Fungal Biology

Series Editors

Vijai Kumar Gupta
ERA Chair of Green Chemistry
Department of Chemistry and Biotechnology
School of Science
Tallinn University of Technology
Tallinn
Estonia

Maria G. Tuohy
Head of the Molecular Glycobiotechnology Group
Biochemistry, School of Natural Sciences
National University of Ireland Galway
Galway
Ireland

About the Series

Fungal biology has an integral role to play in the development of the biotechnology and biomedical sectors. It has become a subject of increasing importance as new fungi and their associated biomolecules are identified. The interaction between fungi and their environment is central to many natural processes that occur in the biosphere. The hosts and habitats of these eukaryotic microorganisms are very diverse; fungi are present in every ecosystem on Earth. The fungal kingdom is equally diverse, consisting of seven different known phyla. Yet detailed knowledge is limited to relatively few species. The relationship between fungi and humans has been characterized by the juxtaposed viewpoints of fungi as infectious agents of much dread and their exploitation as highly versatile systems for a range of economically important biotechnological applications. Understanding the biology of different fungi in diverse ecosystems as well as their interactions with living and non-living is essential to underpin effective and innovative technological developments. This series will provide a detailed compendium of methods and information used to investigate different aspects of mycology, including fungal biology and biochemistry, genetics, phylogenetics, genomics, proteomics, molecular enzymology, and biotechnological applications in a manner that reflects the many recent developments of relevance to researchers and scientists investigating the Kingdom Fungi. Rapid screening techniques based on screening specific regions in the DNA of fungi have been used in species comparison and identification, and are now being extended across fungal phyla. The majorities of fungi are multicellular eukaryotic systems and therefore may be excellent model systems by which to answer fundamental biological questions. A greater understanding of the cell biology of these versatile eukaryotes will underpin efforts to engineer certain fungal species to provide novel cell factories for production of proteins for pharmaceutical applications. Renewed interest in all aspects of the biology and biotechnology of fungi may also enable the development of "one pot" microbial cell factories to meet consumer energy needs in the 21st century. To realize this potential and to truly understand the diversity and biology of these eukaryotes, continued development of scientific tools and techniques is essential. As a professional reference, this series will be very helpful to all people who work with fungi and should be useful both to academic institutions and research teams, as well as to teachers, and graduate and postgraduate students with its information on the continuous developments in fungal biology with the publication of each volume.

More information about this series at http://www.springer.com/series/11224

Bhim Pratap Singh • Lallawmsanga
Ajit Kumar Passari
Editors

Biology of Macrofungi

Springer

Editors
Bhim Pratap Singh
Molecular Microbiology and Systematics Laboratory
Department of Biotechnology
Mizoram University
Aizawl, Mizoram, India

Lallawmsanga
Molecular Microbiology and Systematics Laboratory
Department of Biotechnology
Mizoram University
Aizawl, Mizoram, India

Ajit Kumar Passari
Molecular Microbiology and Systematics Laboratory
Department of Biotechnology
Mizoram University
Aizawl, Mizoram, India

ISSN 2198-7777 ISSN 2198-7785 (electronic)
Fungal Biology
ISBN 978-3-030-02621-9 ISBN 978-3-030-02622-6 (eBook)
https://doi.org/10.1007/978-3-030-02622-6

Library of Congress Control Number: 2018965199

This Springer imprint is published by the registered company Springer Nature Switzerland AG
The registered company address is: Gewerbestrasse 11, 6330 Cham, Switzerland

This volume is dedicated to
Late Shri Ram Prasad
father of Dr. Bhim Pratap Singh,
Senior Editor of this volume,
for his continuous motivation and
encouragement

Late Shri Ram Prasad (1938–2011)

Foreword

It is very important to estimate and document the mushroom (macrofungi) biodiversity as they are well known for the production of bioactive compounds having biotechnological applications and are recognized as supplementary foods due to high nutritional content and medicinal importance. Among the few important features of mushrooms, they can be used as a source of food and medicines besides their key ecological roles. They are also considered as the efficient food for the future as they have high amount of proteins and other nutrients, so they can be used to cope up with the nutrient malnutrition in developing countries. Mushrooms are the only edible source of Vitamin D and considered among the top foods to prevent osteoporosis.

The book volume published by Springer Nature on *Biology of Macrofungi* has received contributions from academicians and scientists from several countries throughout the world including the United Kingdom, the United States, Brazil, China, Serbia, etc. I believe the knowledge shared by the contributors will be very helpful for the readers working in the field of macrofungi. The book has descriptions about the methods used for the identification of macrofungi and how the macrofungi can be exploited for several applications for sustainable development. The book also describes how to cultivate the important types of mushrooms to generate self-employment.

I am sure that the contents given in this volume are a comprehensive coverage of various important aspects of exploitation of macrofungi and their applications in health and pharmaceutical industries. I congratulate the editors and contributors for bringing this volume on *Biology of Macrofungi*.

K. R. S. Sambasiva Rao
Vice-Chancellor
Mizoram University (A Central University)
Aizawl, Mizoram, India

Preface

Mushrooms are often been referred as functional food due to their high nutritional contents. They contain high amount of antioxidants, the chemicals that are required to get rid of free radicals. Free radicals are very dangerous for the body cells as they lead to several deadly diseases including cancer. Mushrooms have high content of selenium, a mineral which is not available by eating most of the fruits and vegetables, and help in detoxifying disease-causing compounds in the body and also play an important role in preventing inflammation. They are high in fiber, potassium, and Vitamin C content which play an important role in preventing cardiovascular health.

The present volume has 19 chapters contributed by the researchers and academicians from several countries including the United Kingdom, the United States, China, Malaysia, Poland, Serbia, Brazil, and India. The volume has covered most of the recent developments in macrofungal biology starting from the latest methods used for the identification of wild macrofungi and several applications of macrofungi for sustainable livelihood.

Editors of the volume express their gratitude to all the contributors for sharing their work in this volume. We are also thankful to Springer Nature publishers for giving us a chance to compile this important volume. We hope that the contents presented in this book will be useful for the readers involved in macrofungi research and all concern.

Mizoram, India

Bhim Pratap Singh
Lallawmsanga
Ajit Kumar Passari

Acknowledgments

My sincere thanks are extended to all the academicians and scientists who have contributed chapters and happily agreed to share their work on various aspects of macrofungi in this volume. At the same time, I also express my deepest gratitude to my family members, especially my wife (Dr. Garima Singh) and my daughter (Aadita Singh), for their kind support which has prompted me to complete the assignment on time. I am also thankful to the Department of Biotechnology (DBT), New Delhi, Government of India, for supporting us financially in the form of several externally funded projects from time to time and for the establishment of DBT Bioinformatics Centre at Mizoram University which was quite useful during the compilation of the book. I am equally thankful to the Springer Publishing for their full cooperation during the production of the volume. In particular, I am thankful to the series editors, Dr. Vijai Kumar Gupta and Prof. Maria G. Tuohy, for accepting our proposal and providing their full support and encouragements. I am also thankful to the production team of Springer Nature for all their efforts for publishing the volume on time. I admit that it is quite possible that some mistakes may have occurred in the text inadvertently, and I take responsibilities for the mistakes, and please feel free to inform me the same.

I am thankful to Prof. KRS Sambasiva Rao, Vice-Chancellor, Mizoram University, for his endeavor and motivations at all stages of the progress.

Mizoram, India Bhim Pratap Singh

Contents

Contributors

João Ricardo Moreira Almeida Graduate Program Microbial Biology, University of Brasilia, Distrito Federal, Brasília, Brazil

Embrapa Agroenergy, Brasília, Distrito Federal, Brazil

Sudheer K. Annepu ICAR-Directorate of Mushroom Research, Solan, India

Madhusmita Borthakur Microbiology Laboratory, Department of Biotechnology & Bioinformatics, North-Eastern Hill University, Shillong, Meghalaya, India

John A. Buswell Institute of Edible Fungi, Academy of Agricultural Sciences, Shanghai, China

Jaime Carrasco Department of Plant Sciences, University of Oxford, Oxford, UK

Peter Chiew Hing Cheong Medicinal Mushroom Research Group (MMRG), Department of Molecular Medicine, University of Malaya, Kuala Lumpur, Malaysia

Aparecido Almeida Conceição Graduate Program of Biosciences, Federal University of Bahia, Vitória da Conquista, Bahia, Brazil

Embrapa Agroenergy, Distrito Federal, Brasília, Brazil

Joice Raisa Barbosa Cunha Graduate Program Microbial Agricultural, Federal University of Lavras, Lavras, Minas Gerais, Brazil

Embrapa Agroenergy, Distrito Federal, Brasília, Brazil

Eustáquio Souza Dias Graduate Program Microbial Agricultural, Federal University of Lavras, Minas Gerais, Brazil

Euziclei Gonzaga de Almeida Graduate Program of Biotechnology and Biodiversity, Federal University of Mato Grosso, Mato Grosso, Grosso, Cuiabá, Brazil

Félix Gonçalves de Siqueira Graduate Program of Biosciences, Federal University of Bahia,Vitória da Conquista, Bahia, Distrito Federal, Brazil

Graduate Program of Biotechnology and Biodiversity, Federal University of Mato Grosso, Mato Grosso, Cuiabá, Brazil

Embrapa Agroenergy, Distrito Federal, Brasília, Brazil

Hesham A. El-Enshasy Institute of Bioproduct Development, Universiti Teknologi Malaysia, Johor Bahru, Johor, Malaysia

School of Chemical Engineering, Faculty of Engineering, Universiti Teknologi Malaysia, Johor Bahru, Johor, Malaysia

City of Scientific Research and Technology Application, New Burg Al Arab, Alexandria, Egypt

Nagib A. Elmarzugi Department of Industrial Pharmacy, Faculty of Pharmacy & Biotechnology Research Center, Tripoli, Libya

Shin Yee Fung Medicinal Mushroom Research Group (MMRG), Department of Molecular Medicine, University of Malaya, Kuala Lumpur, Malaysia

Centre for Natural Products Research and Drug Discovery (CENAR), University of Malaya, Kuala Lumpur, Malaysia

University of Malaya Centre for Proteomics Research (UMCPR), University of Malaya, Kuala Lumpur, Malaysia

Francisco J. Gea Centro de Investigación Experimentación y Servicios del Champiñón (CIES), Quintanar del Rey, Cuenca, Spain

Jasmina Glamočlija Department of Plant Physiology, Institute for Biological Research "Siniša Stanković", University of Belgrade, Belgrade, Serbia

Anna Goyal Department of Microbiology, College of Basic Sciences and Humanities, Punjab Agricultural University, Ludhiana, Punjab, India

Moni Gupta Division of Biochemistry, Sher-e-Kashmir University of Agricultural Sciences and Technology of Jammu, Jammu, India

Sachin Gupta Division of Plant Pathology, Sher-e-Kashmir University of Agricultural Sciences and Technology of Jammu, Jammu, India

Xiaojun Huang State Key Laboratory of Food Science and Technology, Nanchang University, Nanchang, Jiangxi Province, China

S. R. Joshi Microbiology Laboratory, Department of Biotechnology & Bioinformatics, North-Eastern Hill University, Shillong, Meghalaya, India

Katarzyna Kała Department of Pharmaceutical Botany, Faculty of Pharmacy, Jagiellonian University Medical College, Kraków, Poland

Anu Kalia Electron Microscopy and Nanoscience Laboratory, Department of Soil Science, College of Agriculture, Punjab Agricultural University, Ludhiana, Punjab, India

Malarvizhi Kaliyaperumal Centre for Advanced Studies in Botany, Guindy Campus, University of Madras, Chennai, Tamil Nadu, India

Meena Kapahi Department of Chemistry, Manav Rachna University (MRU), Faridabad, India

Gagandeep Kaur Punjab Agricultural University, Ludhiana, Punjab, India

Kezhocuyi Kezo Centre for Advanced Studies in Botany, Guindy Campus, University of Madras, Chennai, Tamil Nadu, India

Marina Kostić Department of Plant Physiology, Institute for Biological Research "Siniša Stanković", University of Belgrade, Belgrade, Serbia

Lallawmsanga Molecular Microbiology and Systematics Laboratory, Department of Biotechnology, Mizoram University, Aizawl, Mizoram, India

Jan Lazur Department of Pharmaceutical Botany, Faculty of Pharmacy, Jagiellonian University Medical College, Kraków, Poland

Ong Mei Leng Harita Go Green Sdn Bhd, Johor Bahru, Johor, Malaysia

Mohamad Fawzi Mahomoodally Department of Health Sciences, Faculty of Science, University of Mauritius, Réduit, Mauritius

Roslinda Abd Malek Institute of Bioproduct Development (IBD), Universiti Teknologi Malaysia (UTM), Skudai, Malaysia

Margarita Department of Plant Sciences, University of Oxford, Oxford, UK

Simone Mendonça Embrapa Agroenergy, Distrito Federal, Brasília, Brazil

Bożena Muszyńska Department of Pharmaceutical Botany, Faculty of Pharmacy, Jagiellonian University Medical College, Kraków, Poland

Sunil A. Nair Dr. Y.S. Parmar University of Horticulture and Forestry, Solan, India

María J. Navarro Centro de Investigación Experimentación y Servicios del Champiñón (CIES), Quintanar del Rey, Cuenca, Spain

Shaoping Nie State Key Laboratory of Food Science and Technology, Nanchang University, Nanchang, Jiangxi Province, China

Ajit Kumar Passari Molecular Microbiology and Systematics Laboratory, Department of Biotechnology, Mizoram University, Aizawl, Mizoram, India

Rubén Darío Romero Pelaéz Graduate Program Microbial Biology, University of Brasilia, Distrito Federal, Brasília, Brazil

Embrapa Agroenergy, Distrito Federal, Brasília, Brazil

Margarita Perez Centro Tecnológico de Investigación del Champiñón de La Rioja (CTICH), Autol, Spain

Gail M. Preston Department of Plant Sciences, University of Oxford, Oxford, UK

Bhim Pratap Singh Molecular Microbiology and Systematics Laboratory, Department of Biotechnology, Mizoram University, Aizawl, Mizoram, India

Marina Soković Department of Plant Physiology, Institute for Biological Research "Siniša Stanković", University of Belgrade, Belgrade, Serbia

Mohammad Soltani Institute of Bioproduct Development (IBD), Universiti Teknologi Malaysia (UTM), Skudai, Malaysia

Gunaseelan Sugantha Centre for Advanced Studies in Botany, Guindy Campus, University of Madras, Chennai, Tamil Nadu, India

Katarzyna Sułkowska-Ziaja Department of Pharmaceutical Botany, Faculty of Pharmacy, Jagiellonian University Medical College, Kraków, Poland

Baby Summuna Sher-e-Kashmir University of Agricultural Sciences and Technology of Jammu, Wadoora, India

Chon Seng Tan Ligno Research Foundation, Balakong Jaya, Selangor, Malaysia

Maria L. Tello Centro Tecnológico de Investigación del Champiñón de La Rioja (CTICH), Autol, Spain

Monika Thakur Amity Institute of Food Technology, Amity University Uttar Pradesh, Noida, India

Davin Uy Research and Innovation Center, Tumnuptek, Chamkarmorn, Phnom Penh, Cambodia

Vandinelma Oliveira Vieira Graduate Program of Biotechnology and Biodiversity, Federal University of Mato Grosso, Cuiabá, Mato, Grosso, Brazil

Embrapa Agroenergy, Brasília, Distrito Federal, Brazil

Yuxiao Wang State Key Laboratory of Food Science and Technology, Nanchang University, Nanchang, Jiangxi Province, China

Chapter 1
Exploration of Macrofungi in Sub-Tropical Semi-Evergreen Indian Forest Ecosystems

Lallawmsanga, Ajit Kumar Passari, and Bhim Pratap Singh

1.1 Introduction

1.1.1 What is a Mushroom?

Mushroom is a term used to define the fruiting body formed by a group of fungi such as **Basidiomycetes** and **Ascomycetes**. They are either epigeous or hypogeous large enough to be seen with the naked eye and can be picked by hand (Chang and Miles 1992). The fruiting bodies of fungi (except for bracket) are normally short lived; and their existence varies from one to another. Mushrooms basically have the specific nutritional and ecological requirements for their growth and development; some of them thrive extremely well in soil and some of them utilize degrading plant residues as saprophytes. Many of the mushrooms exist in symbiotic relationship with roots of higher plant species. Due to the differences in their lifestyle, their ecological influence also varies from one to another (Lallawmsanga et al. 2016). These macrofungi appear at different times in a year and some species do not appear every year (Packham et al. 2002). They are a group of fungi that forms relatively conspicuous sporocarp (Ferris et al. 2000). As the tropical regions are considered rich in biodiversity, most of the new species of mushrooms reported in the recent years are from these regions; these newly reported mushrooms are usually associated with the native trees.

The occurrence of mushroom fruiting bodies varies according to different ecological climate (Arora 1991). They are essential for recycling of nutrients, growth and development of saplings in the forest base while some of them cause damage to the trees as they are in parasitic association with them. Humid climate provides

Lallawmsanga · A. K. Passari · B. P. Singh (✉)
Molecular Microbiology and Systematics Laboratory, Department of Biotechnology, Mizoram University, Aizawl, Mizoram, India
e-mail: bhimpratap@mzu.edu.in

B. P. Singh et al. (eds.), *Biology of Macrofungi*, Fungal Biology,
https://doi.org/10.1007/978-3-030-02622-6_1

more distribution as compared to hot and dry periods (Sibounnavong et al. 2008). Mueller et al. (2007) estimated that there are 53, 000 to 110, 000 mushrooms in the world. However, Hawksworth (2001) estimated the global mushroom is about 140,000 species and approximately 14,000 species of them are known. It is also assumed that the mushrooms which are still unexplored could be of highly beneficial to mankind. **Biodiversity** research has become more important following the Convention on Biological Diversity in 1992 to understand the structure and function of ecosystems (Gadd 2007). Besides flora and fauna, fungi are considered as one of the most significant organisms in the world due to their important role in ecosystem (Mueller et al. 2004). Fungi have great importance ecologically as well as in industrially. They are found in almost all part of human life and ecosystems (Garibay-Orijel et al. 2009).

There are over 27,000 mushroom species recorded in India which makes it the second largest biotic community after insects (Manoharachary et al. 2005). True fungi belong to kingdom **Eukaryota** which has four phyla, 103 orders, 484 families and 4979 genera. True fungi consists of monophyletic groups, of which Basidiomycetes alone accounts for 35% of the fungal species described so far. They are of great importance ecologically as well as industrially. Ascomycetes, Basidiomycetes, some Zygomycetes and Glomeromycetes are grouped under **hypogeous fungi** (Trappe and Castellano 1991). The loss of spore discharge mechanisms in Ascomycetes and Basidiomycetes are the common characteristics observed in the two (Alessandra et al. 2014).

1.1.2 Why do we need to Explore Mushrooms?

Mushrooms are rich in food, medicine and plant growth promotion (Ghate and Sridhar 2015). So they have large number of importance in biodegradation, ecosystems, food industry and pharmacology (Barros et al. 2007). Mushrooms are considered as healthy foods due to their high composition of nutrients like proteins, carbohydrates, vitamins, fatty acids, amino acids and various elements in their fruiting body (Liu et al. 2015; Dimitrijevic et al. 2016). Mushrooms are known as an unlimited source of **bioactive compounds** and their immense diversity provide a massive prospective in the discovery and development of several drugs (Butler 2004). They are also known to produce the most important kinds of bioactive compounds including lectins, peptidoglucans, phenolic compounds, polysaccharides, steroids, etc., that acquire a broad range of pharmacological properties like **antimicrobial** (Alves et al. 2013), **antioxidant** (Ferreira et al. 2009), **anti-tumor** (Popovic et al. 2013), and therapeutic properties such as counteracting diseases including hypertension, hypercholesterolemia and cancer (Gast et al. 1988). These factors made the large scale production of mushroom became an important industry in several countries (Rafique 1996; Izlam et al. 2009).

Rise in food prices due to high biofuel prices have caused scarcity of food and malnutrition throughout the world. To lessen this, cultivation of mushroom is beneficial and applicable for the poor farmers as it is labor intensive, short-duration for production and land saving (Shah et al. 2004). It could be more applicable in developing countries where malnutrition is one of the biggest problems as mushrooms contain high level of nutrients and its high productivity per unit area (Eswaran and Ramabadran 2000).

1.2 Exploration of Wild Mushroom Diversity

1.2.1 Macroscopic Identification

Mushrooms collected from the selected reserve forests were first identified based on their macroscopic characteristics. The followings are the macroscopic features used for the identification up to genus: attachment, cap size, colour, gills, shape, surface texture and moisture, spacing, stem size, the presence or absence of partial and universal veils. The identification was done by using seven mycological characters such as hymenium type, cap shape, gills, stipe character, color of the spore print, ecological type, and edibility (Krishna et al. 2015).

1.2.2 Molecular Characterization

Several numbers of molecular markers have been employed in the rapid identification of mushrooms in modern biotechnological research (Lian et al. 2008). DNA based techniques have eased the knowledge about microbial diversity from the natural ecosystems (Tuckwell et al. 2005). Mullis and Faloona (1987) have developed DNA-based **PCR** and taxon specific primers for the identification and study various kinds of fungi. The internal transcribed spacer region has been widely employed for the identification of fungi at species level (Sanchez-Ballesteros et al. 2000). Molecular fingerprinting offer an efficient way to study genetic variations which further helps in breeding programs due to its properties including high levels of detectable polymorphism and independence of environmental parameters (Yin et al. 2014).

Various molecular techniques such as random amplified polymorphic DNA (**RAPD**), amplified fragment length polymorphism (**AFLP**) analysis, simple sequence repeat (**SSR**) analysis, inter-simple sequence repeat (**ISSR**) analysis have been commonly employed for the genetic characterization of plants, animals, fungi and other organisms (Mei et al. 2014). RAPD analysis is used to reveal genetic similarity and phylogenetic analysis due to simplicity in its technique. RAPD is furthermore employed for classification and study the genetic diversity in newly discovered species that are with agricultural and industrial importance (Shakeel et al. 2013). The advantages of employing RAPD as a molecular marker is that the DNA sequence information is not required, moreover they require only a small amount of DNA template, no harmful contamination and low-cost technology (Mei et al. 2014). Among the molecular markers, ISSR has gained interest as it is known to be abundant, highly informative, highly polymorphic and extremely reproducible. It has been effectively employed in the molecular identification of plants (Zietkiewicz et al. 1994) and fungi (Tang et al. 2010).

1.3 Biodiversity of Mushrooms in two Ecosystems

Mizoram, a state in North Eastern Region of India is a big bio-prospecting area. It is (21°57' - 24°30' North and 92°15' - 93°26' East) located in the extreme southern part of northeastern India, has a geographical area of 2,108,100 ha (0.6% of India's geographical area). Mizoram is bounded on the north by Assam and Manipur, on the east and south by Myanmar and on the west by Bangladesh and Tripura. The terrain is hilly and mostly undulating with the average altitude ranging from 500 to 800 m and the maximum reaching 2157 m in the Blue Mountains. It is an important part of the Indo-Burma Biodiversity Hot Spot. Therefore, existence of agriculturally and industrially potential mushroom strains with diverse genetic resources in this region cannot be ruled out. It is well recognized that the diversity of Mushrooms in North-Eastern region, especially in Mizoram remains underexplored. It is important to explore the genetic resources of this region that will be useful for various purposes in the field of industrial, agricultural and pharmaceutical sectors.

1.3.1 Dampa Tiger Reserve

Dampa Tiger Reserve (Mamit District, Mizoram) is the largest wildlife sanctuary in Mizoram. It covers up an area of about 500 sq.km, falls within 23° 23' 15"N – 23° 42' 20"N latitudes and 92° 16' 25"E – 92° 25' 55"E longitudes. Altitude in the forest ranged between 800 to 1100 m above mean sea level. The annual rainfall ranged between 2000 mm to 2500 mm and maximum rainfall occurs in the month of June to August. Pleasant and warm climate prevails in the reserve all through the year with temperate to chill winter during November to January in higher altitudes. The natural vegetation includes *Michelia champaca*, *Dipterocarpus turbinatus*, and *Terminalia chebula* in the lower elevations while the higher elevations are characterized by *Castanopsis indica, Dendrocalamus longispathus, Gmelina arborea, Lannea coromandelica, Mesua ferrea, Quercus* sp., *Schima wallichii* and *Sterculia villosa*. The reserve hosts several endangered species including *Panthera tigris* (tiger), *Neofelis nebulosa* (clouded leopard) and *Elephus maximus* (Asiatic elephant). It is especially affluent *M. assamensis, M. leonina, M. arctoides, Nycticebus bengalensis, Trachypithecus pileatus* and *T. phayrei, Macaca mulatta*.

1.3.2 Murlen National Park

Murlen Nation Park (Champhai District, Mizoram), located in the neighborhood of the Indo-Myanmar boundary and is important for its propinquity to the Chin Hills. It covers up an area of roughly 100 sq. km. It lies within 23° 32' 42" – 23° 41' 36"N latitudes and 92°13' 12" – 92°27' 24"E longitudes with altitude ranged from 1000–1600 m. Sub-Montane forest and semi-Evergreen forest are important vegetation that covers of the area. Murlen is a host for variety of rich flora and fauna. The prominent plant species that can be seen includes *Arundinaria callosa, Prunus myrica, Rhododendron, Quercus, Betula, Schima wallichai, Arundinaria callosa, Canes, Michelia champaca* and different types of orchids. It is one of the most important bird areas (IBM) supporting several threatened species. The Park provides a habitat for mammals *Callosciurus pygerythrus, Cervus unicolor, C. erythraeus, Hylobates hoolock, Macaca assamensis, M. arctoides, Martes flavigula, Muntiacus muntjak, Ratufa bicolour, Sus scrofa* and *Trachypithecus pileatus*.

1.3.3 Diversity of Macrofungi in Dampa Tiger Reserve and Murlen National Park

The mushroom samples were collected from the abovementioned forests in 2013–2015. The mushroom fruiting bodies were first photographed in their natural habitat before collecting them. The collected fruiting bodies were carefully cleaned to remove soil and other substrates using fine brush. The collected mushroom samples were kept in separate paper bags to avoid mixing and were transferred to the laboratory. The specimens preserved as herbarium were identified with the help of standard literatures (Adhikari 2000).

The assessment of mushroom diversity of the Dampa Tiger Reserve and Murlen National Park (Mizoram, India) was accomplished and we collected 249 mushroom taxa (Fig. 1.1 A-E and Table 1.1) mainly from the genera *Amanita, Boletus, Russula, Psathyrella, Schizophyllum, Calocybe, Lentinus, Polyporus, Pleurotus, Trametes* and *Tricholoma* (Fig. 1.2). In total there are 39 mushroom families recorded from the 249 taxa collected (Table 1.2) and the major families are Agaricaceae, Amanitaceae, Boletaceae, Lyophyllaceae, Marasmiaceae, Polyporaceae, Psathyrellaceae, Russulaceae, Schizophyllaceae and Tricholomataceae (Fig. 1.3).

Fig. 1.1 A-E Wild mushrooms collected from Dampa Tiger Reserve and Murlen National Park, Mizoram, India (A) *Chlorosplenium aeruginascens*; (B) *Collybia cookie*; (C) *Clavulinospsis laeticolor*; (D) *Auricularia auricular-judae*; (E) *Favolaschia cyathea*; (F) *Cookeina tricholoma*; (G) *Coltricia perennis*; (H) *Helvella* sp.; (I) *Marasmius androsaceus*; (J) *Xylaria polymorpha*; (K) *Scleroderma citrinum*; (L) *Pholiota squarrosa*; (M) *Pleurotus* sp.; (N) *Geoglossum cookeanum*; (O) *Oudemansiella mucidula*; (P) *Mycena* sp.; (Q) *Xeromphalina campanella*; (R) *Coprinus disseminates*

Fig. 1.1 (continued)

Table 1.1 Wild mushrooms collected from Mizoram, India

Genus	No. of individuals
Agaricus	4
Lepiota	1
Lycoperdon	1
Leucoagaricus	1
Amanita	15
Auricularia	4
Bolbitius	3
Boletus	41
Xerocomus	1
Cantharellus	2
Clavaria	1
Clavulinopsis	2
Cortinarius	1
Crepidotus	4
Calocera	1
Entoloma	3
Geoglossum	1
Chlorociboria	1
Helvella	1
Laccaria	4
Hygrocybe	2
Coltrichia	5
Hebeloma	2
Inocybe	1
Leotia	1
Calocybe	8
Termitomyces	2
Marasmius	2
Xeromphalina	4
Hydropus	1
Phlebia	1
Merulius	1
Mycena	5
Favolaschia	1
Phallus	2
Hymenopellis	1
Xerula	3
Mucidula	2
Pleurotus	6
Lentinus	8
Trametes	7
Panus	2
Polyporus	6
Pycnosporus	1

Table 1.1 (continued)

Genus	No. of individuals
Coprinellus	3
Coprinopsis	1
Coprinus	1
Psathyrella	12
Lactarius	5
Russula	23
Cookeina	1
Schizophyllum	10
Scleroderma	4
Agrocybe	1
Pholiota	4
Stropharia	1
Thelephora	2
Clitocybe	1
Collybia	1
Lepista	1
Lichenomphalina	1
Tricholoma	6
Flammulina	1
Xylaria	4

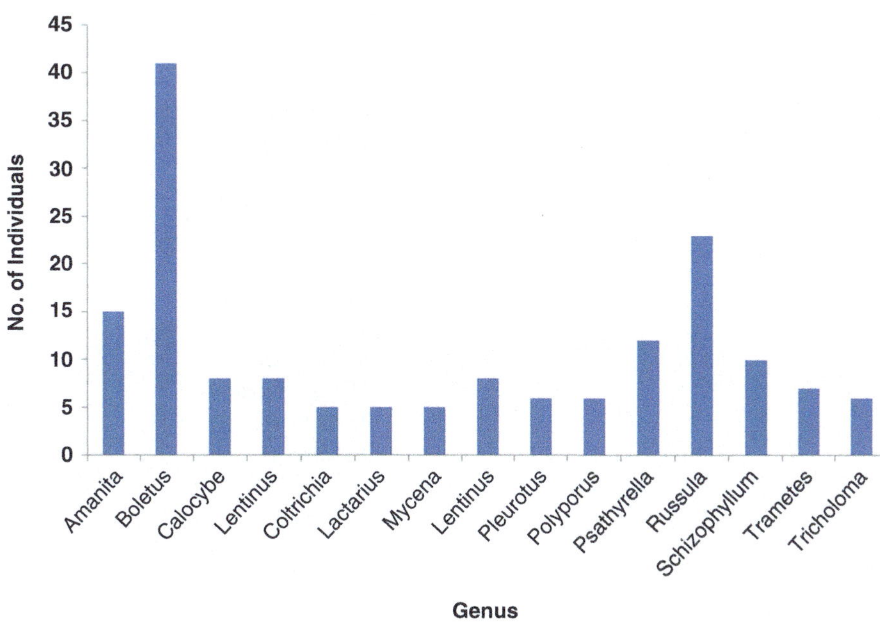

Fig. 1.2 Major genera recorded from Dampa Tiger Reserve and Murlen National Park, Mizoram, India

Table 1.2 Family wise distribution of wild mushrooms collected from Mizoram, India

Family	No. of individual
Agaricaceae	7
Amanitaceae	15
Auriculariaceae	4
Bolbitiaceae	3
Boletaceae	42
Cantharellaceae	2
Clavariaceae	3
Cortinariaceae	1
Crepidotaceae	4
Dacrymycetaceae	1
Entolomataceae	3
Geoglossaceae	1
Helotiaceae	1
Helvellaceae	1
Hydnangiaceae	4
Hygrophoraceae	2
Hymenochaetaceae	5
Hymenogastraceae	2
Inocybaceae	1
Leotiaceae	1
Lyophyllaceae	10
Marasmiaceae	7
Meruliaceae	2
Mycenaceae	6
Phallaceae	2
Physalacriaceae	6
Pleurotaceae	6
Polyporaceae	24
Psathyrellaceae	17
Russulaceae	28
Sarcoscyphaceae	1
Schizophyllaceae	10
Sclerodermataceae	4
Strophariaceae	6
Thelephoraceae	2
Tricholomataceae	11
Xylariaceae	4

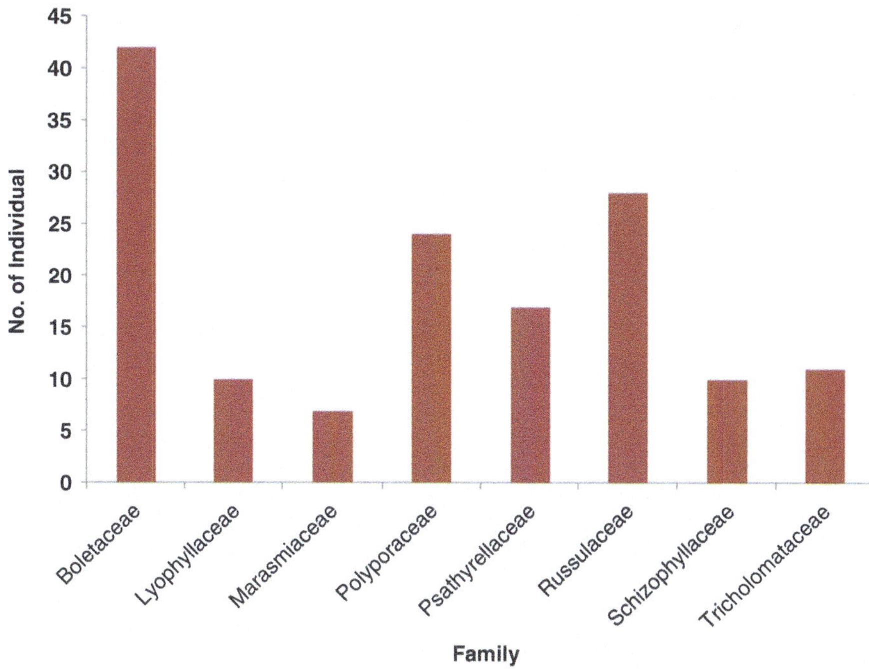

Fig. 1.3 Major families recorded from Dampa Tiger Reserve and Murlen National Park, Mizoram, India

1.4 Future Prospects

Assessment of wild mushroom diversity in the northeast India is essential as it plays a significant role in the socio-economic life of the tribal population. They are imperative for the rural and sub-urban inhabitants through food security and livelihood. They can build important dietary addition through protein and various micronutrients and, together with their medicinal activities. As this region is still underexplored for mushroom biodiversity, there is a very high chance of obtaining mushrooms with high potential in diverse applications.

Acknowledgements The authors are thankful to the Department of Biotechnology, Government of India for DBT sponsored NER-Twinning project (No. BT/320/NE/TBP/2012). We gratefully acknowledge Chief Wildlife Warden, Environment and Forest Department, Government of Mizoram, India. We are grateful for Department of Biotechnology, Government of India for establishing DBT-BIF centre and DBT-State Biotech Hub in the Department of Biotechnology, Mizoram University.

References

Adhikari MK (2000) Mushrooms of Nepal. P.U. Printers, Kathmandu

Alessandra Z, Domizia D, Luigi RG, Simonetta F, Niccol BGM, Mirco I, Asuncion M, Lahsen K, Abdulhakim B, Federica P, Riccardo C, Giuseppe V (2014) Hypogeous fungi in Mediterranean Maquis, Arid and Semi-Arid forests. Mycorrhiza 24(6):481–486

Alves MJ, Ferreira ICFR, Froufe HJC, Abreu RMV, Martins A, Pintado M (2013) Antimicrobial activity of phenolic compounds identified in wild mushrooms, SAR analysis and dochking studies. J Appl Microbiol 115:346–357

Arora D (1991) All that rain promises and more: a hip pocket guide to Western Mushrooms. Ten Speed Press, Berkeley

Barros L, Calhelha RC, Vaz JA, Ferreira ICFR, Baptista P, Estevinho LM (2007) Antimicrobial activity and bioactive compounds of Portuguese wild edible mushrooms methanolic extracts. Eur Food Res Technol 225:151–156

Butler MS (2004) The role of natural product chemistry in drug discovery. J Nat Prod 67:2141–2153

Chang ST, Miles PG (1992) Mushroom biology - A new discipline. The Mycologist 6:64–65

Dimitrijevic MV, Mitic VD, Cvetkovic JS, Stankov Jovanovic VP, Mutic JJ, Nikolic Mandic SD (2016) Update on element content profiles in eleven wild edible mushrooms from family Boletaceae. Eur Food Res Technol 242:1–10

Eswaran A, Ramabadran R (2000) Studies on some physiological, cultural and post harvest aspects of oyster mushroom, *Pleurotus eous*. Trop Agri Res 12:360–374

Ferreira ICFR, Barros L, Abreu RMV (2009) Antioxidants in wild mushrooms. Cur Med Chem 16:1543–1560

Ferris R, Peace AJ, Newton AC (2000) Macro fungal communities of low land Scots pine and Norway spruce plantations in England: relationships with site factors and stand structure. For Ecol Manag 131:255–267

Gadd GM (2007) Geomycology: Biogeochemical transformations of rocks, minerals, metals and radionuclides by fungi bioweathering and bioremediation. Mycol Res 111:3–49

Garibay-Orijel R, Cordova J, Cifuentes J, Valenzuela R, Estrada-Torres E A, Kong A (2009) Integrating wild mushrooms use into a model of sustainable management for indigenous community forests. For Ecol Manag 258:122–131

Gast GH, Jansen E, Bierling J, Haanstra L (1988) Heavy metals in mushroom and their relationship with soil characteristics. Chemosphere 60(4):789–799

Ghate SD, Sridhar KR (2015) Contribution to the knowledge on macrofungi in mangroves of the southwest India. Plant Biosyst 150:977–986

Hawksworth DL (2001) The magnitude of fungal diversity: the 1.5 million species estimate revisited. Mycol Res 105(12):1422–1432

Izlam MZ, Rahman MH, Hafiz F (2009) Cultivation of oyster mushroom (*Pleurotus flabellatus*) on different substrates. Int J Sustain Crop Prod 4(1):45–48

Krishna G, Samatha B, Himabindu SVSSSLN, Prasad MR, Rajitha B, Charaya MAS (2015) Macrofungi in some forests of Telengana State, India. J Mycol 382476:1–7

Lallawmsanga, Passari AK, Mishra VK, Leo VV, Singh BP, Meyyappan GV, Gupta VK, Uthandi S, Upadhyay RC (2016) Antimicrobial potential, identification and phylogenetic affiliation of wild mushrooms from two sub-tropical semi-evergreen Indian forest ecosystems. PLoS One 11(11):e0166368

Lian B, Zang J, Hou W, Yuan S, Smith DL (2008) PCR-based sensitive detection of the edible fungus *Boletus edulis* from rDNA ITS sequences. Electron J Biotechnol 11(3):1–8

Liu B, Huang Q, Cai H, Guo X, Wang T, Gui M (2015) Study of heavy metal concentrations in wild edible mushrooms in Yunnan Province, China. Food Chem 188:294–300

Manoharachary CS, Singh KR, Adholeya A, Suryanarayanan TS, Rawat S, Johri BN (2005) Fungal biodiversity: distribution, conservation and prospecting of fungi from India. Curr Sci 89:58–71

Mei ZQ, Fu SY, Yu HQ, Yang LQ, Duan CG, Liu XY, Gong S, Fu JJ (2014) Genetic characterization and authentication of *Dimocarpus longan* Lour. using an improved RAPD technique. Genet Mol Res 13(1):1447–1455

Mueller GM, Schmit JP, Huhndorf SM, Ryvarden L, O'Dell TE, Lodge JE, Leacock PR, Mata M, Umana L, Wu Q, Czederpiltz DL (2004) Recommended protocols for sampling macrofungi. In: Mueller GM, Bills GF, Foster MS (eds) Biodiversity of Fungi: inventory and monitoring methods. Elsevier Academic Press, San Diego, pp 168–172

Mueller GM, Schmit JP, Leacock PR, Buyck B, Cifuentes J, Desjardin DE, Halling RE, Hjortstam K, Iturriaga T, Larsson KH, Lodge DJ, May TW, Minter D, Rajchenberg M, Redhead SA, Ryvarden L, Trappe JM, Watling R, Wu Q (2007) Global diversity and distribution of macrofungi. Biodivers Conserv 16:37–48

Mullis KB, Faloona FA (1987) Specific synthesis of DNA *in vitro* via a polymerase-catalyzed chain reaction. Methods Enzymol 155:335–350

Packham JM, May TM, Brown MJ, Wardlaw TJ, Mills KA (2002) Macrofungal diversity and community ecology in mature and regrowth wet eucalypt forest in Tasmania: A multivariate study. Austral Ecol 27:149–161

Popovic V, Zivkovic J, Davidovic S, Stevanovic M, Stojkovic D (2013) Mycotherapy of cancer: an update on cytotoxic and antitumor activities of mushrooms, bioactive principles and molecular mechanisms of their action. Curr Top Med Chem 13:279–2806

Rafique AN (1996) Studies on the cultivation of mushroom *Pleurotus species* in Gujarat. Ph. D. Thesis, Department of Microbiology, M.G. Science Institute, Navrangpura, Ahmedabad

Sánchez-Ballesteros J, González V, Salazar O, Acero J, Portal MA, Julián M, Rubio V, Bill GF, Platas G, Mochales S, Peláez F (2000) Phylogenetic study of *Hypoxylon* and related genera based on ribosomal ITS sequences. Mycologia 92:964–977

Shah ZA, Ashraf M, Ishtiq C (2004) Comparative study on cultivation and yield performance of oyster mushroom (*Pleurotus ostreatus*) on different substrates (Wheat straw, leaves, sawdust). Pak J Nutri 3:158–160

Shakeel M, Ilyas M, Kazim M (2013) Evaluation of synthetic hexaploid wheats (derivative of durum wheats and *Aegilops tauschii* accessions) for studying genetic diversity using randomly amplified polymorphic DNA (RAPD) markers. Mol Biol Rep 40:21–26

Sibounnavong P, Cynthia CD, Kalaw SP, Reye RG, Soytong K (2008) Some species of macrofungi at Puncan, Carranglan, Nueva Ecija in the Philippines. J Agri Technol 4(2):105–115

Tang LH, Xiao Y, Li L, Guo Q, Bian YB (2010) Analysis of genetic diversity among Chinese *Auricularia auricular* cultivars using combined ISSR and SRAP markers. Curr Microbiol 61:132–140

Trappe JM, Castellano MA (1991) Keys to genera of truffles (Ascomycetes). McIlvainea 10:47–65

Tuckwell DS, Nicholson MJ, McSweeney CS, Theodorou MK, Brookman JL (2005) The rapid assignment of ruminal fungi to presumptive genera using ITS1 and ITS2 RNA secondary structures to produce group-specific fingerprints. Microbiology 151:1557–1567

Yin Y, Liu Y, Li H, Zhao S, Wang S, Liu Y, Wu D, Xu F (2014) Genetic diversity of *Pleurotus pulmonarius* revealed by RAPD, ISSR, and SRAP fingerprinting. Curr Microbiol 68:397–403

Zietkiewicz E, Rafalski A, Labuda D (1994) Genome fingerprinting by simple sequence repeat (SSR) – anchored polymerase chain reaction amplification. Genomics 20(2):176–183

Chapter 2
A Global Overview of Edible Mushrooms

Malarvizhi Kaliyaperumal, Kezhocuyi Kezo, and Sugantha Gunaseelan

2.1 Introduction

Fungi are one of the most diverse and prominent organisms to inhabit and influence the earth. They are an essential component of ecosystem in recycling the mineral nutrients by acting as agents of decaying. Members of **Ascomycota** and **Basidiomycota**, under a precise combination of various abiotic conditions and surrounding flora are known to produce a detectable fruiting body called as "mushrooms" (Stojchev 1995). According to Chang and Miles (1992), "macrofungus are naked to eyes and are able to grow above ground (epigeous) and underground (hypogeous). They might have originated from ancient lineage ca. 400 million years ago and flourished in association with land plants as both saprobes and parasites (Boyce et al. 2007).

Considering the rich magnitude of fungal diversity, total of estimated mushroom diversity available to science is very less (Hawksworth 1991, 2001). While within reported mushrooms, only 50% (7000 species) acquire varying degrees of edibility; $\leq$ 3000 known species belongs to 31 different genera; ca. 1-10 % are poisonous mushrooms (Miles and Chang 1997). Mushrooms are recognized as rich sources of diverse bioactive principles that make them medically significant as theapeutic agents against pathogens, curing many health disorders and diseases (Wasser and Weis 1999; Lindequist et al. 2005; Ajith and Janardhanan 2007). Nevertheless many edible mushrooms have been integrated with human life since ancient times. **Mycophagy** is the act of consuming mushrooms. Hay (1887), a well-known British mycologist proposed the exclusive terms "mycophilia" and "mycophobia". Mycophilic societies refer to the peoples who like and appreciate mushrooms since ancient time. Mycophobic societies comprise of people showing aversion and fear

M. Kaliyaperumal (✉) · K. Kezo · S. Gunaseelan
Centre for Advanced Studies in Botany, Guindy Campus, University of Madras, Chennai, Tamil Nadu, India
e-mail: malarvizhi@unom.ac.in

B. P. Singh et al. (eds.), *Biology of Macrofungi*, Fungal Biology,
https://doi.org/10.1007/978-3-030-02622-6_2

towards mushrooms (Wasson and Wasson 1957). In general they have been used as food or food products by many tribal and urban peoples for their taste, dietary value with low calories and cholesterol but have high proteins content, minerals, fibres, good amount of vitamins and trace elements (Wani et al. 2010).

2.2 Ethnomycology

Ethnomycology gives details on WEF inculcated in our life since ancient times. Consequently, hallucinogenic mushrooms and their religious significance have been included under ethnomycology (Schultes 1940; Wasson 1968). Diverse areas of mushroom knowledge that includes entheogenics, cultivation, nutraceutics, **mushroom taxonomy**, mycophagy and mycopharmaceuticals are also represented by ethnomycological works (Wasser 2010).

According to Wasson (1971) (father of modern Ethnomycology), *Amanita muscaria* (divine mushroom of immortality) was used for scared rituals by "Aryans" over 4000 years ago (Fig 2.1). They believed that "Soma rasa", the vedic juice is believed to give divine qualities on the spirit of the consumer, even immortality. Traditionally, mushrooms have been used as food, medicine, poison and spiritual practices in religious rituals across the world since 5000 BC (Winkler 2008). The ethnomycological data of mushroom is well documented in various parts of the globe, especially in Asia (Kang et al. 2012, 2013; Pala et al. 2013), Africa (Oso 1975; Kinge et al. 2011), and Central America (Montoya-Esquivel 1998; Montoya-Esquivel et al. 2001).

Fig. 2.1 Stone carving of "'Weeping God" at Quetzalcoatl stone carving of Quetzalcoatl

Fericgla (1994) categorized different European peoples as eminently mycophilic. Ancient Greeks believed that consuming mushrooms offers stamina for soldiers in war and they named it as "sons of the gods," because of their mysterious appearance after thunderstorms. The ancient Romans believed mushrooms as "the food of the gods" and many Romanian writers made an attempt to explain significance of thunderstorms in the life cycle of fungi. Italian peoples are strongly mycophilic, but among the south Catalonia Sapnish community, the tradition of consuming WEF is less common. *Amanita caesarea* is known as "Caesar's mushroom" part of ancient Itlian cuisine, still it exits in many parts of county as a "*ovolo* or *ovolo buono*" or "fungo reale" (Reyna et al. 2002). Nowadays food menus were dominated by embracing diverse species of edible ones such as *Tuber* spp. (truffles) and *Boletus edulis* (porcini) in most of the European countries.

In 500 BC, Theophrastus defined truffles as "a natural phenomenon of great complexity with no stem, root, branch, fibre, leaf, bud or flower". Truffles are hypogeous fungi unique in its appearance served along with dessert only to pharaoh and royals (Trappe 1990). These hypogeous fungi were known to consume by ancient Babylonians, Etruscans, Egyptians, Greek and Romans (Tartufi 2011; Reyna and Garcia-Barreda 2014). At the eighteenth century, truffles regained its status in the cuisines of French and Italian people (Heim 1969). The truffles collected from the Muqattam hills were used only to grace the cuisines of "Fatimid Caliphates" in Egypt. Later, availability of large quantities of truffles in local markets of Cairo has made them cheaper that graced the common people food too (Trappe 1990).

Power et al. (2015) observed the spores from bolete and agaric mushrooms from the dental calculus of an adult woman from Magdalenian population (People of Paleolithic in Western Europe). Both agaric and bolete include many edible and medicinal mushrooms, this discovery could perhaps suggest the intentional consumption of fungi during Old Stone Age era which has not been reported earlier (Fig. 2.2).

Fig. 2.2 *Mushrooms Rock art at Selva Pascuala in Spain Credit: Juan Francisco Ruiz López*

Two fragments of a *Fomitopsis betulina* (formerly *Piptoporus betulinus*) basidiomata was found in Ötzi, a mummified Calcolithic Tyrolean body of an iceman who might have died 5300 years ago in an alpine glacier in the Val Senales glacier, Italy. This mushroom was known for its edibility and medicinal properties (Peintner et al. 1998). Ötzi, revealed the traditional knowledge and knowing of mushroom by ancient tribal people. Russians are well-known mycophilic group; duing weekend they had habit of collecting WEF from the forest (Filipov 1998). Russian's passion for mushroom was described among the Estonians by a saying: "Where there is a mushroom coming up, there is always a Russian waiting for it".

Mushrooms belong to the genus *Agaricus* and *Boletes* are not appreciated among African tribes for consumption, yet Europeans living in Africa habitually consume them (Rammeloo and Walleyn, 1993). From ancient period, ethnic people of Congo-Guinea basin have used edible mushrooms to supplement and diversify their diet (Buyck and Nzigidahera 1995). Dijk et al. (2003) documented three different edible species of cup-fungi (*Cookeina sulcipes* and *C. tricholoma*) and bird-nest's fungus (*Cyathus striatus*) which referred by a same name Tôloñg by Bantu and Bagyeli tribes of south Cameroon Tôloñg. In evident to this, *Peziza* was the common name used for both cup fungi and bird-nest's fungi in the mycological history of Europe. In many Slavic cuisines of Russians, mushroom holds a prominent status. Eskimos (Yupiks and Chukchi) consumed several species of wild *Leccinum, Lactarius, Russula* and *Armillariella* collected from Arctic Tundra (Yamin-Pasternak 2007, 2008).

Long net stinkhorns/ bamboo fungus is traditionally known for its consumption during the Mexican divinatory rituals because of its distinct shape. Tribals from New Guinea considered this fungus as sacred. The Urhobo and the Ibibio ethinic people from Nigeria used stinkhorns to prepare harmful charms (Oso 1975). Yekuana tribes, who are native to Amazon rain forest in Southern Venezuela, consumed two species *viz., Auricularia mesenterica* and *Polyporus* sp. (Chitty 1992). In Chile, WEF were seems to be part the food or cuisine about 13000 years ago (Rojas and Mansur, 1995).

China is a distinctive example of a mycophilic soceity when compared to other countries. According to the archaeological documentation, edibility of wild fungi is first well renowned in China even before the birth of Christ (FAO 1998). Paddy straw mushroom (*Volvariella volvacea*) a common edible mushroom has been cultivated in China during eighteenth century (Chang 1977). Later, it was cultivated in other South Asian countries during 1932 to 1935, (Baker 1934; Chang 1974). Ethnic tribes from Nepal believed that the mushrooms found in high-altitude areas were related to lack of toxicity. Christensen et al. (2008) supported this observation by confirming the frequency of poisonous mushrooms in the *Pinus wallichiana* forest compared to the forest types at lower altitudes was very low. Their study also noted that, there are no known poisonous mushrooms looking like the frequently eaten ones in the high-altitude forests. The first documentation of Yarsagumba (Caterpillar fungus or *Cordyceps*) was by a Tibet physician who described the importance of the mushroom as a "sexual tonic" in his text entitled "An Ocean of Aphrodisiacal Qualities". From ancient times, *Cordyceps sinensis* has been described in old

Chinese and Tibetan medicinal books. Many tribes store mushrooms for future use by drying them in porous baskets or outdoors in structures similar to granaries (Beals 1933; Maniery 1983).

The significant cultural values of wild edible mushrooms diverge across the world. In Indian ayurveda, mushrooms were kept under "tamasika ahara" (Tamasic diet) along with meat of an animal, fish, the fertilized egg, onion, garlic etc., which were believe to cause certain potentially physical conditions and also as a medicine for enhancing energy and vitality (Saddler 2003). Wild edible mushrooms and ethno-mycological practices of the wild mushrooms have been documented by several authors from North-East Indian states (Boruah et al. 1996; Sing et al., 2002). *Morchella* spp. (Ascomycotas) generally recognized as morels and 'Guchhi' in the Indian market are well known for its edibility (Lakhanpal et al. 2010). The local community in Northwestern Himalaya used common names for many of the wild mushrooms which, suggest that they have a practice of utilizing it for a very long period. Semwal et al. (2014) documented some of those wild edible mushrooms from Northwestern Himalaya. Traditionally in India, mushroom collection is known to be the final alternative for poor people in lean periods (Harsh et al. 1993); Chinese and Mexicans often offered mushrooms as gifts owing to their nutritional values (Härkönen 2002; Garibay-Orijel et al. 2007). Because of the fact that habitats include decaying matter and fear of poisoning these wild mushrooms are avoided religiously (Härkönen 2002; Walleyn and Rammeloo 1994). Numerous tribes belonging to the Nahua speaking Indians of Mesoamerica used psychoactive fungus (*Psilocybe*) in magico-religious ceremonies as divinatory sacraments (Schultes 1939, 1940; Wasson and Wasson 1957).

Britain is usually classified as mycophobic. While Castilian or Valencian of Spain also considered as mycophobic society. Mycophilic immigrants and commercial reasons have changed attitudes of **mycophobic community**. For example, now, there are an increasing number of Americans people who collect WEF in the forest (Dyke and Newton 1999). However variations among these societies still exist widely in different parts of the world. Variable traditions also exist in the United Republic of Tanzania (Härkönen et al. 1994a, b). This clearly indicates that mushroom collection was found to be more common among the tribes who lived in high-altitude forest areas which are important part of their diet and provide income when they started selling in local markets.

2.3 Common Wild Edible Mushroom

WEF are non-timber forest resource well documented in many parts of the world. Of the 1.5 million estimated fungi, only 14000 nos. were described across the world (Chang and Miles 2004). About 7000 mushrooms (50%) possess varying degree edibility and 3000 from 31 genera are potentially edible but 10% are known to be lethal (Chang and Miles 1996). Boa (2004) compiled over 200 genera of macrofungi which contains species that are either consumed directly as food or are used indirectly for health benefits. Some common edible species are *Agaricus, Auricularia, Dictyophora, Flammulina, Hericium, Lentinula, Pholiota, Pleurotus, Tremella* and *Volvariella* have been consumed across the world. But there are species that are highly esteemed such as *Cantharellus, Sparassis, Lactarius, Suillus, Tuber* and *Morchella.* Often there are species that are eaten in a region or country which are consider harmful or poisonous by others e.g. *Agaricus arvensis, A. semotus, Amanita gemmate, Coprinus atramentarius,* and *Lenzites elegans* (Lincoff and Mitchel 1977; Logemann et al. 1987; Rammeloo and Walleyn 1993 and Chang and Mao 1995). In developing countries, WEF are the source of food and medicine (Cakilcioglu et al. 2011) and provides additional income.

In the recent years, the nomenclature and identification has step up with the molecular characterization of mushroom, thus knowing the scientific name of a mushroom highly increase the chance of identifying its edibility. Garibay-Orijel et al. (2006) reported about 300 wild edible mushroom has been consumed by rural people of Mexico and ca. 180 species were known to be potentially edible mushrooms (Cordova et al. 2002). In some cases, knowing the genus alone suffices its edibility for example; all known species of *Cantharellus* are edible (Boa 2004). *Tuber* sp., generally known as truffles is considered one of the costliest mushrooms, particularly in the Northern European counties. Few desert truffles like *Terfezia* and *Tirmania* are endemic in arid and semi-arid areas of the Mediterranean (Hall et al. 2007).

Following are some of the important genera that are consumed worldwide (Iordanov et al. 1978; Saenz et al. 1983; Bon 1987; Zang 1988; Rammeloo and Walleyn 1993; Buyck 1994; Chang and Mao 1995; Degreef et al. 1997; Hall et al. 1998a, b; Jordan 2000; Boa 2004; Thawthong et al. 2014; Zhang et al. 2015;):

1. ***Agaricus***: The genus *Agaricus* is an important source of food and medicine, hosting over 434 species worldwide, of which about 60 species are reportedly eaten in 29 countries. *Agaricus* L. includes economically important species like *A. bisporus*, commonly known as the button mushroom (Cappelli 1984; Kerrigan 1986; Largeteau et al. 2011). It's considered as the most widely cultivated edible species of mushrooms with over 32 % of the total mushroom production, worldwide. Some of the edible species includes; *A. arvensis, A. aurantiacus, A. bingensis, A. bisporus, A. bulbillosus, A. blazei, A. campestris, A. comptulus, A. croceolutescens, A. endoxanthus, A. erythrotrichus, A. gennadii, A. goossensiae, A. maculatus, A. micromegethus, A. nivescens, A. placomyces, A. purpurellus, A. rodmani, A. semotus, A. sylvicola, A. silvaticus, A. subedulis, A. subperonatus, A. subrutilescens, A. volvatulus.* Others species were not well accepted but commercially cultivated species includes; *A. arvensis, A. campestris, A. bitorquis* and *A. subrufescens.* Few species are reported to be poisonous for example *A. xanthodermus, A. litoralis.*
2. ***Amanita*:** There are about 83 edible species reported from 31 countries (Boa 2004). *A. caesarea* is highly valued and one of the most highly sought mushroom worldwide (Boa 2004; Wang and Chen 2014) especially in Mexico, Nepal and Turkey. Some of the edible species consumed worldwide are; *A. argentea, A.*

aurea, A. bingensis, A. caesarea, A. calopus, A. calyptratoides, A. calyptroderma, A. ceciliae, A. craseoderma, A. crassiconus, A. crocea, A. flammeola, A. flavoconia, A. gemmate, A. goossensiae, A. hemibapha, A. hovae, A. inaurata, A. loosii, A. masasiensis, A. muscaria, A. perphaea, A. rubescens, A. robusta, A. subviscosa, A. spissa, A. strobilaceovolvata, A. tuza, A. fulva, A. virgineoides, A. vaginata, A. umbonata, A. xanthogala and *A. zambiana* (Simmons et al. 2002; Flores et al. 2002). Few of this species are exported and traded. Some species of *Amanita* have conflicting reports on edibility for example *A. gemmata* reportedly edible in Mexico and Costa Rica while in Guatemala the species was reported with case of poisoning like-wise, in *A. flavoconia* and *A. spissa* (Logemann et al. 1987; Chang and Mao 1995). Some species are highly poisonous for example *A. phalloides,* which is popularly called as "death cap" known for more number of deaths after consuming it (Lincoff and Mitchel 1977; Liu and Yang 1982; Boa 2004).

3. ***Auricularia***: This genus is known by their several common names; ear fungi, Judas's ear, Jew's ear, jelly ear, black jelly etc., and distributed throughout the temperate and subtropical regions worldwide (Ingold 1985; Du et al. 2011). There are about 13 edible species reported from 24 countries. Many Southeast Asian countries like China, Taiwan, Thailand, Philippines Indonesia and Malaysia are into cultivation of this Ear fungus. However, China alone produces about 3.6 million tonnes per year that is 6% of the total world's production. *A. auricula* and *A. polytricha* are widely considered to be the earliest cultivated mushroom dating back 600 A.D. China (Lou 1978; Quimio 1979; Li 2012). Some of the other edible species consumed worldwide are; *A. auricula-judae, A. cornea, A. delicata, A. fuscosuccinea, A. mesenterica, A. polytricha* and *A. tenuis,* (Prance 1984; Flores 2002).
4. ***Boletus:*** About 72 edible species have been reported from over 30 countries (Boa 2004). The common species, *B. edulis,* known by several names (king bolete, *porcini, suilli, penny buny, panza etc.*) is an ecologically and economically important species consumed worldwide (Arora Arora and Dunham 2008; Feng et al. 2012). Yugoslavia was the highest exporter during 1993–1995, producing a maximum of 5186 tonnes in 1993. In the South Africa, *B. edulis* were introduced to the native forest with the plantation of exotic trees; however the locals were skeptical of consuming the unknown variety (Boa 2004). In China, families living in the mountain areas, exploited *B. edulis* and sold them in farmers market to overcome financial insecurity (Zhang et al. 2017). *B. edulis* was popularly consumed in countries like; Europe, North America and Asia (Agueda et al. 2008). Other edible species consumed worldwide are; *B. aereus, B. aestivalis, B. appendiculatus, B. aurantiacus, B. atkinsonii, B. barrowsii, B. bicoloroides, B. bouriqueti, B. bulbosus, B. calopus, B. caudicinus, B. citrifragrans, B. colossus, B. communis, B. crassus, B. cyanescens, B. elegans, B. emodensis, B. erythropus, B. felleus, B. frostii, B. griseus, B. impolitus, B. loyo, B. luridus, B. luridiformis, B. michoacanus, B. nigroviolaceus, B. pseudoloosii, B. pinicola, B. pinetorum, B. pinophilus, B. rubellus, B. regius, B. reticulatus, B. russellii, B. scaber, B. subtomentosus, B. sulphureus, B. speciosus, B. truncatus, B. variega-*

tus, B. variipes, B. violaceofuscus, B. vitellinus and *B. zelleri* (Bouriquet 1970; Vasilèva 1978; Malyi 1987; Adhikari and Durrieu, 1996; Montoya-Esquivel 1998; FAO 1998; Hall et al. 1998a, b; Ereifej and Al-Raddad 2000; Montoya-Esquivel et al. 2001; Sabra and Walter 2001).

5. ***Cantharellus***: *Cantharellus* spp. are ectomycorrhizal fungi, about 22 edible species reported from 45 countries with a good reputation for edibility (Buyck 2008; Arora and Dunham 2008). The species are commonly known for their fruity, apricot-like odour with diverse species distributed throughout the world. Most common species, *C. cibarius* popularly known as golden chanterelle, is a highly commercial species which is harvested from nature alone. As these mushrooms are mycorrhizal and haven't been mass cultivated successfully but procuring from the local sellers will be expensive (Hall and Zambonelli 2012; Yun and Hall 2004). There are no known poisonous species. Some species are very common in the markets of many countries and are sold in mixture of different species. Some edible species spread across the world are; *C. cibarius, C. cinereus, C. cinnabarinus, C. congolensis, C. cyanescens, C. cyanoxanthus, C. densifolius, C. eucalyptorum, C. floccosus; C. floridulus, C. formosus, C. ignicolor, C. incarnatus, C. infundibuliformis, C. isabellinus, C. longisporus, C. luteocomus, C. luteopunctatus, C. lutescens, C. madagascariensis, C. miniatescens, C. minor, C. odoratus, C. platyphyllus, C. platyphyllus, C. pseudofriesii, C. pseudocibarius, C. ruber, C. rufopunctatus, C. splendens, C. symoensii, C. subalbidus, C. subcibarius, C. tenuis* and *C. tubaeformis* (Bouriquet 1970; Buyck 1994; Härkönen et al. 1994a; Adhikari and Durrieu 1996; Adhikari 1999; Tedder et al. 2002; Flores 2002).
6. ***Clitocybe***: This genus is estimated to have ca. 1131 associated species. However, only a few members of this genus considered as edible and others as toxic or poisonous. Although, the genus is better known for its toxicity (not as deathly), some species have proven to be beneficial in medical aspect. For example *Clitocybe nebularis* (Pohleven et al. 2009), *Clitocybe maxima* (Zhang et al. 2010) and *C. alexandri* (Vaz et al. 2010). *C. clavipes, C. fragrans, C. geotropa, C. gibba, C. hypocalamus, C. infundibuliformis, C. nebularis, C. odora, C. squamulosa* and *C. suaveolens* are few examples of edible species reported from various parts of the world including; Australia, Bulgaria, Chile, China, Hong Kong, India, Indonesia, Mexico, Russia and Ukraine (Burkhill 1935; Vasilèva 1978; FAO 1998).
7. ***Cordyceps***: It's a unique Ascomycetous genus grows on the larva of insects. About 37 edible species have been reported from three countries. Although edible species are consumed only for their health benefits, while many species has been described from Japan, they are intensively collected in parts of China and Nepal. *C. cicadicola, C. gunnii, C. liangshanensis, C. ophioglossoides, C. militaris* and *C. sinensis* are some valued species for their medicinal property (Hall et al. 1998a, b; Gong and Peng 1993; Yang et al. 2009). *C. militaris* is medicinally important species with beneficial properties such as antioxidant (Chen et al. 2013; Jiang et al. 2011), **antitumor**, **anti-inflammatory** and **immunomodulatory** and effects (Hsu et al. 2008; Jiang et al. 2011; Bai and

Sheu 2018). Localities collect these mushrooms to overcome the finacinal needs. Due to the anthropological effects and habitat loss, the ecosystem of *C. sinensis* has been affected which ultimately declined the natural yield.

8. ***Cortinarius***: There are about 30 edible species reported from over 11 countries. In Europe and North America, *Cortinarius* spp. is less popular due to the incidences associated with poisonous species. An example, *C. orellanus* was responsible for a total count of 11 dead in the year 1952 (Lincoff and Mitchel 1977; Lampe and Ammirati 1990). Only a few edible species have been reported from countries such as Costa Rica, China, Japan, Russia and Ukraine. The edible species includes; *C. alboviolaceus, C. armeniacus, C. armillatus, C. claricolor, C. claricolor var. turmalis, C. cornucopioides, C. collinitus, C. crassus, C. elatior, C. glaucopus, C. largus, C. mucosus, C. multiformis, C. orichalceus, C. praestans, C. purpurascens, C. rufo-olivaceus, C. varius* (Vasilèva 1978; Liu and Yang 1982; Zang 1984; Chamberlain 1996; Montoya-Esquivel et al. 2001).
9. ***Flammulina***: *F. velutipes,* popularly known "golden needle mushroom" was ranked fifth in the year 1997 for a total worldwide production of edible mushrooms in Southeast Asian counties like China, Japan, Korea, and Taiwan (Kües and Liu 2000; Psurtseva 2005). Till 1977 *Flammulina* was considering monotypic genus which was separated into two species namely; *F. ononides* and *F. velutipes*. Later, several species were reported from across the world and base on their authentic descriptions, 14 species has been accepted under this genus however information on their edibility is very limited (Redhead and Perterson 1999; Perez and Fernández 2007; Bas, 1983; Ge et al. 2008 and Ge et al. 2015).
10. ***Laccaria***: Nine edible spp. are reported from 17 countries. Common species is *L. lacata* found in North temperate countries such as Europe, North America, Mexico and Costa Rica (Chamberlain 1996; Tedder et al. 2002). This genus is mycorrhizal thus cultivation is not promising however wild mushrooms are collected and sold in the local markets (Boa 2004). Some of the edible species consumed worldwide are: *L. amethystea, L. amethystina, L. amethysteo- occidentalis, L. bicolor, L. edulis, L. farinacea, L. laccata, L. proxima* and *L. scrobiculatus* (Lopez et al. 1992; Tedder et al. 2002; Flores 2002).
11. ***Lactarius***: There are about 94 edible species reported from over 39 countries. All species of *Lactarius,* when fresh, are characterized by the unique ability to produce a milky fluid, if cut or broken. The color and taste of the milk varies between the species and are considered of great taxonomical value (Athanasakis et al. 2013). They are widely distributed, from Asia, America to Europe (Flores et al. 2002). In Spain, particularly in Palencia, *L. deliciosus* are a valuable mushroom which is sold at 2 € per kg and about 4000 kg are marketed on a daily basis during season (Roman and Bao 2004). China produces around 308000 tons of *L. deliciosus* annually (Sun and Xu 1999) while Estonia produces over 250 tons of *L. rufus* (Kalamees and Silver 1988). *L. piperatus* (peppery milk-cap; currently placed under the genera *Lactifluus*) and *L. torminosus* (woolly or bearded milk-cap) were reported to be edible and included in Turkish cuisines (Malyi 1987; Caglarirmak et al. 2002) but former is reported as poisonous in China (Liu and Yang 1982). Some edible species include; *L. akahatsu,*

L. angustus, L. annulatoangustifolius, L. camphoratus, L. carbonicola, L. chrysorrheus, L. congolensis, L. controversus, L. corruguis, L. deliciosus, L. denigricans, L. densifolius, L. edulis, L. flavidulus, L. gymnocarpoides, L. gymnocarpus, L. hatsudake, L. heimii, L. indigo, L. insulsus, L. inversus, L. lapponicas, L. kabansus, L. laevigatus, L. laeticolor, L. latifolius, L. luteopus, L. medusae, L. mitissimus, L. necator, L. pelliculatus, L. phlebophyllus, L. piperatus, L. princeps, L. pseudovolemus, L. pubescens, L. pyrogalus, L. quietus, L. resimus, L. rubidus, L. rubrilacteus, L. rubroviolascens, L. rufus, L. salmonicolor, L. sanguifluus L. scrobiculatus, L. sesemotani, L. subdulcis, L. subindigo, L. tanzanicus, L. torminosus, L. trivialis, L. vellereus, L. volemoides, L. volemus, L. xerampelinus, L. yazooensis and *L. zonarius* (Bouriquet 1970; Vasilèva 1978; Härkönen et al. 1994b; Adhikari and Durrieu 1996; Namgyel 2000; Demirbas 2000; Montoya- Esquivel et al. 2001; Deschamps 2002; Caglarirmak et al. 2002; Lian et al. 2007).

12. ***Leccinum***: There are about 14 edible species widely collected and consumed in Europe and New Zealand (Boa 2004). The species *L.* versipelle popularly collected in Poland and highly valued when fresh but not as much when dried (Guminska and Wojewoda 1985). Some edible species reported across the world are; *L. aurantiacum, L. chromapes, L. extremiorientale, L. griseum, L. holopus, L. lepidum, L. manzanitae, L. oxydabile, L. rugosiceps, L. scabrum, L. testaceoscabrum* and *L. versipelle* (Lincoff and Mitchel 1977; Vasilèva, 1978; Malyi 1987; Walleyn and Rammeloo 1994; Maŕtinez et al. 1997).
13. ***Lentinula***: There are only about three edible species reported from six countries. *L. edodes* popularly known as "shiitake" and is one of the most cultivated mushroom worldwide especially South East Asia (Reshetnikov et al. 2001). In 1986, worldwide production of *L. edodes* was 14% (Chang and Miles 2004). Species such as *L. boryana, L. edodes, L. lateritia* are edible, reported from Chile, India, Mexico, Nepal, Papua New Guinea and Thailand (Purkayastha and Chandra 1985; Jones et al. 1994; Sillitoe 1995; Schmeda et al. 1999; Adhikari 1999).
14. ***Lentinus***: There are about 28 edible species reported from over 24 countries (Boa 2004). Most species of *Lentinus* are edible, but few species with their tough texture are of less significant. The edible species such as; *L. sajor-caju* and *L. strigosus,* are important species possessing anti-oxidant property (Yang et al. 2002). *L. araucariae, L. brunneofloccosus, L. crinitus, L. glabratus, L. sajor-caju, L. strigosus, L. squarrosulus, L. tuber-regium, L. velutinus,* are few edible species reported from Benin, Brazil, Burundi, Central Africa, China, Congo, Ethiopia, Gabon, Ghana, India (Zang 1984; Prance 1984; Rammeloo and Walleyn 1993; Buyck 1994; Kalotas 1997; Obodai and Apetorgbor 2001).
15. ***Lepista***: Commonly called "blewit mushroom", a wild edible mushroom (Eyüpoğlu et al. 2011). They are found throughout mainland of Europe and in many other parts of the world including North America. They are very important economically, and are cultivated by Mushroom Research Center in France (Suberville et al. 1996), but cultivated wood blewit mushrooms are not delicious compared to wild wood blewit mushrooms (Barutçiyan 2012). Wood ble-

wits are collected for their medicinal uses. They are highly nutritious containing 44.2% crude protein, 9.0% lipids, 5.4% ash and 41.4 % carbohydrates (Colak et al. 2007). Medicinal uses includes; prevention against beriberi (Dulger et al. 2002), antimicrobial, antioxidant properties (Pinto et al. 2013). Some of the *Lepista* spp. consumed throughout the world are; *L. caespitosa; L. caffrorum, L. dinahouna, L. glaucocana, L. irina, L. luscina, L. nuda, L. personata* and *L. sordida,* (Hall et al. 1998a, b).

16. ***Lycoperdon***: About 22 edible species are reported from over 19 countries (Boa 2004). The genus is distributed worldwide; Common species are *Lycoperdon giganteum, L. pyriforme* and *L. perlatum* popularly known as puffballs, one of the biggest edible mushrooms with size reaching up to 150 Cm in diameter (John et al. 2011). Habitat to woods, grassy areas, and along roads. However, *L. perlatum* looks similar to immature fruit bodies of poisonous *Amanita* spp. (Lassoe et al. 1996). Some edible species are *L. asperum, L. candidum, L. endotephrum, L. echinatum, L. gemmatum, L. marginatum, L. oblongisporum, L. peckii, L. perlatum, L. pyriforme, L. pusillum, L. rimulatum, L. spadiceum, L. umbrinum, L. umbrinum var. floccosum* reported from countries such as Banin, Bhutan, Bulgaria, Canada, China, India, Kyrgyzstan, Madagascar, Mexico (Elćhibaev 1964; Harsh et al. 1996; Namgyel 2000; Lian et al. 2007), *L. asperum, L. pusillum, L. perlatum, L. pyriforme, L. spadiceum,* are used as medicine in China (Chang and Mao 1995).
17. *Macrolepiota*: There are about 13 edible species from over 33 countries. The common species *M. procera* traded in small scale; they have high nutritional values and are consumed all over the world (Boa 2004). *Chlorophyllum molybdites* a species with conflicting report of poisonous mushroom is often confuse with *M. procera*. Few worldwide distributed edible species worldwide are: *M. africana, M. dolichaula, M. excoriata, M. excoriata var. rubescens, M. gracilenta, M. gracilenta var. goossensiae, M. procera, M. prominens, M. procera var. vezo, M. puellaris, M. rhacodes* and *M. zeyheri* (Elćhibaev 1964; Bouriquet 1970; Vasilèva 1978; Saenz et al. 1983; Purkayastha and Chandra 1985; Adhikari and Durrieu 1996; Degreef et al. 1997; FAO 1998; Tedder et al. 2002). *M. neomastoidea,* distributed throughout Korea and other East Asian countries is reported to be poisonous (Kim et al. 2009).
18. ***Morchella***: There are about 62 reported species from over 28 countries (Boa 2004; Negi 2006). *M. esculenta* is well known species which are consumed by the people but are also known for their toxicity when eaten raw forms (Lincoff and Mitchel 1977). The species *M. esculanta* is among the most highly prized and morphologically recognizable fungi in the world (Goldway et al. 2000). In Turkey, this species cost around 130 euro/kg. (Okan et al. 2013). They are diverse and found worldwide. Some of the edible species are: *M. angusticeps, M. conica, M. conica var. rigida, M. costata, M. crassipes, M. deliciosa, M. elata, M esculenta, M. esculenta var. rotunda, M. esculenta var. umbrina, M. esculenta var. vulgaris* and *M. intermedia* (Singh and Rawat 2000; Deschamps 2002).
19. ***Pleurotus***: There are about 40 edible species reported from over 35 countries (Boa 2004). *P. ostreatus* most widely consumed popular species; commonly

known as oyster mushroom. They are cultivated in many parts of the world so *P. ostreatus* production was at 14.2 % of the total WEF produced worldwide (Chang 1999). Some of the species consumed around the world are; *P. abalonus, P. citrinopileatus, P. cornucopiae, P. cystidiosus, P. concavus, P. djamor, P. eryngii, P. floridanus, P. pulmonarius, P. ostreatus, P. rhodophyllus, P. spodoleucus, P. sapidus* and *P. salignus* (Zang 1984; Buyck 1994; Chamberlain 1996; Namgyel 2000).

20. ***Podaxis***: They appear like a stalked-puffball and are secotioid fungi in Agaricaceae which comprises about 44 species (Conlon et al. 2016). *P. pistillaris* is common edible species under this genus commonly reported from Afghanistan, Australia and Hawaiian Islands. Besides their use as food (Abraham et al. 2017), they are also used as hair dye in Australia (Batra 1983) and as baby-powder in West Africa (Gérault and Thoen 1992). In India and Pakistan, *P. pistillaris* species was reported to be edible (Batra 1983) but they were reported poisonous by Nigeriens (Walleyn and Rammeloo 1994).
21. ***Polyporus***: There are about 30 edible species reported across 20 countries. Many species are reportedly used as remedial medicine or are eaten but are relatively of minor importance (Boa 2004). Some of the edible species reported from around the world are; *P. aquosus, P. alveolaris, P. arcularius, P. badius, P. brumalis, P. blanchettianus, P. brasiliensis, P. confluens, P. croceoleucus, P. elegans, P. eucalyptorum, P. fimbriatus, P. grammocephalus, P. indigenus, P. moluccensis, P. mylittae, P. rhizomorphus, P. rugulosus, P. sapurema, P. stipitarius, P. squamosus, P. sanguineus, P. tricholoma, P. tubaeformis, P. tenuiculus, P. tinosus, P. tuberaster* and *P. umbellatus*, (Burkhill 1935; Bouriquet 1970; Prance 1984; Remotti and Colan 1990; Walleyn and Rammeloo 1994; Chang and Mao 1995; Sillitoe 1995; Adhikari and Durrieu 1996; Kalotas 1997; Hall et al. 1998a, b; Adhikari 1999; Härkönen 2002).
22. ***Ramaria***: There are about 44 edible species reported from over 18 countries. Several major species are regularly collected and sold in markets of Nepal and Mexico, of which *R. botrytis* are the most popularly consumed species (Boa 2004). In Nepal, *R. formosa* is considered as edible while in Bulgaria, it was treated as poisonous (Iordanov et al. 1978; Adhikari and Durrieu 1996). Some of the edible species collected worldwide are; *R. araiospora, R. apiculata, P. aurea, R. bonii, R. botrytoides, R. botrytis, R. cystidiophora, R. flava, R. flavobrunnescens, R. mairei, R. ochracea, R. obtusissima, R. rosella, R. rubiginosa, R. rubripermanens, R. subaurantiaca, R. stricta, R. sandaracina, R. sanguinea* and *R. subbotrytis* (Liu and Yang 1982; Walleyn and Rammeloo 1994; Chamberlain 1996; FAO 1998; Montoya-Esquivel 1998; Flores 2002).
23. ***Russula***: Nearly, 128 edible species has been repoted over 28 countries (Boa 2004). *R. emetic* is eaten by Mexicain and Russian but otherwise believe to be poisonous when eaten uncooked (Vasilèva 1978). *Russula* is mycorrhizal fungi, so very difficult to bring into cultivation and it is highly diverse. Some of the edible specie includes: *R. alutacea, R. atrovirens, R. atropurpurea, R. chamaeleontina, R. cyclosperma, R. cyanoxantha, R. cellulata, R. compressa, R. congoana, R. diffusa* var. *diffusa, R. delica, R. depallens, R. emetic, R. lepida, R.*

erythropus, R. grisea, R. hiemisilvae, R. meleagris, R. minutula, R. oleifera, R. olivacea, R. pectinata, R. pseudopurpurea, R. phaeocephala, R. pseudostriato-viridis, R. roseoalba, R. roseostriata, R. rubra, R. sesenagula, R. striatoviridis, R. testacea, R. vesca, R. virescens, R. viscida and *R. xerampelina* (Liu and Yang 1982; Buyck 1994; Degreef et al. 1997; Tedder et al. 2002).

24. ***Suillus***: There are about 27 edible species reported from over 25 countries. *S. luteus* is the common species collected and consumed worldwide; major collectors have been from Argentina, Ecuador and Chile (Hedger 1986). *S. granulatus*is is another species widely recorded edible and a good source of carbohydrate and minerals (FAO 1998). The species *S. placidus* are considered edible in Russia, however poisonous in China (Vasilèva 1978; Chang and Mao 1995). Estonia and Mexico, in the late 80's, were the leading producers of *Suillus* spp. producing about 280 kg/ha in total (Villarreal and Guzmán 1985; Kalamees and Silver 1988). Some of the important species includes: *S. abietinus, S. acidus, S. americanus, S. bovinus, S. brevipes, S. cavipes, S. granulatus, S. grevillei, S. hirtellus, S. luteus, S. lactifluus, S. pictus, S. placidus, S. plorans, S. pungens, S. pseudobrevipes, S. subluteus, S. tomentosus, S. variegatus* and *S. viscidus* (Lincoff and Mitchel 1977; Vasilèva 1978; Namgyel 2000; Montoya-Esquivel et al. 2001).
25. ***Sparassis***: *S. crispa* is popular edible species commonly known as Cauliflower mushroom. They are ectomycorrhizal fungus associated with coniferous forest from the mountains of Eastern Asia, Europe and North America (Humpert et al. 2001; Adhikari et al. 2005). However, they are also collected in countries such as; Canada, China, India, Mexico, Russia, Turkey, Ukraine, USA (Vasilèva 1978; Purkayastha and Chandra 1985; Hall et al. 1998a, b; Tedder et al. 2002). While, *S. crispa* are known for their edibility, they are more popularly known for their medicinal properties (Kawagishi 2007; Kwon et al. 2009; Kimura 2013; Elsayed et al. 2014).
26. ***Terfezia***: They are generally called as "desert truffles" native to arid and semi-aridlands of Mediterranean countries, parts of Asia, Europe, North America, North and South Africa. *T. claveryi* and *T. areanaria* are the popular species and very expensive in Europe. They are served as a major course in high-class restaurants; while in the countries of Middle East and Gulf, North Africa, this mushroom are eaten raw (Fortas and Chevalier 1992; Bradai et al., 2015). There are about 7 edible species reported from over eight countries which includes; *T. boudieri, T. claveryi, T. decaryi, T. pfeilii, T. leonis, T. arenaria, T. leptoderma* (Bouriquet 1970; Al-Naama et al. 1998; Martinez et al. 1997; FAO 2001; Sabra and Walter 2001).
27. ***Termitomyces***: It is known to be extremely esteemed genus with high nutritional values e.g. *T. clypeatus* possess a significant quantity of nutrients (Ogundana and Fagade 1982, Tibuhwa 2012). There are about 27 edible species reported from over 35 countries (Boa 2004) with world largest species, *T. titanicus.* They are more prevalent in usage among Africans and Asians, still poorly documented (Pegler and Vanhaecke 1994). *T. clypeatus* are collected and sold by the local markets of Tibet, Nepal and Northern India (Harsh et al.

1996). Notable edible species include: *T. aurantiacus, T. albuminosus, T. clypeatus, T. cylindricus, T. eurrhizus, T. entolomoides, T. fuliginosus, T. globulus, T. heimii, T. le-testui, T. mammiformis, T. medius, T. microcarpus, T. robustus, T. schimperi, T. striatus, T. titanicus* (Pegler and Vanhaecke 1994).

28. ***Tricholoma***: There are about 57 edible species reported from over 30 countries (Boa 2004). *T. matsutake* is the most valued and expensive species (Hall et al., 1998). *T. matsutake* are exported mostly from countries such as Bhutan, China, Koreas and Russia (Yeh 2000; Namgyel 2000; Winkler 2002) as such in China a '*T. matsutake*' farmers income is slated around 5 – 6 million USD per annum (Winkler 2002). While *T. pessundatum* are believed to be poisonous (Lincoff and Mitchel 1977) but it has been consumed in Hong Kong (Chang and Mao 1995). The edible species includes; *T. caligatum, T. columbetta,, T. equestre, T. flavovirens, T. georgii, T. imbricatum, T. magnivelare, T. matsutake, T. mauritianum, T. mongolicum, T. nauseosum, T. personatum, T. pessundatum, T. portentosum, T. quercicola, T. russula, T. rutilans, T. saponaceum,T. scabrum, T. sejunctum, T. sulphureum, T. terreum, T. tigrinum, T. ustaloides* and *T. vaccinum,* (Liu and Yang 1982; Purkayastha and Chandra 1985; Malyi 1987; Kytovuori 1989; Chang and Mao 1995; Hall et al. 1998b; Namgyel 2000; Tedder et al. 2002; Winkler 2002).
29. ***Tuber*:** There are about 18 edible species reported from eight countries (Boa 2004). Edible species such as; *T. aestivum* (black truffle), *T. borchii* (white truffle), *T. brumale* (black truffle), *T. indicum* (black truffle), *T. magnatum* (white truffle) and *T. melanosporum* (black truffle). These are few popular species which are widely studied. Some of these species are sold at a very high rate costing around 600 to 6000 € per kg (Luard 2006). *T. indicum* is one of the renowned commercial truffles in Yunnan Province, China and it has been exported to Japan, United States, Europe and Australia since the 1980's (Tao and Liu 1990). Truffles are known for their variety of aromatic property and thus appeals differently from person to person for example; *T. melanosporum* have an aroma of 'wet forest' in between the taste of a radish and a tint hazelnut, while the *T. magnatum* gives an aroma of garlicky cheese with subtle methane overtones (Cullere et al. 2009). *Tuber* spp. have been reported from all over the world; edibles species includes; *T. aestivum, T. borchii, T. brumale, T. californicum, T. gibbosum, T. hiemalbum, T. indicum, T. melanosporum, T. magnatum, T. mesentericum, T. moschatum, T. oligospermum, T. rufum* and *T. sinosum* (Zang and Pu 1992; Hall et al. 1998a; Sabra and Walter 2001; Moreno-Arroyo et al. 2001).
30. ***Volvariella***: There are about 12 edible species across 27 countries (Boa 2004). The most common species is *V. volvacea;* it was first cultivated by Buddhist monks for their consumption. Later, in 1875, it was gifted to the royal family as a tribute. Cultivation of these paddy straw mushrooms was first started almost 300 years ago during the eighteenth century (Chang 1977). It was introduced to other parts of the Asian counties during 1932 to 1935 (Baker 1934; Chang 1974) (Reshetnikov et al. 2001). Following are some countries that are either harvesting from the wild or cultivation: Benin, Central Africa, Chile, China, Congo, Costa Rica, Hong Kong, Indonesia, Ghana, India, Israel, Madagascar,

Malawi, Mauritius, Mexico, Nepal, Nigeria, Peru, Russia, Taiwan, Thailand. Some common edible species are: *V. bombycina, V. bakeri, V. diplasia, V. esculenta, V. earlei, V. parvispora, V. speciosa, V. terastria* and *V. volvacea* (Bouriquet 1970; Oso 1975; Vasilèva 1978; Sarkar et al. 1988; Remotti and Colan 1990; Wasser 1995; FAO 1998; Adhikari 1999).

2.4 Nutrional Properties of Edible Mushrooms

With a long history of mushrooms, as food source, they are also reported for favorable nutritional effects on human health. The ancient Romans called "food of the gods" and the first Egyptians called the "gifts from God of Osiris" and Chinese called it "the elixir of life". Chang (1999) stated that over 2000 years mushrooms were consumed for their nutrition and therapeutics properties in China. In vegetarian diets, these nutrients are extremely valuable because they offer all the essential amino acids; they have higher protein content than most vegetables. Among the 2000 edible mushrooms, ca. 850 spp. are available in India. The edible mushrooms have wider range of usage not only as food but also in pharmaceuticals, nutraceuticals and cosmeceuticals. They are in rich carbohydrate, high proteins content (including amino acids), fibre, low fat and calories, and chitin. In addition, trace elements such as calcium, phosphorus, iron, copper, chlorine, sodium, zinc, manganese and bromine are present; it can be recommended for those who have high cholesterol because mushroom has very low fat content (Mattila et al. 2001; Barros et al. 2007). They also possess higher quantities of vit A, B complex (vit B1 (Thiamin), B2 (Rhiboflavin), B3 (pantetonic acid), B5 (Nicotinic acid) that are reported to be good for the nervous system. Besides, their vit C and vit D concentration is 5-10 times higher than vit B3. Mushrooms also possess various metabolites, such as phenolics, flavonoids, polyketides, carotenoids, terpenoids, variegatic acid, quinones and steroids (Teissedre and Landrault 2000; Cheung et al. 2003). Presence of phenolics and flavonoids plays a vital role in antioxidant property of mushrooms since they authorize to be reducing agents and also as singlet oxygen quenchers, respectively (Rice-Evans et al. 1996). Nutritional composition of common edible mushrooms were listed Table 2.1. The nutrient constituents of mushrooms such as protein, amino acids, dietery fiber, carbohydrates, lipids, micro-nutrients, minerals, ash and less fat and nearly no cholesterol are accountable for the medicinal properties (Tsai et al. 2007; Chang and Wasser 2012; Liu et al. 2012; Kalogeropoulos et al. 2013).

2.4.1 Protein

The dietary significance of mushrooms is mainly associated to the protein index. Mushroom is known to posses more dietary protein value when compared to protein from plant origin (FAO 1991). Varying protein constituent of mushrooms is

Table 2.1 Nutritional compositions of important wild edible and cultivated mushrooms

Species	Common names	Carbohydrate	Crude fat	Crude fiber	Crude protein	Ash	Reference
Agaricus bisporus	Button mushroom	74	2.18	-	14.08	9.74	Reis et al. 2012
Agaricus brasiliensis	Almond mushroom	26.74	2.62	-	26.74	6.81	Tsai et al. 2008
Agaricus campestris[b]	Meadow mushroom	58.16	0.11	-	18.57	23.16	Pereira et al. 2012
Amanita caesaria	Caesar's mushroom	55.63 ± 0.06	3.50 ± 0.00	-	34.77 ± 0.06	6.05 ± 0.01	Ouzouni et al. 2009
Armillaria mellea[a]	Honey fungus	65.47 ± 0.15	2.10 ± 0.02	-	24.47 ± 0.12	7.95 ± 0.02	Ouzouni et al. 2009
Armillaria tabesceus[a]	Ringless honey mushroom	66.87 ± 0.06	2.54 ± 0.03	-	22.90 ± 0.20	7.63 ± 0.15	Ouzouni et al. 2009
Boletus aereus[a]	Queen bolete	62.10 ± 0.10	4.47 ± 0.02	-	27.17 ± 0.15	6.25 ± 0.02	Ouzouni et al. 2009
B. armeniacus	-	68.1	1.56	-	18.25	12.09	Pereira et al. 2012
B. edulis	Black headed bolete	70.95	2.45	-	21.07	5.53	Heleno et al. 2011
B. erythropus[b]	Red stemmed bolete	52.43	0.75	-	20.92	25.90	Grangeia et al. 2011
B. reticulates[b]	-	55.16	2.55	-	22.57	19.72	Heleno et al. 2011
Calocybe gambosa	St.George mushroom	69.82	0.83	-	15.46	13.89	Vaz et al. 2011
Calvatia utriformis[b]	Saddle shaped buff ball	59.92	1.92	-	20.37	17.81	Grangeia et al. 2011
Cantharellus cibarius[a]	Chanterelle, Yellow Chanterelle, Girrole	66.07 ± 0.23	2.88 ± 0.02	-	21.57 ± 0.21	9.44 ± 0.01	Ouzouni et al. 2009
Clitocybe odora[b]	Aniseed mushroom	70.66	2.46	-	17.33	6.42	Akata et al. 2012
C. subconnexa[a]	-	27.35 ± 0.13	1.02 ± 0.09	38.74 ± 0.79	7.42 ± 0.25	5.98 ± 0.04	Heleno et al. 2015
Coprinus comatus	Shaggy ink cap	70.35	1.13	-	15.67	12.85	Vaz et al. 2011
Fistulina hepatica[a]	Beefsteak fungus, Ox tongue	66.00 ± 0.10	3.17 ± 0.02	-	22.60 ± 0.20	8.20 ± 0.10	Ouzouni et al. 2009
Flammulina velutipes[b]	Velvet Shank	70.85	1.84	-	17.89	9.42	Pereira et al. 2012
Flammulina velutipes	Velvet Shank	85.99	2.9	-	3.87	7.3	Reis et al. 2012
Hygrophorus russula[a]	Russula Wax- cap	53.33 ± 0.06	6.00 ± 0.10	-	32.47 ± 0.06	8.18 ± 0.02	Ouzouni et al. 2009
Hypsizigus marmoreus	White beech mushroom	68.56	4.9	-	19.6	7.75	Lee et al. 2009
Laccaria laccata[#]	Deceiver	12.77	3.76	-	62.78	20.69	Heleno et al. 2009
Lactarius deliciosus[b]	Saffron milk cap	64.63	8.02	-	20.02	7.15	Akata et al. 2012
Lentinula edodes	Shiitake mushroom	87.14	4.4	-	1.73	4.4	Reis et al. 2012
Lepista nuda[a]	Wood Blewit	56.33 ± 0.15	3.23 ± 0.01	-	34.37 ± 0.15	6.03 ± 0.02	Ouzouni et al. 2009

Lycoperdon echinatum[#]	Spiny puff ball	65.83	1.22	-	23.52	9.43	Grangeia et al. 2011
Macrolepiota dolichaula	Paraol Mushroom	56.2 ± 0.10	3.2 ± 0.20	4.85 ± 0.18	19.95 ± 1.35	7.3 ± 0.15	Atri et al. 2014
M. procera	parasol mushroom	60.82 ± 0.11	3.4 ± 0.08	5.1 ± 0.22	19.95 ± 1.06	1.93 ± 0.06	Atri et al. 2014
Pleurotus eryngii	king trumpet mushroom, French horn mushroom or king oyster mushroom	81.37	1.45	-	11.0	6.18	Reis et al. 2012
P. ostreatus	tree oyster mushroom or the grey oyster mushroom	85.86	1.4	-	7.02	5.7	Reis et al. 2012
P. sajor-caju	Oyster *Mushroom.*	55.3	1.0	-	37.4	6.3	Akyüz and Kirbağ 2010
Ramaria largentii[a]	Orange coral mushroom	87 ± 0.25 6.	67 ± 0.12 58.	-	80 ± 0.46 5.	67 ± 0.12	Ouzouni et al. 2009
Russula cyanoxantha[b]	Chacoal buner	74.65	1.52	-	16.8	7.03	Grangeia et al. 2011
R. delica	milk-white brittlegill	63.87 ± 0.31	4.44 ± 0.04	-	26.10 ± 0.30	5.61 ± 0.03	Ouzouni et al. 2009
Termitomyces badius	Bhatolian+, Baat Koir+	39.0 ± 0.17	2.2 ± 0.10	2.5 ± 0.01	44.00 ± 0.10	6.6 ± 0.03	Atri et al. 2014
T. heimii	Goal Tatmour[c], Joru Koir[c]	36.2 ± 0.72	1.65 ± 0.19	5.0 ± 0.11	40.95 ± 0.84	8.6 ± 0.05	Atri et al. 2014
T. mammiformis	Goal Tatmour[c], Joru Koir[c]	47.65 ± 0.02	3.3 ± 0.17	8.0 ± 0.26	23.45 ± 0.04	9.9 ± 0.09	Atri et al. 2014
T. striatus	Goal Tatmour[c], Joru Koir[c]	60.27 ± 0.20	3.25 ± 0.06	4.1 ± 0.15	12.95 ± 0.05	12.13 ± 0.33	Atri et al. 2014
Volvopluteus gloiocephalus[a]	Big sheath mushroom	13.97 ± 0.34	4.62 ± 0.04	39.12 ± 0.29	19.66 ± 0.14	14.19 ± 0.07	Heleno et al. 2015

[a]g/100g of tissue
[b]Wild edible mushoom
[c]Local names

reliant on both physical and biological causes and it also differs during the fruiting body development and on **genetic structure** of the species (Ragunathan and Swaminathan 2003; Agrahar and Subbulakshmi 2005; Chang et al. 1981). For example, the total protein content of mushrooms is considerably low when boiled but remained relatively constant when air-dried at 40°C (Barros et al. 2007). Xu et al. (2011) published a comprehensive data of bioactive proteins from mushrooms. Mushrooms may not be having protein more than that of animal meats but the amount of crude protein is above than the other foods (Chang and Miles 1993). In cultivated mushrooms such as *A. bisporus*, *L. edodes*, *Pleurotus* spp., and *V. volvacea,* the total protein index ranges from 1.75 to 3.63% (Chang 1980). The cultivated mushrooms like *A. bisporus, P. ostreatus* and *P. sajor-caju* possess higher protein content than the untamed mushroom (Akyüz and Kirbağ 2010). In *P. ostreatus*, highest protein content with 92% digestibility was reported (Vetter and Rimoczi 1993).

Litchfteld et al. (1963) analyzed protein index in the dried mycelium in *Morchella* spp. The commercial morel mushroom powder found to have considerably higher protein content (51 g/100g dm) than the other three cultivated species viz., *M. crassipes* (22.8 g/100g dm), *M. esculenta* (25 g/100g dm) and *M. hortensis* (26.9 g/100g dm), (51 g/100g dm). The total protein in dried mycelium of *Agaricus arvensis* (28.16%), *A. campestris* (30.16%), *Morchella deliciosa* (29.16%) and *M. esculenta* (34.7%) was reported by Samajipati (1978).

The crude proteins of *C. indica*, *L. subnudus* and *V. volvacea* were found to be 14 to 27% (Purkayastha and Chandra 1976). Haddad and Hayes (1978) determined the protein content from the mycelium of *A. bisporus* (32 to 42% dm) but the total protein content of dried *A. bisporus* was found to be 46.5% dry weight basis, which is slightly higher when compared with mycelia protein content (Abou et al. 1987). The protein content in *Lactarius deliciosus* and *L. sanguifluus* was 14.71 to 17.37% and 15.20 to 18.87%, respectively (Sharma et al. 1988).

Bauer-Petrovska (2001), studied the protein profile of 52 different Macedonian edible mushrooms belongs to 17 different genera. His investigation revealed the maximum protein content was 48.81–52.06% dm in *Tricholoma georgii*, *Macrolepiota mastoidea*, and *Calvatia caelata,* while *Laetiporus sulphureus* and *Cantharellus cibarius* contains low protein content (14.00–16.19% dm). Maximum amount of albumins and globulins was observed in WEF but prolamins and glutelins are present extremely less quantity (Bauer-Petrovska 2001). Numerous researchers revealed that truffles acquired more protein than the other edible mushrooms (Singer 1961). Desert truffles composed of 20–27% (dm) protein (Kagan-Zur and Roth-Bejerano 2008). Three truffles collected from Iraqi namely, *Terfezia claveryi*, *Tirmania nivea* and *T. pinoyi* posses 8.02 to 13.84% protein content (Hussan and Al-Ruqaie 1999). Total protein content of Saudi Arabian black (Gibaah and Kholeissi) and white (Zubaidi) desert truffles ranged from 19.59 to 27.18% (Sawaya et al. 1985). So mushroom is a promising food that possibly helps to overcome the malnutrition crisis in the world.

2.4.2 Essential Amino Acids

Based on essentiality it is classified as essential amino acids (cannot be made by the body, supplemented to diet) and non-essential amino acids (can be synthesized by our body) (Young 1994). The proteins from commercially cultured mushrooms possess amino acids necessary for us and lysine is the important one among them, whereas tryptophane and methionine are the least required essential amino acids (Hughes et al. 1958; Altamura et al. 1967). The amino acid composition in some mushroom varieties can be equivalent to that of hen's egg and several species of mushroom is nearly equal to or superior than soy proteins (Yin and Zhou 2008). Hence, addition of mushrooms in vegan diet aid to achieve the essential amino acids, where intake of animal based protein is restricted (Galante and de Araujo 2014). Available literature suggests that the amino acid constituents of mushroom protein were meagerly studied even though they posses more nutrients than plants (Kalač 2009). The unique umami savour of mushrooms is due to the presence of aspartic and glutamic acid (Phat et al. 2016). Five different wild edible *Lentinus* spp. were rich in aspartic acid; maximum amount is present in *L. squarrosulus* (0.25 - 0.37%) (Sharma et al. 2012). Sawaya et al. (1985) were the first to report sulphur amino acids such as cystine, lysine, methionine and tryptophan in *Terfezia claveryi*, *Tirmania nivea* and *T. pinoyi*. Later they have observed sulphur rich amino acids in European truffles. These amino acids also limit the assimilation of mushroom protein (Dabbour and Takruri 2002). According to Bano and Rajarathnam (1982) *Pleurotus* spp. contains the lowest essential amino acids (tryptophane and methionine).

Many mushroom lectins have been discovered in past few years. The first known fungal lectin was from fly agaric mushroom (*Amanita muscaria*). The lectin activity reported to be related with the toxicity of the fungi (Ford 1910). Soon lectins from many common edible mushrooms including *B. edulis*, *L. deliciosus* and *L. edodes* reported to posses autonomy toxicity (Guillot et al. 1991; Tsivileva et al. 2005; Vetchinkina et al. 2008). Few mushroom lectins known to tolerate extensive variation in pH and temperature. Lectin derived from *V. volvacea* was found to be stable even at 80 °C and wide range of pH (Lin and Chou 1984). Lyophilized powder of *A. bisporus* lectin is commercialized and marketed by Sigma Aldrich Co.,

The exogenous amino acid content of frozen *P. ostreatus* (798 mg/100g of fm) is high when compared with *A. bisporus* (651 mg/100g fm) (Bernaś and Jaworska 2010). The amino acid of canned *A. bisporus* was higher (913.6 mg/100 g fm) than the *P. ostreatus* (769.3 mg/100g fm) (Jaworska and Bernaś 2011).

2.4.3 Fats/ Lipids

Crude fat (total lipids) of mushrooms constitute wide range of lipid complexes with free fatty acids, sterols, sterol esters, glycerides (mono-, di-, and tri) and phospholipids. According to Crisan and Sands (1978), the crude fat in mushroom ranges between 1% to 20% dm. Chang and Miles (2004) reported total lipids, occurrence of unsaturated fatty acids and abundance of linoleic acid from several mushroom species such *Agaricus, Auricularia, Boletus, Flammulina, Lentinula, Pleurotus* and *Volvariella* which vary from 1.1 to 8.3% dm. The total lipid of cultivated and wild strains of *P. ostreatus* was 3-5% (Hiroi 1982). The cap region contains more lipids when compared with the stalk. Total fatty acids in wild and cultivated strains of *P. ostreatus* constitute 20-30% neutral lipid, about 10% of glycolipid, 60-70% of phospholipid and 70-80% of linoleic acid. About 10 saturated, 6 monoenic and 4 polyunsaturated fatty acids were identified in *Boletus* spp.; linoleic, oleic and palmitic acids were primary that constitutes about 86–94% of total fatty acids (Hanuš et al. 2008). Phosphatidyl ethanolamine and phosphatidyl choline to be the most important individual phospholipids present in *A. bisporus* (Holtz and Schisler 1971) but few strains lack these fatty acids. Of the 58 edible mushroom screened, phosphatidylcholine was the major phospholipid found in 55 edible species (Vaskovsky et al. 1998). Huang et al., (1985) study shows the unusual elevated level of ergosterol and provitamin D2 interfered saponifiable lipid production in *V. volvacea.* Occurrence of high content of unsaturated fatty acids made WEF as nutritional dietary supplement and food.

2.4.4 *Fiber and Carbohydrates*

Fiber Still many WEF are underutilized and less explored in the view as a source of dietary fibre. A cluster of indigestible carbohydrates is crude fibre in other words **carbohydrate polymers** with ten or more monomeric units are called as dietary fibre, which cannot be hydrolyzed by the endogenous enzymes in humans (Codex 2010). *Boletus* spp. contains higher quantity of insoluble fibre (22-30% dm) than soluble (4–9% dm) (Manzi et al. 2004). In general, mushrooms reported to have 40% dm of crude fibre except *Craterellus aureus* and *Sarcodon aspratus* (5% dm). In *Pleurotus* spp. fibre content ranges from 7.4 to 27.6% dm but comparatively less in *V. volvacea* (4 to 20% dm) (Li and Chang 1982).

Carbohydrates In general, mushrooms constitute less amount of carbohydrate. It is an ideal diet for diabetic people, since its showing extremely slight consequence on human blood glucose level. They are having unique carbohydrates that can be stored as glycogen, which is common in human and animals but not as starch as in case of plants (Kalač 2013). Mushroom carbohydrates may comprise of hexoses, methylpentoses, pentoses, amino sugars, disaccharides, sugar acids and sugar alcohols (Crisan and Sands 1978). Mushroom polysaccharides are also best known for its **antitumor** and **immunomodulating** properties. These properties are reported to be possessed by many higher basidiomycetes because of the presence of some specific carbohydrates including, arabinose, fructose, fucose, glucose, maltose, mannitol, mannose, rhamnose, sucrose, trehalose and xylose (Zaidman et al. 2005; Zhang et al. 2007).

Structurally polysaccharide is composed of a backbone of β (1, 3)-linked glucose residues with acidic sugars, galactose and mannose residues in branches (Yoshioka et al. 1975). β-glucans are the important cell wall component of fungi which is the key polysaccharides found in mushrooms. They act as ligand and activate the membrane receptors to induce signaling pathways including defence against pathogenic microbes (Falch et al. 2000; Ishibashi et al. 2001; Kataoka 2002). They also stimulate the human immune system from detrimental contaminants and mutagens and provoke adaptive and innate resistant together (Vetvicka 2004). β-glucans of both wild and cultivated mushrooms are accountable for the anticancer, anticholesterolemic, antioxidant, immunomodulating and neuroprotective activities. *Pleurotus* spp. contains higher carbohydrates i.e. 46.6 to 81.8% when compared to *A. bisporus* (60% dm) (Bano and Rajarathnam 1982).

About 80–90% dm of mushroom cells consists of chitin. Eight *Boletus* spp. were reported to have 6.8–10.2% (dm) of carbohydrate (Manzi et al. 2004). When compared with water soluble polysaccharides, it known to have less bioactivities. **Chitin** is indigestible for humans and act as an important dietary fibre (Tao et al. 2006). Due to the presence α and β-glucans, chitin, galactans, hemicellulose, mannans and xylans mushrooms are known to be a prospective candidate for a potential source of **prebiotics**.

Mannitol is responsible for mass of texture of mushroom. Kalač (2012) analyzed the free sugar content in 27 species of WEF belong to 19 different genera. He observed the average amount of mannitol and trehalose ranges from 28.9 to 39.2 g kg^{-1} dm whereas glucose, fructose, mannose, ribose, sucrose and xylose occur at a low level. Trehalose and mannitol were reported at higher level in cultivated and mycorrhizal edible mushrooms (Reis et al. 2011) but these are not easy for humans to digest. During the course of processing there is slight decrease in the quantity of mannitol and trehalose (Barros et al. 2007). Grangeia et al., (2011) investigation revealed that the content of sugar in edible mycorrhizal species is higher (160 to 420 g kg^{-1} dm) than the edible saprotrophic mushrooms (up to 150 g kg^{-1} dm).

2.4.5 Mineral Composition

Mushrooms are also source minerals, possess highest amount of potassium (K), subsequently calcium (Ca), magnesium (Mg), phosphorus (P) and sodium (Na). These are known as major and minor mineral elements constitutes cadmium (Cd), copper (Cu), iron (Fe), manganese (Mn), molybdenum (Mo) and zinc (Zn) etc (Bano and Rajarathnam 1982; Li and Chang 1982). The Cu content is higher in *Pleurotus* spp. and varies from 12.2 to 21.9 ppm (Bano and Rajarathnaum 1982). Singer (1961) reported important minerals including Al, Ca, Cu, Fe, K, Mg, Mn, Na, P, S, Si and Zn from truffles. The European truffles contain abundant potassium, phosphorus, iron, and calcium (Saltarelli et al. 2008).

Mushrooms studied from metropolitan and industrial areas are contaminated with lethal compounds such as As, Cd, Hg, and Pb (Falandysz and Borovička 2013). The content of Cd, Cr, Ni and Zn from the mushrooms collected in rural areas is comparatively high, which may be because of quarries and industrial activities (Zhang et al. 2015). Silver (Ag) is one of the metallic elements with no dietary importance; due to its high affinity for proteins it is toxic and consequently gets collected in mushrooms (Falandysz and Borovička 2013).

Ash: The amount of ash in edible mushrooms is the poorly studied factor or this is not measured as a component for the estimation to analyze the mushroom quality (Falandysz et al. 2007; Falandysz et al. 2008). Estimation of ash requires high sophisticated laboratory infrastructure and instruments that may not be feasible for all (Falandysz et al. 2001). But the ash profile will give only general indication of mineral constituents of the mushrooms. The total ash content of mushroom is very less, only about 5–12% of dry matter.

2.4.6 Vitamins

Mushrooms composed of a number of primary vitamins including **vit B** complex, **vit C** and **vit D** (Cheung 2010; Kalač 2013). Information about vitamin composition of WEF has been lacking when compared with cultivated species (Mattila et al. 2001). The most potent provitamin A, *β*-carotene, is very low, ≥ 6 mg per kg dm found in Portuguese WEF (Pereira et al. 2012). The primary vitamins of mushroom constitute ascorbic acid, niacin, thiamine, tocopherols and riboflavin (Quan et al. 2007; Zhu et al. 2007; Yin and Zhou 2008; Zhou and Yin 2008; Xu et al., 2012). *Boletus edulis*, *B. speciosus* and *Thelephora ganhajun* were reported to possess tocopherol and vit D2 at a range of 8.9–45 and 4.7–194 mg/100g dm, respectively (Wu et al. 2005; Zhou and Yin 2008). In *Agaricus* spp. ascorbic acid content are relatively lower, unlike *B. edulis* and *C. cibarius*, which have higher content of ascorbic acid. Vit B complex have been reported from *A. bisporus* (white and brown), *L. edodes* and *P. ostreatus* (Caglarirmak 2011). Jaworska and Bernas (2009) stated that the levels of niacin and riboflavin decreased during mushroom processing. A type of vit B9 (total folates) have been quantified from several cultivated mushroom such as; crimini, chanterelle, enoki, maitake, morel, oyster, portabella, shiitake, UV-treated Portabella and white button mushrooms (Phillips et al. 2011). Tang et al. (2012) described ergosterol (640–1770 mg kg^{-1} dm) and also several phytosterols, especially brassicasterol from several *Tuber* spp. The amount of vitamins was found to have definite effects on the cooking and industrial processing of mushroom. In canned *Boletus*, vit B1 lost at a rate of 21–57% and vit B2 at a rate of 8–74% and at the worst case, reaching up to 76–99% lost in vit B complex (Yin and Zhou 2008; Zhou and Yin 2008).

2.4.7 Other Aromatic Metabolites

Each mushroom species posses a very characteristic aroma, which helps to determine them distinctly from other mushrooms (Cronin and Wada 1971). This unique characteristic aroma of mushrooms can be differentiated into volatile and nonvolatile components (Maga 1981). Some of the C8 aliphatic components are responsible for the unique flavour of mushroom are; 1-octen-3-ol, 2-octen-1ol, 3-octanol, 1-octanol, 1-octen-3-one and 3-octanone (Cho et al. 2006). While, 1-octene and 2-octene (often 3-octanone) are responsible for the typical aroma in mushrooms (Combet et al. 2006).

Truffles are popularly known for its unique aroma. The distinctive feature of the truffles is they do not share same desirable aroma even if morphology is same. More than 200 VOCs have been reported from truffles (Kanchiswamy et al. 2015). Various VOCs were reported in six different species namely *Tuber aestivum, T. borchii, T. brumale, T. dryophilum, T. magnatum* and *T. mesentericum* in various ratios (Federico et al. 2015). Thiophene, sulfur containing volatiles is the characteristic of *Tuber borchii* (Splivallo and Ebeler 2015). Accordingly, aroma of truffles may vary from cheesey, creamy, dusty, earthy, gasoline-like, garlicky, leathery, pungent and vanilla-like (Xiao et al. 2015). Bis (methylthio) methane, Dimethyl sulphide, 3-Ethyl-5-methylphenol, Hexadecanoic acid, 5-Methyl-2-propylphenol and B-Phenylethanol are example of few aromatic compounds reported from truffle (Omer et al. 1994; Buzzini et al. 2005; Cullere et al. 2009). Various preservation methods including refrigeration (4 °C) (Saltarelli et al. 2008), irradiation (Nazzaro et al. 2007) and modified atmosphere packing (MAP) (Rivera et al. 2010) are employed for industrial preservations of these aroma compounds. Pinho et al. (2008) studied the volatile components of eleven WEF (*Amanita rubescens, Boletus edulis, Cantharellus cibarius, Fistulina hepatica, Hygrophorus agathosmus, Russula cyanoxantha, Suillus bellini, Suillus granulatus, Suillus luteus, Tricholoma equestre* and *Tricholomopsis rutilans*) and concluded with 65 such compounds are responsible for the odor of mushroom which could be a key character in identifying these mushroom.

2.5 Hallucinogenic Mushrooms

Hallucinogenic or magic mushrooms have been widely consumed by indigenous groups in Mexicans. It came into the public attention in 1957 and then gained more popularity since then. Comprehensive detail of magic mushrooms consumption and its role in rituals among Mexican tribes and others across the world gave a spark among the psychoactive mushrooms consumers (Wasson et al. 1978). The first record of hallucinogenic mushroom was credited to the Yoruba tribe of Nigeria in Africa. It was traced back to the Paleolithic period (7000 – 9000 years ago). Of the 180 type magic mushrooms in the world, *Psilocybe* spp. is the "true" magic mushrooms, generally called as “shrooms”. They possess psychoactive indole of tryptamines called psilocybin and psilocin that has low level of physiological toxicity and never give addiction except low to acute psychedelic effects (Johnson et al. 2008; Tylš et al. 2014). More than 3000 years, the **psychoactive** fungus belongs to the genera *Psilocybe* and perhaps *Panaeolus* have been used conventionally. Besides the shrooms, there are many mushrooms such as *Conocybe, Copelandia, Galerina, Gymnopilus, Inocybe, Lycodperdon, Mycena, Panaeolus, Panaeolina, Pholotina* and *Pluteus* etc., are known to posses tryptamine derivates. The use of magic mushrooms alone or with alcohol

Fig. 2.3 *Ancient origin of magic mushrooms Credit:* Robert Brusco

was comparatively safe (van Amsterdam et al. 2011) but with mild to adverse effects like psychological distress, dangerous behaviour and enduring psychological problems (Carbonaro et al. 2016). *Psilocybe* spp. can be easily mistaken in wild with morphologically similar and non- *Psilocybe* or inedible or poisonous mushrooms. But at times illegal selling of "*Psilocybe* like poisonous mushrooms" has become lethal to an individual leading to death. In many countries including Australia, America and Europe illicit growing, possession and sale of magic mushrooms is punishable. *Psilocybe* have been cultivated sacredly because of the special kind of neurotropic (hallucinogenic) chemical constituent Psilocybin. These mushrooms are recognized as little saints or flesh of the gods among the native religious people (Fig. 2.3). French mycologist Heim (1969) documented this neurotropic species as *Psilocybe* species, which has been traditionally used during spiritual practices (Guzman 2008). This provides evidence that ancient Egyptians were not an exclusive group to exploit this substance in rituals (Guzman 2008).

While in India, people are well aware of these mushrooms either consume it with omelette or along with bread/ butter jam or with bread/banana or with honey (personnel survey; unpublished data). Even though it is under punishable act, many law enforcement officials are least aware of it. So there is a dire need to have a study on systematic research and on the abuse of the same among the youth. Scientists have adapted simple screening techniques to discriminate psilocybin and non-psilocybin mushrooms (Marumaya et al. 2006). DNA-based approach and **LC/MS** has been adapted to detect hallucinogenic mushroom and psychedelic drugs, respectively in grow kits from illegal market (Gambaro et al. 2015).

Nevertheless, psilocybins have been prescribed by the physicians in treating neurotic disorders in humans. These studies are looking at psilocybin and other hallucinogens to treat a number of psychiatric and stress disorders including chronic depression, post-traumatic stress disorder, and drug or alcohol dependency.

However, the mechanism and pharmacological profile of pure drug has to be compared with mushroom preparations. Studies reveal that psilocybin may decrease the depression and death anxiety along with increased the positive attitude in life of cancer patients. Pure psilocybin, "Sandbox" was marketed by Novartis has been recommended by physicians for psychedelic psychotherapy. Potentials of psilocybin in curing obsessive compulsive disorder (Wilcox 2014) and cluster headaches (Sewell et al. 2006) have also been investigated.

2.6 Cutivation of Edible Mushroom

The cultivation practice of WEF started several centuries ago; *Auricularia auricula* was probably the first mushroom to be purposely cultivated around A.D. 600, followed by *Flammulina velutipes* (ca. 800 AD). According to the "*Chinese Book of Agriculture*" (1313), the first historical record on cultivated mushroom was *Lentinula edodes* (Chang and Hayes 1978). Later, in France (1600 AD) *Agaricus bisporus* were first cultivated in outdoor later followed by *V. volvacea* (1700 AD) and *Tremella fuciformis* (1800 AD). This shows that, earlier practice of mushroom cultivation included only of outdoor that implies the limited knowledge and understanding of developing spawn, substrate and composts. It was only at the later part of the seventeenth century, the spawning technique for *Agaricus* was developed (Treshaw 1944). Elliott (1985) described the method to achieve pure culture. Later this was successfully accomplished in United Kingdom, followed by France in 1894 and United States in 1902 (Chang and Miles 2004). Since then the progress in mushroom cultivation technique has improved by miles, cultivating over 100 of species and producing millions of tons worldwide (Chang and Mao 1995; Stamets 2000; Boa 2004) (Table 2.2).

Table 2.2 A global overview of wild edible mushroom production with respect to countries according to Food and Agriculture Organization (FAO), United Nations.

Sl. No.	Countries	Production in tonnes (2000)	Production in tonnes (2005)	Production in tonnes (2010)	Production in tonnes (2015	Production in tonnes (2016)
1	Albania	100	100	100	100	100
2	Algeria	113	170	211	2526	1890
3	Armenia	0	0	80	278	361
4	Australia	36000	47992	41295	42777	50387
5	Austria	1000	900	1300	1200	1400
6	Azerbaijan	0	0	1900	1515	1562
7	Belarus	5000	6851	7000	7568	10135
8	Belgium	46300	41420	39154	30440	29450
9	Bosnia and Herzegovina	1200	2000	1200	1228	1203
10	Brunei Darussalam	8	9	12	13	14
11	Bulgaria	11500	1427	1619	2520	1473

(continued)

Table 2.2 (continued)

Sl. No.	Countries	Production in tonnes (2000)	Production in tonnes (2005)	Production in tonnes (2010)	Production in tonnes (2015	Production in tonnes (2016)
12	Canada	80241	80071	78452	118642	133935
13	China, Hong Kong SAR	31	31	35	31	31
14	China, mainland	2400000	3400000	4826000	826152	7786368
15	Taiwan	8196	9643	7689	9939	11530
16	Cyprus	1730	1014	790	711	761
17	Czechia	1000	350	526	551	561
18	North Korea	5745	6030	5906	5927	5868
19	Denmark	8686	10946	3000	3930	3930
20	Estonia	0	0	0	130	51
21	Finland	1536	1996	1645	1248	1345
22	France	203861	138541	119373	101135	101949
23	Germany	62000	50000	60000	62594	72141
24	Greece	845	2292	1397	4400	3601
25	Hungary	16926	19734	14026	28621	32311
26	Iceland	447	438	579	550	585
27	India	24000	40000	40600	33699	29992
28	Indonesia	28000	30854	61376	33485	40906
29	Iran (Islamic Republic of)	0	0	74500	132331	150063
30	Ireland	59800	62400	54500	72200	70000
31	Israel	7500	9500	9500	11000	11000
32	Italy	72492	88361	684401	594835	683620
33	Japan	67224	66000	65764	65711	65579
34	Jordan	500	688	764	841	856
35	Kazakhstan	500	503	507	515	513
36	Kyrgyzstan	264	200	200	231	226
37	Latvia	500	530	135	64	62
38	Lithuania	6000	4087	10434	13824	15785
39	Luxembourg	16	5	5	5	5
40	Madagascar	1000	1487	1882	2269	2262
41	Malta	898	989	1088	2021	1676
42	Mongolia	0	200	253	310	326
43	Montenegro	0	0	600	600	600
44	Morocco	1800	1924	1996	2087	2105
45	Netherlands	265000	245000	266000	310000	300000
46	New Zealand	8500	8600	5687	2110	1740
47	Philippines	568	508	526	556	580
48	Poland	109273	160000	230000	252944	260140
49	Portugal	1196	1377	1500	10754	12093
50	Republic of Korea	20659	28375	26250	26292	26158

(continued)

Table 2.2 (continued)

Sl. No.	Countries	Production in tonnes (2000)	Production in tonnes (2005)	Production in tonnes (2010)	Production in tonnes (2015	Production in tonnes (2016)
51	Republic of Moldova	2000	2000	2000	2034	2135
52	Romania	5000	5630	9973	10955	14519
53	Russian Federation	6000	5000	5373	8660	9682
54	Serbia	12000	12521	5000	5365	5403
55	Singapore	0	3	117	32	30
56	Slovakia	600	1100	2335	2000	2074
57	Slovenia	1269	1200	1131	1074	1063
58	South Africa	7278	8385	12217	18267	18803
59	Spain	63254	137764	133000	218795	197010
60	Switzerland	7148	7440	8465	7307	7089
61	Thailand	9500	9800	5746	1147	960
62	Macedonia (FYROM)	2000	3000	2900	2876	2866
63	Tunisia	99	116	136	150	153
64	Turkey	7000	17000	21559	39495	40272
65	Ukraine	3500	6000	11000	12480	14740
66	United Kingdom	89900	74000	69300	103197	99813
67	United States of America	383830	386984	359469	429562	419630
68	Uzbekistan	397	464	600	673	694
69	Viet Nam	20500	17702	19934	22854	23701
70	Zimbabwe	230	350	526	697	684
71	China	2408227	3409674	4833724	8036122	7797929
72	European Union	1030582	1051064	1706632	1830148	1906833
73	Least Developed Countries	1000	1487	1882	2269	2262
74	Land Locked Developing Countries	5391	6717	8966	9129	9367
75	Small Island Developing States	0	3	117	32	30
76	Low Income Food Deficit Countries	31636	48531	49715	43496	39726
77	Net Food Importing Developing Countries	3399	4415	5031	5657	5702

Mushroom production has increased gradually in the agricultural related industries ever since the end of World War II. Initially, *Agaricus* production was at greater rate and subsequently there was a greater raise in production of *Lentinula*, *Flammulina*, and *Pleurotus* (Chang and Buswell 2008). In twentieth century, the wide uses of industrialized cultivation techniques were applied for the mushroom productions. The development of mushroom farming skills has been principally responsible for the raise in mushroom production in recent years.

2.6.1 Major steps in Mushroom Cultivation

Cultivation of mushroom is relatively a primitive process however with modernization and since the recognition of its important health benefits and upliftment of economy, it has become an industrial venture in most nation producing hundreds and thousands of tonnes every year. Nonetheless, production in small scales industries plays a major role in smaller markets in developing nations. In either case, the concept of cultivation focuses on increasing the yield within a short stipulated time. This requires proper understanding on selection of high-yielding strains and media for spawn making, improved management of the mushroom beds, including pest and dieases management. Besides, continous supply of mushrooms to the consumers and marketing are also vital progression in the mushroom farming. Thus, there are a number of factors involved in mushroom production and a successful grower requires scientific knowledge, training and practice. Mushroom cultivation generally occur in the following six phases that follows (Buswell 1984; Nair 1991; Dawit 1998; Chang and Miles 2004): (1) selection of a mushroom species, (2) selection of a fruiting culture, (3) development of spawn, (4) preparation of compost, (5) spawn running, and (6) mushroom development.

Many strategies have to be adapted for successful production of WEF. Principally, selection of mushroom strains which have high demand and market value has to be studied. Maintenance cost, influence of other environmental factors on mushroom growth and accessibility of substrate for cultivation are the other factors. The cultures of edible ones are capable of producing fruiting bodies under suitable growing conditions. Strain improvement techniques like "mating with other isolates" is not necessary in case of heterothallic or a homothallic species since they can able to form fruit bodies. To avoid the spore density in the air of mushroom houses, sporeless strains of *Pleurotus* spp. have gained great commercial interest than non-sporeless strains. The latter may lead to respiratory tract problem and allergy to the mushroom workers. Quality of the mushroom spawn is mainly depends on quality and combination of the substrates and genetic constitution of the mushroom.

There is synthetic compost which is used for growing most of the mushroom; they are made up of agricultural and chemical materials but without animal manure. The mycelium grows at geate rate when larger quantity of spawn is used but it may also increase production cost. The requiment of temperature, humidity, pH and aeration varies at every stages of mushroom poduction. "Flushes" or appearance of mushrooms is in periodical cycles and they can be picked at different stages of development in accordance to consumer preference or market value. Nevertheless, harvesting varies among the species; *V. volvacea* and *P. pulmonarius* equires only simple farming activity than *A. bisporus, F. velutipes*, and *H. marmoreus* which needs a high-technology industry.

Among the hundred species of cultivated fungi, commercial markets are still dominated by *A. bisporus*, *L. edodes* and *Pleurotus* spp. and this account for nearly three quarters of the cultivated mushrooms grown around the world (Chang 1999; Boa 2004). Whether it is large scale industrial or small scale, cultivation edible fungi is profitable as well as they are highly nutritional as seen in countries such as Africa, Brazil, China, Mexico (Pauli 1999; Mshigeni and Chang 2000; Martinez-Carrera et al. 2001). On the other hand, cultivation of some species of mushroom such as *L. edodes* may lead decline in forests trees. Qingyuan of China is known as "mushroom capital of the world", suffering extensive deforestation from wood exploited for mushroom cultivation (Pauli 1998).

The number of cultivated species is ever growing as the technology and practical advice are easily available (Stamets 2000). Aside from saprobic species, **ectomycorrhizal** species can also be cultivated, where the tree are inoculated with the inoculum species which is allowed to infect the roots and form ectomycorrhizae, after which this tree are carefully tented for the production of fruiting body. Cultivation of ectomycorrhizal species are not fully developed and are constantly being refined and improved, cultivation of truffle mushroom is an example (Hall et al. 2007).

Some of the cultivated species, cultivated worldwide are given below (Stamets 2000; Chang and Mao 1995):

Agaricus arvensis, A. augustus, A. bisporus, A. bitorquis, A. blazei, A. campestris, A. subrufescens, Amanita brunnescens, Auricularia auricula-judae, A. fuscosuccinea, A. polytricha, Coprinus comatus, Flammulina velutipes, Laetiporus sulphureus, Lentinula edodes, Lentinus strigosus, L. tigrinus, L. tuber-regium, Morchella angusticeps, M. esculenta, Pleurotus cornucopiae, P. cystidiosus, P. eryngii, P. euosmus, P. ostreatus, P. pulmonarius, P. rhodophyllus, Pluteus cervinus, Polyporus indigenus, P. saporema, Volvariella bombcyina and *V. volvacea.*

References

Adhikari MK (1999) Wild relatives of some arable mushrooms found in Nepal. National Conference on Wild Relatives of Cultivated Plants in Nepal, Kathmandu, pp 149–155

Adhikari MK, Durrieu G (1996) Ethnomycologie Nepalaise. Bulletin Societé Mycologique de France 112:31–41

Adhikari MK, Devkota S, Tiwari RD (2005) Ethnomycolgical knowledge on uses of wild mushrooms in western and central Nepal. Our Nat 3:13–19

Agrahar M, Subbulakshmi G (2005) Nutritional value of edible wild mushrooms collected from the Khasi hills of Meghalaya. Food Chem 89:599–603

Agueda B, Parladé J, Fernández-Toirán LM, Cisneros O, De Miguel AM, Modrego MP, Martínez-Pena F, Pera J (2008) Mycorrhizal synthesis between *Boletus edulis* species complex and rock-roses (*Cistus* sp.). Mycorrhiza 18:443–449

Ajith TA, Janardhanan KK (2007) Indian medicinal mushrooms as a source of antioxidant and antitumor agents. J Clin Biochem. Nutr 40:157–162
Akata I, Ergonul B, Kalyoncu F (2012) Chemical compositions and antioxidant activities of 16 wild edible mushroom species grown in Anatolia. Int J Pharmacol 8:134–138
Akyüz M, Kirbağ S (2010) Nutritive value of wild edible and cultured mushrooms. Turk J Biol 34:97–102
Al-Naama MM, Ewaze JO, Nema JH (1998) Chemical constituents of Iraqi truffles. Iraq J Agric Sci 6:1–56
Altamura MR, Robbins FM, Andreotti RE, Long L Jr, Hasselstrom T (1967) Mushroom ninhydrin-positive compounds, amino acids, related compounds and other nitrogen substances found in cultivated mushroom, *Agaricus compestris*. J Agric Food Chem 15:1040–1043
Arora D, Dunham SM (2008) A new, commercially valuable chanterelle species, *Cantharellus californicus* sp. nov., associated with live Oak in California, USA. Econ Bot 62(3):376–391
Athanasakis G, Aligiannis N, Zagou GZ, Skaltsounis AL, Fokialakis N (2013) Antioxidant properties of the wild edible mushroom *Lactarius salmonicolor*. J Med Food 16(8):760–764
Atri NS, Kumari B, Upadhyay RC (2014) Taxonomy, sociobiology, nutritional and nutraceutical potential of termitophilous and lepiotoid mushrooms from North West India. In: proceedings of 8th International conference on mushroom biology and mushroom products. World Society of Mushroom Biology and Mushroom Products & ICAR Directorate of Mushroom Research, Chambaghat, Solan. 2:479–489
Bai K-C, Sheu F (2018) A novel protein from edible fungi *Cordyceps militaris* that induces apoptosis. J food drug anal 26:21–30
Baker JA (1934) Mushroom growing in Wellesley and Penang Provinces. Malay Agric J 22:25–28
Bano Z, Rajarathnam S (1982) In Tropical Mushrooms – Biological Nature and Cultivation Methods. Chinese University Press, Hong Kong, pp 363–380
Barros L, Baptista P, Correia DM, Casal S, Oliveira B, Ferreira ICFR (2007) Fatty acid and sugar compositions, and nutritional value of five wild edible mushrooms from Northeast Portugal. Food Chem 105(1):140–145
Barutçiyan J (2012) Türkiye'nin Mantarları-1. Oğlak Yayınları, İstanbul
Bas C (1983) *Flammulina* in Western Europe. *Persoonia* 12:51–66
Batra LR (1983) Edible Discomycetes and Gasteromycetes of Afghanistan, Pakistan and Northwestern India. Biologia 29:293–304
Bauer PB (2001) Protein fraction of edible Macedonian mushrooms. Eur Food Sci Technol 212:469–472
Beals RL (1933) Ethnology of the Nisenan. University of California Publications in American Archaeology and Ethnology Berkeley, 31(6):335–414
Bernaś E, Jaworska G (2010) Comparison of amino acid content in frozen *P. Ostreatus* and *A. Bisporus* mushrooms. Acta Sci Pol 9(3):295–303
Boa E (2004) Wild edible fungi: a global overview of their use and importance to people. Food and Agricultural organisation of the United Nations, Rome
Bon M (1987) The mushrooms and toadstools of Britain and North–Western Europe, vol 278. Hodder & Stoughton, London
Boruah P, Adhikary RK, Kalita P, Bordoloi D (1996) Some edible fungi growing in the forest of East Khasi Hills (Meghalaya). Adv For Res Ind 14:214–219
Bouriquet G (1970) Les principaux champignons de Madagascar. Terre Malagache 7:10–37
Boyce Kevin C, Carol L, Hotton Marilyn L, Fogel, George D, Cody, Hazen RM, Knoll AH, Hueber FM (2007) Devonian landscape heterogeneity recorded by a giant fungus. Geol Soc Am:399–402
Bradai L, Bissati S, Chenchouni H, Amrani K (2015) Effects of climate on the productivity of desert truffles beneath hyper. Int J Biometeorol 59(7):907–915
Burkhill IH (1935) A dictionary of the economic products of the Malay Peninsula. Crown Agents for the Colonies, London
Buswell JA (1984) Potentials of spent mushroom substrates for bioremediation purposes. Compost 2:31–35

Buyck, B. (1994) Ubwoba: Les charnpignons comestibles de l'ouest du Burundi. Brussels, Administration Generale de la Cooperation au Dcveloppement. 123.

Buyck B (2008) Wild edible mushrooms in Madagascar. Econ Bot 62:509–520

Buyck B, Nzigidahera B (1995) Ethnomycological notes from western Burundi. Belg J Bot 128:131–138

Buzzini P, Gasparetti C, Turchetti B, Cramarossa MR, Vaughan-Martini A, Martini A (2005) Production of volatile organic compounds (VOCs) by yeasts isolated from the ascocarps of black (*Tuber melanosporum* Vitt.) and white (*Tuber magnatum* Pico) truffles. Arch Microbiol 184:187–193

Caglarirmak N (2011) Chemical composition and nutrition value of dried cultivated culinary-medicinal mushrooms from Turkey. Int J Med Mushrooms 13(4):351–356

Caglarirmak N, Unal K, Otles S (2002) Nutritional value of wild edible mushrooms collected from the Black Sea region of Turkey. Micol Aplicada Int 14(1):1–5

Cakilcioglu U, Khatun S (2011) Nitrate, moisture and ash contents of edible wild plants. J Cell Plant Sci 2(1):1–5

Cappelli A (1984) Agaricus L: Fr. incl. colour plates. Libreria editrice Biella Giovanna, Saronno Italy, p 558

Carbonaro TM, Bradstreet MP, Barrett FS, MacLean KA, Jesse R, Johnson MW, Griffiths RR (2016) Survey study of challenging experiences after ingesting psilocybin mushrooms: Acute and enduring positive and negative consequences. 2016. *J Psychopharmacol*:1–11

Chamberlain M (1996) Ethnomycological experiences in South West China. Mycologist 10:13–16

Chang ST (1974) Production of straw mushroom (*Volvariella volvacea*) from cotton wastes. Mushroom J 21:348–354

Chang ST (1977) The origin and early development of straw mushroom cultivation. Econ Bot 31:374–376

Chang ST (1980) Mushrooms as human food. BioScience 30:399–401

Chang ST (1996) Mushroom research and development - equality and mutual benefit. Mush. Biol Mush Prod 2:1–10

Chang ST (1999) World production of cultivated edible and medicinal mushrooms in 1997 with emphasis on *Lentinus edodes* in China. Int J Med Mushrooms 1:291–300

Chang ST, Buswell JA (2008) Safety, quality control and regulational aspects relating to mushroom nutriceuticals. Proc. 6th Intl. Conf. Mushroom biology and mushroom products 2008:188–195

Chang ST, Hayes WA (eds) (1978) The biology and cultivation of edible mushrooms. Academic Press, New York

Chang ST, Mao XL (1995) Hong Kong Mushrooms [in Chinese]. Chinese University Press, Hong Kong

Chang ST, Miles PG (1992) Mushroom biology: a new discipline. Mycologist 6:64–65

Chang ST, Miles PG (1993) "The nutritional attributes and medicinal value of edible mushrooms," in Edible Mushrooms and Their Cultivation 27–39.

Chang ST, Miles PG (2004) Mushrooms: Cultivation, nutritional value, medicinal effect, and environmental impact, 2nd edn. CRC Press, Boca Raton

Chang ST, Wasser SP (2012) The role of culinary-medicinal mushrooms on human welfare with pyramid model for human health. Int J Med Mushrooms 14:95–134

Chang ST, Lau DW, Cho KY (1981) The cultivation and nutritional value of *Pleurotus sajor-caju*. Eur J Appl Microbiol Biotechnol 12:58–62

Chen X, Wu G, Huang Z (2013) Structural analysis and antioxidant activities of polysaccharides from cultured *Cordyceps militaris*. Int J Biol Macromo 58:18–22

Cheung PCK (2010) The nutritional and health benefits of mushrooms. Nutr Bull 35:292–299

Cheung LM, Cheung PCK, Ooi VEC (2003) Antioxidant activity and total phenolics of edible mushroom extracts. Food Chem 8:249–255

Chitty DF (1992) Vegetacion y etnobotanica del Valle de Culebra (Mawadianejodo), Estado Amazonas, Venezuela. Acta Terramaris 5:1–42

Cho IH, Choi HK, Kim YS (2006) Difference in the volatile composition of pine-mushrooms (*Tricholoma matsutake* Sing.) according to their grades. J Agric Food Chem 54:4820–4825

Codex Alimentarius (2010) Report of the 31st session of the Codex Committee on nutrition and foods for specific dietary uses. ALINORM

Colak A, Kolcuoglu Y, Sesli E, Dalman O (2007) Biochemical composition of some Turkish fungi. Asian J Chem 19:2193–2199

Combet E, Henderson J, Eastwood DC, Burton KS (2006) Eight carbon volatiles in mushrooms and fungi: properties, analysis, and biosynthesis. Mycoscience 47:317–326

Conlon BH, De Beer ZW, Henrik H, Aanen DK, Poulsen M (2016) Phylogenetic analyses of diverse *Podaxis* specimens from Southern Africa reveal hidden diversity and new insights into associations with termites. Fungal Biol 120:1065–1076

Cordova J, Garibay-Orijel R, Valenzuela R, Cifuentes J (2002) Inventario de las especies de hongos comestibles del bosque de Pino-Encino de Ixtlán de Juárez, Oaxaca (México). In: Nanacatepec: Studies on the Latin American Fungi (eds. G. Guzmán and G. Mata). Universidad Veracruzana, Mexico: 540.

Crisan EV, Sands A (1978) In: Chang ST, Hayes WA (eds) Nutritional value, in the biology and cultivation of edible mushrooms. Academic Press, New York, pp 137–168

Cronin DA, Wada S (1971) J Sci Food Agric 22:477–479

Cullere L, Ferreira V, Chevret B, Venturini ME, Sanchez-Gimeno AC, Blanco D (2009) Characterisation of aroma active compounds in black truffles (*Tuber melanosporum*) and summer truffles (*Tuber aestivum*) by gas chromatography– olfactometry. Food Chem 122(1):300–306

Dabbour IR, Takruri HR (2002) Protein digestibility using corrected amino acid score method (PDCAAS) of four types of mushrooms grown in Jordan. Plant Foods for Hum Nutr 57:13–24

Dawit A (1998) Mushroom cultivation: a practical approach. Berhanena Selam printing press, Addis Ababa

Degreef J, Malaisse E, Rammeloo J, Baudart E (1997) Edible mushrooms of the Zambezian woodland area: a nutritional and ecological approach. BASE (Biotechnologie, Agronomie, Societe et Environnement) 1:221–231

Demirbas A (2000) Accumulation of heavy metals in some edible mushrooms from Turkey. Food Chem 68:415–419

Deschamps JR (2002) Hongos silvestres comestibles del Mercosur con valor gastronómico, Documentos de trabajo. No. 86. Universidad de Belgrano, Argentin 25

Dijk HV, Onguene NA, Kuyper TW (2003) Knowledge and utilization of edible mushrooms by local populations of the rain forest of south Cameroon AMBIO. J Hum Environ 32(1):19–23

Du P, Cui BK, Dai YC (2011) Genetic diversity of wild *Auricularia polytricha* in Yunnan province of South-western China revealed by sequence-related amplified polymorphism (SRAP) analysis. *J Med Plants Res* 5:1374–1381

Dulger B, Ergul CC, Gucin F (2002) Antimicrobial activity of the macrofungus *Lepista nuda*. Fitoterapia 73:695–697

Dyke AJ, Newton AC (1999) Commercial harvesting of wild mushrooms in Scottish forests is it sustainable. Scott For 53(2):77–85

Elćhibaev AA, (1964) *S'edobnye griby Kirgizii [Edible mushrooms of the Kirghiz SSR]*, Kirgizskoi SSR. Izdatel'stvo Akademii Nauk 44

Elliott TJ (1985) Spawn-making and spawns. In: Flegg PB, Spencer DM, Wood DA (eds) *The Biology and Technology of the Cultivated Mushroom*. Wiley, New York

Elsayed EA, Enshasy EH, Wadaan MA, Aziz R (2014) Mushrooms: a potential natural source of anti-inflammatory compounds for medical applications. Mediat Inflamm 805841

Ereifej KI, Al-Raddad AM (2000) Identification and quality evaluation of two wild mushrooms in relation to *Agaricus bisporus* from Jordan. In L.Van Griensven, ed. Science and cultivation of edible fungi 721-724. Proceedings of the 15th International Congress on the Science and Cultivation of Edible Fungi. Maastricht, Netherlands 15-19 May 2000

Eyüpoğlu OE, Ozan V, Atacı N, Arısan (2011) Determination of some enzymes, which have industrial importance by lignolitic enzymes, from white saprophyte mushrooms and role of acidic conditions effect mechanism in production. Biyoloji Bilimleri Araştırma Dergisi 4 (2): 93–98.

Falandysz J, Borovička J (2013) Macro and trace mineral constituents and radio nuclides in mushrooms-health benefits and risks. Appl Microbiol Biotechnol 97:477–501

Falandysz J, Szymczyk K, Ichihashi H, Bielawski L, Gucia M, Frankowska A (2001) ICP/MS and ICP/AES elemental analysis (38 elements) of edible wild mushrooms growing in Poland. Food Addit Contam 18:503–513

Falandysz J, Kunito T, Kubota R, Bielawski L, Mazur A, Falandysz JJ (2007) Selected elements in Brown Birch Scaber Stalk *Leccinum scabrum*. J Environ Sci Health Part A 42:2081–2088

Falandysz J, Kunito T, Kubota R, Bielawski L, Frankowska A, Falandysz J (2008) Multivariate characterization of elements accumulated in King Bolete *Boletus edulis* mushroom at lowland and high mountain regions. J Environ Sci Health 43:1692–1699

Falch BH, Espevik T, Ryan L, Stokke BT (2000) The cytokine stimulating activity of (1→3)-β-D-glucans is dependent on the triple helix conformation. Carbohydrate Research. Int J Microbiol 329(3):587–596

FAO (1991) Protein quality evaluation. Food and Agricultural Organization of the United Nations, Rome

FAO (1998) In: Ciesla WM (ed) Non-wood forest products from conifers. Non-wood Forest Products 12, Rome, p 138

FAO (2001) In: Wong J, Thornber K, Baker N (eds) Resource assessment of non-wood forest products: experience and biometric principles. Non-wood Forest Products 13, Rome, p 126

Federico V, Cosimo T, Antonio P, Nadia B, Valentina L (2015) Volatile organic compounds in truffle (Tuber magnatum Pico): comparison of samples from different regions of Italy and from different seasons. Scientific reports 5:12629

Fericgla JM (1994) El hongo y la génesis de las culturas. Duendes y gnomos: Ambitos culturales forjados por el consumno de la seta enteógena *Amanita muscaria*. Los Libros de la Liebre de Marzo. Barcelona, España

Filipov D (1998) Mushroom season has Russians in fungi frenzy. Boston Globe

Flores R, Bran MDC, Honrubia M (2002) Edible mycorrhizal mushrooms of the west Highland Guatemala. In: Hall IR, Wang Y, Zambonelli A, Danell E (eds) Edible ectomycorrhizal mushrooms and their cultivation. Proceedings of the second international conference on edible mycorrhizal mushrooms. July 2001, Christchurch. CD-ROM. New Zealand Institute for Crop and Food Research Limited, Christchurch

Ford WW (1910) The distribution of haemolysins, agglutinins and poisons in fungi, especially the *Amanitas*, the *Entolomas*, the *Lactarius* and the *Inocybes*. J Pharmacol Exp Ther 2:285–318

Fortas Z, Chevalier G (1992) Effet des conditions de culture sur la mycorrhization de l'Helianthemum guttatum par trois espèces de terfez des genres Terfezia et Tirmania d'Algérie. Can J Bot 70:2453–2460

Galante F, de Araujo MVF (2014) Fundamentos de Bioqu ımica (2a). Editora Rideel, Sao Paulo

Gambaro V, Roda G, Visconti GL, Arnoldi S, Casagni E, Ceravolo C, Acqua LD, Farè F, Rusconi C, Tamborini L, Arioli A, Mora D (2015) Taxonomic Identification of Hallucinogenic Mushrooms Seized on the Illegal Market Using a DNA-Based Approach and LC/MS-MS Determination of Psilocybin and Psilocin. J Anal Bioanal Tech 6:6

Garibay-Orijel R, Cifuentes J, Estrada-Torres A, Caballero J (2006) People using macro-fungal diversity in Oaxaca. Mexico. Fungal Divers 21:41–67

Garibay-Orijel R, Caballero J, Estrada-Torres A, Cifuentes J (2007) Understanding cultural significance, the edible mushrooms case. J Ethnobiol Ethnomed 3(4):1–18

Ge ZW, Yang ZL, Zhang P, Matheny PB, Hibbett DS (2008) *Flammulina* species from China inferred by morphological and molecular data. Fungal Divers 32:59–68

Ge ZW, Liu XB, Zhao K, Yang ZL (2015) Species diversity of *Flammulina* in China: new varieties and a new record. Mycosystema 34(4):589–603

Gérault A, Thoen D (1992) Les champignons dans les pharmacopees traditionelles de l'Afrique de L'Ouest. Revue de Médecine et de Pharmacie Africa 1(1):45–53

Goldway M, Amir R, Goldberg D, Hadar Y, Levanon D (2000) *Morchella conica* exhibiting a long fruiting season. Mycol Res 104(8):1000–1004

Gong CL, Peng GP (1993) Culture of *Cordyceps rnilitaris* on Chinese silkworms and the analysis of its components. Zhongguo Shiyongjun (Edible Fungi of China: abimonthly journal) 12(4):21–23

Grangeia C, Heleno SA, Barros L, Martins A, Ferreira ICFR (2011) Effects of trophism on nutritional and nutraceutical potential of wild edible mushrooms. Food Res Int 44:1029–1035

Guillot J, Giollant M, Damez M, Dusser M (1991) Isolation and characterization of a lectin from the mushroom, *Lactarius deliciosus*. J Biochem 109:840–845

Gumińska B, Wojewoda W (1985) Grzyby i ich oznaczanie. PWRiL, Warszawa

Guzman G (2008) Diversity and use of traditional mexican medicinal fungi:A review. Int J Med Mush 10(3):209–217

Haddad NA, Hayes WA (1978) Nutritional factors and the composition of the *Agaricus bisporus* mycelium. Mushroom Sci 10:715–722

Hall IR, Zambonelli A (2012) Laying the foundations. Chapter 1. In: Zambonelli A, Bonito G (eds) Edible Mycorrhizal Mushrooms. Dordrecht, Springer, pp 3–16

Hall IR, Zambonelli A, Primavera E (1998a) Ectomycorrhizal fungi with edible fruiting bodies 3. *Tuber magnatum* Tuberaceae. Econ Bot 52(2):192–200

Hall IR, Buchanan PK, Wang Y, Cole ALJ (1998b) Edible and poisonous mushrooms: an introduction. New Zealand Institute for Crop and Food Research Limited, Christchurch, p 248

Hall IR, Brown GT, Zambonelli A (2007) Taming the truffle. The history, lore, and science of the ultimate mushroom. Timber Press Inc, Portland, p 304

Hanuś LO, Shkrob I, Dembitsky VM (2008) Lipids and fatty acids of wild edible mushrooms the genus *Boletus*. J Food Lipids 15:370–383

Härkönen M (2002) Mushroom collecting in Tanzania and Hunan (southern China): inherited wisdom and folklore of two different cultures. In: Watling R, Frankland JC, Ainsworth AM, Isaac S, Robinson CH (eds) *Tropical mycology*, Vol. 1 *Macromycetes*. Wallingford, UK, CAB International, pp 149–165

Härkönen M, Saarimäki T, Mwasumbi L (1994a) Edible and poisonous mushrooms of Tanzania. Afr J Mycol Biotechnol 2(2):99–123

Härkönen M, Saarimäki T, Mwasumbi L (1994b) Tanzanian mushrooms and their uses. 4. Some reddish edible and poisonous *Amanita* species. Karstenia 34:47–60

Harsh NSK, Rai BK, Ayachi SS (1993) Forest fungi and tribal economy - a case study in Baiga tribe of Madhya Pradesh, India. J Trop For 9:270–279

Harsh NSK, Tiwari CK, Rai BK (1996) Forest fungi in the aid of tribal women of Madhya Pradesh, India. Sustain For 1:10–15

Hawksworth DL (1991) The fungal dimension of biodiversity: magnitude, significance, and conservation. Mycol Res 95:641–655

Hawksworth DL (2001) The magnitude of fungal diversity: the 1.5 million species estimate revisited. Mycol Res 105:1422–1432

Hay WD (1887) An Elementary Text-Book of British Fungi. S. Sonnenschein, Lowrey, London

Hedger J (1986) Suillus luteus on the Equator. Bulletin of the British Mycological Society 20:53–54

Heim R (1969) Champignons d'Europe. Généralités. Ascomycètes. Basidiomycètes. 2ème édition. Paris: N. Boubée & Cie

Heleno SA, Barros L, Sousa MJ, Martins A, Ferrerira ICFR (2009) Study and characterization of selected nutrients in wild mushrooms from Portugal by gas chromatography and high performance liquid chromatography. Microchem J 93:195–199

Heleno SA, Barros L, Sousa MJ, Martins A, Santos-Buelga C, Ferreira ICFR (2011) Targeted metabolites analysis in wild *Boletus* species. LWT 44:1343–1348

Heleno SA, Barros L, Martins A, Morales P, Ruiz VF, Glamoclija J, Sokovic M, Ferreira ICFR (2015) Nutritional value, bioactive compounds, antimicrobial activity and bioaccessibility studies with wild edible mushrooms. LWT - Food Science and Technology 63:799–806

Hiroi M (1982) Fatty acid composition of mushroom lipids. Part 5. Comparison of lipid components of wild and cultivated mushroom (*P. ostreatus*). J Kumayama Womans Univ 18:1–8

Holtz RB, Schisler LC (1971) Lipid metabolism of *Agaricus bisporus* (Lange) sing.: I. Analysis of sporophore and mycelial lipids. Lipids 6(3):176–180

Hsu CH, Sun HL, Sheu JN, Ku MS, Hu CM, Chan Y, Lue KH (2008) Effects of the immunomodulatory agent *Cordyceps militaris* on airway inflammation in a mouse asthma model. Pediatr Neonatol 49:171–178

Hughes DH, Lynch DL, Somers GF (1958) Chromatographic identification of the amino acids and carbohydrates in the cultivated mushroom *Agaricus campestris* L. ex. Fries. *J Agric Food Chem.* 6:850–853

Humpert AJ, Giachini AJ, Castellano MA, Spatafora JW (2001) Molecular phylogenetics of Ramaria and related genera: evidence from nuclear large subunit and mitochondrial small subunit rDNA sequences. Mycologia 93(3):465–477

Hussain G, Al-Ruqaie IM (1999) Occurrence, chemical composition, and nutritional value of truffles: an overview. Pakistan J Biological 2:510–514

Ingold CT (1985) Water and spore discharge in Ascomycetes and Hymenomycetes. *Trans Br Mycol Soc* 85:575–583

Iordanov D, Vanev SG, Fakirova (1978) Gubite *v* Bulgariya: Opredelitel na nai- prostranenite yadlivi i otrovni gubi (Fungi of Bulgaria: keys to the identification of the most widely distributed edible and poisonous fungi). Sofiya, Izd-vo na Bulg Akad na Naukite

Ishibashi KI, Miura NN, Adachi Y, Ohno N, Yadomae T (2001) Relationship between solubility of grifolan, a Fungal 1,3-β-D-glucan, and production of tumor necrosis factor by macrophages *in vitro*. Biosci Biotechnol Biochem 65(9):1993–2000

Jaworska G, Bernas E (2009) The effect of preliminary processing and period of storage on the quality of frozen *Boletus edulis* (Bull:Fr.) mushrooms. Food Chem 113:936–943

Jaworska G, Bernaś E (2011) Comparison of amino acid content in canned *Pleurotus ostreatus* and *Agaricus bisporus* mushrooms. veg crop res bull 74:107–115

Jiang Y, Wong JH, Fu M, Ng TB, Liu ZK, Wang CR, Li N, Qiao WT, Wen TY, Liu F (2011) Isolation of adenosine, isosinensetin and dimethylguanosine with antioxidant and HIV-1 protease inhibiting activities from fruiting bodies of *Cordyceps militaris*. Phytomedicine 18:189–193

John C, Leffingwell1 ED, Alford (2011) Volatile constituents of the giant puffball mushroom (*Calvatia gigantea*) Leffingwell Reports 4.

Johnson MW, Richards WA, Griffiths RR (2008) Human hallucinogen research: guidelines for safety. J Psychopharmacol 22:603–620

Jones EBG, Whalley AJS, Hywel-Jones NL (1994) A fungus foray to Chiang Mai market in Northern Thailand. Mycologist 8(2):87–90

Jordan P (2000) The Mushroom guide and identifier: The ultimate guide to identifying, picking and using mushrooms. Hermes House, London, p 100

Kagan-Zur V, Roth-Bejerano N (2008) Dessert truffles. Truffles 1:32–37

Kalač P (2009) Chemical composition and nutritional value of European species of wild growing mushrooms: a review. Food Chem 113:9–16

Kalač P (2012) Chemical composition and nutritional value of European species of wild growing mushrooms, In Mushrooms: Types, Properties and Nutrition, ed. by Andres S and Baumann N. Nova Science. New York 129–152

Kalač P (2013) A review of chemical composition and nutritional value of wildgrowing and cultivated mushrooms. J Sci Food Agric 93:209–218

Kalamees K, Silver S (1988) Fungal productivity of pine heaths in North-West Estonia. Acta Botanica Fennica 136:95–98

Kalogeropoulos N, Yanni A, Koutrotsios G, Aloupi M (2013) Bioactive microconstituents and antioxidant properties of wild edible mushrooms from the island of Lesvos, Greece. Food Chem Toxicol 55:378–385

Kalotas A (1997) Aboriginal knowledge and use of fungi. In: Fungi of Australia, vol 1B. Introduction Fungi in the environment, Canberra, Australian Biological Resources Study, pp 269–295

Kanchiswamy CN, Malnoy M, Maffei ME (2015) Chemical diversity of microbial volatiles and their potential for plant growth and productivity. Front Plant Sci 6:151

Kang Y, Łuczaj Ł, Ye S, Zhang S, Kang J (2012) Wild food plants and wild edible fungi of Heihe valley (Qinling Mountains, Shaanxi, central China): Herbophilia and indifference to fruits and mushrooms. Acta Soc Bot Pol 81:405–413

Kang Y, Łuczaj Ł, Kang J, Zhang S (2013) Wild food plants and wild edible fungi in two valleys of the Qinling Mountains (Shanxi, central China). J Ethbobiol Ethnomed 9(1):26

Kataoka K, Muta T, Yamazaki S, Takeshige K (2002) Activation of macrophages by linear (1→3)-*β*-D-glucans. Implications for the recognition of fungi by innate immunity. J Biol Chem 277(39):36825–36831

Kawagishi H, Hayashi K, Tokuyama S, Hashimoto N, Kimura T, Dombo M (2007) Novel bioactive compound from the *Sparassis crispa* mushroom. Biosci Biotechnol Biochem 71:1804–1806

Kerrigan RW (1986) Agaricales of California. Agaricaceae, vol vol 6. Mad River Press, Eureka

Kim KH, Park KM, Choi KM SU, Lee KR KR (2009) Macrolepiotin, a new indole alkaloid from *Macrolepiota neomastoidea*. J Antibiot 62:335–338

Kimura T (2013) Natural products and biological activity of the pharmacologically active Cauliflower mushroom *Sparassis crispa*. BioMed Res Int:156–167

Kinge TR, Tabi EM, Mih AM, Enow EA, Njouonkou L, Nji TM (2011) Ethnomycological studies of edible and medicinal mushrooms in the Mount Cameroon region (Cameroon, Africa). Int J Med Mushrooms 13(3):299–305

Kues U, Liu Y (2000) Fruiting body production in basidiomycetes. Appl Microbiol Biotechnol 54:141–152

Kwon AH, Qiu Z, Hashimoto M, Yamamoto K, Kimura T (2009) Effects of medicinal mushroom (*Sparassis crispa*) on wound healing in streptozotocin-induced diabetic rats. Am J Surg 197:503–509

Kytovuori I (1989) The *Tricholoma caligatum* group in Europe and North Africa. Karstenia 28:65–77

Lakhanpal TN, Shad O, Rana M (2010) Biology of Indian morels. I K International Publ, New Delhi

Lampe KF, Ammirati JF (1990) Human poisoning by mushrooms in the genus *Cortinarius*. McIlvainea 9(2):12–25

Largeteau ML, Llarena-Hernández RC, Regnault-Roger C, Savoie JM (2011) The medicinal *Agaricus* mushroom cultivated in Brazil: bilology, cultivation n non-medicinal valorization. Appl Micobiol Biotehnol 92:897–907

Lassoe T, Del Conte A, Lincoff G (1996) The mushroom book. Kindersley Publishers, New York

Lee YL, Jian AY, Mau JL (2009) Composition and non-volatile taste components of *Hypsizigus marmoreus*. LWT – Food Sci Technol 42:594–598

Li Y (2012) Present development situation and tendency of edible mushroom industry in China. *Mushroom Sci* 18:3–9

Li GSF, Chang ST (1982) Nutritive value of *Volvariella volvacea*. In: Chang ST, Quimio TH (eds) *Tropical Mushrooms æ Biological Nature and Cultivation Methods*. Chinese University Press, Hong Kong, pp 199–219

Lian B, Dong Y-R, Hou W-G, Tong L-H, Yuan S (2007) Ectomycorrhizal Fungi in Jiangsu Province, China. Pedosphere 17:30–35

Lin JY, Chou TB (1984) Isolation and characterization of a lectin from edible mushroom, *Volvariella volvacea*. J Biochem 96:35–40

Lincoff G, Mitchel DH (1977) Toxic and hallucinogenic mushroom poisoning. A Hand book for physicians and mushroom hunters. Van Nostrand Reinhold Company, New York, p 267

Lindequist U, Niedermeyer THJ, Julich WD (2005) The pharmacological potential of mushrooms. eCAM 2(3):285–299

Litchfield JH, Vely VG, Overbeck RC (1963) Nutrient content of morel mushroom mycelium: aminoacid composition of the protein. J Food Sci 28:741

Liu WP, Yang HR (1982) An investigation of mushroom poisoning in Ninghua county during the last 20 years. Chin J Prev Med 16:226–228

Liu Y, Sun J, Luo Z, Rao S, Su Y, Xu R, Yan Y (2012) Chemical composition of five wild edible mushrooms collected from Southwest China and their antihyperglycemic and antioxidant activity. Food Chem Toxicol 50:1238–1244
Logemann H, Argueta J, Guzman G, Montoya-Bello L, Bandala-Munoz VM, de Leon Chocooj R (1987) Lethal poisoning by mushrooms in Guatemala. Revista Mexicana de Micología 3:211–216
Lopez GA, Cruz JMM, Zamora-Martinez MC (1992) Evaluación de la produccion de hongos comestibles silvestres en San Juan Tetla, Puebla. Ciclo 1992. In Reunion Cientzfica Forestal y Agropecuaria 182–191.
Lou LH (1978) *The Cultivation and Processing of Edible Mushrooms*. Peking Agricultural University, Peking, pp 96–106
Luard E (2006) Truffles. Frances Lincoln, London
Maga JA (1981) Mushroom flavor. J Agric Food Chem 1981(*29*):1–4
Malyi LP (1987) Resources of edible fungi in. Belorussia (Belarus) and the possibility of their utilization Rastitelo'nye Resursy 23(4):532–536
Maniery J (1983) A chronicle of murphys rancheria (Mol-Pee-So): a historic central sierra
Manzi P, Marconi S, Aguzzi A, Pizzoferrato L (2004) Commercial mushrooms: Nutritional quality and effect of cooking. Food Chem 84:201–206
Martínez A, Oria de Rueda JA, Martínez P (1997) Estudio sobre la potencialidad de los diferentes usos del bosque para la creación de empleo y actividad económica en el medio rural de Castilla León. Universidad de Report for the Junta de Castilla y León y Fondo Social Europeo. 348 pp.
Martínez-Carrera D, Bonilla M, Martinez W, Sobal M, Aguilar A, Gonzalez E (2001) Characterization and cultivation of wild *Agaricus* species in Mexico. Micol Aplicada Int 13:9–24
Maruyama T, Kawahara N, Yokoyama K, Makino Y, Fukiharu T, Goda Y (2006) Phylogenetic relationship of psychoactive fungi based on rRNA gene for a large subunit and their identification using the TaqMan assay (II). Forensic Sci Int 163:51–58
Mattila P, Konko K, Eurola M (2001) Contents of vitamins, mineral elements, and some phenolic compounds in cultivated mushrooms. J Agric Food Chem 49(5):2343–2348
Medina-Ortiz AJ, Herrera T, Marco A, Vásquez-Dávila, Raja HA, Figueroa M (2017) The genus *Podaxis* in arid regions of Mexico: preliminary ITS phylogeny and ethnomycological use. MycoKeys 20:17–36
Miles PG, Chang ST (1997) Mushroom Biology- concise basics and current developments. World Scientific, Singapore
Montoya-Esquivel A (1998) Ethnomycology of Tlaxcala Mexico. McIlvainea 13(2):6–12
Montoya-Esquivel A, Estrada-Torres A, Kong A, Juarez-Sanchez L (2001) Commercialization of wild mushrooms during market days of Tlaxcala, Mexico. Micol Aplicada Int 13:31–40
Moreno-Arroyo B, Recio JM, Gomez J, Pulido E (2001) *Tuber oligospermum* from Morocco. Mycologist 15:41–42
Mshigeni KE, Chang ST (2000) A guide to successful mushroom farming: with emphasis on technologies appropriate and accessible to Africa's rural and peri-urban communities. UNDP/UNOPS regional project RAF/99/021. University of Namibia, Windhoek
Nair MC (1991) Indian mushroom. Proceedings of the national symposium on mushrooms. Kerala Agricultural University, Vellanikkara
Namgyel P (2000) The story of Buddha mushroom. *Tricholoma matsutake*. Unpublished manuscript, Thimpu. 14 pp.
Nazzaro F, Fratianni F, Picariello G, Coppola R, Reale A, Luccia DA (2007) Evaluation of gamma rays influence on some biochemical and microbiological aspects in black truffles. Food Chem 103:344–354
Negi CS (2006) Morels (Morchella sp.) in Kumayun Himalaya Nat Prod Rad 5(4): 306–310
Obodai M, Apetorgbor M (2001) An ethnobotanical study of mushroom germplasm and its domestication in the Bia Biosphere Reserve of Ghana. Report presented to UNESCO through Environmental Protection Agency of Ghana, Accra

Ogundana SK, Fagade OE (1982) Nutritive value of some Nigerian edible mushrooms. Food Chem 8(4):263–268
Okan OT, Sibel Y, Ayşenur Y, Barutçiyan J, Deniz I (2013) Wild Edible Mushrooms Having an Important Potential in East Black Sea Region 673–680
Omer EA, Smith DL, Wood KV, El-Menshawi BS (1994) The volatiles of desert truffle: *Tirmania nivea*. Plant Foods Hum Nutr 45:247–249
Oso BA (1975) Mushrooms and the Yoruba people of Nigeria. Mycologia 67(2):311–319
Ouzouni PK, Petridis D, Koller W, Riganakos KA (2009) Nutritional value and metal content of wild edible mushrooms collected from West Macedonia and Epirus, Greece. Food Chem 115:1575–1580
Pala SA, Wani AH, Bhat MY (2013) Ethnomycological studies of some wild medicinal and edible mushrooms in the Kashmir Himalayas (India). Int J Med Mushrooms 15:211–220
Pauli G (1998) Qingyuan: the mushroom capital of the world (available at www.zeri.org/news/1998/august/aug_chin.htm).
Pauli G (1999) Sustainable development in the Amazon forest (available at www.zeri.org)
Pegler DN, Vanhaecke M (1994) *Termitomyes* of Southeast Asia. Kew Bull 49:717–736
Peintner U, Poder R, Pumpel T (1998) The iceman's fungi. Mycol Res 102(10):1153–1162
Pereira E, Barros L, Martins A, Ferreira ICFR (2012) Towards chemical and nutritional inventory of Portuguese wild edible mushrooms in different habitats. Food Chem 130:394–403
Perez-Butron JL, Fernández-Vicente J (2007) Una nueva especie de *Flammulina* P. Karsten, *F. cephalariae* (Agaricales) encontrada en España. *Revista Catalana de Micologia* 29:81–91
Phat C, Moon B, Lee C (2016) Evaluation of umami taste in mushroom extracts by chemical analysis, sensory evaluation, and an electronic tongue system. Food Chem 192:1068–1077
Phillips KM, Ruggio DM, Horst RL, Minor B, Simon RR, Feeney MJ, Byrdwell WC, Haytowitz DB (2011) Vitamin D and sterol composition of 10 types of mushrooms from retail suppliers in the United States. J Agric Food Chem 59:7841–7853
Pinho PG, Ribeiro B, Gonçalves RF, Bapista P, Valentão P, Seabra RM, Andrade PB (2008) Aroma compounds in eleven edible mushroom species: Relationship between volatile profile and sensorial characteristics. In: Blank I, Wüst M, Yeretzian C (eds) Expression of multidisciplinary flavour science. Proceedings of the 12th Weurman Symposium. Zürich University of Applied Sciences, Wädenswil
Pinto S, Barros L, Sousa MJ, Ferreira CFRI (2013) Chemical characterization and antioxidant properties of *Lepista nuda* fruiting bodies andmycelia obtained by in vitro culture: Effects of collection habitat and culture media. Food Res Int 51:496–502
Pohleven J, Obermajer N, Sabotic J, Anzlovar S, Sepcic K, Kos J, Kralj B, S ˇ trukelj B, Brzin J (2009) Purification, characterization and cloning of a ricin B-like lectin from mushroom *Clitocybe nebularis* with antiproliferative activity against human leukemic T cells. Biochim Biophys Acta 1790:173–181
Power RC, Salazar-García DC, Straus LG, González Morales MR, Henry AG (2015) Microremains from El Mirón Cave human dental calculus suggest a mixed plant-animal subsistence economy during the Magdalenian in Northern Iberia. J Archaeol Sci 60:39–46
Prance G (1984) The use of edible fungi by Amazonian Indians. Adv Econ Bot 1:127–139
Psurtseva NV (2005) Modern taxonomy and medical value of the *Flammulina* mushrooms. Int J Med Mushrooms 7:449–451
Purkayastha RP, Chandra A (1976) Amino acid composition of protein of some edible mushroom growth in synthetic medium. J Food Sci Technol 3:13–17
Purkayastha RP, Chandra A (1985) Manual of edible mushrooms. Today and tomorrow's Printers and Publishers, New Delhi
Quan XL, Wang HJ, Shi TY, Zhang MS (2007) Nutritive components comparison between Tricholoma matsutake and Tricholoma bakamatsutake. Edible fungi 2:54–55
Quimio TH (1979) Taxonomic consideration of Auriculariales imported into the Philippines. Philipp J Biol 6:69–72
Ragunathan R, Swaminathan K (2003) Nutritional status of *Pleurotus* spp. grown on various agro-wastes. Food Chem 80:371–375

Rammeloo J, Walleyn R (1993) The edible fungi of Africa south of the Sahara: a literature survey. Scripta Botanica Belgica 5:1–62

Redhead SA, Petersen RH (1999) New species, varieties and combinations in the genus *Flammulina*. Mycotaxon 71:285–294

Reis FS, Heleno SA, Barros L, Sousa MJ, Martins A, Santos-Buelga C (2011) Toward the antioxidant and chemical characterization of mycorrhizal mushrooms from northeast Portugal. J Food Sci 76:C824–C830

Reis FS, Barros L, Martins A, Ferreira ICFR (2012) Chemical composition and nutritional value of the most widely appreciated cultivated mushrooms: an inter-species comparative study. Food Chem Toxicol 50:191–197

Remotti CD, Colan JA (1990) Identification of wild edible fungi in Dantas Forest, Huanuco. Revista Forestal del Peru 17:21–37

Reshetnikov SV, Wasser SP, Tan KK (2001) Higher basidiomycota as a source of antitumour and immunostimulating polysaccharides. Rev Int J Medi Mushrooms 3:361–394

Reyna S, García-Barreda S (2014) Black truffle cultivation: a global reality. For Sys 3(2):317–328

Reyna S, Rodriguez-Barreal J, Folch L, Perez-Badia R, Garcia S, Jimenez E (2002) Truffle silviculture in Mediterranean forests. In: Hall IR, Wang Y, Zambonelli A, Danell E (eds) Edible ectomycorrhizal mushrooms and their cultivation. Proceedings of the second international conference on edible ectomycorrhizal mushrooms. July 2001, Christchurch. CD-ROM. New Zealand Institute for Crop and Food Research Limited, Christchurch

Rice-Evans C, Miller NJ, Paganga G (1996) Structure-antioxidant activity relationship of flavonoids and phenolic acids. Free Radical Bio Med 20:933–956

Rivera CS, Blanco D, Salvador ML, Venturini ME (2010) Shelf life extension of fresh *Tuber aestivum* and *Tuber melanosporum* truffles by modified atmosphere packaging with microperforated films. J Food Sci 75(4):225–233

Rojas C, Mansur E (1995) Ecuador: Informaciones Generales Sobre Productos Non Madereros en Ecuador. In Memoria, Consulta De Expertos Sobre Productos Forestales no Madereros Para America Latina Y el Caribe, pp. 208-223. Serie Forestal #1. Santiago, Chile, FAO Regional Office for Latin America and the Caribbean.

Roman DM, Boa E (2004) Collection, marketing and cultivation of edible fungi in Spain. Micol Aplicada Int 16(2):25–33

Sabra A, Walter S (2001) Non-wood forest products in the Near East: a regional and national overview. Working paper FOPW/01/2. Rome, FAO. 120 pp.

Sadler Michele J (2003) Nutritional properties of edible fungi. Br Nutr Found Nutr Bull 28:305–308

Saenz JA, Lizano AVM, Nassar MC (1983) Edible, poisonous and hallucinatory fungi in Costa Rica. Revista de Biologia Tropical 31:201–207

Saltarelli R, Ceccaroli CP, Barbieri E, Stocchi V (2008) Effect of storage on biochemical and microbiological parameters of edible truffle species. Food Chem 109:8–16

Samajipati N (1978) Nutritive value of Indian edible mushrooms. Mushroom Sci 10:695–703

Sarkar BB, Chakraborty DK, Bhattacharjee A (1988) Wild edible mushroom flora of Tripura. Indian Agriculturist 32:139–143

Sawaya WN, Al-Shalhat A, Al-Sogair A, Mohammad M (1985) Chemical composition and nutritive value of truffles of Saudi Arabia. J Food Sci 50:450–453

Schmeda-Hirschmann G, Razmilic I, Reyes S, Gutierrez MI, Loyola JI (1999) Biological activity and food analysis of *Cyttaria* spp. (Discomycetes). Econ Bot 53(1):30–40

Schultes RE (1939) Plantae Mexicanae II. The Identification of Teonanácatl, a Narcotic Basidiomycete of the Aztecs. Botanical Museum Leaflets, Harvard University 7:37–56

Schultes (1940) Teonanácatl: the narcotic mushroom of the aztecs. Am Anthropol 42:429–444

Sewell RA, Halpern J, Pope HG (2006) Response of cluster headache to psilocybin and LSD. Neurology 66(12):1920–1922

Sharma RP, Kaisth KR, Lakhanpal TN (1988) Protein and mineral content of two edible *Lactarius* species. Ind. J Mushrooms 14:4447

Sharma SK, Atri NS, Joshi R, Ashu G, Arvind G (2012) Evaluation of wild edible mushrooms for amino acid composition. Acad J Plant Sci 5(2):56–59

Sillitoe P (1995) Ethnoscientific observations on entomology and mycology in the southern highlands of Papua New Guinea. Sci N G 21(1):3–26
Simmons C, Henkel T, Bas C (2002) The genus *Amanita* in the Pakaraima mountains of Guyana. Persoonia 17(4):563–582
Sing, Sing NI, Sing SM, Th C (2002) In: Vij SP, Kondo K, Sharma ML, Gupta A (eds) Fleshy Fungi of Manipur in Plant Genetic Diversity: Exploaration, Evaluation, Conservation. Afficiated East West Press Pvt. Ltd, New Delhi
Singer R (1961) Mushrooms and truffles. Leonard Hill Ltd., Aberdeen
Singh SK, Rawat GS (2000) Morel mushroom industry in India. Plant Talk 21:36–37
Splivallo R, Ebeler SE (2015) Sulfur volatiles of microbial origin are key contributors to human-sensed truffle aroma. Appl Microbiol Biotechnol 99(6):2583–2592
Stamets P (2000) Techniques for the cultivation of the medicinal mushroom royal sun *Agaricus - Agaricus blazei* Murr. (Agaricomycetideae). Int. J Med Mushrooms 2:151–160
Stojchev G (1995) New fungi for Bulgaria. Higher institute of agriculture-plovdiv, Jubilee scientific session. IV(1): 229–232
Suberville NC, Cruz C, Guinberteau J, Montury M (1996) Correlation between fatty acid content and aromatic compound release in Fresh Blewit (*Lepista nuda*). J Agric Food Chem 44:1180–1183
Sun WS, Xu JY (1999) Cultivation of edible fungi has become one of the backbone industries in rural economy of China. Edible Fungi of Chin 18(2):5–6
Tang Y, Li HM, Tang YJ (2012) Comparison of sterol composition between *Tuber* fermentation mycelia and natural fruiting bodies. J Agric Food Chem 132:1207–1213
Tao K, Liu B (1990) Ecology and nutritive value of *Tuber sinese*. J Shan xi Univ (Nat Sci Ed) 13:319–329
Tao Y, Zhang L, Cheung PCK (2006) Physicochemical properties and antitummor activities of water-soluble native and sulphated hyperbranched mushroom polysaccharides. Carbohydr Res 341:2261–2269
Tartufi RS (2011) Frutti della terra, figli degli dei. Series: I preziosi della gastronomia. Genova: Sagep
Tedder S, Mitchell D, Farran R (2002) Property rights in the sustainable management of non-timber forest products. Victoria, British Columbia, British Columbia, Ministry of Forests 140 pp.
Teissedre PL, Landrault N (2000) Wine phenolics: contribution to dietary intake and bioavailability. Food Res Int 33:461–467
Thawthong A, Karunarathna SC, Thongklang N, Chukeatirote E, Kakumyan P, Chamyuang S, Rizal M, Mortimer PE, Xu JC, Callac P, Hyde KD (2014) Discovering and Domesticating Wild Tropical Cultivatable Mushrooms. Chiang Mai J Sci 41:731–764
Tibuhwa DD (2012) Folk taxonomy and use of mushrooms in the communities around Ngorongoro and Serengeti National Park, Tanzania. J Ethnobiol Ethnomed 8:36–10
Trappe JM (1990) Use of truffles and false truffles around the world. In: Bencivenga M, Granetti B (eds) Proceedings, Atti del Secondo Congresso Internazionale sul Tartufo. Comunita Montana dei Monti Martini edel Serano, Spoleto, Italy
Treshaw C (1944) Nutrition of the cultivated mushroom, *Dansk Bot*. Arkiv 11(6):1–180
Tsai H, Tsai H, Mau JL (2007) Nonvolatile taste components of fruit bodies and mycelia of shaggy ink cap mushroom *Coprinus comatus* (O.F. Müll.: Fr.) Pers. (Agaricomycetideae). Int J Med Mushrooms 9:47–55
Tsai SY, Tsai HL, Mau JL (2008) Non-volatile taste components in *Agaricus blazei*, *Agrocybe cylindracea* and *Boletus edulis*. Food Chem 107:977–983
Tsivileva OM, Nikitina VE, Garibova LV (2005) Effect of culture medium composition on the activity of extracellular lectins of *Lentinus edodes*. Appl Biochem Microbiol 41:174–176
Tylš F, Páleníček T, Horáček J, Psilocybin (2014) summary of knowledge and new perspectives. Eur Neuropsychopharmacol 24:342–356
van Amsterdam J, Opperhuizen A, van den Brink W (2011) Potential of magic mushroom use: A review. Regul Toxicol Pharmacol 59:423–429

Vasilèva LN (1978) Edible mushrooms of the Far East. Far Eastern Publishing House, Vladivostock
Vaskovsky VE, Khotimchenko SV, Boolugh EM (1998) Distribution of diacylglycerotrimethylhomoserine and phosphatidylcholine in mushrooms. Phytochemistry 47(5):755–760
Vaz JA, Heleno SA, Martins A, Almeida GM, Vasconcelos MH, Ferreira ICFR (2010) Wild mushrooms *Clitocybe alexandri* and *Lepista inversa*: in vitro antioxidant activity and growth inhibition of human tumour cell lines. Food Chem Toxicol 48:2881–2884
Vaz JA, Barros L, Martins A, Santos-Buelga C, Vasconcelos MH, Ferreira ICFR (2011) Chemical composition of wild edible mushrooms and antioxidant properties of their water soluble polysaccharidic and ethanolic fractions. Food Chem 126:610–616
Vetchinkina EP, Nikitina VE, Tsivileva OM, Garibova LV (2008) Activity of *Lentinus edodes* intracellular lectins at various developmental stages of the fungus. Appl Biochem Microbiol 44:66–72
Vetter J, Rimoczi I (1993) Crude, digestible and indigestible protein in fruiting bodies of Pleurotus ostreatus. Zeitschrift fur Lebensmittel Untersuchung und Forschung 197:427–428
Vetvicka V, Yvin JC (2004) Effects of marine β-1,3 glucan on immune reactions. Int Immunopharmacol 4(6):721–730
Villarreal L, Guzmán G (1985) Producción de lost hongos comestibles silvestres en los bosques de México I. Revista de la Sociedad Mexicana de Micología 1:51–90
Walleyn R, Rammeloo J (1994) The Poisonous and Useful Fungi of Africa South of the Sahara: A Literature Survey. Scripta Botanica Belgica 10:1–56
Wang Y, Chen YL (2014) Recent advances in cultivation of edible mycorrhizal mushrooms. Verlag Berlin Heidelberg. 23: 375 – 397
Wani BA, Bodha RH, Wani AH (2010) Nutritional and medicinal importance of mushrooms. J Med Plants Res 4(24):2598–2604
Wasser SP (1995) Edible and poisonous mushrooms of Israel. Modan Press, Tel-Aviv
Wasser SP (2010) Current findings, future trends and unsolved problems in studies of medicinal mushrooms. Appl Microbiol Biotechnol 89:1323–1332
Wasser SP, Weis AL (1999) Medicinal properties of substances occurring in higher Basidiomycetes mushrooms: current perspectives (Review). Int J Med Mushrooms 1:31–62
Wasson RG (1968) Soma-Divine Mushroom of Immortality. Harcourt, Brace & World, New York
Wasson RG (1971) The Soma of the Rig Veda: what was it? J Am Orient Soc 91(2):169–187
Wasson VP, Wasson RG (1957) Mushroom, Russia and History. Pantheon Books, New York
Wasson RG, Ruck CAP, Hoffman A (1978) The Road to Eleusis: Unveiling the Secret of the Mysteries. Harcourt Brace Jovanovich, New York
Wilcox J (2014) Psilocybin and Obsessive Compulsive Disorder. J psychoactive drugs 46(5):393–395
Winkler D (2002) Forest use and implications of the 1998 logging ban in the tibetan prefectures of Sichuan: case study on forestry, reforestation and NTFP in Litang County, Ganzi TAP, China. In Z. Ziang, M. Centritto, S. Liu & S. Zhang, cds. The ecological basis and sustainable management of forest resources. Informatore Botanico Italiano 134 (Supplemento 2
Winkler D (2008) Present and historic relevance of Yartsa Gunbu (*Cordyceps sinensis*). An ancient myco-medicinal in Tibet. Fungi 1:6–7
Wu SX, Wang BX, Guo SY, Li L, Yin JZ (2005) Yunnan wild edible *Thelephora ganhajun* Zang nutrients analysis. Mod Prev Med 32:1548–1549
Xiao DR, Liu RS, He L, Li HM, Tang YL, Liang XH (2015) Aroma improvement by repeated freeze-thaw treatment during Tuber melanosporum fermentation. Sci Rep 5:17120
Xu X, Yan H, Chen J, Zhang X (2011) Bioactive proteins from mushrooms. Biotechnol Adv 29(6):667–674
Xu DX, Lin J, Duan ZM, Wan YP, Bai B, Sun C (2012) Detection of chemical compositions of wild *Lactarius volemus* from Yunnan province. Edible Fungi 4:60–61
Yamin-Pasternak S (2007) An ethanomycologicl approach to land use values in Chukotka. Ètues/Inuit/Studies 31(1-2):121–111

Yamin-Pasternak S (2008) From disgust to desire: Changing the attitude towards mushrooms among the peoples of Beringian mushrooms. Econ Bot 62(3):214–222

Yang JH, Lin HC, Mau JL (2002) Antioxident properties of several commercial mushrooms. Food Chem 77:229–235

Yang FQ, Feng K, Zhao J, Li SP (2009) Analysis of sterols and fatty acids in natural and cultured *Cordyceps* by one-step derivatization followed with gas chromatography-mass spectrometry. J Pharm Biomed Anal 49:1172–1178

Yeh E (2000) Forest claims, conflicts and commodification: the political ecology of Tibetan mushroom-harvesting villages in Yunnan Province, China. China Q 161:225–278

Yin JZ, Zhou LX (2008) Analysis of nutritional components of 4 kinds of wild edible fungi in Yunnan. Food Res Dev 29:133–136

Yoshioka Y, Emori M, Ikekawa J, Fukuoka F (1975) Isolation, purification and structure of components from acidic polysaccharides of *Pleurotus ostreatus* (Fr.) Quel. Carbohydrate Res 43:305–320

Young VR (*1994*) Adult amino acid requirements: the case for a major revision in current recommendations. J Nutr *124*(*8*):*1517S–1523S*

Yun W, Hall IR (2004) Edible ectomycorrhizal mushrooms: challenges and achievements. Can J Bot 82:1063–1073

Zaidman YM, Mahajna J, Wasser SP (2005) Medicinal mushroom modulators of molecular targets as cancer therapeutics. Appl Microbiol Biotechnol 67(4):453–468

Zang M (1984) Mushroom distribution and the diversity of habitats in Tibet, China. McIlvainea 6(2):15–20

Zang DC (1988) Collybia albuminosa at Lianshan District. Zhongguo Shiyongjun (Edible Fungi of China: a bimonthly journal) 7(1):28–31

Zang M, Pu C (1992) Confirmatory Tuber indica distributed in China. Zhongguo Shiyongjun (Edible Fungi of China: a bimonthly journal) 11(3):19

Zhang M, Cui SW, Cheung PCK, Wang Q (2007) Antitumor polysaccharides from mushrooms: a review on their isolation process, structural characteristics and antitumor activity. Trends Food Sci Technol 18(1):4–19

Zhang GQ, Wang YF, Zhang XQ, Ng TB, Wang HX (2010) Purification and characterization of a novel laccase from the edible mushroom *Clitocybe maxima*. Proc Biochem 45:627–633

Zhang JX, Chen Q, Huang CY, Gao W, Qu JB (2015) History, current situation and trend of edible mushroom industry development. Mycosystema 34:524–540

Zhang C, Mingxia H, Liu J, Xu X, Cao Y, Gao F, Yiwei F, Wang W, Yun W (2017) Brief introduction to a unique edible Bolete—*Phlebopus portentosus* in Southern China. J Agric Sci Technol 7:386–394

Zhou LX, Yin JZ (2008) Yunnan wild edible Boletus nutrition analysis and evaluation. Edible Fungi 4:61–62

Zhu XQ, Wang XJ, Xiong Z (2007) Nutrient analysis of the wild *Lentinula edodes*. Forest By-Product and Speciality in China 2:9–11

Chapter 3
Molecular Characterization of Wild Mushrooms: A Paradigm Shift from Morphotyping

Madhusmita Borthakur and S. R. Joshi

3.1 Introduction

Being close to nature, **ethnic communities** have proximate association with natural resources and mushrooms are one resource that has been harvested and used by mankind since time immemorial. This has resulted in acquisition of immense knowledge about the benefits and utilization of different bioresources for consumption and medicinal purposes (Sawian et al. 2007; Singh et al. 2012). In the Asian sub-continent, mycophilic societies are associated among the indigenous people of North Eastern India, Western Ghats and North Western India, China. The local inhabitants are often seen collecting mushrooms from their neighboring localities, meadows and forests for consumption and for selling them to earn partial livelihood for their family during monsoon season when other forest products are unavailable for harvest.

Ethnomycological knowledge among Indian tribal population has been profound and broad with about 283 wild types of mushrooms being consumed out of 2000 edible species which are recorded worldwide (Purkayastha and Chandra 1985). These mushrooms which are often consumed by the ethnic community are harvested and picked from the forest area based on the traditional knowledge about their morphology and the toxicity. This knowledge on the edibility and the medicinal importance of the mushrooms has been passed from one generation to the other in the form of word of mouth without much record and scientific documentation.

The primary identification of mushroom involves its phenotypic characteristics and its **palynological** study. The common morphology of a mushroom is its umbrella shape structure but few resemble toads sitting atop on the cap. The common oyster mushroom and the family of *Amanita* possess distinct gills which are important for its identification. Genus *Russula* is known to possess hollow white stalk while

M. Borthakur · S. R. Joshi (✉)
Microbiology Laboratory, Department of Biotechnology & Bioinformatics,
North-Eastern Hill University, Shillong, Meghalaya, India
e-mail: srjoshi@nehu.ac.in

B. P. Singh et al. (eds.), *Biology of Macrofungi*, Fungal Biology,
https://doi.org/10.1007/978-3-030-02622-6_3

Boletus consists of a bulbous base at its stalk considered as important features for their identification. The dimension and the colour of the spores (when stained with **Lactophenol cotton blue** and **Melzer staining**) are also an important observable features in identification of mushrooms (Figs. 3.1a, 3.1b and 3.1c) Since a single

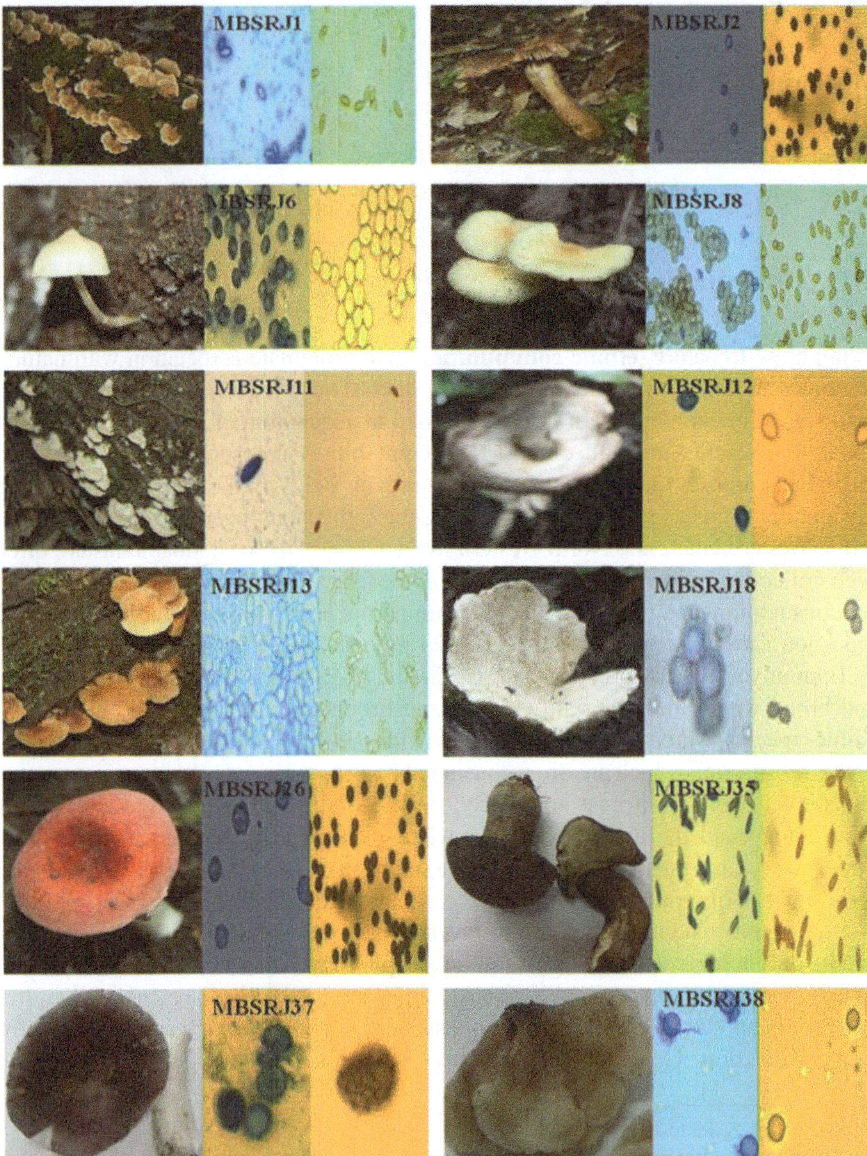

Fig. 3.1a Morphotyping and photomicrographical analysis of mushrooms of Meghalaya, India using Lactophenol cotton blue and Melzer staining (100xmagnifications)

Fig. 3.1b Morphotyping and photomicrographical analysis of mushrooms of Meghalaya, India using Lactophenol cotton blue and Melzer staining (100xmagnifications)

spore is almost invisible to naked eye, spore prints in a dark or lighter background have often been considered for identification purposes. The colour of the spore prints varies during different growth stages of a mushroom. They are mostly whitish to creamish during younger stage to brownish lilac at maturity. The preliminary identification of mushrooms is solely based on morphological features during field

Fig. 3.1c Morphotyping and photomicrographical analysis of mushrooms of Meghalaya, India using Lactophenol cotton blue and Melzer staining (100xmagnifications)

observation and microscopic features of its spores, pileus, stipe, volva (if present) and host. Few mushroom specimens are described and identified till the genus level solely based on the morphotypic features (Table 3.1). The palynological parameters can also be considered as important tool for their identification (Table 3.2).

Table 3.1 Morphological characterization of few mushrooms of Meghalaya, India

Specimen identity number	Fruiting body (Pileus)	Stalk (Stipe)	Gills (Lamellae)	Morphological identification
MBSRJ1	5-11 cm wide;Perennial asci shaped with wavy lobed edges, yellow with white margin, leathery tough to velvety in texture	Absent, the asci shaped cap attach to bark of the tree	The pores are mostly small whitish to tan in colour	*Formitopsis*sp. Order: Polyporales Family: Formitopsidaceae
MBSRJ2	4.5–8.3 cm wide, convex with margin tucked when young but broadly convex to flat at maturity, pale brownish yellow	3–8.1 cm long,0.8–2.2 cm thick, white hollow dry in texture nearly tapering base	Close distinct gills attached to the stalk pale yellowish to white	*Russulafoetens* Order: Russulales Family: Russulaceae
MBSRJ6	1-5 cm wide, conical to bell shapedwith a central bump, inrolled margin with radial lined, grayish to dirty tan in colour with a dark brown center, bald tacky texture	4.3–7.8cmn long1-2 cm thick, hollow, bald, slender with few tiny fiber like filaments colour varies from white to tan yellowish	Whitish attached to the stalk, prominentpinkish to brownish cross vein at maturity	*Mycenagalericulata* Order: Agaricales Family: Mycenaceae
MBSRJ8	2-5 cm convex wide to nearly flat, dry, velvety in touch, yellowishwith a darker pointed center, the margin sometimes inrolled to wavyin nature	2.5–8.5 cm long, 3.5–6.5 cm thick, tapering at the base, bright brownish to rusty brown from the upward base, faint ring like zone at the end of the stalk	Close crowded often attached to the stalk yellowish to olive yellow in colour, short gills are often connected to one another	*Hypholomafasciculare* Order: Agaricales Family: Strophariaceae
MBSRJ11	Upto 3.5 cm wide, irregularbracket shaped, flattened convexthick leathery, with marginal zone of whitish to pale lilac in colour	Absent, cap attached to tree bark	Pores are angular to erodic spines atmaturity lilac tobrownish spores are often seen at the upper surface of the pores	*Trichaptumbioforme* Order:Polyporales Family:Polyporaceae
MBSRJ12	4.5–8.5 cm wide, broadly convex,with a hollow depression continues to the stalk, dry fuzzy wrinkled purplish to purplish brown, the margin is inrolled wavy, purplishto brownish in colour	6 cm long and 1.5 cm wide, slightly broaden towards the end, colour similar to the cap, presence of tiny hairs at the basalend	Frequent short closer gills continuing to the stalk whitish to pale brownish in colour	*Panus*sp. Order: Polyporales Family: Panaceae

(continued)

Table 3.1 (continued)

Specimen identity number	Fruiting body (Pileus)	Stalk (Stipe)	Gills (Lamellae)	Morphological identification
MBSRJ13	1.5-3 cm wide, convex to nearly flat at maturity, slightly bell shaped sticky moist to tacky in touch, yellowish to brownish orange in colour, margin inrolled finely lined	2–5.4 cm long, brownish to reddish brown at the basal stalk, dry to moist at maturity	Gills attached to the stalk short frequent close or nearly distant, gills turn into rusty brown atmaturity	*Galerina* sp. Order: Agaricales Family: Hymenogastraceae
MBSRJ18	9.5 cm wide, broadly convex, shallowly depressed, continuing the stalk, wavy edges, soft whitish incolour	7.5 cm long, white with a tapering end, solid	Runs down the stem, crowded, forking, pale creamy	*Lactariusglaucescens* order: Agaricomycetes Family: Russulaceae
MBSRJ26	5.5 cm wide, convex to broadlyflattened, fairly smooth to slight depression at the center, often seendarker at the center, the margin is peeled off to slightly lined at maturity	5.5 cm long, white hollow, 2.5 cm thick, firm, fairly smoothen in touch	Gills attached running slightly down to the stalk whitish to creamish with frequently closer to each other	*Russulalepida* Order: Russulales Family: Russulaceae
MBSRJ35	17.8 cm wide, convex, broadly flattened, greasy to tacky, browncap, margin lobed, wavy, irregular with thick white flesh and brownish fruiting body	9.5 cm long, 5.5 cm thick, broaden at the base of the stalk, pale yellowish irregular stalk tapering at the base	Pores are stuffed long cylindrical with regular dimension, yellowish at the surface. Pores are seen attached to the stalk.	*Boletus* sp. Order: Boletales Family: Boletaceae
MBSRJ37	7.7 cm wide, round to convex,shallowly depressed at the center, dry smooth to touch,purplish pink outer layer, the margin slightly lined to inward rolled	8.4 cm long to 2.2 cm thick, whitish, hollow, brittle to dry in touch	Gills slightly running down the stem, closed to crowded near the stem, whitish in colour to spotted brown near the stalk, soften in touch	*Russula*sp. Order: Russualales Family: Russulaceae
MBSRJ38	2.5–3.9 cm wide, asci with wavy lobed inrolled margin, creamishto dark brown, surface slightlyumbonated, sticky, moist, smooth fibrillose in texture	Stipe is absent	Lamellae sinuate, crowded, whitish to pale yellowish at the time of maturity	*Campanophyllumproboscideum* Order: Agaricales Family: Cyphellaceae

MBSRJ39	7.3 cm wide, white to creamishin colour, funnel shaped withcap continuing to the stalk, white firm flesh crumbly fruiting body with the presence of milky latex like substance on breakage	6.4 cm long hollow stalk, whitish in colour	Gills are decurent and narrow and distantly present.	*Lactifluusvellereus*
				Order: Russulales
			Milky white latex like substance are often seen on breakage	Family: Russulaceae
MBSRJ45	Elongated spike like with a capdiameter of 6 cm, fragile, rough to pitted and ridge, olive brown dark brown in colour	Stipe hollow elongated withlength range from 7.9–10.7 cm,thick with the presencepotholes, presence of a lateralpale brown volval remnantaround the stalk base		*Phallus* sp.
				Order: Phallales
				Family: Phallaceae
MBSRJ48	1.8–3.9 cm across, broadly convex to bell shaped, with a central depression, soften in touch, dry, pale brown with the surface wrinkled or ribbed	1.7–2.9 cm long, equal, dry, presence of fine hair, pale at the apex, reddish brown to blackish at the stalk base	Gills broadly attached to the stalk, buff to whitish in colour, presence of collar that attach the gills to stalk	*Micromphalefoetidum*
				Order: Agaricales
				Family: Marasmiaceae
MBSRJOY	Upto 12 cm wide, broadly convexto flat at maturity, kidney shape with inrolledmargin,whitish in colour	Stalk is absent or lateral in few, 1 cm long, whitish, soften in touch	Gills continuing the lateral stalk, whitish to creamish depends on age with spores at upper layer	*Pleurotusostreatus*
				Order: Agaricales
				Family: Pleurotaceae
MBSRJ51	6.2 cm wide, broadly convex to flat,dry to sticky, yellowish to dark brownish in colour	6 cm long, slender, equal,tough, with enlarge base dark brown to blackish in color	Yellowish to orange pores, depressed at the center of stalk, uniform tube size	*Boletus pseudocalopus*
				Order: Boletales
				Family: Boletaceae
MBSRJ54	12.1 cm wide, stipitate, funnelshape, colour varies formingtwo zone, the inner darker yellow to brownish to outerlighter to pale yellowish to whitish, edges wavy, the cap continues the stalk	Stalk 4.8 cm long, tough, hollow, darker zone just beneath the cap	The pores are yellowish, small uniform	*Amaurodermarugosum*
				Order: Polyporales
				Family: Ganodermataceae

(continued)

Table 3.1 (continued)

Specimen identity number	Fruiting body (Pileus)	Stalk (Stipe)	Gills (Lamellae)	Morphological identification
MBSRJ55	13.5 cm wide, broadly convex with a slight central bump, cap surface covered with dark brown scaly fibers over beige surface with dark center	7.8 cm long with 1 cm wide, hairy surface, ragged, white ring with brown scales undersurface, the colour similar to the cap	Gills are crowded, whitish, few are scaly similar to cap	*Echinodermaaspera* Order: Agaricales Family: Agaricaceae
MBSRJ57	6.5 cm wide, convex, wavy edges, moist surface, withyellowish flesh, the cap colour is yellowish brown	The stalk is wavy with a tapering end, fibrous surface, 6 cm long, reddish brown surface	Fine angular pores, yellowish, long, decurrent, coarse, moist	*Suillus*sp. Order: Boletales Family: Suillaceae
MBSRJ58	5.5 cm wide, broadly flat with a depressed center,yellowish with striated edges	White hollow stalk, 3.5 cm long, equal attach from the cap center	White decurent gills, forked at the stalk	*Gymnomyces* sp. Order: Russulales Family: Russulaceae
MBSRJ59	16.5 cm wide, wavy edges,broadly convex, dry surfacerugulose, faded brownishyellow	7 cm long, 5 cm wide, apex tapering, dark brown, fibrous scales at the stalk, apically sub reticulate	Long slender pores, yellowish, abundant, close, depressed at the stipe	*Leccinumrugosiceps* Order: Boletales Family: Boletaceae
MBSRJ61	5.5 cm wide, convex,margininvolute, black tough scalous fibers on beige surface	3 cm long with tapering end, hairy fibrous surface similar to the pileus	Whitish to grayish pores underneath the cap	*Strobilomycesfloccopus* Order: Boletales Family: Boletaceae
MBSRJ62	Wood decaying with a capdiameter of 7.5 cm, brown toblackish asci, wavy margin,tough surface,cap continuingthe attached stalk	Thin slender stalk with a tapering end, colour similar to the fruiting cap	Highly polyporous attached undersurface the cap, pale yellowish to white	*Amauroderma* sp. Order: Polyporales Family: Ganodermataceae
MBSRJ63	3.5 cm convex cap, smooth, moist, reddish brown, sticky,remains of partial vein on the edges	Pale yellowish hollow stalk of 3.5 cm length, 1 cm wide attach to the cap, soft smooth surface	Pale yellowish, attached to stem, nearly distant, close with a vein like structure	*Agrocybeochracea* Order: Agaricales Family:Agaricomycetes

MBSRJ64	7.25 cm convex flat with acentral depression, sticky surface, the margin lined,bruising green to beige, bald	6.5 cm long, hollow,pitted somewhat, color similar to cap	Gills continues to stalk, nearly distant, spotting green in colour	*Lactariuspurpureus* Order: Russulales Family: Russulaceae
MBSRJGN	15 cm wide, kidney shaped, wavy, rough surface, cracked, sometimes velvety, brownish with white margin, attached to tree bark	Stalk absent	Highly polyporous, tough, circular creamish to yellowish pores	*Heterobasidionannosum* Order: Russulales Family: Bondarzewiaceae
MBSRJ67	Broadly convex 3.5 cm, withshallow depressed center, dry, brownish, Somewhat wrinkled with lined margin with radial splits	5 cm long, equal withslight flaring at the end velvety to fine hairy in touch, brownish to nearly blackish at the base	Gills whitish to yellowish distantly present, attached to stalk	*Gymnopussubnudus* Order: Agaricales Family: Marasmiaceae
MBSRJ69	Cap wide of 5 cm, brownish, tough with shallow depressed at the center, curvy edges	Short stalk of 2.5 cm long, colour similar to the cap	Gills attached to stem, distantly apart, adnate, creamish in colour	*Collybia* sp. Order: Agaricales Family: Tricholomataceae
MBSRJ70	Convex tongue shape withcap diameter of 6 cm, fuzzysurface, the margin wavy, translucent whitish in colour	4.25 cm long, vertical, gelatinous, colour similar to cap	Whitish spines running down the stem, fuzzy in touch	*Pseudohydnumgelatinosum* Order: Tramellales Family: Tramellaceae
MBSRJMR1	Globular, 10 cm wide, rough scaly surface, cracked, yellow to brownish in colour	Stalk absent	Purplish black dust like substance "Gleba" are found inside	*Scleroderma citrinum* Order: Boletales Family: Sclerodermatacea
MBSRJ74	7.35 cm wide, semicircular, irregular edges, zones furrowed, properly distinct, brownish	Stubby lateral stalk	Yellowish to brownish circular pores, separated by layers	*Ganodermaaustrale* Order: Polyporales Family: Ganodermataceae
MBSRJ77	6 cm convex undecorated, darker at the centre, margin tuberculate-striate, a darker zone at the inner end of the margin	8.25 cm long, dirtywhite, undecorated withtiny fibrils, hollow, cylindrical, presence of awhite membranous volvalremnant around the stalk	Gills are free, crowded, whitish to creamish in colour	*Amanita lignitincta* Order: Agaricales Family: Amanitaceae

(continued)

Table 3.1 (continued)

Specimen identity number	Fruiting body (Pileus)	Stalk (Stipe)	Gills (Lamellae)	Morphological identification
MBSRJ79	7.2 cm plano convex, slimywrinkled, dark brown to reddish brown surface	4.25 cm long, 1 cm thick dotted rough surface, moist gelatinous dots, creamish to dirty white surface	Whitish to yellowish pores covered with white partial veins young	*Suillusbrevipes* Order: Boletales Family: Suillaceae
MBSRJPSI	5.85 cm broadly conical, with a pointed tip at the center, brownish, stratiated margin	Cylindrical hollow stalk of 3.2 cm wide, white in uur, tiny hairy fibrils on the surface	Gills distantly separated, thick beneath the cap attach to the stalk, whitish in colour	*Inocybe* sp. Order: Agaricales Family: Inocybaceae
MBSRJ82	6.45 cm campanulate cap, slightly umbonate, Plano convex, grayish brown darker at the center with tuberculate striated margin	8 cm long, sub-cylindricalwith slightly expanded apex,grayish tiny fibrillose scalesover the stalk surface,incomplete ring like projection at the stalk base	Gills are free, crowded, subtruncate, grayish to somewhat darker at the edges	*Amanita* sp. Order: Agaricales Family: Amanitaceae
MBSRJ83	5.5 cm, flat, brown, striated margin with the presence of adpressed scales, tiny fibrils at the cap surface	6.28 cm long, slender, hollow, ring and volva present with white floccose squamules	Gills are free, crowded, attached to the stalk	*Amanita spissacea* Order: Agaricales Family: Amanitaceae
MBSRJ84	Cap surface of 7.2 cm, white conical to applanate with volval remnants of 0.2 cm wide, tiny fibrilous appendiculate over the surface	Stipe length of 12 cm, attenuated upwards, covered with white floccose squamules with a basal bulb of 3 cm wide, subglobular, with conical warts	Gills are free, attenuated, whitish to creamish in colour	*Amanita virgineoides* Order: Agaricales Family: Amanitaceae
MBSRJ86	5.2 cm flat, convex, greyish brown with darker center, smooth velvety in touch with lined margin	5.5 cm long, tapering at the apex, whitish to light brownish in colour, shaggy with a skirt like ring, enlarge sac like white volva present at the base	Gills are free from the stem, crowded to close, whitish in colour	*Amanita spreta* Order: Agaricales Family: Amanitaceae

Table 3.2 Microscopical analysis of few mushrooms from Meghalaya, India

Mushroom specimen	Spore print	Lactophenol cotton blue staining	Melzer's staining	Spore morphology	Spore size (µm)
MBSRJ1	Creamy	Cyanophilic	Pseudoamyloid	Symmetry:Symmetrical	Face view: 5.8–9.06 × 3.5–4.7
				Shape: Cylindrical	
				Ornamentatioon: Smooth	Mean: 7.43 × 4.1
					Mean L/B ratio: 1.81
MBSRJ2	White	Cyanophilic	Amyloid	Symmetry:Symmetrical	Face view: 6.8–7.06 × 5.3–6.23
				Shape:Globose	
				Ornamentation: Spiky warts	Mean: 6.93 × 5.77
					Mean L/B ratio: 1.2
MBSRJ6	White	Cyanophilic	Inamyloid	Symmetry:Symmetrical	Face view: 8.1–9.76 × 5.5–6.7
				Shape:Subglobose	
				Ornamentation: Smooth	Mean: 8.93 × 6.1
					Mean L/B ratio: 1.46
MBSRJ8	Purple brown	Amphophilic	Inamyloid	Symmetry:Symmetrical	Face view: 5.6–7.9 × 3.2–4.7
				Shape: Ellipsoidal	
				Ornamentation: Smooth	Mean: 6.75 × 3.95
					Mean L/B ratio: 1.7
MBSRJ11	White	Cyanophilic	Pseudoamyloid	Symmetry:Symmetrical	Face view: 6.8–8.06 × 2.8–3.7
				Shape:Allantoid	
				Ornamentation: Smooth	Mean: 7.43 × 3.25
					Mean L/B ratio: 2.28
MBSRJ12	White	Cyanophilic	Inamyloid	Symmetry:Symmetrical	Face view: 4.8–6.06 × 2.5–4.2
				Shape: Ellipsoidal	
				Ornamentation: Smooth	Mean: 5.43 × 3.35
					Mean L/B ratio: 1.62

(continued)

Table 3.2 (continued)

Mushroom specimen	Spore print	Lactophenol cotton blue staining	Melzer's staining	Spore morphology	Spore size (μm)
MBSRJ13	Rusty Brown	Cyanophilic	Inamyloid	Symmetry: Asymmetrical Shape:Subellipsoidal Ornamentation: Smooth	Face view: 7.8–10.8 × 4.8–6.4 Mean: 9.3 × 5.6 Mean L/B ratio: 7.45
MBSRJ18	Creamy	Cyanophilic	Amyloid	Symmetry: Symmetrical Shape: Ellipsoidal Ornamentation: Warts	Face view: 5.8–7.06 × 5.5–6.7 Mean: 6.43 × 6.1 Mean L/B ratio: 1.05
MBSRJ26	Creamy	Cyanophilic	Amyloid	Symmetry:Symmetrical Shape:Globose Ornamentation: Warts	Face view: 6.8–8.06 × 5.5–6.7 Mean: 7.43 × 6.1 Mean L/B ratio: 1.21
MBSRJ35	Brown	Cyanophilic	Pseudoamyloid	Symmetry:Symmetrical Shape:Subfusoid Ornamentation: Smooth	Face view: 10–12.4 × 3.5–4.5 Mean: 11.2 × 4 Mean L/B ratio: 3.55
MBSRJ37	White	Cyanophilic	Amyloid	Symmetry:Symmetrical Shape:Subglobose Ornamentation: Isolated Warts	Face view: 6.8–9.06 × 5.5–5.7 Mean: 7.93 × 5.6 Mean L/B ratio: 1.41
MBSRJ38	White	Cyanophilic	Pseudoamyloid	Symmetry:Symmetrical Shape: Ellipsoidal Ornamentation: Smooth	Face view: 4.8–5.06 × 2.4–2.7 Mean: 4.93 x 2.55 Mean L/B ratio: 1.93

MBSRJ39	White	Cyanophilic	Amyloid	Symmetry:Symmetrical	Face view: 7.2–8.06 × 4.5–6.7
				Shape: Ellipsoidal	
				Ornamentation: Isolated warts	Mean: 7.63 x 5.6
					Mean L/B ratio: 1.36
MBSRJ45	Greenish brown	Cyanophilic	Pseudoamyloid	Symmetry:Symmetrical	Face view: 8.2–8.66 × 3.5–3.7
				Shape: Ellipsoidal	
				Ornamentation: Smooth	Mean: 7.63 × 3.6
					Mean L/B ratio: 2.11
MBSRJ48	Whitish buff	Cyanophilic	Pseudoamyloid	Symmetry:Symmetrical	Face view: 8.1–9.06 × 3.5–4.0
				Shape: Ellipsoidal	
				Ornamentation: Smooth	Mean: 8.58 × 3.75
					Mean L/B ratio: 2.288
MBSRJOY	Creamy	Cyanophilic	Pseudoamyloid	Symmetry:Symmetrical	Face view: 6.8–8.1 × 3.5–4.2
				Shape: Ellipsoidal	
				Ornamentation: Smooth	Mean: 7.45 × 3.85
					Mean L/B ratio: 1.93
MBSRJ51	Yellow	Cyanophilic	Inamyloid	Symmetry:Symmetrical	Face view: 8.8–9.06 × 4.5–5.2
				Shape:Fusoid	
				Ornamentation: Smooth	Mean: 8.93 × 4.85
					Mean L/B ratio: 1.84
MBSRJ54	Brownish Yellow	Cyanophilic	Pseudoamyloid	Symmetry:Symmetrical	Face view: 9.8–10.1 × 5.5–5.7
				Shape:Fusoid	
				Ornamentation: Smooth	Mean: 9.95 × 5.6
					Mean L/B ratio: 1.77

(continued)

Table 3.2 (continued)

Mushroom specimen	Spore print	Lactophenol cotton blue staining	Melzer's staining	Spore morphology	Spore size (µm)
MBSRJ55	White	Cyanophilic	Pseudoamyloid	Symmetry:Symmetrical	Face view: 7.8–8.06 × 3.5–3.7
				Shape:Fusoid	
				Ornamentation: Smooth	Mean: 8.94 × 4.59
					Mean L/B ratio: 1.94
MBSRJ57	Greenish yellow	Amphophilic	Inamyloid	Symmetry:Symmetrical	Face view: 8.8–10.1 × 3.5–3.7
				Shape: Spindle shape	
				Ornamentation: Smooth	Mean: 9.45 × 3.6
					Mean L/B ratio: 2.6
MBSRJ58	Creamy	Cyanophilic	Amyloid	Symmetry:Symmetrical	Face view: 5.8–6.06 × 4.5–5.7
				Shape:Globose	
				Ornamentation: Warts	Mean: 5.93 × 5.1
					Mean L/B ratio: 1.16
MBSRJ59	Pale Yellow	Amphophilic	Inamyloid	Symmetry:Symmetrical	Face view: 8.8–9.5 × 2.5–4.7
				Shape:Subfusoid	
				Ornamentation: Smooth	Mean: 9.15 × 3.6
					Mean L/B ratio: 2.54
MBSRJ61	Grayish Brown	Amphophilic	Pseudoamyloid	Symmetry:Symmetrical	Face view: 6.8–7.06 × 5.5–5.7
				Shape:Globose	
				Ornamentation: Ridges and lines forming reticulum	Mean: 6.93–5.6
					Mean L/B ratio: 1.23
MBSRJ62	Creamy	Cyanophilic	Amyloid	Symmetry:Symmetrical	Face view: 6.1–6.7 × 4.5–5.7
				Shape:Subglobose	
				Ornamentation: Thick double wall with Warts	Mean: 6.4–5.1
					Mean L/B ratio: 1.94

MBSRJ63	Brown	Amphophilic	Pseudoamyloid	Symmetry:Symmetrical	Face view: 6.8–9.65 × 5.5–5.7
				Shape: Ellipsoidal	
				Ornamentation: Smooth	Mean: 8.22 × 5.6 Mean L/B ratio: 1.46
MBSRJ64	Creamy	Cyanophilic	Amyloid outerwarts	Symmetry:Symmetrical	Face view: 6.8–7.06 × 5.5–5.7
				Shape:Subglobose	
				Ornamentation: Warts	Mean: 6.93 × 5.6 Mean L/B ratio: 1.23
MBSRJGN	Pale creamy	Cyanophilic	Inamyloid	Symmetry:Symmetrical	Face view: 6.5–6.8 × 4.5–5.7
				Shape:Subglobose	
				Ornamentation:Smoothe	Mean: 6.65 × 5.1 Mean L/B ratio: 1.3
MBSRJ67	Creamy	Cyanophilic	Inamyloid	Symmetry:Symmetrical	Face view: 8.5–10.1 × 3.5–4.1
				Shape: Ellipsoidal	
				Ornamentation:Smooth	Mean: 9.3–3.8 Mean L/B ratio: 2.44
MBSRJ69	White	Cyanophilic	Inamyloid	Symmetry:Symmetrical	Face view: 8.3–9.66 × 3.5–3 .7
				Shape: Ellipsoidal	
				Ornamentation:Smooth	Mean: 8.98–3.6 Mean L/B ratio: 2.49
MBSRJ70	White	Cyanophilic	Amyloid	Symmetry:Symmetrical	Face view: 4.5–5.2 × 4.1–4.8
				Shape:Subglobose	
				Ornamentation:Smooth	Mean: 4.85–4.45 Mean L/B ratio: 1.08

(continued)

Table 3.2 (continued)

Mushroom specimen	Spore print	Lactophenol cotton blue staining	Melzer's staining	Spore morphology	Spore size (μm)
MBSRJMR1	Dark brown	Cyanophilic	Pseudoamyloid	Symmetry:Symmetrical	Face view: 8–12.06 × 7.5–9.7
				Shape:Globose	
				Ornamentation:Reticulate	Mean: 10.03–8.6
					Mean L/B ratio: 1.16
MBSRJ74	Yellow	Cyanophilic	Inamyloid	Symmetry:Symmetrical	Face view: 8.8–10.1 × 4.5–5.7
				Shape: Ovoid	
				Ornamentation:Smooth	Mean: 8.94–4.59
					Mean L/B ratio: 1.94
MBSRJ77	White	Cyanophilic	Inamyloid	Symmetry:Symmetrical	Face view: 9.5–12.5 × 7.5–8.9
				Shape:Subglobose	
				Ornamentation:Smooth	Mean: 11–8.2
					Mean L/B ratio: 1.34
MBSRJ79	Yellow	Cyanophilic	Pseudoamyloid	Symmetry:Symmetrical	Face view: 7.8–9.02 × 2.8–3.7
				Shape:Subfusoid	
				Ornamentation:Smooth	Mean: 8.41–3.25
					Mean L/B ratio: 2.58
MBSRJPSI	Brown	Cyanophilic	Inamyloid	Symmetry:Symmetrical	Face view: 12.8–12.9 × 9.5–10.7
				Shape:Fusoid	
				Ornamentation:Tiny Nodules	Mean: 12.85–10.1
					Mean L/B ratio: 1.27

MBSRJ82	White	Cyanophilic	Inamyloid	Symmetry:Symmetrical	Face view: 9–12.1 × 8.5–11.8
				Shape:Subglobose	
				Ornamentation:Smooth	Mean: 10.55–10.15
					Mean L/B ratio: 1.03
MBSRJ83	White	Cyanophilic	Amyloid	Symmetry:Symmetrical	Face view: 7.8–10.1 × 6.5–9.8
				Shape:Fusoid	
				Ornamentation:Smooth	Mean: 8.95–8.15
					Mean L/B ratio: 1.09
MBSRJ84	White	Cyanophilic	Amyloid	Symmetry:Symmetrical	Face view: 9.8–12.1 × 4.5–4.8
				Shape: Ellipsoidal to Subglobose	Mean: 10.93–4.65
				Ornamentation:Smooth	Mean L/B ratio: 2.35
MBSRJ86	White	Cyanophilic	Inamyloid	Symmetry:Symmetrical	Face view: 7.8–9.5 × 5.5–6.8
				Shape: Ellipsoidal	
				Ornamentation:Smooth	Mean: 8.65–6.15
					Mean L/B ratio: 1.4

Although morphological and microscopic features have been the basis for identification of mushrooms, there are many unsolved dilemma left in the typing and classification at lower taxon level because of the paucity of morphological features (Hibbett et al. 2007). This is further complicated by other factors which include the convergent evolution and cryptic speciation.

Epigenetic factors have made the morphological features unstable at both intra- and interspecies level leading to high incongruency thus making the use of molecular marker as a reliable confirmatory identification alternative (Feng et al. 2012). Grouping or characterizing the species based on morphotyping has led to convergent evolution (Brun and Silar 2010). Apart from these, the concepts of exploiting molecular approaches have unraveled the knowledge on the nature of evolution and the phylogenetic concept of the life forms in the last decade. The upgradation of various statistical methods along with bioinformatics tools has aided and simplified the process of evaluation of the evolutionary clade of a species. Approaches such as amplification of specific hypervariable gene loci by polymerase chain reaction (**PCR**) or restriction digestion of a specific gene sequence using restriction fragment length polymorphism (**RFLP**) have provided deeper understanding on the typification and diversity of species.

Till recently, the absence of a universally accepted DNA barcode for fungi, the second most specious eukaryotic kingdom (Blackwell 2011; Mora et al. 2011), has been a serious limitation for **biodiversity** studies. For proper identification and discovery of true cryptic species and classifying them in the pre-existent evolutionary clade in fungi, there is a felt need for developing molecular barcoding tools and markers in mushrooms.

3.2 Paradigm Shift from Classical to Advanced Molecular Approaches

Mycology is mostly familiarized to the lower species of fungi without much attention being paid to describe and identifies cryptic species. Mushrooms studies have been revolving around wood rotten fungi and the poisonous 'toadstools'. It is well known that fungi are known to be found everywhere presenting morphological similarity with one another thus making the differentiation among the closely related species very difficult. Till the last decade, the gross identification of these cryptic species was solely dependent on the morphotypic features of the individual which were later supplemented with the *in vitro* micro graphical characters. The advancement in molecular biology tools in the recent era has resulted in shift to a new level of nucleic acid and its sequence analysis leading to nucleic acid revolution in mushroom identification. The levels of genetic information contained at the cellular level are far more advanced than that of the phenotypic features. Several variations observable at the genetic level can be incorporated to the phenotypic features and are collectively considered under "epigenetic" factors.

The taxonomic study in fungi is associated with 80,000 species, traditionally differentiated into six main domain based on the sexual reproductive units, ultrastructural

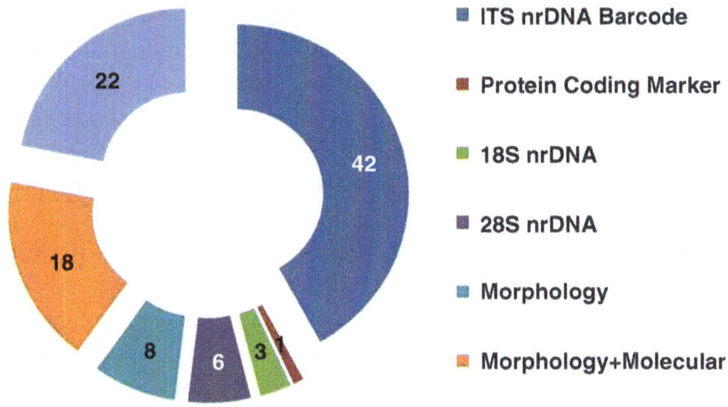

Fig. 3.2 Percentage of using various markers for the identification of the fungal species (Adapted from Raja et al. 2017)

components and biochemical analysis. These domains include Myxomycetes or Protists, Oomycetes and the true fungi which are classified into Ascomycetes, Basidiomycetes, Zygomycetes and Chytrids (McLaughlin et al. McLaughlin et al. 2001a, 2001b). A pseudo-division includes the Deuteromycetes or Fungi Imperfecti commonly referred to as moulds or rust have asexually reproducing unit and some known sexual reproductive stages. With an estimate of 2.2 to 3.8 million fungal species (Hawksworth and Lacking 2017), the problem with cataloguing of fungi in the database has already become a challenging task without the contribution of molecular information.

In the last decades, taxonomist have classified the species based on morphology, niche types, growth behavior, their interaction with the host and to the ecological factors (Fig. 3.2). Thus the classification and identification of the cryptic species from the taxonomical point of view is far from being complete. It is well accepted that less than 1% of the microbiological species are culturable in *in-vitro*. So, the need of the biological markers for easy identification of the unculturable environmental samples are of utmost importance developing barcodes for forms like mushrooms is expected to answer the difficult questions posed in their classification approaches.

3.3 Molecular Markers for Identification

With the advancement of the barcode in the prokaryotic system, amplification of the nucleic acids and the initial construction of the phylogenetic map work on fungi began with nuclear ribosomal genes. The structural gene sequence of rRNA is known for its well conserved regions at both genus and species level. Initially **18SrRNA**, a homolog to **16S rRNA** in bacteria was used for the phylogenetic analysis but the region has limitations of having fewer hypervariable domains. The large ribosomal subunit **28S rRNA** gene discriminates the species on its own or combines with the ITS region. The internal transcribed spacer region (**ITS**) of the ribosomal region of DNA (rDNA) is commonly used as a marker for resolution at or below the genus level (Seifert 2009; Begerow et al. 2010) and is a recommended marker by the International Fungal Barcoding Consortium (Schoch et al. 2012). In fungi, the ITS ranges from 550–600 bp comprising of two hypervariable spacer regions, ITS1 and ITS2 which are separated by highly conserved 5.8S rRNA gene (White et al. 1990). The region is flanked by 18S rRNA and 28S rRNA genes towards the 5′-end and 3′-end of ITS-1 and ITS-2 spacer region respectively. In fungi, ITS is found in multiple tandem copies in each haploid genome making it highly desirable for use as a fungal barcode aided by its easier amplification even in trace amounts of biological samples and is extensively used for the genetic distance mapping in fungi. ITS has been used as a fungal universal barcode for the identification of the cryptic species and has been cited in more than 500 literature (Schoch et al. 2012). Apart from the ITS, the small ribosomal subunit (SSU) region when combined with the NS1 and NS4 region gives an information about the phylogenetic placement of higher fungi in taxonomical level. Similarly for the identification of the species at the intermediate taxonomical level, large ribosomal subunit (LSU) primers in combination with LROR and LR6 are being used (Vilgalys and Hester 1990). These genes are widely used as markers in the fungal systematic nomenclature such as Assembling the Fungal Tree of Life (AFTOL).

In mycological identification, the sequence obtained for a species of fungi is compared with the unknown sequence from database available in Sequence Database such as GenBank at **NCBI** (National Centre for Biotechnology Information) and **EMBL**, The European Nucleotide Sequence Archive of European Molecular Biology Laboratory.

Apart from the ITS region, the mitochondrial gene, which codes for the cytochrome oxidase I (COX I) used as a biomarker in the animal kingdom have been explored for fungal systems but are found to be inappropriate for the identification at species level for fungi (Schoch et al. 2012). The protein encoding genes like RNA polymerase II (RPB1, RPB2) are being widely used for the phylogenetic analysis of fungi but due to it ubiquitous and single copy number and slow rate of sequence diversion it is not highly recommended (Tanabe et al. 2002). For species level identification of fungi in general, the short ribosomal subunit (SSU) has no such barcode gap and is not recommended as an identification marker. The large ribosomal subunit (LSU) sometimes show no such amplification in diverse species adding to editing error and are not considered at par with the usage of SSU marker. However, the hypervariable ITS regions have much superior barcode gap and due to its tandem repeats and resolving power to distinguish the mushrooms at its species level, it is a defined barcode for the phylogenetic analysis of the mushrooms at its basic level of species discrimination phylogenetically.

3.4 ITS Marker Assisted Identification of Mushrooms

Polyphasic characterization gives a preliminary identification of a species but genus level identification is always left to a dilemma when considering the macroscopic and microscopic features. Few mushroom species look morphologically alike to other species as observed in a study carried to characterize wild mushrooms by the authors` group. For example, the specimen MBSRJ1and MBSRJ11 is a common wood fungus alike to *Trametes* species but identification based on the molecular marker using ITS identified the specimens as *Formitopsis ostreoformis* and *Trichaptum biforme* respectively (Fig. 3.3). In the same study, the morphological identification of few mushroom species could not be completed due to the lack of information and relevant data in the existing databases. The paucity in the morphological data can also be highly affected by the epigenetic factors. The mushroom species MBSRJ18 and MBSRJ39 were morphologically alike but confirming the molecular data resembled them as two different species, *Lactarius glaucescens* and

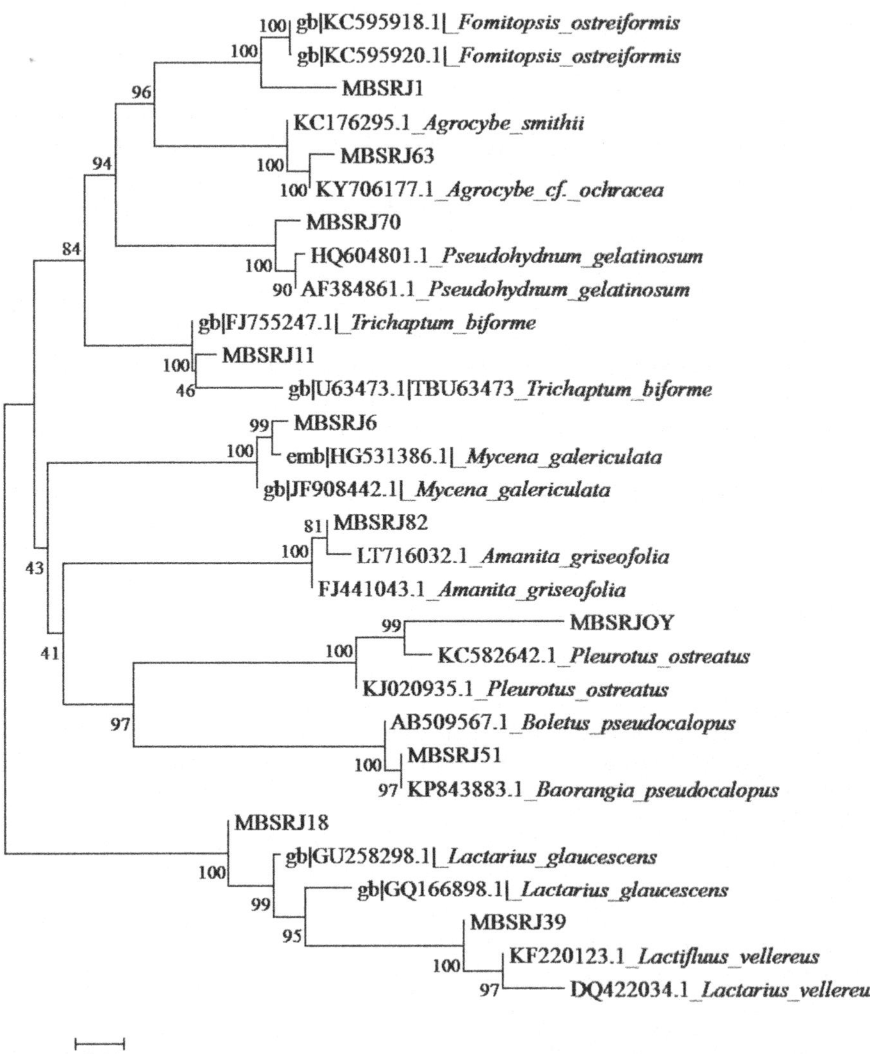

Fig. 3.3 Evolutionary distance relationship of few mushroom species based on ITS sequence similarity

Lactifluus vellereus respectively. The molecular identification using the ITS loci was first time reported in the study for MBSRJ82 which was identified to be *Amanita griseofolia*. The morphological trait in these cryptic species has always leaded to misidentification which may be the effect of convergent evolution. Thus confirming the use of molecular marker such as ITS for correct identification is necessary along with the traditional technique of identifying the mushrooms. Hence employing the phylogenetic analyses becomes a useful tool for classification of unknown sequence and incorporating such emerging information into the existing evolutionary framework.

3.5 Conclusion

Though a large number of the universal biomarkers for the phylogenetic analysis of fungi in general and macrofungi in particular have been in use, the markers are not the only restricted as genetic markers in phylogenetic analysis. There is a need for exploration of additional biomarkers for **fungal taxonomy**. Compared to the earlier approaches of identifying the fungal species with the basic morphological and microscopic features, there is a need towards improving the molecular markers along with the algorithms of various statistical tools which can provide an appropriate authentication for the identification of the fungi. The combination of both morphotyping and molecular genetic marker studies will relatively assure a full proof system for the phylogenetic analysis of the higher eukaryotes like mushrooms. The need for biomarker is not only essential to delineate the existing morphotyping but also to authenticate and classify the novel undescribed forms of macrofungi by linking them with available sequences in various databases and genebanks.

References

Begerow D, Nilsson H, Unterseher M, Maier W (2010) Current state and perspectives of fungal DNA barcoding and rapid identification procedures. Appl Microbiol Biotechnol 87:99–108

Blackwell M (2011) The Fungi: 1, 2, 3 … 5.1 million species? Am J Bot 8:426–438

Brun S, Silar P (2010) Convergent evolution of morphogenetic processes in Fungi. In: Evolutionary biology-concepts, molecular and morphological evolution. Springer Berlin Heidelberg, Berlin/Heidelberg, pp 317–328

Feng B, Xu J, Wu G, Zeng NK, Li YC, Tolgor B, Kost GW, Yang ZL (2012) DNA sequence analyses reveal abundant diversity, endemism and evidence for Asian origin of the porcini mushrooms. PLoS One 7(5):e37567

Hawksworth DL, Lacking R (2017) Fungal diversity revisited: 2.2 to 3.8 million species. Microbiol Spectr 5(4). https://doi.org/10.1128/microbiolspec.FUNK-0052-2016

Hibbett DS, Binder M, Bischoff JF et al (2007) A higher-level phylogenetic classification of the Fungi. Mycol Res 111:509–547

McLaughlin DJ, McLaughlin EG, Lemke PA (eds) (2001a) The Mycota. Vol. VII. Part A. systematics and evolution. Springer-Verlag, Berlin

McLaughlin DJ, McLaughlin EG, Lemke PA (eds) (2001b) The Mycota. Vol. VII. Part B. systematics and evolution. Springer-Verlag, Berlin

Mora C, Tittensor DP, Adl S, Simpson AG, Worm B (2011) How many species are there on earth and in the ocean? PLoS Biol 9(8):e1001127

Purkayastha RP, Chandra A (1985) Manual of Indian edible mushrooms. Today and Tomorrow's Publication, New Delhi

Raja HA, Miller AN, Pearce CJ, Oberlies NH (2017) Fungal Identification Using Molecular Tools: A Primer for the Natural Products Research Community. Journal of Natural Products 80(3):756–770

Sawian JT, Jeeva S, Lyndem FG, Mishra BP, Laloo RC (2007) Wild edible plants of Meghalaya, North East India. Nat Prod Rad 6:410–426

Schoch CL, Seifert KA, Huhndorf S, Vincent R, Spouge JL, Levesque CA, Chen W (2012) Fungal barcoding consortium. Nuclear ribosomal Internal Transcribed Spacer (ITS) region as a universal DNA barcode marker for Fungi. PNAS 109(16):6241–6246

Seifert KA (2009) Barcodi Fungi: progress towards DNA barcoding of fungi. Mol Ecol Resour 9(1):83–89

Singh B, Sinha BK, Phukan SJ, Borthakur SK, Singh VN (2012) Wild edible plants used by Garo tribes of Nokrek biosphere reserve in Meghalaya. Indian J Tradit Know 11(1):166–171

Tanabe Y, Watanabe MM, Sugiyama J (2002) Are *Microsporidia* really related to Fungi? A re-appraisal based on additional gene sequences from basal fungi. Mycol Res 106:1380–1391

Vilgalys R, Hester M (1990) Rapid genetic identification and mapping of enzymatically amplified ribosomal DNA from several *Cryptococcus* species. J Bacteriol 172(8):4238–4246

White TJ, Bruns T, Lee S, Taylor JW (1990) Amplification and direct sequencing of fungal ribosomal RNA genes for phylogenetics. In: Innis MA, Gelfand DH, Sninsky JJ, White TJ (eds) PCR protocols: a guide to methods and applications. Academic Press Inc, New York, pp 315–322

Chapter 4
Antimicrobial and Hepatoprotective Activities of Edible Mushrooms

Jasmina Glamočlija, Marina Kostić, and Marina Soković

4.1 Introduction

People in many countries of the world consume mushrooms in their daily diet and this habit has been present in the society for thousands of years. Mushrooms consumption has increased in recent years, due to the scientifically supported facts that a stable and balanced nutrition has an important role in overall human well-being. Mushrooms are widely consumed for their food properties, excellent nutritional characteristics and curative effects; accordingly they belong to functional foods with many medicinal effects (Valverde et al. 2015).

Mushrooms have a great nutritional value since they are rich in protein content, essential amino acids and dietary fibre, and on the other side they contain a low fat content. Different species of edible fungi obtain a high moisture value up to 95%, depending on environmental circumstances and phase of growth. It is important to highlight the fact that edible species of fungi contain different vitamins such as K, H, E, D, B1, B2, B12, as well as mono- and poly-unsatturated fatty acids with various biological effects, phenolic compounds, organic acids, sterols, terpenes etc (Barroetavena and Toledo 2017).

More than 3000 mushrooms are marked and consumed as edible; only one third of them are cultivated for commercial purposes with just ten species being grown on the industrial level (Chang and Miles 2004). Many edible mushrooms and number of macromycetes possessed therapeutic properties with edible and medicinal value, approximately 2500 species (FAO 2004; FAO 2009). Global economic value of mushrooms is increased in the past few decades, since mushrooms are not only consumed for their edibility and respected for their nutritional properties, but also because they have many confirmed medicinal uses. Mushrooms production is

J. Glamočlija · M. Kostić · M. Soković (✉)
Department of Plant Physiology, Institute for Biological Research "Siniša Stanković", University of Belgrade, Belgrade, Serbia
e-mail: mris@ibiss.bg.ac.rs

B. P. Singh et al. (eds.), *Biology of Macrofungi*, Fungal Biology,
https://doi.org/10.1007/978-3-030-02622-6_4

continuously rising together with the rise in global human population. Production patterns vary according to the geographic region; according to the reports from 2014, Asian countries produce more than 74.64% of world mushroom markets, followed by Europe (19.63%) (Rosmiza et al. 2016). In the USA the total value of mushrooms produced and solid during 2015–2016 is about 880 million US $ (Gargano et al. 2017). The major producer is China with estimated production of more than 64% of the world mushroom production in 2012 (FAO 2015).

Mushrooms are attractive as a source of **therapeutic compounds**, and are classified as excellent functional food. They played an important role in the healthcare systems of Asia and South America and have been traditionally used for the management of important diseases and conditions related to health (Roupas et al. 2012). Different mushrooms species have a long history of ethnological use in folk medicine, especially in Asian countries (China, India, Japan and Korea), but also in eastern European countries and Africa (Wasser 2014). Biological activity is expected to be explored from fungal mycelium or the culture fluid where mycelium was grown and fungal fruiting bodies. Natural products obtained from mushrooms primary and secondary metabolism have introduced a revolution in the area of prevention and treatment of various diseases (De Silva et al. 2013). Different bioactive compounds of edible mushrooms are linked with their immunomodulatory, antioxidant, anthelmintic, antitumor and anticancer, antiatherogenic, antimicrobial, and hypoglycemic properties (Sánchez 2017a). Therefore, natural products from edible mushrooms have received attention from the scientific community since they present a valuable source of molecules with potential medical application (David et al. 2015).

The liver is one of the vital organs of the human body with the ability of glandular secretion, which has a number of important physiological roles necessary for the body to maintain its physiological functions. The primary functions of the liver are: excretion of hormones, cholesterol; metabolism of proteins, fats, and carbohydrates; enzyme activation; storage of vitamins, glycogen, and minerals; production of bile and excretion bilirubin; synthesis of plasma proteins, such as albumin, and clotting factors and contributes in blood detoxification and purification (Seif 2016).

The **hepatocytes** are responsible for many of the important metabolic processes in the body:

- Absorbs overbalanced glucose from digested food and stores it in the form of glycogen for later use;
- Hepatocytes take in fatty acids from the blood passing through the liver and metabolize them to produce energy in the form of ATP. Glycerol is converted into glucose by hepatocytes. They can also produce phospholipids, cholesterol, and lipoproteins that are used by other cells in the body. Excess cholesterol gets excreted from the body as a component of bile;
- Hepatocytes first remove the amine groups of the amino acids and convert them into ammonia and eventually urea. Urea is less toxic than ammonia and can be excreted in urine;
- The hepatocytes of the liver remove many potentially toxic substances and monitor the contents of the blood. Enzymes in hepatocytes metabolize drugs and alcohol into their inactive forms. The liver also metabolizes hormones and removes them from circulation.

- The liver provides storage of vitamins (A, D, E, K, and B12), many essential nutrients (glycogen and fatty acids from digested triglycerides), and minerals (copper and iron) obtained from hepatic portal system;
- Production of protein components (prothrombin, fibrinogen, and albumins) included in blood plasma is an important feature of hepatic cells;
- Kupffer cells of the liver contribute in the **inflammatory** response. They play an important role in capturing and digesting various parasites including bacteria and fungi as well as worn-out blood cells and cellular debris.

Liver is susceptible to many diseases and conditions that may occur: cancer, cirrhosis, hepatitis and damage by toxins and medications (Seif 2016). Alcohol in high doses can be hepatotoxic and long-term alcohol abuse is a common cause of liver disfunction. Infections and general inflammation by different viruses or medications may affect the liver (side effect of antibiotics, statins and others) (Wang et al. 2010).

It is estimated that adult population of more than fifty million people in the world, would be affected with chronic liver disease due to viral hepatitis, alcohol, and nonalcoholic steatohepatitis (NASH) as the most common causative factors. Liver disease was the 9th leading digestive disease diagnosed at ambulatory care visits in USA during 2004 (Everhart and Ruhl 2009).

The risks of developing fatty liver disease could be prevented by maintaining an adequate body weight with balanced diet. As mentioned above, edible mushrooms are excellent sources of biologically active compounds and nutraceuticals (Reis et al. 2017). It has been more then 30 years since the introduction of the term functional foods. Consumers recently have been more interested food derived from natural production as well as their use as health promoting agents.

Mushrooms present a nutritious alternative and a valuable source of various healty compounds (Chang and Buswell 1996) which are extractable from either the mushroom mycelium or the fruiting body. Continous intake of either mushrooms or mushroom dietary supplements in the form of pills can improve individuals overall health.

The edible mushrooms which demonstrate useful **antimicrobial** and **hepatoprotective** activities about which data will appear in this chapter include species belonging to genus *Agaricus, Coprinus, Flammulina, Grifola, Laetiporus, Lentinus, Meripilus, Morchella, Pleurotus* and *Polyporus*.

Other species (*Ganoderma, Inonotus* and *Trametes* species) mentioned in this chapter are inedible due to their coarse and hard texture or bitter taste, but are consumed in different forms as liquors, concentrates, tinctures, hot water extracts, tonics, powders, soups and teas. (Quang et al. 2006).

4.2 Antimicrobial Activity of Edible Mushrooms

Human society has been confronted, since the beginig, with the problem of infectious diseases caused by pathogenic microorganisms. In addition, other pathogens can cause damage to crops and many food sources and are considered primary causes of food spoilage (Soković et al. 2013). The treatment of infectious diseases caused by bacteria has been dealt with, with the discovery of **antibiotics** (Van Hoek et al. 2011).

However, their excessive and inappropriate during last fifty years has led to the increase in the number of resistant strains of bacteria which today represent a health threat of global proportions (Wilson et al. 2002). In addition, commercially available antibiotics can lead to induced hepatotoxicity, which may have serious consequences in patients with hepatitis or HIV infection (Sharma et al. 2015; Andrade and Tulkens 2011).

On the other hand, opportunistic phatogenic micromycetes cause serious infections and in recent years have increased the number of immunocompromised or hospitalized patients especially. The fungal infections are so widespread and frequent that they cause at least 1.5 million deaths worldwide each year (Brown et al. 2012).

Human and fungi have the same cellular organization umans (both are eukaryotic organisms) and have a similar cellular machinery. Therefore toxic compounds used to treat pathogenic fungi could probably be toxic for human cells. Antifungal drugs used in clinical practice have shown selective biological activity, which enables the appearance of resistant strains. For all these reasons, the search for new antimicrobial compounds ever present.

Fungi contain many components that may have antimicrobial activity which allow them to survive in the environment. The antimicrobial properties of both edible and inedible mushrooms are frequently discussed in numerous papers (Alves et al. 2012 a, b). Suay et al. (2000) tested 204 species and found that 45% extracts have antimicrobial properties, and show primarily antibacterial activity.

During the last decade, a large number of species of subdivisions Basidiomycotina and some Ascomycotina were tested in various studies which showed their antimicrobial potential (Barros et al. 2008 a, b; Ramesh and Ramesh and Pattar 2010; Ochoa-Zarzosa et al. 2011; Schwan 2012; Alves et al. 2013; Kostić et al. 2017; Soković et al. 2017). Results of these studies indicate that macromycetes are rich with compounds that possess various biological activities (Boucher et al. 2009). Crude extracts and isolated compounds (terpenoids, sterols, organic acid, derivatives of anthraquinone, the peptides and proteins, ribonuclease) exhibit direct antimicrobial activity, while polysaccharides exhibit indirect antimicrobial activity. It is believed that those compounds interact in unison, although individual activity should not be neglected (Alves et al. 2013). The methods used for rapid detection of antimicrobial activity are numerous: microdilution and disk diffusion methods which include the incorporation of the extract into the culture media. However the use of different methods deter in the compilation and interpretation of their antimicrobial activity.

Furthermore, the use of various methods of extraction and different types of the solvent also pose difficulty in the determination of antimicrobial activity. Petrović et al. (2014a) in their study confirm that the methods of extraction have a significant effect on the determination of antimicrobial activity. They determined that the extract of *Laetiporus sulphureus* obtained by the ultrasonic apparatus had a better activity than the extract obtained by classical extraction. Chemical macromycetes profile is determined by geographical factors, native to the specific species, as well the phase of the life cycle of the **macromycetes** (whether it was collected in the form of mycelium, young or old fruit body). Macromycetes are rich in compounds which are known in the literature to have antimicrobial activity: phenol and organic acids, peptides, proteins and others (Barros et al. 2008a; Alves et al. 2013).

The results of the antimicrobial activity indicate that antimicotics showed better activity against the different pathogenic microorganisms then natural antimicrobial agents; however, commercial antibiotics also exhibit a series of follow-up of adverse effects (e.g. hepatotoxicity induced). Using edible macromycetes which have antimicrobial properties and also don't display adverse effects definitely attract attention.

4.2.1 Antimicrobial Activity of the Most Extensively Cultivated Edible Mushroom

The commercial production of mushrooms is continuously increasing, 7,959,979 tonnes grown in 2012, 1,869,091 tonnes were harvested in Europe while in China accounting biggest production, 5,150,000 tonnes (Grujičić et al. 2015). Human population increased and so did the consumption of mushrooms as well as their mushroom production, which is estimated to grow 15% per year (Kamarudzaman et al. 2015). *Agaricus bisporus* (J. E. Lange) Emil J. Imbach. is the most extensively cultivated mushroom worldwide, followed by *Lentinula edodes* (Berk.) Pegler and *Pleurotus ostreatus* (Jacq. ex Fr.) P. Kumm.

4.2.1.1 Antimicrobial Activity of the *Agaricus bisporus*

White button mushroom, *Agaricus bisporus* is the most widely cultivated (Fig. 4.1) and consumed species of mushroom in the world (Gargano et al. 2017). The methanolic extracts of three *Agaricus* species showed antimicrobial activity against six species of **Gram-positive** bacteria, seven species of **Gram-negative** bacteria and two species of yeast (Ozturk et al. 2011). *A. bisporus* extract exhibited better antimicrobial activity compared to other mushroom species especially against Gram-positive bacteria (Table 4.1).

Fig. 4.1 *Agaricus bisporus* (J.E.Lange) Imbach Photo by: Ivanka Milenković

Table 4.1 Antimicrobial activity of selected edible mushrooms

Mushroom	Type of extract	Antimicrobial activity	References
Agaricus bisporus **(J.E.Lnge) Imbach**	Methanolic	**antibacterial**	Abah and Abah (2010), Ozturk et al. (2011), Sharma et al. (2015), Waithaka et al. (2017))
		IZ 4-23mm	
		antifungal	Ozturk et al. (2011), Sharma et al. (2015), Waithaka et al. (2017)
		IZ 8-21mm	
	Acetone	**antibacterial**	Sharma et al. (2015)
		IZ 10-17.77mm	
Agaricus blazei **Murill**	Methanolic	**antibacterial**	Stojković et al. (2014)
		MIC 0.1-2.3 mg/mL	
		antifungal	
		MIC 0.1-1.25 mg/mL	
	Ethanolic	**antibacterial**	
		MIC 0.04-1.15 mg/mL	
		antifungal	
		MIC 0.15-3.125 mg/mL	
A. campestris L.	Methanolic	**antibacterial**	Glamočlija et al. (2015)
		MIC 0.58-2.34 μg/mL	
		antifungal	
		MIC 0.39-6.25 μg/mL	
	Ethanolic	**antibacterial**	
		MIC 0.03-2.34 μg/mL	
		antifungal	
		MIC 0.1-3.12 μg/mL	
A. macrosporus **(F. H. Moller & Jul. Schff) Pilat**	Methanolic	**antibacterial**	
		MIC 0.4-1.15 μg/mL	
		antifungal	
		MIC 0.4-3 μg/mL	
	Ethanolic	**antibacterial**	
		MIC 0.35-1.7 μg/mL	
		antifungal	
		MIC 0.5-2.34 μg/mL	
A. bitorquis **(Quelet) Sacc.**	Methanolic	**antibacterial**	
		MIC 0.29-2.34 μg/mL	
		antifungal	
		MIC 0.78-3.12 μg/mL	
	Ethanolic	**antibacterial**	
		MIC 0.23-1.17 μg/mL	
		antifungal	
		MIC 0.39-3.12 μg/mL	

(continued)

Table 4.1 (continued)

Mushroom	Type of extract	Antimicrobial activity	References
Coprinus comatus **(O.F.Müll.) Pers.**	Methanolic	**antibacterial** MIC 0.75-3 mg/mL **antifungal** MIC 0.75-3 mg/mL	Stojković et al. (2013)
	Aqueous	**antibacterial** MIC 13-52 mg/mL **antifungal** MIC 13-52 mg/mL	Klančnik et al. (2017)
Grifola frondosa **(Dickson: Fries) S. F. Gray**	Hot alkaline	**antibacterial** MIC 0.02-2.50 mg/mL	Klaus et al. (2015)
Laetiporus sulphureus **(Bull.) Murill**	Methanolic	**antibacterial** MIC 0.9-3.6 mg/mL $^{-1}$ **antifungal** MIC 1.25-4.5 mg/mL^{-1}	Petrović et al. (2014b)
	Polysaccharide	**antibacterial** MIC 0.4-3.1 mg/mL $^{-1}$ **antifungal** MIC 0.5-4 mg/mL^{-1}	Petrović et al. (2014b)
Lentinula edodes(**Berk.) Pegier**	Aqueous	**antibacterial** MIC 5- >50 mg/mL IZ 15-21.2 mm	Hirasawa et al. (1999, Venturini et al. (2008)
	Ethyl acetate	**antibacterial** MIC 0.1- 2 mg/mL	Hirasawa et al. (1999)
	Chloroform	**antibacterial** MIC 0.01- > 1.5 mg/mL	Hirasawa et al. (1999)
Morchella esculenta **(L.) Pers.**	Methanolic	**antibacterial** MIC 0.02- >10 mg/mL IZ 6.16-8.34mm	Heleno et al. (2013)
	Ethanolic	**antibacterial** MIC 0.8- 50 mg/mL	Dimitrijevic et al. (2015)
Morchella conica **Pers.**	Methanolic	**antibacterial** MIC 0.70-7.5 mg/mL **antifungal** MIC 0.78-12.5 mg/mL	Vieira et al. (2015)
	Ethanolic	**antibacterial** IZ 4-29mm	Turkoglu et al. (2006)
Meripilus giganteus **Karst.**	Methanolic	**antibacterial** MIC 0.012-2.5 mg/mL IZ 8.5-17.5 mm **antifungal** MIC 0.025-0.3 mg/mL	Karaman et al. (2010, Karaman et al. (2014), Stojković et al. (2017)

(continued)

Table 4.1 (continued)

Mushroom	Type of extract	Antimicrobial activity	References
Pleurotus ostreatus **(Jacq.: Fr.) Kumm.**	Aqueous	**antibacterial**	Bawadekji et al. (2017), Owaid et al. (2017)
		IZ 1-30.66 mm	
		antifungal	
		IZ 20.66-33.33 mm	
	Methanolic	**antibacterial**	Akyüz et al. (2010), Chowdhury et al. (2015)
		MIC 5-8 mg/mL	
		IZ 5-10.5 mm	
		antifungal	
		MIC 4mg/mL	
		IZ 8-15.5 mm	
	Ethanolic	**antibacterial**	Ahmad et al. (2014)
		IZ 6.75-21.83 mm	
	Ethyl acetate	**antibacterial**	Uddin et al. (2015), Roy et al. (2016)
		IZ 7.1-13.43 mm	
	Hexane	**antibacterial**	Uddin et al. (2015)
		IZ 11.09-24.56 mm	
	Chloroform	**antibacterial**	Uddin et al. (2015)
		IZ 10-17 mm	
Polyporus squamosus **(Huds.) Fr**	Methanolic	**antibacterial**	Fernandes et al. (2016), Mocan et al. (2018)
		MIC 0.20-20.4 mg/mL	
		antifungal	
		MIC 0.40-3.13 mg/mL	
	Ethanolic	**antibacterial**	Dimitrijevic et al. (2015)
		MIC 6.3-50 mg/mL	

MIC- minimal inhibitory concentration, IZ- inhibition zone

In previous studies the extracts of *A. bisporus* prepared with methyl alcohol showed antimicrobial activities against some food borne and clinically isolated pathogenic bacteria, yeasts, and dermatomycetes using the **agar well diffusion** method (Table 4.1) (Akyüz et al. 2010; Abah and Abah 2010). Antimicrobial potential of *A. bisporus* extracts isolated from different part of fruit body has been reported also by Ndungutse et al. (2015). They suggest that the stipes of *A. bisporus* could be used as source of natural antimicrobials. Tehrani et al. (2012) examined the effects of the total protein aqueous extract of the cultivated *A. bisporus* against *Staphylococcus aureus* and methicillin-resistant *S. aureus* and demonstrated their significant antibacterial activity.

The freeze-dried extract of ethanol extract of fruiting bodies of *A. bisporus* displays activity towards *Escherichia coli* and *S. aureus*. *Pseudomonas aeruginosa* has proved to be the most resistant strain in bacteria tests (Vamanu 2012).

Chitin and chitosan from *A. bisporus* are **polysaccharides** with high molecular weight. The antimicrobial effect of thisese compounds depends on molecular weight and their activity is based on decreasing bacterial adhesion to the culture medium (Rajewska and Bałasińska 2004).

The antimicrobial activities of methanolic and ethanolic extracts of fruiting bodies of *A. bisporus* and *A. brasiliensis* were assessed and compared. *A. brasiliensis* revealed the highest antioxidant potential, and its ethanolic extract displayed the highest antibacterial potential, while the methanolic extract of *A. bisporus* revealed the highest antifungal activity. The eight Gram-positive and Gram-negative bacteria and eight micromycetes were used in order to investigate the antimicrobial activity of mushroom extract *in vitro* by the **microdilution** method. The ethanolic extract of both species was tested *in situ* for inhibition of *Listeria monocytogenes* growth in yoghurt. *A. brasiliensis* ethanolic extract proved to be more efficient than that of *A. bisporus* in controlling *L. monocytogenes* growth (Stojković et al. 2014).

The extracts isolated from three wild *Agaricus* species collected from Serbia showed antimicrobial and **anti-quorum sensing** (AQ) properties. Sub-inhibitory concentrations of the ethanolic extracts demonstrated reduction of virulence factors, AQ inhibition zones, twitching and swimming motility of the bacteria. The biofilm forming ability of *P. aeruginosa* PAO1 was also reduced in a concentration-dependent manner at sub-MIC values. Extracts of *A. bisporus* possessed the lowest antibacterial activity compared to other mushroom species. Ethanolic extracts of all the tested species showed once more the highest antifungal activity. The methanolic and ethanolic extracts of the tested *Agaricus* species are a promising source of antimicrobial and antiquorum sensing compounds (Glamočlija et al. 2015).

Sharma et al. (2015) and Waithaka et al. (2017) has been investigating the antimicrobial properties of *Agaricus bisporus* by the agar disc diffusion method by testing a panel of pathogens: *Escherichia coli, Staphylococcus aureus, Pseudomonas aeruginosa, Enterococcus faecalis, Streptococcus pneumonia, Proteus vulgaris, Candida albicans, Aspergillus niger, Fusarium oxysporum, Ustilago maydis, Microsporum gypseum* and *Malassezia furfur* (Table 4.1). The extracted metabolites did not inhibit the growth of Gram-positive bacteria.

Atila et al. (2017) published a study determinating the antimicrobial activity of *Agaricus bisporus* with special emphasis on combining the use of *A. bisporus* exctract with **silver nanoparticles** (AgNPs). Sudhakar et al. (2014) combined the use of the AgNPs with the *A. bisporus* extract using a different method. They suggested that *A. bisporus*-AgNPs may have an important advantage over conventional antibiotics. Ul-Haq et al. (2015) determined that combinig AgNPs with *A. bisporus* extract have shown a higher zone of inhibition against Methicillin-Resistant *Staphylococcus aureus* strains compared to other medicinal mushrooms. The same authors determined the synergistic effect of *A. bisporus*-AgNPs with gentamicin and ceftriaxone on selected bacteria. Mirunalini et al. (2012) where investigated the antibacterial activity of the synthesized *A. bisporus* - AgNPs against Gram-positive bacteria like *S. aureus* and Gram-negative bacteria *S. typhimurium*, *Proteus* sp., *Enterobacter* sp. and *Klebsiella* sp. (Dhanasekaran et al. 2013).

These studies have shown that edible mushroom *A. bisporus* is a promising source of new therapeutics against many diseases.

4.2.1.2 Antimicrobial Activity of the *Lentinula edodes*

Lentinula edodes (Berk.) Pegier (shiitake) is the most cultivated species of mushrooms worldwide being the world's second largest cultivated fungi (Li and Hu 2014). *L. edodes* is called a medicinal mushroom for its long traditional use in especially in oriental medicine. (Bisen et al. 2010). Various extracts and isolated compounds of this mushroom are presented in the list together with ther notable pharmacological properties (Finimundy et al. 2014).

Extracts of *L. edodes* possessed antimicrobial activity towards pathogenic fungi, Gram-positive and Gram-negative bacteria, as well as antiviral activity. Ethyl acetate extract of this species was observed for several bacteria: *S. aureus, B. subtilis, Bacillus cereus*, and *S. epidermidis*. Water extract of this species exhibited notable activity against MRSA strain. *Streptococcus pyogenes* was shown to be susceptible to the effect of chloroform extract of *L. edodes*. Strong bactericidal potential on *Streptococcus mutans* was also recorde (Alves et al. 2012b). Rincão et al. (2012) investigated antiviral activity of *L. edodes* ethanol, aqueous extracts, and a polysaccharide towards bovine herpes virus type 1 and poliovirus type 1.

Lentinan a polysaccharide isolated from the shiitake mushroom is one of the most medically important compounds. It was shown that it is active against *Listeria monocytogenes* and *Mycobacterium tuberculosis* (Ooi and Liu 2000). Oxalic acid is a compound obtained from *L. edodes* that was responsible for bacteriostatic effects against *Bacillus megaterium, S. pyogenes* and *S. aureus* (Nikitina et al. 2007). Lentin, a protein isolated from this species was able to inhibit growth of mycelia in various microfungi species including *Botrytis cinerea, Mycosphaerella arachidicola* and *Physalospora piricola* (Ngai and Ng 2003).

Lentinan was able to enhace host resistance to bacterial, fungal and viral infections including AIDS as studied by Chihara (1992). Lentinan also reduced the toxicity of a drug commonly used to treat HIV and AIDS patients. There is ongoing research on lentinan activity towards suppression of symptoms arised in patients suffering from HIV virus infection.

Bactericidal activity against *Prevotella intermedia* and *Streptococcus mutans* was confirmed for ethyl acetate and chloroform extracts of shiitake mushroom (Hirasawa et al. 1999) (Table 4.1) A cyclic organosulfur compound namely lenthionine, that is partially responsible for the specific taste of shiitake, showed bacteriostatic effects on *Bacillus subtilis, S. aureus*, and *Escherichia coli* (Hatvani 2001).

The antibacterial activity of *L. edodes* against *B. subtilis* was evaluated in cell-free filtrates obtained after growth in 14 different culture media (Hassegawa et al. 2005). The antimicrobial activity of aqueous, ethyl acetate, hexane and methanol extracts from wild growing and cultivated samples of *L. edodes* was recorded in **disk-diffusion** assay against nine foodborne pathogenic bacterial strains. Aqueous extracts from *L. edodes* had the highest antimicrobial activity in one study where 48 species was tested against various bacteria (Table 4.1) (Venturini et al. 2008).

It has been reported that extracts of shiitake possess antibacterial activity enhancing host immunity against infections (Hearst et al. 2009). Rao et al. (2009) identified 34 compounds namely new antimicrobial metabolites in shiitake mushroom, as a result of the demonstration of its antibacterial and antifungal activity. The lower level of activity of polar compounds in water-soluble extracts indicated that the antibacterial properties of shiitake mushrooms may have arisen largely from less polar organic compounds and the data also suggest that the higher activity may be due to the presence of disulphides, lenthionine compounds in the organic extracts. Fungal metabolite **bostrycoidin** has been shown to possess antimicrobial activity against *M. tuberculosis in vitro.*

It has been shown that the extract of *L. edodes* inhibits oral pathogenic bacteria which were marked as main causative agents of of gingivitis (Lingstrom et al. 2012; Ciric et al. 2011). According to Spratt et al. (2012) the fraction of low molecular weight compounds isolated from *L. edodes* aqueous extract had potential activity against oral pathogens *in vitro*. Concerning all previous studies on *L. edodes* it could be concluded that this mushroom presents magnificent source of antimicrobial compounds that could be further used in medical practice, after detailed preclinical and clinical studies.

4.2.1.3 Antimicrobial Activity of the *Pleurotus ostreatus*

Pleurotus is a genus of wood-rotting fungi which became one of the most cultivated species throughout the world; with oyster mushroom being the third most produced species on the industrial level, after *Agaricus* and shiitake (Chang 2006).

P. pulmonarius, *P. ostreatus*, *P. djamor* and *P.eryngii* are species that are usually cultivated for a larger industrial production (Gargano et al. 2017).

Pleurotus ostreatus was to be active against single and **multi-drug resistant** bacterial strains of *S. aureus, Staphylococcus epidermidis* and *E. coli* (Akyüz et al. 2010) and was shown to possess antifungal activity on *Candida* species (Wolff et al. 2008). *Pleurotus* spp. methanol extracts possessed activity on different microorganisms including: *Bacillus megaterium, E.coli, Klebsiella pneumoniae, S. aureus, C. glabrata, C. albicans, Epidermophyton* and *Trichophyton* (Table 4.1) (Akyüz et al. 2010).

Antimicrobial activity of *P. ostreatus* extracts was depended on the polarity of solvents used for the extraction; petroleum ether extract was more active against Gram negative bacteria when compared to the activity of acetone extract (Iwalokun et al. 2007). Petroleum ether and acetone extracts of *P. ostreatus* were effective against *B. subtilis, E. coli* and *S. cerevisiae.* The study Nehra et al. (2012) was designed to evaluate the antimicrobial potential extracts fruiting body powder of *P. ostreatus* different polarity (benzene, chloroform, acetone, ethanol, methanol and distilled water) tested in agar well diffusion assay. The extracts effectively inhibited the growth of Gram-positive four and seven Gram-negative species of bacteria. Among the microorganisms tested, *P. aeruginosa, Aeromonas hydrophila, S. aureus* and *S. typhimurium* were found to be more susceptible to the extracts as compared to the other microbes.

The study Oyetayo and Ariyo (2013) antimicrobial properties of *P. ostreatus* cultivated on three different substrates carried out by the agar diffusion method. The ethanol, water and methanol extracts of *P. ostreatus* are also effectively inhibited the growth of a large number of pathogenic microorganisms. Uddin et al. (2015) have shown that a water extract of the cultivated sample *P. ostreatus* effective to inhibit growth of *B. cereus*, *B. subtilis*, and *P. aeruginosa* (Table 4.1).

Gemmotherapic extract of *P. ostreatus* was investigated for its antimicrobial activity against the following bacteria: *B. cereus* var. *mycoides*, *B. subtilis*, *P. aeruginosa*, *Serratia marcescens* and *Streptococcus faecalis* by using the well diffusion assay. The extract was prepared from young parts of *P. ostreatus*, in accordance with gemmotherapic principles. The extract had a significant inhibitory activity against *B. cereus* var. *mycoides* and *B. subtilis* (Pauliuc and Dorica 2013).

The antimicrobial potential of ethanol extract of *P. ostreatus* was tested for antimicrobial potential against Gram-negative bacteria (*Agrobacterium tumifaciens*, *E. coli*, *Erwinia carotovora K. pneumonia*, *P. aeroginosa* and *S. typhi*), Gram-positive bacteria (*B. atrophaeus*, *B. subtilis* and *S. aureus*) and a fungus (*C. albicans*). The results of antimicrobial tests were compared with the activitz of standard therapeutic antibiotic drugs. *E. coli* was not susceptible to the effect of the ethanol extract which was exhamined (Table 4.1) (Ahmad et al. 2014).

The antimicrobial activity of methanolic extract *P. ostreatus* was determined by the agar well diffusion method. The oyster mushroom has a broad-spectrum of antibacterial and antifungal activity against *B. subtilis*, *E. coli*, *K. pneumonia*, *P. aeruginosa*, *S. aureus* and *S. typhi* and fungal species *Saccharomyces cerevisiae* and *C. albicans*. *S. cerevisiae* and *C. albicans* were found to be very susceptible to the effect of the extract, while *P. aeruginosa* was more resistant to mushrooms extract when compared with other microbial species investigated (Table 4.1) (Chowdhury et al. 2015).

Ethyl acetate extract of the *P. ostreatus* is tested for their *in vitro* growth inhibitory activity against five Gram-positive bacteria, five Gram-negative bacteria and three fungi by disc diffusion method using. The extract of *P. ostreatus* has moderate antibacterial effect and no antifungal activity was observed on the experimental fungi (Table 4.1) (Roy et al. 2016).

Antimicrobial effect depends on the nature of solvent in which fungal material were dissolving to make test samples. The ability of *P. ostreatus* (methanolic extract obtained from commercially cultivated fruiting bodies) to reduce *in vitro* growth of selected pathogenic microorganisms – causative agents of chronic diseases: *P. aeruginosa*, *Aspergillus niger* and *A. fumigatus* is presented by Petrović et al. (2016). The Gram negative bacteria *P. aeruginosa* was inhibited strongly. As for the antifungal activity, the extract possessed prominent antifungal effect on wide variety of *Aspergillus* species – common causative agents of aspergillosis.

The screening of 20 strains of *P. ostreatus* from a bank of edible mushrooms bought from international strain banks in the order to detect activity could possess health benefit related properties; 13 strains of have been cultivated on different wood logs; six strains of oyster mushroom were able to produce sporocarps. The bacteriostatic/bactericidal activity from dried mushrooms in water extracts investigated against the model microorganisms *S. aureus* and *P. aeruginosa* (Parola et al. 2017).

The Bawadekji et al. (2017) investigated antifungal and antibacterial potential of cold crude extract of basidiocarps of *P. ostreatus* against five bacterial strains and fungus *Candida albicans*. All strains were tested by well diffusion technique. Crude extract of *P. ostreatus* fruiting bodies showed an important zone of inhibition only toward *C. albicans*, *P. aeruginosa* and *S.aureus* (Table 4.1). The fruiting body powder of *P. ostreatus* was extracted with ethanol by using the agar well diffusion technique and tested for their antifungal activity against six fungi: *A. fumigatus, Aspergillus flavus, Humicola grisea, Penicillium chrysogenum, Sporotricum carnis* and *Thermoascus aurantiacus*. The extract of *P. ostreatus* showed the maximum inhibition against *P. chrysogenum* and minimum against *A. flavus* (Kumar and Yadav 2014).

The antifungal activity *P. ostreatus* and three other *Pleurotus* species were tested against three fungal pathogens: *Pythium* sp., *Trichoderma harzianum* and *Verticillium* sp. by using *in vitro* disc diffusion method. The most sensitive species was shown to be *T. harzianum* - an important plant pathogen (Table 4.1) (Owaid et al. 2017).

4.2.2 Antimicrobial Activity of Selected Wild Edible Mushrooms

Edible wild mushrooms are considered as a source of bioactive compounds, especially phenolic compounds and nutrients. Over the past years, many wild growing mushrooms have been explored in order to demonstrate their diverse biological activity precisely because of their potential uses as functional food and/or in medical purposes (Heleno et al. 2015; Soković et al. 2017). Bacterial and fungal infections represent the global problem for human health. During time, many bacteria have developed resistance to some commercial antibiotics and therefore we investigated antimicrobial potential of wild edible mushrooms which can help us in battle against pathogenic microorganisms. The following text will described some of the wild species that have a significant antimicrobial activity and some of them, in addition to numerous other roles, also have **hepatoprotective** activity.

Aqueous extract of *A. blazei* Murill (*Agaricus brasiliensis*) showed **antibiofilm** and QS activity against *P. aeruginosa* (Soković et al. 2014), while is proven for methanolic and etanolic extracts to have antimicrobial effect against Gram positive (*B. cereus, , Micrococcus flavus, S. aureus, L. monocytogenes*), Gram negative (*P. aeruginosa, Salmonella typhimurium, Enterobacter cloacae, E. coli*) bacteria and some fungal species (*A. fumigatus, Aspergillus ochraceus, Aspergillus versicolor, A. niger, Penicillium funiculosum, Penicillium ochrochloron, Penicillium verrucosum* var. *cyclopium, Trichoderma viride*) (Table 4.1) (Stojković et al. 2014). Also, Glamočlija et al. (2015) reported antimicrobial activity of extracts of other species from genus *Agaricus* against above mentioned bacteria and fungi (Table 4.1).

Coprinus comatus (O.F.Müll.) Pers., shaggy ink cap, (Fig. 4.2), is mushroom with wide range of biological activity (hypoglycemic, hypolipidemic, immunomodulatory, antitumor, antioxidant and antibacterial potential) (Yu et al. 2009). It's reported antibacterial and antifungal activity of wild *C. comatus* methanolic and aqueous extracts against ten bacteria, eight molds and one yeast species (Table 4.1) (Stojković et al. 2013; Klančnik et al. 2017). Other species from the same genus, *Coprinus atramentaria*, also showed antibacterial activity (Khan et al. 2016).

Fig. 4.2 *Pleurotus ostreatus* (Jacq.: Fr.) Kumm. Photo by: Ivanka Milenković

Hot alkalis extract and partially purified polysaccharides from fruiting bodies of *Grifola frondosa* (Dicks.) Gray (Fig. 4.2), inhibited growth of nine ATCC isolates (Table 4.1) (Klaus et al. 2015). He et al. (2010), described that polysaccharides can cause damage on the cell wall and cytoplasm membrane which can results in cell death.

Laetiporus sulphureus (Bull.) Murill (Fig. 4.2) is well known edible mushroom rich with polysaccharides, proteins, fibers, etc. (Khatua et al. 2017). Many researchers investigated biological activity of this mushroom and proved that *L. sulphureus* have antioxidant and antimicrobial activity, hypoglycemic and hepatoprotective effect (Soković et al. 2018). Antimicrobial activity of different types of extracts was reported (Table 4.1) (Petrović et al. 2014b). Methanolic and polysaccharide exctracts showed the best inhibitory activity against *S. typhimurium*, *S. aureus*, *P. ochrochloron* and *A. ochraceus* (Petrović et al. 2014b). Extracts obtained from different solvent, posessed good activity against pathogenic microorganisms, despite on extraction methodology (Petrović et al. 2014a).

Methanolic extract of ***Meripilus giganteus*** (Pers.) P. Karst. (Fig. 4.2) exhibited antioxidant, antitumor and antimicrobial activity (Karaman et al. 2014; Stojković et al. 2017). Antimicrobial activity was examinate by microdilution method (Table 4.1) and showed that the most sensitive bacteria and fungi were *S. aureus*, *B. cereus* and *A. versicolor*, *P. funiculosum* (Stojković et al. 2017). While, other authors, exanimate only antibacterial activity of methanolic extracts originating from Serbia, against large number of bacteria, by the same methods (Table 4.1) and showed that the most sensitive bacteria were *Salmonella enteritidis* (by disk diffusion method) and *S. aureus* and *B. subtilis* (by microdillution method) (Karaman et al. 2010; Karaman et al. 2014).

Morchella esculenta (L.) Pers. (Fig. 4.2) is highly appreciated wild edible mushroom. Morel extracts demonstrated antimicrobial, antioxidant, anti-inflammatory, hepatoprotective, and antitumor activities (Mau et al. 2004; Nitha et al. 2007; Alves et al. 2012a; Nitha et al. 2013; Heleno et al. 2013; Dimitrijevic et al. 2015). Steroids as well as polysaccharides isolated from this mushrooms showed high *in vitro* antioxidant potential, while galactomannan demonstrated immunostimulatory activities (Duncan et al. 2002). Heleno et al. (2013) reported antibacterial activity of methanol extract from wild species *M. esculenta* collected in Portugal and Serbia, against five bacteria by disk diffusion and broth dilution methods (Table 4.1). Extract showed the best activity against *S. aureus*. Besides antibacterial effect, same authors demonstrated demelanizing activity of methanolic extracts against *A. fumigatus*, *A. flavus*, *P. funiculosum*, *P. ochrochloron* (Heleno et al. 2013).

Ethanolic extract showed slightly weaker activity than methanolic (Table 4.1) with the best activity against *K. pneumoniae* (Dimitrijevic et al. 2015). Methanolic extracts of fruiting bodies *Morchella conica* Pers. (Fig. 4.3) from different regions demonstrated antimicrobial potential against various bacteria and fungi. The highest antibacterial potential, extracts showed against *B. cereus*, and the highest antifungal potential was archived against *T. viride* (Vieira et al. 2015). Turkoglu et al. (2006) demonstrated antibacterial activity of ethanolic extract of *M. conica*, with the strongest activity against *M. flavus*.

Polyporus squamosus (Huds.) Fr. (Fig. 4.3), edible mushroom commonly used as spice, but also has been shown to possess antioxidant (Akata et al. 2012), immunomodulating (Babakhin et al. 1996), antimicrobial, antibiofilm and anti-quorum sensing activities (Fernandes et al. 2016; Mocan et al. 2018). The antimicrobial activity of different types of extract of wild edible *P. squamosus* from different regions was investigated. Methanolic extracts from Serbia and Portugal exhibited better antibacterial potential (MIC range 0.20-3.13 mg/mL), than the one from Romania (MIC range 0.61-20.4 mg/mL) (Table 4.1) (Fernandes et al. 2016; Mocan et al. 2018). Methanolic extracts originating from Serbia and Portugal mushrooms showed antifungal activity, with *A. versicolor* as the most sensitive

Fig. 4.3 (**a**) *Agaricus macrosporus* (F.H. Møller & Jul. Schäff.) Pilát; (**b**) *Coprinus comatus* (O.F.Müll.) Pers.; (**c**) *Grifola frondosa* (Dicks.) Gray; (**d**) *Laetiporus sulphureus* (Bull.) Murill; (**e**) *Meripilus giganteus* (Pers.) P. Karst.; (**f**) *Morchella esculenta* (L.) Pers.; (**g**) *Morchella conica* Pers.; (**h**) *Polyporus squamosus* (Huds.) Fr.. Photos by: Jasmina Glamoćlija

isolate (Fernandes et al. 2016). Etanolic extract from wild *P. squamosus,* collected in Serbia showed moderate antibacterial activity (Table 4.1), with *S. aureus* as the most sensitive species (Dimitrijevic et al. 2015).

4.3 Hepatoprotective Activity of Edible Mushrooms

In many countries (dates back thousands of years), natural products were used in prevention and/or treatment of different liver diseases (Zhang et al. 2013). There are treatments for most of the liver diseases, but many types remain untreatable and one of the reasons can be emergence of drug resistance. It has been shown that many mushroom extracts possess hepatoprotective activities (Soares et al. 2013a). In the near future, the new strategy for the prevention and/or treatment of liver diseases through applications of mushroom extracts can be developed.

4.3.1 Antioxidant Properties of Compounds from Edible Mushrooms and their Hepatoprotective Activity

The drugs and toxins could be a reason for accumulation of **reactive oxygen species** (hydrogen peroxide, hydroxyl radical, and superoxide) in hepatocytes, inducing tissue injury (Poli 1993). Metabolism of normal cells produces reactive oxygen and nitrogen species and normally host cells are protected from oxygen derived radical injury by antioxidants and free radical scavengers, but in liver diseases, redox potential is increased, thereby damaging the hepatic tissue. (Coskun et al. 2005; Muriel 2009).

Antioxidants play a vital role in hepatoprotective capability by scavenging free radicals. For this reason, the exploration of substances of natural origin with antioxidant property has become a fundamental focus of the study of hepatoprotection these days. Numerous reports on edible mushrooms have presented the chemical, nutritional, and medicinal characteristics and demonstrated the presence of compounds with antioxidant activity isolated from, mycelium and fruiting bodies. Ascorbic acid, carotenoids, tocopherols and phenolics which has been isolated from the different species of mushrooms are compounds for which it is shown to increase the antiviral and antihypercholesterolaemic activity, and meliorate the toxic effect of radio- and chemotherapy (Ferreira et al. 2009; Reis et al. 2012; Stajić et al. 2013; Tel et al. 2015; Kozarski et al. 2015; Sánchez et al. 2017b).

Numerous wild mushrooms have antioxidant activity, which is associated with their phenolic content (Ferreira et al. 2009). There are limited numbers of *in vivo* studies about the antioxidant potential of mushrooms, because the most of studies used *in vitro* assays. Gan et al. (2013), presented antioxidant effect of total flavonoid content (**TFC**) and total phenolic content (**TPC**) of *A. bisporous* and *A. brasiliensis* in aqueous and 60% ethanol extract *in vitro*. Results showed that ethanol extract of mushrooms possess high antioxidant activity due to presence of phenolics. Liu et al. (2013) reported a *in vitro* study in which *A. bisporus* ethanolic extract was found to have moderate hydrogen peroxide scavenging activity and strong superoxide, hydroxyl, 2,2- diphenyl-1-picrylhydrazyl (**DPPH**) radical scavenging activity and reducing power. *In vivo* antioxidant assays were performed with *A. bisporus* ethanolic extract for 30 days via gavage. Ethanolic extract significantly enhanced the activities of antioxidant enzymes in livers, hearts and serums of mice.

Several authors evaluated the antioxidant activity of mushroom extracts isolated by dichloromethane, cyclohexane, methanol, ethanol, and water from *A. bisporus*, *P. ostreatus*, *Boletus edulis*, *Cantharella cibarius* and *Lactarius piperatus* (Barros et al. 2008b; Keles et al. 2011) and were assessed by α-diphenyl-β-picrylhydrazyl (DPPH), oxygen radical absorbance capacity (**ORAC**), ferric reducing antioxidant power (**FRAP**), and 2,2′-azino-bis-3-ethylbenzthiazoline-6-sulphonic acid (**ABTS**) methods. To evaluate the TPC of the selected species Folin–Ciocalteu test was performed (Arnao et al. 2017).

Antioxidant activity of mushroom extracts and dry mushroom substances sequentially isolated by methanol, dichloromethane, cyclohexane, and water from *A. bisporus*, *Phaeolus schweinitzii*, *P. ostreatus*, *Inonotus hispidus*, *Tricholoma caligatum*, *Tricholoma columbetta*, *Xerocomus chrysenteron* and *Hydnellum ferrugineum*, were evaluated by ABTS þ scavenging capacity, DPPH, FRAP, ORAC, and FolineCiocalteu TPC methods. The highest antioxidant capacity values were found for methanolic fractions of *I. hispidus* and *P. schweinitzii*. Extracts from other species had lower antioxidant activity (Smolskaite et al. 2015).

Recent published data revealed total polyphenols content for thirteen frequently consumed mushroom species from Poland. Concentration of flavonoids, polyphenols, β-carotene, lycopene and their reducing power and ability to scavenge ABTS cation radical in aqueous and methanolic extracts of dried fruiting bodies were determined. Robaszkiewicz et al. (2010) founded that the concentration of antioxidants varies depending on the species and parts of the fruiting body. Authors observed a strong correlation between the total phenolics and reducing power/scavenging effects in both types of extracts, while for flavonoids this correlation was moderate. Lycopene had a significant contribution to the scavenging activity of methanolic extracts while β-carotene didn't contribute to the antioxidative activity of the extracts- (Robaszkiewicz et al. 2010). Other authors shown that the scavenging activity depends on the harvesting stage and climate soil (Heleno et al. 2013; Smolskaite et al. 2015).

Dimitrijević et al. (2016) described antioxidant properties of six wild mushrooms: *Boletus edulis*, *B. impolitus*, *B. regius*, *Leccinum pseudoscaber*, *Lactarius deliciosus* and *L. volemus* which were collected in Serbia. The antioxidant properties were determinated through the evaluation of the radical scavenging ability and reducing power of the samples. The *B. regius* sample showed the highest, while the *L. volemus* sample showed the lowest scavenging and reducing power activity.

Due to their bioactive compounds, cultivated or wild edible mushrooms have been related to significant antioxidant properties.

4.3.2 Effects of Bioactive Compounds Isolated from Edible Mushrooms in Control of Cholesterol and Triglycerides

Major causes of death across many industrialized countries are cardiovascular diseases. Mensink et al. (2003) have underlined that cardiovascular diseases have a multifactorial aetiology being mostly caused by **atherosclerosis**, and have identified potential risk biomarkers such as high blood pressure, **oxidative damage**, lipid and lipoprotein metabolism, haemostatic function, and metabolism of homocysteine. The edible mushrooms as a healthy food that could be used as protective agents in mentioned diseases received attention of scientists and consumers. Also other properties of the mushrooms, such as anti-inflammatory, hypocholesterolemic and hypoglycemic which result in the prevention of cardiovascular diseases have also been attributed to them (Guillamón et al. 2010; Valverde et al. 2015).

In many examinations of cholesterol lowering properties of edible mushrooms (*Auricularria polytricha, A. bisporus, L. edodes, Flammulina velutipes* and *P. ostreatus)* the hypocholesterolemic activity has been reported (Guillamón et al. 2010). The beneficial effects of edible mushrooms, particularly of *P. florida, P. sajor-caju* and *P. ostreatus* were investigated on normocholestolemic and hypercholesterolemic animal models. Hypercholesterolemic rats were feeding with 5% of oyster mushroom powder which led to significant repression of the increment of plasma cholesterol in rats (Alam et al. 2009). Animal studies were focused on the examination of the hypocholesterolemic effect of *P. ostreatus,* oyster mushroom especially, which revealed a suppression of the activity of HMG-CoA reductase and inhibition of lipid peroxidation in hypercholesterolemic and normocholestolemic conditions. The lovastatin can be found in large amounts in the fruiting bodies of different cultured *Pleurotus* species. Therefore, for consumption as a natural cholesterol-lowering agent could be recommended mature fruiting bodies of *P. ostreatus*, but a lovastatin as a drug was approved by FDA in 1987. Some inhibitors of HMG-CoA reductase have been isolated from different mushrooms, such as meviolin from *Pleurotus* spp. and eritadenine from *L. edodes* (Guillamón et al. 2010).

Mevinolin (a statin polysaccharide), is a known pharmacological inhibitor of the HMG-CoA reductase. A high amount of mevinolin has been found in *P. ostreatus* fruiting bodies, but also in sporocarps of *P. eryngii* and *P. cornucopiae*. Also, the mycelia from three *Pleurotus* species (*P. ostreatus, P. saca and P. sapidus*) produced mevinolin (Guillamón et al. 2010). The shiitake mushroom *L. edodes* can lower free cholesterol in plasma and blood pressure, as well as accelerate accumulation of lipids in the liver, by removing them from circulation. The shiitake is able to lower blood serum cholesterol via a factor known as eritadenine (lentysine, lentinacin). Mechanism of action is the acceleration of excretion of ingested cholesterol and its metabolic decompositions. Wasser and Weis (1999) shown that the eritadenine reduces serum cholesterol in animals. A diet containing 5% shiitake fruiting bodies given to hypercholesterolemic rats, reduced plasma total lipids, phospholipids, cholesterol, triglycerides, low-density lipoprotein (**LDL**), and the high-density lipoprotein (**HDL**) (Yoon et al. 2011). Enman (2007) demonstrated that the shiitake mushrooms can lower the serum cholesterol in humans; giving them 90 g of fresh shiitake daily per week.The serum cholesterol was lowered by 12%.

Xu et al. (2007) demonstrated that the 5% chitosan from *A. bisporus* treatment despite a greatly increased intake of cholesterol, moderate cholesterol metabolism in rats. In clinical research on human volunteers, it was founded that the other species of this genus, *A. brasiliensis,* decreased body mass index, body weight, blood glucose level blood cholesterol level, and percentage body fat, visceral fatand neutral fat level, and reduced obesity thus activated the immune function and normalized liver function in people with poor health (Liu et al. 2008).

The data presented in the study of Jeong et al. (2010), the *A. bisporus* fruiting bodies shown a positive influence on liver function and lipid metabolism on rats. The rats were under the hypercholesterolemic diet and the other group of rats was with type 2 diabetes which was induced by streptozotocin. Animals were treating, orally, with the *A.bisporus* powder (200 mg/kg of body weight) per three weeks. Based on studies of the other mushrooms, the cholesterol- and glucose- lowering effects of *A. bisporus* are probably the result of a number of mechanisms involving dietary fibers and other active components from mushroom which can act individually or in combination.

Caz et al. (2016) investigated the hypocholesterolemic activity of lard functionalized (LF) with mushroom extracts and examined influence of the LF on transcriptional mechanisms involved in metabolism of cholesterol. Their research suggest that LF may have potential as a dietary supplement for neutralizing diet-induced hypercholesterolemia and may be a source for the development of novel functional foods with low cholesterol-levels.

The analysis the influence of some hypocholesterolemic compounds obtained from edible mushrooms and the molecular reactions occurring during synthesis and absorption of cholesterol suggested that the hypocholesterolemic effect of extracts could be due to transcriptional and post-transcriptional modulations besides other side effects. *In vitro* and *in vivo* studies indicated that the mushroom extracts (β-glucans and other water-soluble compounds) and fungal sterols down-regulated genes involved in homeostasis of cholesterol also stimulated transcriptional profiles on the similar way like some hypocholesterolemic drugs (Gil-Ramírez et al. 2018).

4.3.3 Hepatoprotective Activity of Edible Mushrooms Against Experimentally Induced Liver Injury

Injury of the liver associated with damaged liver function caused by using of drug or another non-infectious agent could be defined as hepatotoxicity. Agents may cause changes in antioxidant enzyme systems and activity of some liver enzymes: alanine transaminase, aspartate aminotransferase, alkaline phosphatase and can affect the bilirubin concentration in blood (Navarro and Senior 2006). Despite of increasing demand to protect liver, modern medicine still needs reliable drugs and searc for it (Lu et al. 2017) and this is a great challenge and acute liver injury and its protection are unexplained field in clinical medicine (Auzinger and Wendon 2008).

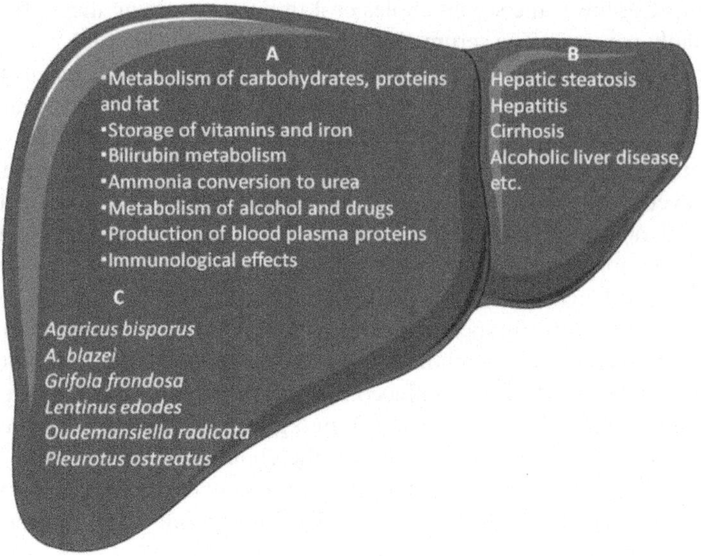

Fig. 4.4 (**a**) Liver functions; (**b**) Liver diseases; (**c**) Selected edible mushrooms with hepatoprotective activity. Picture of liver is downloaded from: smart.servier.com

In a long history and experience of using natural products in Japan and China, edible mushrooms treatment of a variety of diseases showed very little side effects and this practise has low cost (Zhu 2009). Therefore, exploring of this new strategy of prevention and treatment of liver diseases with different mushrooms species become intensive among scientists (Fig. 4.4).

In the study Ooi (1996) the hetatoprotective activities of water extracts obtained from cultured mycelia or fruit bodies of seven edible mushroom species *Volvariella volvacea, L. edodes, Flammulina velutipes, Auricularia auricula, Tremella fuciformis, G. frondosa* and *Tricholoma lobayense* were tested in vivo experiments in rats, toward paracetamol-induced liver damage. The extracts of the *L. edodes* and *G. frondosa* possessed a great hepatoprotective potential administered at both doses (100 mg/kg and 300 mg/kg body weight) while *T. lobayense* showed hepatoprotective effect at dose of 300 mg/kg. All the other species of mushrooms did not have effect on enzyme activities of serum transaminases that are measured in the blood.

The activity of aqueous extract obtained from *A. blazei* was tested as a pretreatment on paracetamol injuried rats (e.g., levels of enzymes were measured in the plasma) and in terms of metabolic and functional index (e.g. gluconeogenesis). *A. blazei* extract possessed a little or it doesn't showed protective potential at all at the paracetamol injury in the hepatic tissue (Soares et al. 2013b).

Selective hepatotoxic synthetic agent, carbon tetrachloride (CCl_4) represents one of the most used toxins in experimentally induced liver fibrosis in laboratory animals. In liver, CCl_4 is activated by oxidases in endoplasmic reticulum and then form CCl_3 radical which combines in the presence of oxygen, with cellular protein and lipid and could induce lipid peroxidation though hydrogen abstraction. The results of this are changes of structures of endoplasmic reticulum or other membranes and loss of metabolic enzyme activation leading to damage and lossing of liver function (Sherry et al. 2018). The several mushroom crude extracts used testing the impact carbon tetrachloride as hepatotoxic agents: *A. blazei (A.brasiliensis), Armillariella tabescens, Macrocybe gigantea, M. esculenta, Calocybe indica, C. comatus, Ganoderma lucidum, G. tsugae, P. ostreatus, P. cornucopiae, Phellinus rimosus* (Soares et al. 2013a). However, there are no many data about its protective effects on liver function.

Wu et al. (2011) employed CCl_4 to provoke hepatic fibrosis at rats and to examine the protective potential of low (200 mg/kg) and high dose of (2000 mg/k) *A. brasiliensis* extract on the liver. They concluded that high-dose of AB treatment showed lower hepatic necrosis and fibrosis induced by CCl_4 in comparison with the control animals.

Pleurotus genus comprises approximately 40 species (Patel et al. 2012). Extract of *Pleurotus* spp. could influence ameliorate or abrogate CCl_4-induced liver injury in rats. In the study Jayakumar et al. (2006), acute hepatotoxicity at Wistar rats was induced by intraperitoneal administration of CCl_4. **Hepatotoxicity** was noticed as direct evidence in the occurrence of alterations in serias of hepatic parameters. The ethanol extract of *P. ostreatus* was administered in a concentration of 200 mg/kg body was able to protect liver against acute hepatotoxicity.

The aqueous extract of *Pleurotus tuber regium* showed antioxidant effect and hepatoprotective activities at CCl_4-induced hepatotoxicity in animal model. According to this results *P. tuber regium* could be used for treatment of liver diseases (alcoholic liver injury, intrahepatic cholestatis, hepatotoxin exposure, liver ischemia and hepatitis casued by viruses, cirrhosis, fibrosis, biliary disease, steatohepatitis, necrosis and inflammation of liver (Dandapat et al. 2015).

The lectin isolated from the other mushroom belonging to family Pleurotaceae and its usage in future in fighting aganist arsenic mediated damage of liver is tested. Rana et al. (2012) showed that *Pleurotus florida* lectin and ascorbic acid possessed protective activity in **oxidative stress** induced by arsenic in liver of rats. This study also showed that orally applied lectins of *P. florida* could significantly reduce the enhanced liver tissue proteins malondialdehyde and level of protein carbonyl in liver tissues. Also, it was shown that constant usage of mushroom lectins and ascorbic acid could significantly reduce the level of hepatic of lipid peroxide and protein carboxylation.

The *Oudemansiella radicata* is wild edible mushroom and has high economic value displayed antioxidative and hepatoprotective activity. Liu et al. (2013) investigated protective potential of *O. radicata* polysaccharides on CCl_4-induced liver injury. During administration of soluble polysaccharides, aques or alkal, isolated from *O. radicata exhibited* suppression of malondialdehyde formation and activation of hepatic **superoxide dismutase** and **glutathione peroxidise**, and *i*ncreaseing in serum alanine aminotransferase and aspartate aminotransferase activities in a CCl_4-induced acute damage of liver.

The studies on the hepatoprotective activities of polysaccharides isolated from *Laetiporus sulphurous* spent mushroom substrates (SMS) in acute alcohol induced liver disease (ALD) mice were investigated. SMS contains residual mycelia and fruiting bodies of mushrooms; trace elements, active enzymes and biomacromolecules (Zhu et al. 2012). Polysaccharides extracted from SMS have received attention owing to their abundant pharmacological activities. The *in vitro* and *in vivo* study demonstrated polysaccharides from *L. sulphurous* as a potential natural and functional food supplement for the prevention and alleviation of ALD and its complication (Zhao et al. 2017).

Commercial cultivation of some mushrooms has not been successful until now and according to that the cultivation of mushroom's mycelium became extensively used for investigation. The aqueous ethanolic extract of mycelium obtained by cultivation of *Morchella esculenta* was tested against CCl_4 and chronic hepatotoxicity induced by ethanol. Extract showed **hepatoprotective** effect by measuring of enzymes responsible for liver function (Nitha et al. 2013).

Maitake (*G. frondosa*) is an edible mushroom used in prevention and treatment of various diseases and possess great medically importance. Product containing ß-glucan extracted from this mushroom has been tested for its antihepatitis effects. Kubo and Nanba (1998) were examined for potential of maitake mushroom to influence on disorders of liver in animal models using hepatic-damaged mice. It was noticed that autoimmune chronic hepatitis is more often in control animals than in mushroom-treated mice.

4.3.4 The Edible Mushrooms as Alternative Methods of Hepatitis Treatment and Liver Damage Induced with Path Ogens - Viruses

The liver has special functions in humanr body, and this organ is very often exposed to may intestinal antigens, including tumor cells, pathogens (bacteria, viruses, and parasites), toxins, and harmless dietary antigens (Mowat 2003). Medical therapies for chronic liver disease are limited and difficult to handle. Patients suffers from these diseases often need other treatments, alternatives to replace or to help to standard care. There are many types of alternative and complementary medical (CAM) approaches. Many natural compounds orgianated from plants were tested toward liver diseases (Seeff et al. 2001). Medicinal mushrooms could be used as novel promising substances in tretamnet of hepatitis.

In order to find novel **therapeutic agents**, Mlinarič et al. (2005) performed screenings potential therapeutic activity of some Slovenian fungi. The dichloromethane and methanol extracts obtained from 57 mushrooms were tested for inhibitory potential against HIV-1 reverse transcriptase activity. Methanolic extracts of 13 musrooms possessed more tehn 40% inhibition effect and only two showed inhibition potential in moe than 80%. *Laetiporus sulphureus* and *Poria monticola* were the most effective. *L. sulphureus* was tested in more detailes, using few different isolates, where preliminary results confirmed that the most active fraction was the one which contained acidic compound with the amino group.

One of the main reasons of liver diseases is viral infections due to hepatitis, in B (HBV) and C (HCV) forms. Approximately 350 million people are living with chronic infection of liver and around 2 billion people are infected with HBV, and 600 000 peopel die per year due to the acute or chronic infection of HBV and its consequences. Data showed that 180 million people have HCV chronic infection and 3 to 4 million people per year become infected. The treatment of chronically infected patients by hepatitis B/C is very costly. Drugs ussualy used for treatment of chronic HBV infection are cytokines, nucleoside analogues (e.g., lamivudine, ganciclovir), vaccines and antibodies against hepatitis B surface antigen (HBsAg) (Papatheodoridis et al. 2002). On the other hand, long-standing treatment with nucleoside analogues is conected with escalating rates of viral resistance attributable to the emergence of resistant HBV strains. For that reason, new efficient hepatoprotective drugs having lower toxicity are indispensable. Nonetheless, there has been a fresh ascent of mercantile interest in favour of the edible mushrooms.

A. blazei is promising candidate in hepatitis complementary treatment. Inuzuka and Yoshida (2002) investigated extract of *Agaricus* spp. and its activity on C-hepatitis. They showed that this extract is useful in treatment in light hepatopathy in patients with C-type hepatitis in clinical examination activity of this mushroom at humans.

Chen et al. (2004) showed that extract of *A. brasiliensis* could be useful in treatment of **hepatitis B** as adjuvant to vaccines *in vivo*. The results showed that this extract has increased HBcAg-specific antibody response, and higher T cell proliferation was notified in mice which were treated with combination of *A. brasiliensis* extract and HBcAg DNA vaccine.

Grinde et al. (2006) noticed the effect on expression of gene in peripheral blood cells at pateinets suffer from chronic **hepatitis C**. Fraction with *β*-glucan isolated from *A. brasiliensis* extract applied to pateient but was not transported in appreciable quantities into the blood.

4.3.5 Clinical Utility of Mushroom Extract in Treatment Liver Diseases

Clinical studies of testing the activity of mushroom preparations on disorders at pathological levelare are very rare. There are data that approximately 30 mushroom exstracts and compounds are beeing investigated in clinical trials by Hospital Clinic of Barselona, National institutes of Health in US; and others hospital, moustly for the immunological use in oncology (Morris et al. 2017).

A. brasiliensis is the most frequently mentioned species in clinical trials. Around 100 000– 300 000 kg of the dried frut body of *A. blazei* is cultivated per year in

Japan. Data showed that 300 000–500 000 people used 3–5 g three times per day of *A. blazei* extract (Wang et al. 2013).

Safety and clinical potential on **γ-GTP** activity of *A. blazei* extract was tested in 20 volunteers suffer from C-hepatitis. Extract showed decreasing activity for serum γ-GTP in 80% of the tested patients, treated orally, two times per day, 8 weeks. There are no any negative toxicological results or any other side effects were observed (Inuzuka and Yoshida 2002).

Hsu et al. (2008) performed a 1 pilot study to test potential of *A. brasiliensis* extract to improve function of liver in hepatitis B patients. They applied *A. brasiliensis* extract of 1500 mg per day, one year, and measured the level of aspartate aminotransferase (AST) and alanine aminotransferase (ALT). The result showed that *A. brasiliensis* extract can improve liver function of the 4 patients, but still this study represents only a small sample research.

Orally applied *A. brasiliensis* (AbM) was tested on chronic hepatitis C (Grinde et al. 2006). The viral load decreased after 7 days of treatment. The treatment also helped in regulation of gene for IFN-α-receptor. Examination of AbM intake in combination with standard IFN- α treatment could be of interest. AbM extract may help in activity of IFN alpha treatment in chronic hepatitis patients. Results from all these clinical data have shown that *A. brasiliensis* extracts are promising agent in treatment of hepatitis, especially in fighting with the resistant variety of the disease. Further clinical studies are necessary to confirm the positive activity of this mushroom in treatment of hepatitis, and to exclude toxic side effects (Mukai et al. 2006).

Mukai et al. (2006) reported that three patients with cancer and hepatic damage, taking *A. brasiliensis* extract, after intake of this extract two died of fulminant hepatitis. It is reported that a causal connection between the *A. brasiliensis* extract and damage of liver was suggested. This made the clinical process more complicated.

Toshiro et al. (2003) observed the effect of **g-aminobutyric acid** (GABA)-enriched *A. blazei* (AG-GABA) to human volonters suffer from mild hypertensy. During the AG-GABA intake period, blood pressure decreased to significant levels, in comparation with the pretest period or placebo. The values of cholesterol and in hepatic transaminases and g-GTP were the same.

The polysaccharides (A-PBP and L-PBP) fractions of *A. blazei* mycelia on serum cholesterol and body weight were observed in 90 volunteers (female), for 8 weeks. The reduction in weight was 11.8% and hypocholesterolemic effect 11.0%. The results of this study showed that the body weight controlling and hypolipidemic effect protein-bound polysaccharides are connected and involved, partly, in absorption of cholesterol, which has a role of dietary fiber and in cholesterol metabolism (Kweon et al. 2002).

4.4 Concluding Remarks

The present study is a part of our ongoing studies about bioactive fungi, their use in ethnomedicine but also their potential as source for new drugs. Owing to the increasing development of novel natural originated drugs in medicine and pharmacy, discovery of novel substances from mushrooms is an important research aim. The desire and needs for safer drugs with lower toxicity is a major concern and these compounds isolated from edible mushrooms could possibly be a good alternative.

Acknowledgement The authors thank Serbian Ministry of Education, Science and Technological Development for financial support (grant number 173032).

References

Abah SE, Abah G (2010) Antimicrobial and antioxidant potentials of *Agaricus bisporus*. Adv Biol Res 4(5):277–282

Ahmad N, Mahmood F, Akbar Khalil S, Zamir R, Fazal H, Abbasi BH (2014) Antioxidant activity via DPPH, gram-positive and gram-negative antimicrobial potential in edible mushrooms. Toxicol Ind Health 30(9):826–834

Akata I, Ergonul B, Kalyoncu F (2012) Chemical compositions and antioxidant activities of 16 wild edible mushroom species grown in Anatolia. Int J Pharm 8:134–138

Akyüz M, Onganer AN, Erecevit P, Kirbağ S (2010) Antimicrobial activity of some edible mushrooms in the Eastern and Southeast Anatolia region of Turkey. J Sci 3(2):125–130

Alam N, Amin R, Khan A, Ara I, Shim MJ, Lee MW, Lee UY, Lee TS (2009) Comparative effects of oyster mushrooms on lipid profile, liver and kidney function in hyper cholesterolemic rats. Mycobiology 37(1):37–42

Alves MJ, Ferreira ICFR, Martins A, Pintado M (2012a) Antimicrobial activity of wild mushroom extracts against clinical isolates resistant to different antibiotics. J Appl Microbiol 113(2):466–475

Alves M, Ferreira ICFR, Dias J, Teixeira V, Martins A, Pintado M (2012b) A Review on antimicrobial activity of mushroom (basidiomycetes) extracts and isolated compounds. Planta Med 78(16):1707–1718

Alves MJ, Ferreira ICFR, Dias J, Teixeira V, Martins A (2013) A review on antifungal activity of mushroom extracts and isolated compounds. Curr Top Med Chem 13(21):2648–2659

Andrade RJ, Tulkens PM (2011) Hepatic safety of antibiotics used in primary care. J Antimicrob Chemother 66:1431–1446

Arnao MB, Cano A, Alcolea JF, Acosta M (2017) Estimation of free radical-quenching activity of leaf pigment extracts. Phytohem Anal 12(2):281–286

Atila F, Owaid MN, Shariati MA (2017) The nutritional and medical benefits of *Agaricus bisporus*: a review. J Microbiol Biotechnol Food Sci 7(3):281–286

Auzinger G, Wendon J (2008) Intensive care management of acute liver failure. Curr Opin Crit Care 14(2):179–188

Babakhin AA, Logina NY, Nolte H, DuBuske LM (1996) Immunomodulating activity of the extract from high mycelium fungus *Polyporus squamosus*. J Allergy Clin Immunol 97:229

Barroetavena C, Toledo CV (2017) The nutritional benefits of mushrooms. In: Ferreira ICFR, Morales P, Barros L (eds) Wild plants, mushrooms and nuts: functional food properties and applications. Wiley-Blackwell, Chichester, pp 65–83

Barros L, Cruz T, Baptista P, Estevinho LM, Ferreira IC (2008a) Wild and commercial mushrooms as source of nutrients and nutraceuticals. Food Chem Toxicol 46:2742–2747

Barros L, Venturini BA, Baptista P, Estevinho LM, Ferreira ICFR (2008b) Chemical composition and biological properties of Portuguese wild mushrooms: a comprehensive study. J Agric Food Chem 56(10):3856–3862

Bawadekji A, Mridha MAU, Al Ali M, Jamith Basha W (2017) Antimicrobial activities of oyster mushroom *Pleurotus ostreatus* (Jacq. ex. Fr.) Kummer. J Appl Environ Biol Sci 7(10):227–231

Bisen PS, Baghel RK, Sanodiya BS, Thakur GS, Prasad GBKS (2010) *Lentinus edodes*: a macrofungus with pharmacological activities. Curr Med Chem 17:2419–2430

Boucher HW, Talbot GH, Bradley JS, Edwards JE, Gilbert D, Rice LB, Scheld M, Spellberg B, Bartlett J (2009) Bad bugs, no drugs: no ESKAPE! An update from the infectious diseases society of America. Clin Infect Dis 48(1):1–12

Brown GD, Denning DW, Gow NAR, Levitz SM, Netea MG, White TC (2012) Hidden killers: human fungal infections. Sci Transl Med 4:165–213

Caz V, Gil-Ramírez A, Santamaría M, Tabernero M, Soler-Rivas C, Martín-Hernández R, Marín FR, Reglero G, Largo C (2016) Plasma Cholesterol-lowering activity of lard functionalized with mushroom extracts is independent of niemann–pick c1-like 1 protein and abc sterol transporter gene expression in hypercholesterolemic mice. J Agr Food Chem 64(8):1686–1694

Chang ST (2006) The world mushroom industry: trends and technological development. Int J Med Mushrooms 8:297–314

Chang ST, Buswell JA (1996) Mushroom nutriceuticals. World J Microbiol Biotechnol 12(5):473–476

Chang ST, Miles P (2004) Mushrooms, 2nd edn. CRC Press LLC, Boca Raton

Chen L, Shao HJ, Su YB (2004) Coimmunization of *Agaricus blazei* murill extract with hepatitis B virus core protein through DNA vaccine enhances cellular and humoral immune responses. Int Immunopharmacol 4(3):403–409

Chihara G (1992) Immunopharmacology of lentinan, a polysaccharide isolated from *Lentinus edodes*: its application as a host defence potentiator. Int J Oriental Med 17:55

Chowdhury M, Kubra K, Ahmed S, Barros L, Ferreira M, Queiros B et al (2015) Screening of antimicrobial, antioxidant properties and bioactive compounds of some edible mushrooms cultivated in Bangladesh. Ann Clin Microbiol Antimicrob 14(1):8

Ciric L, Tymon A, Zaura E, Lingstrom P, Stauder M, Papetti A, Signoretto C, Pratten J, Wilson M, Spratt D (2011) *In Vitro* assessment of shiitake mushroom (*Lentinula edodes*) extract for its antigingivitis activity. J Biomed Biotechnol:507908

Coskun O, Kanter M, Korkmaz A, Oter S (2005) Quercetin, a flavonoid antioxidant, prevents and protects streptozotocin-induced oxidative stress and β-cell damage in rat pancreas. Pharmacol Res 51:117–123

Dandapat S, Sinha MP, Kumar M, Jaggi Y (2015) Hepatoprotective efficacy of medicinal mushroom *Pleurotus tuber-regium*. Environ Exp Biol 13:103–108

David B, Wolfender JL, Dias DA (2015) The pharmaceutical industry and natural products: historical status and new trends. Phytochem Rev 14:299–315

De Silva DD, Rapior S, Sudarman E, Stadler M, Xu J, Alias SA, Hyde KD (2013) Bioactive metabolites from macrofungi: ethnopharmacology, biological activities and chemistry. Fungal Divers 62:1–40

Dhanasekaran D, Latha S, Saha S, Thajuddin N, Panneerselvam A (2013) Extracellular biosynthesis, characterisation and *in-vitro* antibacterial potential of silver nanoparticles using *Agaricus bisporus*. J Exp Nanosci 8(4):579–588

Dimitrijevic M, Stankov Jovanovic V, Cvetkovic J, Mihajilov-Krstev T, Stojanovic G, Mitic V (2015) Screening of antioxidant, antimicrobial and antiradical activities of twelve selected Serbian wild mushrooms. Anal Methods 7:4181

Dimitrijević MV, Mitić VD, Stankov-Jovanović VP, Nikolić JS, Stojanović GS (2016) Comprehensive evaluation of the antioxidant activity of six wild edible mushroom species. Adv Technol 5(2):53–59

Duncan C, Pugh J, Pasco G, David N, Ross S, Samir A (2002) Isolation of a galactomannan that enhances macrophage activation from the edible fungus *Morchella esculenta*. J Agric Food Chem 50:5683–5685

Enman J (2007) Production and quantification of eritadenine, a cholesterol reducing compound in shiitake (*Lentinus edodes*) [dissertation]. Luleå University of Technology Department of Chemical Engineering and Geosciences Division of Biochemical and Chemical Process Engineering

Everhart JE, Ruhl CE (2009) Burden of digestive diseases in the United States Part III: liver, biliary tract, and pancreas. Gastroenterology 136:1134–1144

FAO (2004) Wild edible fungi: a global overview of their use and importance to people. In: Non-wood forest products. Food and Agriculture Organization of the United Nations, Rome

FAO (2009) Make money by growing mushrooms (Elaine Marshall and N. G. (Tan) Nair) rural infrastructure and agro-industries division. Food and Agriculture Organization of the United Nations, Rome

FAO Stat (2015) http://faostat.fao.org/site/567/pageID567

Fernandes A, Petrović J, Stojković D, Barros L, Glamočlija SM, Anabela M, Isabel F (2016) *Polyporus squamosus* (Huds.) Fr from different origins: chemical characterization, screening of the bioactive properties and specific antimicrobial effects against *Pseudomonas aeruginosa*. LWT-Food Sci Technol 69:91–97

Ferreira ICFR, Barros L, Abreu RMV (2009) Antioxidants in wild mushrooms. Curr Med Chem 16(12):1543–1560

Finimundy TC, Dillon AJP, Henriques JAP, Ely MR (2014) A Review on general nutritional compounds and pharmacological properties of the *Lentinula edodes* mushroom. Int J Food Sci Nutr 5:1095–1105

Gan CH, Nurul Amira B, Asmah R (2013) Antioxidant analysis of different types of edible mushrooms (*Agaricus bisporous* and *Agaricus brasiliensis*). Int Food Res J 20(3):1095–1102

Gargano ML, van Griensven LJLD, Isikhuemhen OS, Lindequist U, Venturella G, Wasser SP, Zervakis GI (2017) Medicinal mushrooms: valuable biological resources of high exploitation potential. Plant Biosyst 151:548–565

Gil-Ramírez A, Morales D, Soler-Rivas C (2018) Molecular actions of hypocholesterolaemic compounds from edible mushrooms. Food Funct 9:53–69

Glamočlija J, Stojković D, Nikolić M, Ćirić A, Barros L, Ferreira IC, Sokovic M (2015) Comparative study on edible agaricus mushrooms as functional foods. Food Funct 6:1900–1910

Grinde B, Hetland G, Johnson E (2006) Effects on gene expression and viral load of a medicinal extract from *Agaricus blazei* in patients with chronic hepatitis C infection. Int Immunopharmacol 6(8):1311–1314

Grujičić BD, Potočnik I, Duduk B, Vujčić Z (2015) Spent mushroom compost as substrate for the production of industrially important hydrolytic enzymes by fungi *Trichoderma* spp. and *Aspergillus niger* in solid state fermentation. Int Biodeterior Biodegrad 104:290–298

Guillamón E, García-Lafuente A, Lozano M, Arrigo D, Rostagno M, Villares MA et al (2010) Edible mushrooms: role in the prevention of cardiovascular diseases. Fitoterapia 81(7):715–723

Hassegawa RH, Kasuya MCM, Vanetti MCD (2005) Growth and antibacterial activity of *Lentinula edodes* in liquid media supplemented with agricultural wastes. Electron J Biotechnol 8(2):212–217

Hatvani N (2001) Antibacterial effect of the culture fluid of *Lentinus edodes* mycelium grown in submerged liquid culture. Int J Antimicrob Agents 17:71–74

He F, Yang Y, Yang G, Yu L (2010) Studies on antibacterial activity and antibacterial mechanism of a novel polysaccharide from *Streptomyces virginia* H03. Food Control 21(9):1257–1262

Hearst R, Nelson D, McCollum G, Millar BC, Maeda Y, Goldsmith CE, Rooney PJ, Loughrey A, Rao JR, Moore JE (2009) An examination of antibacterial and antifungal properties of constituents of shiitake (*Lentinula edodes*) and oyster (*Pleurotus ostreatus*) mushrooms. Complement Ther Clin Pract 15:5–7

Heleno SA, Stojković D, Barros L, Glamočlija J, Soković M, Martins A, Queiroz MJ, Ferreira I (2013) A comparative study of chemical composition, antioxidant and antimicrobial properties of *Morchella esculenta* (L.) Pers. from Portugal and Serbia. Food Res Int 51(1):236–243

Heleno SA, Barros L, Martins A, Morales P, Fernández Ruiz V, Glamoclija J, Sokovic M, Ferreira I (2015) Nutritional value, bioactive compounds, antimicrobial activity and bioaccessibility studies with wild edible mushrooms. LWT-Food Sci Technol 63:799–806

Hirasawa M, Shouji N, Neta T, Fukushima K, Takada K (1999) Three kinds of antibacterial substances from *Lentinus edodes* (Berk.) Sing. (shiitake, an edible mushroom). Int J Antimicrob Agents 11(2):151–157

Hsu CH, Hwang KC, Chiang YH, Chou P (2008) The mushroom *Agaricus blazei* murill extract normalizes liver function in patients with chronic hepatitis B. J Altern Complement Med 14(3):299–301

Inuzuka H, Yoshida T (2002) Clinical utility of ABCL (Agaricus Mushroom Extract) treatment for C-type hepatitis. Jpn Pharmacol Ther 30:103–107

Iwalokun BA, Usen UA, Otunba AA, Olukoya DK (2007) Comparative phytochemical evaluation, antimicrobial and antioxidant properties of *Pleurotus ostreatus*. Afr J Biotechnol 6:1732–1739

Jayakumar T, Ramesh E, Geraldine P (2006) Antioxidant activity of the oyster mushroom, *Pleurotus ostreatus*, on CCl4-induced liver injury in rats. Food Chem Toxicol 44:1989–1996

Jeong SC, Jeong YT, Yang BK, Islam R, Koyyalamudi SR, Pang G, Cho KY, Song CH (2010) White button mushroom (*Agaricus bisporus*) lowers blood glucose and cholesterol levels in diabetic and hypercholesterolemic rats. Nutr Res 30:49–56

Kamarudzaman AN, Chay TC, Amir A, Talib SA (2015) Biosorption of Mn (II) ions from aqueous solution by *Pleurotus* spent mushroom compost in a fixedbed column. Procedia Soc Behav Sci 195:2709–2716

Karaman M, Jovin E, Malbasa R, Matavuly M, Popović M (2010) Medicinal and edible lignicolous fungi as natural sources of antioxidative and antibacterial agents. Phytother Res 24:1473–1481

Karaman M, Stahl M, Vulić J, Vesić M, Canadanović-Brunet J (2014) Wild-growing lignicolous mushroom species as sources of novel agents with antioxidative and antibacterial potentials. Int J Food Sci Nutr 65(3):311–319

Keles A, Koca I, Genccelep H (2011) Antioxidant properties of wild edible mushrooms. J Food Process Technol 2:130

Khan AA, Gani A, Ahmad M, Masoodi FA, Amin F, Kousar S (2016) Mushroom varieties found in the Himalayan regions of India: antioxidant, antimicrobial, and antiproliferative activities. Food Sci Biotechnol 25(4):1095–1100

Khatua S, Ghosh S, Acharya K (2017) *Laetiporus sulphureus* (Bull.: Fr.) Murr. as food as medicine. Pharmacogn J 9:1–15

Klančnik A, Megušar P, Sterniša M, Jeršek B, Bucar F, Smole Možina S, Kos J, Sabotič J (2017) Aqueous extracts of wild mushrooms show antimicrobial and antiadhesion activities against bacteria and fungi. Phytother Res 31(12):1971–1976

Klaus A, Kozarski M, Vunduk J, Todorovic N, Jakovljevic D, Zizak Z et al (2015) Biological potential of extracts of the wild edible Basidiomycete mushroom *Grifola frondosa*. Food Res Int 67:272–283

Kostić M, Smiljković M, Petrović J, Glamočilija J, Barros L, Ferreira ICFR, Ćirić A, Soković M (2017) Chemical, nutritive composition and wide-broad bioactive properties of honey mushroom *Armillaria mellea* (Vahl: Fr.) Kummer. Food Funct 8:3239–3249

Kozarski M, Klaus A, Jakovljević D, Todorović N, Vunduk J, Petrović P et al (2015) Antioxidants of edible mushrooms. Molecules 20(10):19489–19525

Kubo K, Nanba H (1998) Modifications of cellular immune responses in experimental autoimmune hepatitis in mice by maitake (*Grifola frondosa*). Mycoscience 39:351–360

Kumar V, Yadav U (2014) Screening of antifungal activity of *Pleurotus ostreatus* and *Agaricus bisporus*. Biolife 2(3):918–923

Kweon MH, Kwon ST, Kwon SH, Ma MS, Park YI (2002) Lowering effects in plasma cholesterol and body weight by mycelial extracts of two mushrooms: *Agaricus blazei* and *Lentinus edodes*. Korean J Microbiol Biotechnol 30:402–409

Li M, Hu JL (2014) Study on survival strategies of farmers engage in small-scale household cultivation of edible mushrooms: take Shandong Province as an example. In: Modern economy, vol 5, pp 1092–1100. https://doi.org/10.4236/me.2014.512100

Lingstrom P, Zaura E, Hassan H, Buijs MJ, Hedelin P, Pratten J et al (2012) The anticaries effect of a food extract (shiitake) in a short-term clinical study. J Biomed Biotechnol:217164

Liu Y, Fukuwatari Z, Okumura K, Takeda K, Ishibashi K, Furukawa M et al. (2008) Immunomodulating activity of *Agaricus brasiliensis* KA21 in mice and in human volunteers. eCAM 5(2):205–209

Liu J, Jia L, Kan J, Jin CH (2013) *In vitro* and *in vivo* antioxidant activity of ethanolic extract of white button mushroom (*Agaricus bisporus*). Food Chem Toxicol 51:310–316

Liu Q, Zhu M, Geng X, Wang H, Ng TB (2017) Characterization of polysaccharides with antioxidant and hepatoprotective activities from the edible mushroom *Oudemansiella radicata*. Molecules 22:234–250

Mau JL, Chang CN, Huang SJ, Chen CC (2004) Antioxidant properties of methanolic extracts from *Grifola frondosa*, *Morchella esculenta* and *Termitomyces alduminosus* mycelia. Food Chem 87:111–118

Mensink RP, Aro A, Den Hond E, German JB, Griffin BA, Ter Meer H-U et al (2003) PASSCLAIM-Diet-related cardiovascular disease. Eur J Nutr 42(11):1/6–1/2

Mirunalini S, Arulmozhi V, Deepalakshmi K, Krisnaveni M (2012) Intracellular biosynthesis and antibacterial activity of silver nanoparticles using edible mushrooms. Not Sci Biol 4(4):55

Mlinarič A, Kac J, Pohleven F (2005) Screening of selected wood-damaging fungi for the HIV-1 reverse transcriptase inhibitors. Acta Pharm 55(1):69–79

Mocan A, Fernandes Â, Barros L, Crişan G, Smiljković M, Soković M, Ferreira ICFR (2018) Chemical composition and bioactive properties of the wild mushroom *Polyporus squamosus* (Huds.) Fr: a study with samples from Romania. Food Funct 9(1):160–170

Morris HJ, Llaurado G, Beltran Y, Lebeque Y, Bermudez RC, Garcia N (2017) The use of mushrooms in the development of functional foods, drugs, and nutraceuticals. In: ICFR F, Morales P, Barros L (eds) Wild plants, mushrooms and nuts: functional food properties and applications. Wiley-Blackwell, Chichester, pp 123–159

Mowat AM (2003) Anatomical basis of tolerance and immunity to intestinal antigens. Nat Rev Immunol 3:331–341

Mukai H, Watanabe T, Ando M, Katsumata N (2006) An alternative medicine, *Agaricus blazei*, may have induced severe hepatic dysfunction in cancer patients. Jpn J Clin Oncol 36(12):808–810

Muriel P (2009) Role of free radicals in liver diseases. Hepatol Int 3:526–536

Navarro VJ, Senior JR (2006) Drug-related Hepatotoxicity. N Engl J Med 354:731–739

Ndungutse V, Mereddy R, Sultanbawa Y (2015) Bioactive properities of mushroom (*Agaricus bisporus*) stipe extracts. J Food Process Preserv 39:2225–2233

Nehra K, Kumar M, Yadav A (2012) Evaluation of antimicrobial potential of fruiting body extracts of *Pleurotus ostreatus* (oyster mushroom). Int J Microbial Resour Technol 1(4):391–400

Ngai PHK, Ng TB (2003) Lentin, a novel and potent antifungal protein from shiitake mushroom with inhibitory effects on activity of human immunodeficiency virus-1 reverse transcriptase and proliferation of leukemia cells. Life Sci 73:3363–3364

Nikitina EV, Tsivileva MO, Pankratov NA, Bychkov AN (2007) *Lentinula edodes* biotechnology-from lentinan to lectins. Food technol biotechnol 45:230–237

Nitha D, Meera CR, Janardhanan KK (2007) Anti-inflammatory and antitumour activities of cultured mycelium of morel mushroom, *Morchella esculenta*. Curr Sci 92:235–239

Nitha B, Fijesh PV, Janardhanan KK (2013) Hepatoprotective activity of cultured mycelium of Morel mushroom, *Morchella esculenta*. Exp Toxicol Pathol 65(1–2):105–112

Ochoa-Zarzosa A, Vázquez-Garcidueñas MS, Robinson-Fuentes VA, Vázquez-Marrufo G (2011) Antibacterial and cytotoxic activity from basidiocarp extracts of the edible mushroom *Lactarius indigo* (Schw.) Fr. (Russulaceae). Afr J Pharm Pharmacol 5(2):281–288

Ooi VEC (1996) Hepatoprotective effect of some edible mushrooms. Phytother Res 10:536–538

Ooi CEV, Liu F (2000) Immunomodulation and anticancer activity of polysaccharide-protein complexes. Curr Med Chem 7:715–719

Owaid MN, Al-Saeedi SSS, Al-Assaffii IAA (2017) Antifungal activity of cultivated oyster mushrooms on various agro-wastes. Summa Phytopathol 43(1):9–13

Oyetayo VO, Ariyo OO (2013) Antimicrobial and antioxidant properties of *Pleurotus ostreatus* (Jacq: Fries) cultivated on different tropical woody substrates. J waste conversion, Bioprod Biotechnol 1(2):28–32

Ozturk M, Duru ME, Kivrak S, Dogan NM, Turkoglu A, Ozler MA (2011) *In vitro* Antioxidant, anticholinesterase and antimicrobial activity studies on three *Agaricus* species with fatty acid compositions and iron contents: a comparative study on the three most edible mushrooms. Food Chem Toxicol 49:1353–1360

Papatheodoridis GV, Dimou E, Papadimitropoulos V (2002) Nucleoside analogues for chronic hepatitis B: antiviral efficacy and viral resistance. Am J Gastroenterol 97:1618–1628

Parola S, Chiodaroli L, Orlandi V, Vannini C, Panno L (2017) *Lentinula edodes* and *Pleurotus ostreatus:* functional food with antioxidant -antimicrobial activity and an important source of Vitamin D and medicinal compounds. FFHD 7(6):773–794

Patel Y, Naraian R, Singh VK (2012) Medicinal properties of *Pleurotus* species (Oyster Mushroom): a review. World J Fungal Plant Biol 3(1):1–12

Pauliuc I, Dorica B (2013) Antibacterial activity of *Pleurotus ostreatus* gemmotherapic extract. J Hortic Forestry Biotechnol 17(1):242–245

Petrović J, Papandreou M, Glamočlija J, Ćirić A, Baskakis C, Proestos C, Lamari F, Zoumpoulakis P, Sokovic M (2014a) Different extraction methodologies and their influence on the bioactivity of the wild edible mushroom *Laetiporus sulphureus* (Bull.) Murrill. Food Funct 5:2948–2960

Petrović J, Stojković D, Reis FS, Barros L, Glamočlija J, Ćirić A, Ferreira I, Sokovic M (2014b) Study on chemical, bioactive and food preserving properties of *Laetiporus sulphureus* (Bull.: Fr.) Murr. Food Funct 5:1441–1451

Petrović J, Ćirić A, Heleno SA, Barros L, Ferreira ICFR, Glamočlija J, Soković M (2016) *Pleurotus ostreatus* (Jacq.ex Fr.) P. K.: a highly nutritious edible mushroom cultivated in Serbia as a potential remedy for chronic infectious diseases. Proceedings of the 13th congress of nutrition, food and nutrition – a roadmap to better health; 2016 October 26–28; Belgrade

Poli G (1993) Liver damage due to free radicals. Br Med Bull 49:604–600

Quang DN, Hashimoto T, Asakawa Y (2006) Inedible mushrooms: a good source of biologically active substances. Chem Rec 6:79–99

Rajewska J, Bałasińska B (2004) Biologically active compounds of edible mushrooms and their beneficial impact on health. Postepy Hig Med Dosw 58:352–357

Ramesh CH, Pattar MG (2010) Antimicrobial properties, antioxidant activity and bioactive compounds from six wild edible mushrooms of Western Ghats of Karnataka, India. Pharmacognosy Res 2(2):107–112

Rana T, Bera AK, Das S, Bhattacharya D, Pan D, Bandyopadhyay S, Mondal DK, Samanta S, Bandyopadhyay S, Das SK (2012) *Pleurotus florida* lectin normalizes duration dependent hepatic oxidative stress responses caused by arsenic in rat. Exp Toxicol Pathol 64(7–8):665–671

Rao JR, Smyth TJ, Millar BC, Moore JE (2009) Antimicrobial properties of shiitake mushrooms (*Lentinula edodes*). Int J Antimicrob Agents 3:591–592

Reis FS, Martins A, Barros L, Ferreira ICFR (2012) Antioxidant properties and phenolic profile of the most widely appreciated cultivated mushrooms: a comparative study between *in vivo* and *in vitro* samples. Food Chem Toxicol 50:1201–1207

Reis FS, Martins A, Vasconcelos MH, Morales P, Ferreira ICFR (2017) Functional foods based on extracts or compounds derived from mushrooms. Trends Food Sci Technol 66:48–62

Rincão VP, Yamamoto KA, Ricardo NMPS, Soares SA, Meirelles LDP, Nozawa C, Linhares RE (2012) Polysaccharide and extracts from *Lentinula edodes*: structural features and antiviral activity. Virol J 9:37

Robaszkiewicz A, Bartosz G, Lawrynowicz M, Soszynski M (2010) The role of polyphenols, β-carotene, and lycopene in the antioxidative action of the extracts of dried, edible mushrooms. J Nutr Metab 42:1–9

Rosmiza M, Davies W, Aznie RC, Jabil M, Mazdi M (2016) Prospects for increasing commercial mushroom production in malaysia: challenges and opportunities. Mediterr J Soc Sci 7:406–415

Roy DN, Azad AK, Sultana F, Anisuzzaman ASM (2016) *In-vitro* antimicrobial activity of ethyl acetate extract of two common edible mushrooms. J Pharmacol 5(2):79–82

Roupas P, Keogh J, Noakes M, Margetts C, Taylor P (2012) The role of edible mushrooms in health: evaluation of the evidence. J Funct Foods 4(4):687–689

Sánchez C (2017a) Bioactives from mushroom and their application. In: Puri M (ed) Food Bioactives. Springer International Publishing, Cham, pp 23–27

Sánchez C (2017b) Reactive oxygen species and antioxidant properties from mushrooms. Synth Syst Biotechnol 2(1):13–22

Schwan WR (2012) Mushrooms: an untapped reservoir for nutraceutical antibacterial applications and antibacterial compounds. Curr Top Nutraceutical Res 10(1):75–82

Seeff LB, Lindsay KL, Bacon BR, Kresina TF, Hoofnagle JH (2001) Complementary and alternative medicine in chronic liver disease. Hepatology 34(3):595–603

Seif HAS (2016) Physiological changes due to hepatotoxicity and the protective role of some medicinal plants. Beni–Suef Univ J Basic Appl Sci 5:134–136

Sharma MV, Sagar A, Joshi M (2015) Study on antibacterial activity of *Agaricus bisporus* (Lang.) Imbach. Int J Curr Microbiol App Sci 4(2):553–558

Sherry D, McCulloch A, Liang Q, Reimann S, Newman PA (2018) Current sources of carbon tetrachloride (CCl4) in our atmosphere. Environ Res Lett 13:024004

Smolskaite L, Petras RV, Thierry T (2015) Comprehensive evaluation of antioxidant and antimicrobial properties of different mushroom species. LWT-Food Sci Technol 60:462–471

Soares AA, de Sá-Nakanishi AB, Bracht A, da Costa SMG, Koehnlein EA, de Cristina Giatti Souza M, Peralta RM (2013a) Hepatoprotective effects of mushrooms. Molecules 18:7609–7630

Soares AA, de Oliveira AL, de Sá-Nakanishi AB, Comar JF, Rampazzo APS, Vicentini FA, Natali MRM, Costa SM, Bracht A, Peralta RM (2013b) Effects of an *Agaricus blazei* aqueous extract pretreatment on paracetamol-induced brain and liver injury in rats. Biomed Res Int 469180

Soković M, Glamočlija J, Ćirić A (2013) Natural products from plants and fungi as fungicides. In: Nita M (ed) Fungicides-showcases of integrated plant disease management from around the world. InTech, New York, pp 185–232

Soković M, Ćirić A, Glamočlija J, Nikolić M, van Griensven LJLD (2014) *Agaricus Blazei* hot water extract shows anti quorum sensing activity in the nosocomial human pathogen *Pseudomonas aeruginosa*. Molecules 19(4):4189–4199

Soković M, Ćirić A, Glamočlija J, Stojković D (2017) The bioactive properties of mushrooms. In: Ferreira ICFR, Morales P, Barros L (eds) Wild plants, mushrooms and nuts: functional food properties and applications. Wiley-Blackwell, Chichester

Soković M, Glamočlija J, Ćirić A, Petrović J, Stojković D (2018) Mushrooms as sources of therapeutic foods. In: Holban AM, Grumezescu AM (eds) Therapeutic foods, handbook of food bioengineering. Elsevier Inc, London

Spratt DA, Daglia M, Papetti A, Stauder M, O'Donnell D, Ciric L, Tymon A, Repetto B, Signoretto C, Houri-Haddad Y, Feldman M, Steinberg D, Lawton S, Longstrom P, Pratten J, Zaura E, Gazzani G, Pruzzo C, Wilson M (2012) Evaluation of plant and fungal extracts for their potential antigingivitis and anticaries activity. J Biomed Biotechnol:510198

Stajić M, Vukojević J, Knežević A, Duletić Laušević S, Milovanović I (2013) Antioxidant protective effect of mushroom metabolites. Curr Top Med Chem 13:2660–2676

Stojković D, Reis FS, Barros L, Glamočlija J, Ćirić A, van Griensven LJ et al (2013) Nutrients and non-nutrients composition and bioactivity of wild and cultivated *Coprinus comatus* (O.F. Müll.) Pers. Food Chem Toxicol 59:289–296

Stojković D, Reis FS, Glamočlija J, Ćirić A, Barros L, JLD van Griensven L et al (2014) Cultivated strains *of Agaricus bisporus* and *A. brasiliensis*: chemical characterization and evaluation of antioxidant and antimicrobial properties for final healthy product - natural preservatives in yoghurt. Food Funct 5:1602–1612

Stojković DS, Kovačević-Grujičić N, Reis FS, Davidović S, Barros L, Popović J, Petrovic I, Pavic A, Ciric A, Stevanovic M (2017) Chemical composition of the mushroom *Meripilus giganteus* Karst. and bioactive properties of its methanolic extract. LWT-Food Sci Technol 79:454–462

Suay I, Arenal F, Asensio FJ, Basilio A, Cabello MA, Diey MT, Garcia JB, Del Val AG, Gorrochategui J, Hernandez P, Pelaez F, Vicente MF (2000) Screening of basidiomycetes for antimicrobial activities. Antonie van Leeuwenhoek 78:129–139

Sudhakar T, Nanda A, Babu SG, Janani S, Evans MD, Markose TK (2014) Synthesis of silver nanoparticles from edible mushroom and its antimicrobial activity against human pathogens. Int J Pharmtech Res 6(5):1718–1723

Tehrani MHH, Fakhrehoseini E, Nejad MK, Mehregan H, Hakemi-Vala M (2012) Search for proteins in the liquid extract of edible mushroom, *Agaricus bisporus*, and studying their antibacterial effects. Iran J Pharm Res 1(1):145–150

Tel G, Ozturk M, Duru ME, Turkoglu A (2015) Antioxidant and anticholinesterase activities of five wild mushroom species with total bioactive contents. Pharm Biol 53:824–830

Toshiro W, Ayako K, Satoshi I, Kumar MT, Shiro N, Keisuke T (2003) Antihypertensive effect of gamma-aminobutyric acid-enriched *Agaricus blazei* on mild hypertensive human subjects. J Jpn Soc Food Sci Technol 50:167–173

Turkoglu A, Kivrak I, Mercan N, Duru ME, Gezer K, Turkoglu H (2006) Antioxidant and antimicrobial activities of *Morchella conica* Pers. Afr J Biotechnol 5(11):1146–1150

Uddin GM, Hossain S, Monirul Islam M, Asaduzzaman M, Bulbul J, Amin R (2015) Evaluation of antimicrobial, antioxidant and cytotoxic property of *Pleurotus ostreatus* mushroom. Int Res J Biol Sci 4(1):29–33

Ul-Haq M, Rathod V, Singh D, Singh AK, Ninganagouda S, Hiremath J (2015) Dried Mushroom Agaricus bisporus mediated synthesis of silver nanoparticles from Bandipora District (Jammu and Kashmir) and their efficacy against Methicillin Resistant *Staphylococcus aureus* (MRSA). Nanosci Nanotechnol Int J 5(1):1–8

Valverde ME, Hernandez-Perez T, Paredes-Lopez O (2015) Edible mushrooms: improving human health and promoting quality life. Int J Microbiol:376387

Vamanu E (2012) Determination of antioxidant and antimicrobial properties of *Agaricus bisporus* from Romanian markets. Ovidius University Annals of Chemistry, Constanta, pp 47–52

Van Hoek AHAM, Mevius D, Guerra B, Mullany P, Roberts AP, Aarts HJM (2011) Acquired antibiotic resistance genes: an overview. Front Microbiol 28:203

Venturini ME, Rivere CS, Gonsalez C, Blanco D (2008) Antimicrobial activity of extracts of edible wild and cultivated mushrooms against foodborne bacterial strains. J Food Prot 71(8):1701–1706

Vieira V, Fernandes A, Barros L, Glamočlija J, Ćirić A, Stojković D, Martins A, Soković M, Ferreira ICFR (2015) Wild *Morchella conica* Pers. from different origins: a comparative study of nutritional and bioactive properties. J Sci Food Agr 96:90–98

Waithaka PN, Gathuru EM, Githaiga BM, Onkoba KM (2017) Antimicrobial activity of mushroom (*Agaricus bisporus*) and fungal (*Trametes gibbosa*) extracts from mushrooms and fungi of Egerton Main Campus. J Biomed Sci Eng 6:3

Wang HJ, Zakhari S, Jung MK (2010) Alcohol, inflammation, and gut-liver-brain interactions in tissue damage and disease development. World J Gastroenterol 16(11):1304–1313

Wang H, Fu Z, Han C (2013) The medicinal values of culinary-medicinal royal sun mushroom (*Agaricus blazei* Murrill). Evid Based Complement Alternat Med:842619

Wasser SP (2014) Medicinal mushroom science: current perspectives, advances, evidences, and challenges. Biomed J 37:345–346

Wasser SP, Weis AL (1999) Therapeutic effects of substances occurring in higher Basidiomycetes mushrooms: a modern perspective. Crit Rev Immunol 19:65–66

Wilson JW, Schurr MJ, LeBlanc CL, Ramamurthy R, Buchanan KL, Nickerson CA (2002) Mechanisms of bacterial pathogenicity. Postgrad Med J 78:216–224

Wolff ERS, Wisbeck E, Silveira MLL, Gern RMM, Pinho MSL, Furlan SA (2008) Antimicrobial and antineoplasic activity of *Pleurotus ostreatus*. Appl Biochem Biotechnol 151:402–412

Wu MF, Hsu YM, Tang MC, Chen HC, Chung JG, Lu HF, Lin JP, Tang NY, Yeh C, Yeh MY (2011) *Agaricus blazei* Murill extract abrogates CCl14-induced liver injury in rats. In Vivo 225(1):35–40

Xu G, Huang X, Qiu L, Wu J, Hu Y (2007) Mechanism study of chitosan on lipid metabolism in hyperlipidemic rats. Asia Pac J Clin Nutr 16(1):313–317

Yoon KN, Alam N, Lee JS, Cho HJ, Kim HY, Shim MJ, Lee MW, Lee TS (2011) Antihyperlipidemic effect of dietary *Lentinus edodes* on plasma, feces and hepatic tissues in hypercholesterolemic rats. Mycobiology 39:96–102

Yu J, Cui PJ, Zeng WL, Xie XL, Liang WJ, Lin GB, Zeng L (2009) Protective effect of selenium-polysaccharides from the mycelia of *Coprinus comatus* on alloxan-induced oxidative stress in mice. Food Chem 117:42–47

Zhang A, Sun H, Wang X (2013) Recent advances in natural products from plants for treatment of liver diseases. Eur J Med Chem 63:570–577

Zhao H, Lan Y, Liu H, Zhu Y, Liu W, Zhang J, et al. (2017) Aantioxidant and hepatoprotective activities of polysaccharides from spent mushroom substrates (*Laetiporus sulphureus*) in acute alcohol-induced mice. Oxi Med Cell Longev 5863523, 12

Zhu P (2009) The present status and prospects of medicinal fungal research and development in China. In: Proceeding of the 5th international medicinal mushroom conference, Nantong, China

Zhu HJ, Sheng K, Yan EF, Qiao JJ, Lv F (2012) Extraction, purification and antibacterial activities of a polysaccharide from spent mushroom substrate. Int J Biol Macromol 5(3):840–843

Chapter 5
Mushroom-Mediated Protection from Oxidative Damage to DNA

John A. Buswell

5.1 Introduction

Reactive molecules derived from molecular oxygen are constantly produced as part of normal cellular aerobic metabolism. These '**reactive oxygen species**' (ROS) exist in many forms including superoxide, hydrogen peroxide, hydroxyl radical, hydroxyl ion and nitric oxide, which have crucial and often specific functions in normal physiological processes. For example, they play important roles in host cell defence mechanisms, in gene expression and the activation of cell signalling cascades (Kehrer and Klotz 2015) and in apoptosis (Ray et al. 2012). However, due to the harmful effects of ROS, normal organism functioning requires that cellular antioxidant systems maintain equilibrium between ROS production and amelioration. Antioxidant defence mechanisms comprise enzymic systems such as **superoxide dismutase**, **catalase** and **glutathione peroxidase**, and low molecular weight cellular metabolites that include glutathione, thioredoxins, alpha-lipoic acid, ascorbic acid and vitamin E. When this equilibrium leans towards an excess production of ROS, the organism is subject to 'oxidative stress' resulting in damage to cellular lipids, proteins and DNA. These adverse effects of ROS are thought to be the cause of numerous medical ailments including diabetes, cancers, atherosclerosis, chronic inflammatory and autoimmune diseases, accelerated ageing, and age-related neuropathological conditions such as dementia, Alzheimers and Parkinson's disease (Halliwell and Gutteridge 2000; Kehrer and Klotz 2015).

In recent years, numerous wild and cultivated mushroom species, including many kinds of edible mushrooms that form part of the diet of populations worldwide, have been reported to exhibit varying degrees of antioxidant activity. Mushroom antioxidant components have been found in aqueous, ethanolic and methanolic extracts of fruit bodies, fungal mycelium and spent culture broth, and

J. A. Buswell (✉)
Institute of Edible Fungi, Shanghai Academy of Agricultural Sciences, Shanghai, China

B. P. Singh et al. (eds.), *Biology of Macrofungi*, Fungal Biology,
https://doi.org/10.1007/978-3-030-02622-6_5

include polysaccharides, tocopherols, phenolics, carotenoids, ergosterol and derivatives, flavonoids, selenium and ascorbic acid. A variety of *in vitro* assays for measuring total antioxidant activity have been adopted including **DPPH•** (2,2-diphenyl-1-picrylhydrazyl), ABTS• + (2,2-azobis-3-ethylbenzthiazoline-6-sulfonic acid), hydroxyl and superoxide radical scavenging activity, inhibition of lipid peroxidation, the **FRAP** (ferric reducing antioxidant power) method, and the inhibition of beta-carotene bleaching. Several comprehensive reviews describing these studies have appeared in the literature during the past decade (Ruiz-Rodriguez et al. 2009; Stajić et al. 2013; Kozarski et al. 2015; Sanchez 2017).

5.2 ROS-Mediated Damage to DNA and the 'Comet Assay'

Although low levels of ROS serve key functions in signal transduction and in regulating gene expression, higher concentrations are highly detrimental to DNA and other biological macromolecules (Ames et al. 1993). DNA damage stemming from excess production and accumulation of intracellular ROS and/or inadequate antioxidant defence mechanisms have been studied extensively given the link between such damage and the etiology of cancer and various age-related disorders (Halliwell and Gutteridge 2000). Model studies have revealed that exposure to different ROS gives rise to numerous types of base lesions and 2-deoxyribose modifications. Oxidative damage in the form of strand breaks, inter- and intra-strand cross links and DNA-protein cross links also adversely affect the DNA structure (Cadet and Wagner 2013).

A popular technique for evaluating oxidative **DNA damage** and amelioration *in vivo* and *ex vivo* at the level of the individual eukaryotic cell is the single cell gel electrophoresis assay (**SCGE**, often referred to as the Comet assay) (Wong et al. 2005). First described by Östling and Johanson (1984) and subsequently revised by Singh et al. (1988), the method is a simple and accurate procedure for determining strand breaks in the DNA of single cells embedded in low-melting-point agarose layered on to a microscope slide. Following cell lysis under neutral or alkaline (pH > 13) conditions and removal of the membrane, the resultant nucleoids are subjected to electrophoresis. As the DNA migrates towards the anode, strand breakage results in a comet-like feature that can be viewed using ethidium bromide staining and **fluorescent microscopy** (Fig. 5.1). The quantity of DNA in the comet tail relative to the head reflects the level of strand breakage and can be quantified using appropriate software.

Because basal levels of cellular DNA damage are normally too low to reveal any protective effects, cells are usually exposed to an oxidative damage-inducing agent (for example, hydrogen peroxide) either before or after treatment with the test compound depending on the nature of the experiment. Embedded cells can also be lysed before antioxidant pre-treatment or oxidative challenge using the lysed cell comet (LCC) assay, thereby eliminating the possibility of the cell membrane blocking access to the DNA and negating evaluation of any direct effects a putative protective

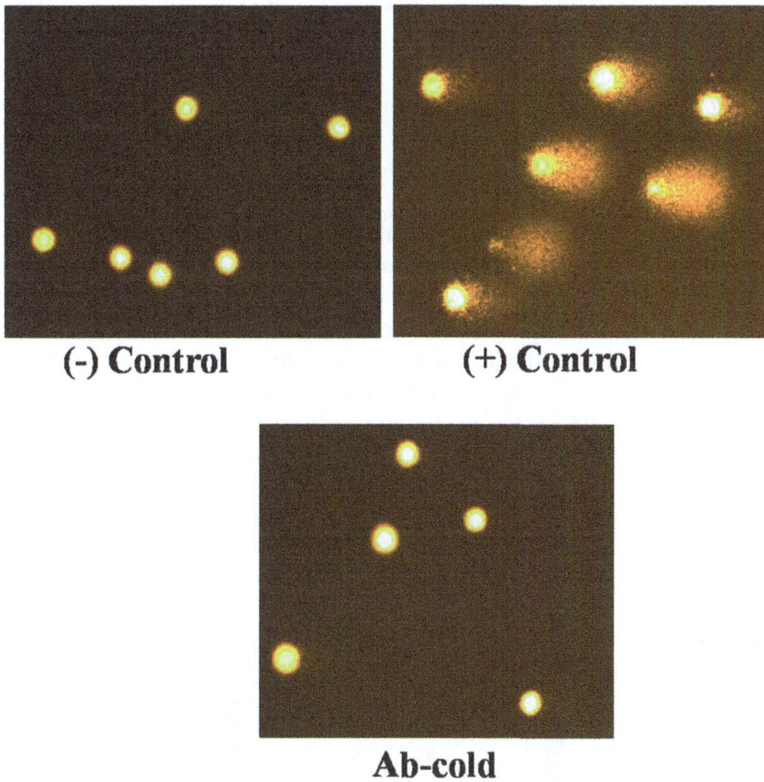

Fig. 5.1 Representative Comet images of Raji (human B-lymphocyte) cells
C (−): negative control (no H_2O_2 challenge); C (+): positive control (H_2O_2 challenge without pre-exposure to *A. bisporus* extract; pre-treated with 0.5 mg/mL*A. bisporus* extract

agent may exert. This procedure also excludes the involvement of cellular responses or adaptation (Szeto et al. 2002). The **Comet assay** is highly flexible and has been adopted to determine diverse forms and different levels of DNA damage using a wide range of cell types.

5.3 Mushroom-Derived Preparations in the Prevention of H_2O_2-induced Oxidative Damage to Cellular DNA

In an early report, hot (100 °C) and cold (20 °C) aqueous extracts of eight mushroom species (*Agaricus bisporus, Auricularia auricula, Flammulina velutipes, Ganoderma lucidum, Hypsizygus marmoreus, Lentinula edodes, Pleurotus sajor-caju* and *Volvariella volvacea*) were evaluated for their capacity to prevent oxidative damage to cellular DNA using cultured human B-lymphocyte cells (Raji) and the

Comet assay (Shi et al. 2001; Shi et al. 2002c) (Fig. 5.1). Significant genoprotection was afforded by cold (20 °C) and hot (100 °C) water extracts of *A. bisporus* and *G. lucidum* fruit bodies, respectively. Whereas no damage amelioration was recorded at 0.1 mg ml^{-1}concentrations or less, crude extracts of both mushrooms provided almost total protection at 0.5 mg ml^{-1}concentrations. This **genoprotective** effect was not the result of an endogenous catalase-like activity associated with the extracts.

Intraperitoneal administration of the *A. bisporus* extract also protected rat lymphocyte DNA against H_2O_2-induced damage in an *ex vivo* assay. No protection was recorded with extracts obtained from *F. velutipes, A. auricula, H. marmoreus, L. edodes, P. sajor-caju* and *V. volvacea.* (Fig. 5.2). There was no evidence of **cytotoxicity** when cells were treated with 1 mg ml^{-1} concentrations of the two DNA protective mushroom extracts for up to 24 h. More recently (Al-A'adhmi et al. 2013), aqueous extracts of *A. bisporus* fruit bodies and mycelium were shown to protect human lymphocyte DNA against the harmful effects of $\mathbf{H_2O_2}$. Cells were pretreated with different concentrations of the extracts for 1 h at 37 °C and then exposed to 100 μM H_2O_2 for five min. Evaluation of oxidative damage as the DNA tail moment

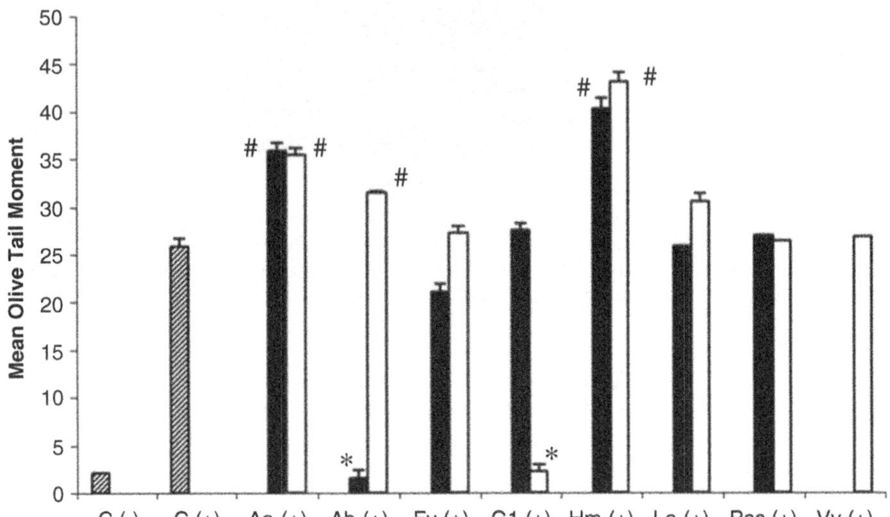

Fig. 5.2 Genoprotective effects of mushroom extracts against H_2O_2-induced damage to Raji (human B-lymphocyte) cells
Aa – *A. auricula*; Ab – *A. bisporus*; Fv – *F. velutipes*; Gl – *G. lucidum*; Hm – *H. marmoreus*; Le – *L. edodes*; Psc – *P. sajor-caju*; Vv – *V. volvacea*. Data are derived from two separate experiments and values are the mean ± SE of the Olive Tail Moment (200 cells). C(−): negative control (no H_2O_2 challenge); C(+): positive control (H_2O_2 challenge without pre-exposure to extract). Closed bars – pre-exposure to cold water extracts (0.5 mg ml^{-1} except Fv where 0.1 mg ml^{-1} was used); Open bars – pre-exposure to hot water extracts (0.5 mg ml^{-1}). *Significant ($P < 0.05$) genoprotective effects found compared with stressed cells without exposure to extract. #Significant ($P < 0.05$) increased damage to DNA compared with stressed cells without exposure to extract (Reproduced from Reference 15 with permission)

using the Comet assay revealed 30% and 38% decreases in **DNA fragmentation** following treatment with 500 μg/mL of fruit body and mycelial extract, respectively compared with the positive control (100 μmol H_2O_2 treatments).

The heat-stable protective activity associated with *G. lucidum* was not further delineated by Shi et al. (2002c). However, Kim and Kim (1999) had earlier reported that protection against oxidative damage to isolated DNA induced by metal-catalyzed Fenton reactions was linked to a water-soluble polysaccharide released from the mushroom fruit body by hot water extraction. More recently, an ethanolic extract of *G. lucidum* (EGL) was shown to protect the DNA of mouse myoblast C_2C_{12} cells against H_2O_2-induced **oxidative damage** (Lee et al. 2016b). Cells were treated with 50 μg/mL EGL for 1 h followed by incubation with and without 1.0 mM H_2O_2 for 24 h. H_2O_2-mediated DNA damage in the C_2C_{12} cells was evaluated using the alkaline Comet assay and Western blot analysis. Exposure to H_2O_2 alone greatly increased the number of DNA breaks, whereas EGL pretreatment afforded approximately 90% protection. Furthermore, compared to untreated controls, C_2C_{12} fibroblasts exposed to 1.0 mM H_2O_2 for 30 min showed markedly increased intracellular ROS levels. The increases were not recorded in EGL-treated cells indicating that the mushroom extract possesses ROS scavenging activity. Moreover, the levels of phosphorylation of nuclear histone H2AX at serine 139, a sensitive marker of DNA double-strand break formation (Bonner et al. 2008), were significantly higher in H_2O_2-treated C_2C_{12} fibroblasts compared to EGL-pretreated cells.

Aqueous extracts of the edible mushrooms *Agrocybe cylindracea* (ACE) and *Lentinus edodes* (LEE) inhibited oxidative damage to the DNA of pUC18 plasmids and HepG2 cells induced by •OH radicals generated from Cu (II) and H_2O_2 (Wang et al. 2004). Levels of DNA damage were determined by measuring the decrease in super coiled DNA and DNA migration after oxidative attack using the Comet assay. The latter revealed that ACE at concentrations of 50, 100 and 200 μg/mL provided 29%, 51% and 67% protection, respectively against cupric-mediated •OH formation. Corresponding values for LEE were 29%, 41% and 47%. Based on the IC_{50} values, the protective effect of ACE was 2.3-fold higher compared to LEE. It was suggested that protein components of ACE contributed to associated free radical scavenging activity. Data demonstrating that a 10^7 KDa protein dimer extracted from *L. edodes* fruit bodies exhibited scavenging activity and protective effects against DNA damage induced by ROS (Wang et al. 2004). A separate study showed that extracts of both unprocessed and sun-dried *L. edodes* fruit bodies prepared using acetone, ethanol, methanol and hot water also protected against H_2O_2-induced damage to human leucocyte DNA (Kim et al. 2012).

Park et al. (2005) used the Comet assay to test the effect of ethanolic extracts of powdered *Inonotus obliquus* (Chaga mushroom) fruit bodies on **oxidative DNA** damage in human lymphocytes taken from two male volunteers. Increased oxidative DNA damage (quantified as % tail DNA) induced by exposure to 200 μM H_2O_2 was significantly inhibited by pretreatment with 6.25, 12.5, 25, 50 and 100 μg/mL of *I. obliquus* extracts by 46, 52, 69, 70 and 80%, respectively. Similar levels of inhibition were recorded when lymphocytes were treated with *I. obliquus* extracts for 30 min following H_2O_2 challenge or when incubated with mushroom extracts and

H_2O_2 simultaneously. Extracts had negligible effects (<5%) on cell viability. Extracts of *I. obliquus* prepared using pepsin also protected the DNA of PC-12 cells from H_2O_2-induced damage (Kim et al. 2011).

Some level of protection from MMS-induced DNA damage was observed when Chinese hamster V79 cells were treated with 'teas' prepared by aqueous extraction of powdered *Agaricus blazei* fruit bodies at 2–8 °C for 1 h, 20–25 °C for 2 h, and 60 °C for five min (Menoli et al. 2001). Chloroform:methanol (3:1) extracts of fruit bodies of six strains of *A. blazei* in various combinations were also assessed for ability to protect against methyl methane sulfonate (MMS)-induced damage to the DNA of V79 cells (Chinese hamster lung cells) (Guterres et al. 2005). The cells were exposed to three different extract concentrations prior to, concurrent with, and following MMS challenge. Decreases in DNA damage (< 50%) were recorded with all the extracts although, in some cases, they were not statistically significant. None of the extracts exhibited genotoxicity under the test conditions. In a more recent study, dried and powdered mycelium of *A. blazei* was tested for its ability to protect human peripheral blood cells (collected from three healthy female participants aged between 20 and 35 years) against 50 μM H_2O_2-induced DNA damage (Živković et al. 2017). **Antigenotoxic** effects were determined following exposure of the blood cells to the mushroom both before and after H_2O_2 challenge. Pretreatment with 500 μg/mL of *A. blazei* resulted in a significant reduction (~45%) in H_2O_2-induced DNA damage. However, a 1000 μg/mL concentration was less effective although the mushroom had no endogenous genotoxic effect at any of the tested concentrations. Conversely, treatment of blood cells with 250, 500 and 1000 μg/mL mushroom material post-oxidative challenge all resulted in fewer cells exhibiting DNA damage ($p < 0.05$).

Clitocybin A (4,6-dihydroxy-2-*p*-hydroxyphenyl-isoindol-1-one), purified from spent culture broth of *Clitocybe aurantiaca* using solvent extraction, silica gel column chromatography, Sephadex LH-20 column chromatography and preparative **HPLC**, exhibited strong free radical scavenging activity against superoxide, ABTS, and DPPH radicals. The antioxidant at 10 mg/mL concentrations also inhibited (~42%) oxidative damage to the DNA of Chinese hamster lung fibroblast cells (V79–4) induced by exposure to 1.0 mM H_2O_2 (Kim et al. 2008).

Park et al. (2009) used a DNA fragmentation assay to demonstrate that hot (100 °C) aqueous extracts of *Phellinus linteus* (PLE) effectively inhibited the formation of strand breaks in *Escherichia coli* ColE1 plasmid DNA induced by hydroxyl radicals produced by exposure to H_2O_2 and $FeCl_2$. The Comet assay was used to show that PLE protected the DNA of rat hepatocytes against H_2O_2-induced damage in a dose-dependent manner. PLE alone did not damage hepatocyte nuclear DNA, whereas a reaction mixture of PLE and red ginseng extract exhibited a synergistic protective effect against DNA damage (Park et al. 2009).

The DNA protective capacity of a *Coriolus versicolor* extract (Chinese Medicine Council of Hong Kong Transitional Registration Number HKP-01010) was determined using both the **standard alkaline comet** (SAC) and the **lysed cell comet** (LCC) assays in parallel (Szeto et al. 2013). Under the former protocol, lymphocytes harvested from three healthy subjects (two males and one female, aged 35–54) were incubated with five concentrations (10^1–10^5 μg/L) of mushroom extract for 30 min at 37 °C before challenge with 45 μM H_2O_2 for 5 min. In the LCC assay, lymphocytes were lysed prior to pretreatment with extract and H_2O_2 challenge. U-shaped dose–responses were recorded in both cases with a 25% decrease in oxidative damage observed with 10^4 μg/L extract in the SAC assay. A genoprotective effect was also observed in the LCC assay but was not statistically significant (Szeto et al. 2013).

5.4 Mechanisms Operative in Mushroom-Mediated Inhibition of H_2O_2-induced Oxidative Damage to Cellular DNA

Subsequent research (Shi et al. 2002b) demonstrated that a heat-labile protein present in the mushroom fruit body, identified as tyrosinase, was responsible for the genoprotective effect of *A. bisporus* (Fig. 5.3). Genoprotection required enzyme-catalysed hydroxylation of tyrosine to L-DOPA and further oxidation of this metabolite to dopaquinone. This was confirmed when addition of increasing concentrations of tropolone, a tyrosinase inhibitor, to the test system was shown to cause parallel decreases in genoprotection (Shi et al. 2002b).

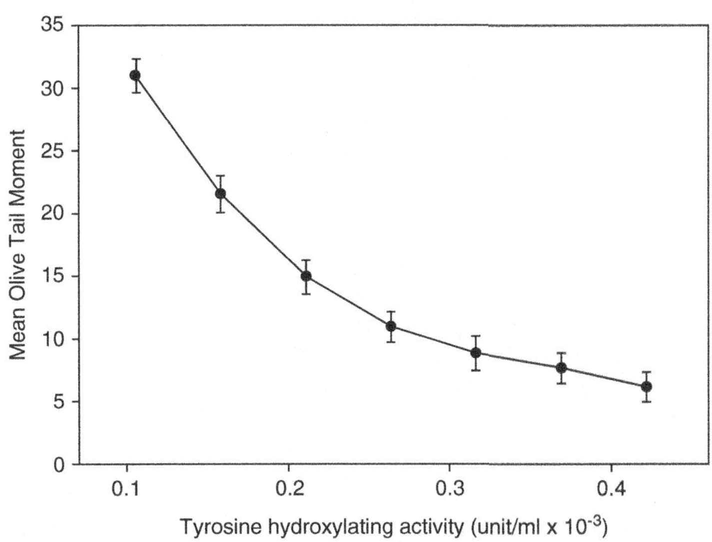

Fig. 5.3 Relationship between genoprotective effect and tyrosine hydroxylating activity
DNA damage is expressed as the Mean Olive Tail Moment (integrated value of DNA density of the comet tail multiplied by the migration distance) ± SE (indicated by error bars) obtained from two replicate experiments (50 cells). Tyrosine hydroxylating activity was varied byincorporating into the assay different quantities of enzyme fraction isolated from *A. bisporus* fruit bodies. (Reproduced from Shi et al. 2002b with permission)

The genoprotective effect of the L-DOPA-tyrosinase system was also conditional on the time allowed for the enzymic reaction to proceed before oxidative challenge (Fig. 5.4). Little genoprotection was recorded when Raji cells were pre-treated with 40 μM L-DOPA and tyrosinase for time periods of up to 30 min before exposure to H_2O_2. However, 40–120 min pre-treatment periods resulted in progressively increasing degrees of genoprotection (Fig. 5.4) (Shi et al. 2002a). Although approximately 50% more oxidation products were generated in the former case, significant genoprotection was evident only on the latter occasion. This delayed response led the authors to conclude that L-DOPA oxidation products (and/or cellular conversion products) were triggering cellular processes which, in turn, were directly responsible for the genoprotective effect. Low concentrations of L-DOPA enhanced the antioxidant status and ROS scavenging capacity of brain cells by increasing intracellular concentrations of the antioxidant glutathione (Mytilineou et al. 1993; Han et al. 1996), and L-DOPA auto-oxidation products enhanced the synthesis of cellular GSH by up-regulating the expression of γ-glutamylcysteine synthetase and γ-glutamyl transpeptidase, enzymes which play important roles in GSH synthesis (Shi et al. 1994). However, no significant fluctuations in GSH levels or catalase activity were detected in Raji cells following pre-treatment with the tyrosinase-tyrosine-L-DOPA system.

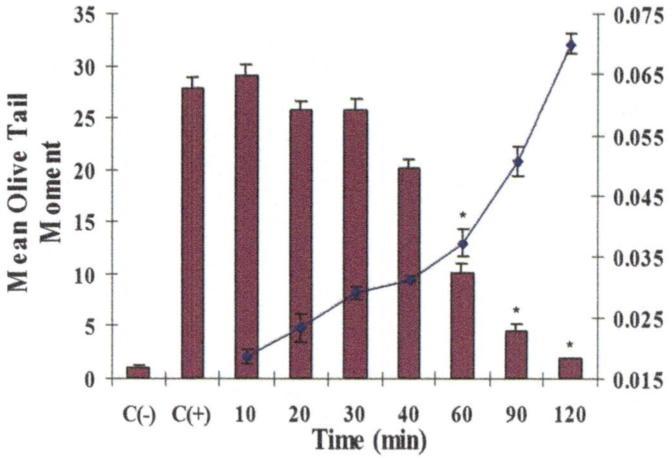

Fig. 5.4 Effect of different exposure times on genoprotective activity of tyrosinase-generated L-DOPA oxidation products
Raji cells were suspended for different times in reaction mixtures containing *A. bisporus* tyrosinase (5.2 mUnits), 40 μM L-DOPA and 10% v/v FBS, prior to oxidative challenge with H_2O_2. DNA damage is expressed as the mean Tail DNA Content ± standard deviation of data from 2–4 slides (100–200 cells). Error bars for dopaquinone hydrazone formation represent the standard deviation of three to six replicates. C (−) negative control (no H_2O_2 challenge); C(+) positive control (no pre-treatment prior to H_2O_2 challenge). *Differences in Tail DNA Content values are significant at $p < 0.01$ compared with the C(+) value using Dunnett's test. (Reproduced from Shi et al. 2002a with permission)

Lee et al. (2016b) undertook a detailed study of the mechanisms underlying the protective effect of an ethanolic extract from *G. lucidum* (EGL) on H_2O_2-induced DNA damage. Protection against oxidative stress by *G. lucidum* is reported to involve, at least in part, the activation of the nuclear factor erythroid 2-related factor 2 (Nrf2), mediated by the expression of heme oxygenase-1 (HO-1) (Lee et al. 2016b). Nrf2, encoded by the *NFE2L2* gene, is a transcription factor that regulates the cellular anti-oxidant response through the expression of antioxidant proteins (Ishii et al. 2000; Kobayashi and Yamamoto 2005). Under oxidative stress conditions, Nrf2 migrates from the cytoplasm, where it is bound to Kelch-like ECH-associated protein 1 (Keap1), to the nucleus. In the nucleus, Nrf2 binds with and activates the anti-oxidant response element (ARE) and initiates transcription of **antioxidative genes** and their proteins such as quinone reductase, glutathione S-transferase and glutathione reductase (Ishii et al. 2000; Kobayashi and Yamamoto 2005). Treatment of C2C12 cells with EGL induced the expression of the Nrf2 protein in a concentration- and time-dependent manner. EGL treatment also caused concentration- and time-dependent increases in phosphorylated Nrf2expression levels and Nrf2 translocation from the cytosol to the nucleus. Phosphorylation of Nrf2 at Ser40 is essential for its stabilization and nuclear translocation (Ishii et al. 2000; Pi et al. 2007; Apopa et al. 2008). In turn, activation of Nrf2 signalling triggered the expression of the HO-1 protein but had no effect on other anti-oxidant enzymes such as nicotinamide adenine dinucleotide phosphate (**NADPH**)-quinone oxidoreductase-1 and thioredoxin reductase-1. HO-1 protects cells exposed to oxidizing agents by catalyzing the oxidation of heme to biologically active products (Scapagnini et al. 2004) that neutralize intracellular ROS (Katori et al. 2002; Motterlini and Foresti 2014). ZnPP, a selective inhibitor of HO-1, significantly reversed the inhibition of ROS generation by EGL in the H_2O_2-stimulated C2C12 cells. Furthermore, silencing of Nrf2 using specific **siRNA** resulted in greatly attenuated levels of EGL-induced expression of Nrf2 and up-regulation of HO-1 compared with those recorded in untransfected cells and siRNA-transfected controls.

5.5 Protection from Oxidative DNA Damage - Dietary Impact

Wild and cultivated edible mushrooms are important dietary components, especially in Asian countries such as China, Japan and South Korea where this special group of fungi has long been recognised as possessing a wide range of medicinal properties. Modern scientific techniques have identified numerous bioactive mushroom components which have been variously reported to exhibit a range of medicinal attributes including anti-cancer, anti-tumour, anti-viral, immunomodulatory, hypocholesterolaemic and hepatoprotective activities (Chang and Wasser 2012). In a recent study on the antioxidant effects of components of the Korean diet, protection against DNA damage in human lymphocytes, assessed using the Comet assay, was highest in mushrooms, followed by vegetables, fruits, seaweeds and kimchi (Lee et al. 2016a). A protective role against DNA oxidative damage for mushrooms consumed as part of the diet was also implied by the results of a comparative study of Korean and American diets (Lee et al. 2016a). This revealed that the Korean diet afforded significantly ($P < 0.01$) greater protection against *ex vivo* lymphocyte DNA damage compared with the latter.

5.6 Closing Comments

Approximately 98,000 fungal species have so far been described (Kirk et al. 2008). Of these, about 64,163 species representing 6355 genera are members of the Ascomycota, while 31,515 species belonging to 1589 genera are assigned to the Basidiomycota (Kirk et al. 2008) the two phyla into which mushrooms are classified. Over 70% of the latter species produce fruiting bodies of sufficient size and suitable structure to be considered as mushrooms according to the definition given by Chang and Miles (1992). Kozarski et al. (2015) listed over 100 species of mushrooms shown to exhibit antioxidant properties as determined by a variety of *in vitro* assays for measuring total antioxidant activity. However, relatively few species that have been shown to exert protection against oxidative damage to macromolecular components (nucleic acids, proteins, lipids) of the cell (Table 5.1). Where such protection has been demonstrated, little is known about the mechanisms involved.

In view of the links between oxidative stress and numerous medical ailments, particularly those associated with a growing ageing population, greater research emphasis on investigating the potential of mushrooms to ameliorate these adverse effects should be highly rewarding.

Table 5.1 Mushroom-based protection against oxidative damage to DNA

Mushroom	Sample	Cell line	Oxidant	Max protection	References
Agaricus bisporus	Cold H_2O extract (FB)[a]	Human B-lymphocytes	H_2O_2	>90% (0.5 mg/mL)	Shi et al. (2002a, b, c)
Agaricus bisporus	H_2O extract (FB)	Human lymphocytes	H_2O_2	38% (500 μg/mL)	Al-A'adhmi et al. (2013)
Agaricus bisporus	H_2O extract (M)[b]	Human lymphocytes	H_2O_2	30% (500 μg/mL)	Al-A'adhmi et al. (2013)
Agaricus blazei	$CHCl_3$:MeOH (3:1) (FB)	Chinese hamster lung cells	MMS[c]	<50%	Guterres et al. (2005)
Agaricus blazei	Mycelium	Human peripheral blood cells	H_2O_2(PoT) #	~57% (250 μg/mL)	Živkovic̆ et al. (2017)
Agaricus blazei	Mycelium	Human peripheral blood cells	H_2O_2(PrT) ##	~45% (500 μg/mL)	Živkovic̆ et al. (2017)
Agrocybe cylindracea	H_2O extract (FB)	HepG2 (human liver cancer cells)	cupric-mediated •OH	67% (200 μg/mL)	Wang et al. (2004)

continued

Table 5.1 (continued)

Mushroom	Sample	Cell line	Oxidant	Max protection	References
Clitocybe aurantiaca	Clytocybin	Chinese hamster lung fibroblasts	H_2O_2	42% (10 μg/mL)	Kim et al. (2008)
Coriolus versicolor	Commercial extract	Human lymphocytes	H_2O_2	25% (10 μg/mL)	Szeto et al. (2013)
Ganoderma lucidum	H_2O (100 °C) extract (FB)	Human B-lymphocytes	H_2O_2	90% (0.5 mg/mL)	Shi et al. (2002c)
Ganoderma lucidum	EtOH (25%) extract (FB)	C2C12 (mouse myoblast cells)	H_2O_2	90% (50 μg/mL)	Lee et al. (2016b)
Inonotus obliquus	EtOH extract (FB)	Human lymphocytes	H_2O_2	100% (100 μg/mL)	Park et al. (2005)
Lentinus edodes	H_2O extract (FB)	HepG2 (human liver cancer cells)	cupric-mediated •OH	47% (200 μg/mL)	Wang et al. (2004)
Phellinus linteus	H_2O (100 °C) extract (FB)	Rat hepatocytes	H_2O_2	100% (100 μg/mL)	Park et al. (2009)

[a]*FB* fruit body, [b]*M* Mycelium, [c]*MMS* Methyl methanesulfonate, # *PoT* post-treatment, ## *PrT* Pre-treatment

References

Al-A'Adhmi MA, Al-Awadi BQ, Abass AAA (2013) White mushroom *Agaricus bisporus* extract inhibits oxidative DNA damage in human lymphocytes as assessed by comet assay. J Gen Environ Res Conserv 1:153–157

Ames BN, Shigenaga MK, Hagen TM (1993) Oxidants, antioxidants, and the degenerative diseases of aging. Proc Natl Acad Sci USA 90:7915–7922

Apopa PL, He X, Ma Q (2008) Phosphorylation of Nrf2 in the transcription activation domain by casein kinase 2 (CK2) is critical for the nuclear translocation and transcription activation function of Nrf2 in IMR-32 neuroblastoma cells. J Biochem Mol Toxicol 22:63–76

Bonner WM, Redo CE, Dickey JS, Nakamura AJ, Sedilnikova OA, Solier S, Pommier Y (2008) γH2AX and cancer. Nat Rev Cancer 8:957–967

Cadet J, Wagner JR (2013) DNA base damage by reactive oxygen species, oxidizing agents, and UV radiation. Cold Spring Harb Perspect Biol 5:a012559

Chang ST, Miles PG (1992) Mushroom biology: a new discipline. Mycologist 6:64–65

Chang ST, Wasser SP (2012) The role of culinary-medicinal mushrooms on human welfare with a pyramid model for human health. Int J Med Mushr 14:95–134

Guterres ZR, Mantovani MS, da Eira AF, Ribeiro LR, Jordão BQ (2005) Genotoxic and antigenotoxic effects of organic extracts of mushroom *Agaricus blazei* Murrill on V79 cells. Genet Mol Biol 28:458–463

Halliwell B, Gutteridge JMC (2000) Free radicals in biology and medicine. Oxford University Press, Oxford

Han SK, Mytilineou C, Cohen G (1996) L-DOPA up-regulates glutathione and protects mesencephalic cultures against oxidative stress. J Neurochem 66:501–510

Ishii T, Itoh K, Takahashi S, Sato H, Yanagawa T, Katoh Y, Bannai S, Yamamoto M (2000) Transcription factor Nrf2 coordinately regulates a group of oxidative stress-inducible genes in macrophages. J Biol Chem 275:16023–16029

Katori M, Anselmo DM, Busuttil RW, Kupiec-Weglinski JW (2002) A novel strategy against ischemia and reperfusion injury: cytoprotection with hemeoxygenase system. Transpl Immunol 9:227–233

Kehrer JP, Klotz L-O (2015) Free radicals and related reactive species as mediators of tissue injury and disease: implications for health. Crit Rev Toxicol 45:765–798

Kim KC, Kim IG (1999) *Ganoderma lucidum* extract protects DNA from strand breakage caused by hydroxyl radical and UV irradiation. Int J Mol Med 4:273–277

Kim Y-H, Cho S-M, Hyun J-W, Ryoo I-J, Choo S-J, Lee S, Seok S-J, Hwang JS, Sohn ED, Yun B-S, Bae KH, Yoo I-D (2008) A new antioxidant, clitocybin A, from the culture broth of *Clitocybe aurantiaca*. J Antibiot 61:573–576

Kim E-K, Lee S-J, Hwang J-W, Kim CG, Choi DK, Lim BO, Kang H, Moon S-H, Jeon B-T, Park P-J (2011) *In vitro* investigation on antioxidative effect of *Inonotus obliquus* extracts against oxidative stress on PC12 cells. J Korean Soc Appl Biol Chem 54:112–117

Kim MJ, Chu WM, Park EJ (2012) Antioxidant and antigenotoxic effects of Shiitake mushrooms affected by different drying methods. J Korean Soc Food Sci Nutr 41:1041–1048

Kirk PM, Cannon PF, Minter DW, Stalpers JA (2008) Dictionary of the Fungi, 10th edn. CABI, Wallingford

Kobayashi M, Yamamoto M (2005) Molecular mechanisms activating the Nrf2-Keap1 pathway of antioxidant gene regulation. Antioxid Redox Signal 7:385–394

Kozarski M, Klaus A, Jakovljevic D, Todorovic N, Vunduk J, Petrović P, Niksic M, Vrvic MM, van Griensven L (2015) Antioxidants of edible mushrooms. Molecules 20:19489–19525

Lee MY, Han J-H, Kang M-H (2016a) Protective effect of Korean diet food groups on lymphocyte DNA damage and contribution of each food group to total dietary antioxidant capacity (TDAC). J Nutr Health 49:277–287 (In Korean with English abstract)

Lee Y-H, Kim J-H, Song C-H, Jang K-J, Kim C-H, Kang J-S, Choi Y-H, Yoon H-M (2016b) Ethanol extract of *Ganoderm alucidum* augments cellular anti-oxidant defense through activation of Nrf2/HO-1. J Pharmacopuncture 19:59–69

Menoli RCRN, Mantovani MS, Ribeiro LR, Speit G, Jordão BQ (2001) Antimutagenic effects of the mushroom *Agaricus blazei* Murrill extracts on V79 cells. Mutat Res 496:5–13

Motterlini R, Foresti R (2014) Heme oxygenase-1 as a target for drug discovery. Antioxid Redox Signal 20:1810–1826

Mytilineou C, Han SK, Cohen G (1993) Toxic and protective effects of L-DOPA on mesencephalic cell cultures. J Neurochem 61:1470–1478

Ostling O, Johanson KJ (1984) Microelectrophoretic study of radiation-induced DNA damages in individual mammalian cells. Biochem Biophys Res Commun 123:291–298

Park E, Jeon K-I, Byun B-H (2005) Ethanol extract of *Inonotus obliquus* shows antigenotoxic effect on hydrogen peroxide induced DNA damage in human lymphocytes. Cancer Prev Res 10:54–59

Park BJ, Lim YS, Lee HJ, Eum WS, Park J, Han KH, Choi SY, Lee KS (2009) Anti-oxidative effects of *Phellinus linteus* and red ginseng extracts on oxidative stress-induced DNA damage. BMB Rep 42:500–505

Pi J, Bai Y, Reece JM, Williams J, Liu D, Freeman ML, Fahl WE, Shugar D, Liu J, Qu W, Collins S, Waalkes MP (2007) Molecular mechanism of human Nrf2 activation and degradation: role of sequential phosphorylation by protein kinase CK2. Free Radic Biol Med 42:1797–1806

Ray PD, Huang B-W, Tsuji Y (2012) Reactive Oxygen Species (ROS) homeostasis and redox regulation in cellular signaling. Cell Signal 24:981–990

Ruiz-Rodriguez A, Santoyo S, Soler-Rivas C (2009) Antioxidant properties of edible mushrooms. Funct Plant Sci Biotechnol 3:92–102

Sanchez C (2017) Reactive oxygen species and antioxidant properties from mushrooms. Syn Sys Biotechnol 2:13–22

Scapagnini G, Butterfield DA, Colombrita C, Sultana R, Pascale A, Calabrese V (2004) Ethyl ferulate, a lipophilic polyphenol, induces HO-1 and protects rat neurons against oxidative stress. Antioxid Redox Signal 6:811–818

Shi MM, Iwamoto T, Forman HJ (1994) γ-Glutamyl cysteine synthetase and GSH increase in quinone-induced oxidative stress in BPAEC. Amer J Physiol 267:414–421

Shi YL, James AE, Benzie IFF, Buswell JA (2001) Prevention of oxidative damage to cellular DNA by mushroom-derived components. Int J Med Mushrooms 3:100

Shi Y-L, Benzie IFF, Buswell JA (2002a) L-DOPA oxidation products prevent H_2O_2-induced oxidative damage to cellular DNA. Life Sci 71:3047–3057
Shi Y-L, Benzie IFF, Buswell JA (2002b) Role of tyrosinase in the genoprotective effect of the edible mushroom, *Agaricus bisporus*. Life Sci 70:1595–1608
Shi YL, James AE, Benzie IFF, Buswell JA (2002c) Mushroom-derived preparations in the prevention of H_2O_2-induced oxidative damage to cellular DNA. Teratog Carcinog Mutagen 22:103–111
Singh NP, McCoy MT, Tice RR, Schneider EL (1988) A simple technique for quantitation of low levels of DNA damage in individual cells. Exp Cell Res 175:184–191
Stajić M, Vukojević J, Knežević A, Laušević SD, Milovanović I (2013) Antioxidant protective effects of mushroom metabolites. Curr Top Med Chem 13:2660–2676
Szeto YT, Collins AR, Benzie IFF (2002) Effects of dietary antioxidants on DNA damage in lysed cells using a modified comet assay procedure. Mutat Res 500:31–38
Szeto YT, Lau PC, Kalle W, Pak SC (2013) Direct human DNA protection by *Coriolus versicolor* (Yunzhi) extract. Pharm Biol 51:851–855
Wang LH, Tsai SJ, Lin SY (2004) Aqueous extract from Taiwanese *Agrocybe cylindracea* strain B protects DNA against •OH-mediated strand breaks. J Food Drug Anal 12:277–285
Wong VWC, Szeto YT, Collins AR, Benzie IF (2005) The comet assay: a biomonitoring tool for nutraceutical research. Curr Top Nutraceutical Res 3:1–14
Živkovic̈ L, Borozan S, Čabarkapa A, Topalovic̈ D, Ciptasari U, Bajic̈ V, Spremo-Potparevic̈ B (2017) Antigenotoxic properties of *Agaricus blazei* against hydrogen peroxide in human peripheral blood cells. Oxidative Med Cell Longev 2017:1–9. https://doi.org/10.1155/2017/8759764

Chapter 6
Chemical and Bioactive Profiling of Wild Edible Mushrooms

Katarzyna Sułkowska-Ziaja, Katarzyna Kała, Jan Lazur, and Bożena Muszyńska

6.1 Introduction

6.1.1 Regulation of Secondary Metabolites Production in Fungi

Metabolism can be divided into two separate levels, basic metabolism and secondary metabolism. Basic metabolism is universal and applies to all species, because it is associated with substances that enable the implementation of basic life functions: carbohydrates, fats, proteins, nucleic acids-compounds that allow division, growth, breathing and reproduction, or necessary life processes. Secondary metabolism leads to the formation of compounds that may be useful for survival in adverse environmental conditions such as, for example, intense **UV radiation** or other (Brakhage and Schroeckh 2011). Mushrooms produce a number of secondary metabolites that play an important role in cellular processes such as transcription or cell-to-cell communication. Many of them have now been used as medicines; inter alia, **antibiotics** and immune suppressants. Examples are *Penicillum chrysogenum* used for the production of penicillin or immunosuppressive drugs, and lovastatin produced by *Tolypocladium inflatum* and *Aspergillus terreus*. Other examples are ergotamine synthesized by the genus *Claviceps* and gibberellins–plant growth hormones produced by *Fusarium fujikuroi* (Brakhage 2013).

Biotechnological methods, including mycelial cultures, are increasingly used in the studies of **secondary metabolites** of mushroom origin. They are a convenient alternative in acquiring biomass compared to sometimes hard-to-find material from the natural environment.

K. Sułkowska-Ziaja · K. Kała · J. Lazur · B. Muszyńska (✉)
Department of Pharmaceutical Botany, Faculty of Pharmacy, Jagiellonian University Medical College, Kraków, Poland

B. P. Singh et al. (eds.), *Biology of Macrofungi*, Fungal Biology,
https://doi.org/10.1007/978-3-030-02622-6_6

The intensity with which they are produced, and consequently accumulated secondary metabolites in *in vitro* cultures, can be regulated, inter alia, with the substrate composition. The composition, which is optimal for the growth of the *in vitro* culture, does not always turn out to be suitable for the synthesis of secondary metabolites. For this reason, the optimization of the composition is usually performed separately in the event of the growth of culture and the synthesis of metabolites (Fox and Howlett 2008).

In numerous experiments, the accumulation of secondary metabolites may be affected by the concentration of sugars in the medium. Increased concentration of this group of compounds causes osmotic stress in the cells, which may result in increased production of metabolites (Brakhage 2013).

Another factor that affects the accumulation of metabolites is the total nitrogen content in the culture medium. This element in the medium is present in the form of $NO^{3-}$and NH^{4+} ions and the selection of their respective concentrations may have a positive effect on the number of secondary metabolites accumulated. Phosphorus is another important element important during the accumulation of metabolites. High concentration of phosphate ions favors an increase of **biomass**, but in many cases limits the accumulation of secondary metabolites. An effective method of increasing the production of secondary metabolites is the addition of precursors and elicitation.

Biosynthesis pathways of secondary metabolites are usually multi-step, and the individual steps run at different efficiencies. Increasing the synthesis of a given compound in *in-vitro* cultures can be achieved by the exogenous addition of precursors. According to this theory, all intermediate metabolite compounds found at the beginning of each pathway enables the biosynthesis of its products to be increased. The addition of a precursor to the culture medium often increases the biosynthesis of metabolites, using the feedback or shortening of the metabolic pathway. The given precursor must be a factor limiting the efficiency of the process, only then will it cause an increase in biosynthesis. An ideal precursor compound should be inexpensive and easily available to make its use cost-effective. Precursors that are often used to increase the synthesis of secondary metabolites are the amino acids from which tropane and indole alkaloids are synthesized. In the case of mycelial cultures, commonly used precursors of indole compounds are anthranilic acid and serine (Opoka et al. 2017).

The content of phenolic compounds, including phenolic acids, can be regulated by the addition of phenylalanine or tyrosinase.

The exact mechanism of elicitor's action has not yet been known. It is assumed that the activation of secondary metabolic pathways allows the body to survive under stressful conditions. Elicitors may, for example, be responsible for the induction of genes associated with synthetic pathways and the accumulation of secondary metabolites. Elicitors increase the production of secondary metabolites by using one or several mechanisms such as modulation of synthesis pathways, accumulation of compounds or limitation of their degradation rate (Palazón et al. 2003).

An example of an elicitor used in mycelial cultures is methyl jasmonate. It has been tested, among others, for the production of aflatoxin b-a carcinogenic metabolite synthesized by some *Aspergillus* species. Concentration of methyl jasmonate had no significant effect on mycelial growth; however, it significantly influenced the increase in aflatoxin synthesis (Meimaroglou et al. 2009).

6.1.2 Main Groups of Bioactive Substances Which Are Occurring in Edible Mushroom

6.1.2.1 Carbohydrates

Trehalose is a disaccharide which is a reserve material in mushrooms and also has properties that protect cells against protein denaturation. In a subarachnoid hemorrhage model using RAW264.7 macrophages and vascular endothelial cells (HUVEC), it was shown that trehalose can inhibit the expression of **proinflammatory** proteins i.e., cyclooxygenase-2, inducible nitric oxide synthase iNOS, and inhibits IκB-α subunit degradation, the NF-κB nuclear transcription factor inhibitor (Echigo et al. 2012). *Agaricus bisporus* dried fruiting bodies contain about 1–3% of trehalose (Wannet et al. 1998).

6.1.2.2 Polysaccharides

Polysaccharides contained in the cell wall of mushrooms hypha, which include glucans, chitin and chitosans, are particularly valuable for the improvement of immunity and modulation of the defense response of the human body (Wu et al. 2004; Kozarski et al. 2014).

β-Glucans, due to their broad spectrum of activity in the immune system, are called biological response modifiers (Novak and Vetvicka 2009). The effect of β-glucans on pro-and anti-inflammatory cytokines has been demonstrated. One of the mechanisms of β-glucans action is binding to pattern recognition receptors (PRR) of immune cells as pathogen-associated molecular patterns (Muta 2006). Such receptors include dectin-1, complementary receptor 3 (CR3, CD11b/CD18), or TLRs (toll-like receptors). In this way, β-glucans activate cell proliferation and maturation of the immune system; stimulate the activation of macrophages and NK cells (Akramiene et al. 2007).

Anti-inflammatory effects are attributed to glucans obtained from *Inonotus obliquus* which *in vitro* demonstrated inhibitory activity of the NF-κB, COX-2 and iNOS signaling pathway in RAW 264.7 cells (Ma et al. 2013). One of the better-known, derived from *A. bisporus*, (1 → 6)-β-D-glucan with a linear structure, showed repressive action of pro-inflammatory genes in *in vitro* studies using the THP-1 cell line, and also inhibited inflammatory reaction induced by **lipopolysaccharide** (LPS), by repressing the expression of COX-2 and interleukin 1 (IL-1) proteins (Smiderle et al. 2013).

Edible mushrooms can be used in the treatment of cancer or as a support for conventional therapy, and even as agents that combat the side effects of cancer therapy (Patel and Goyal 2012; Zong et al. 2012). This has been confirmed in clinical studies in recent years (Kosanić et al. 2016; Meng et al. 2016).

Extracts and compounds isolated from mushrooms exhibit many mechanisms of **anticancer** activity, for example, inhibiting the kinase, and hence the cell cycle, or inhibiting angiogenesis. They are also inducers of reactive oxygen species, antimitotic agents, and topoisomerase inhibitors stimulating apoptosis of cancer-transformed cells (Patel and Goyal 2012; Kosanić et al. 2016). In the effective anticancer therapy, it is necessary to apply comprehensively acting factors, because cells that have undergone mutation have a unique ability to multiply and spread, creating metastases that cause high mortality (Zong et al. 2012).

The main and best-known compounds responsible for the anticancer effect are polysaccharides and their combinations with peptides (proteoglycans) or steroids (Ruthes et al. 2015; Kosanić et al. 2016; Meng et al. 2016; Singdevsachan et al. 2016). These compounds can be used in the treatment of cancer or viral diseases (e.g., AIDS), and used as prebiotics. Polysaccharides isolated from edible mushrooms activate the immune response in *in vitro* and *in vivo* studies, acting as biological stimulants. The most important anticancer polysaccharides are polysaccharide-protein complexes, fiber, (1 → 3) α-glucans and (1 → 3), (1 → 6)-β-glucans. In turn, heteropolysaccharides showing antiproliferative effect in relation to cancer cells have been shown to be substances that inhibit the development of cancers (carcinostatic), mainly after intraperitoneal or oral administration (Byerrum et al. 1957; Zong et al. 2012; Singdevsachan et al. 2016).

The importance of mushroom polysaccharides in modulating the function of the immune system and thus the potential inhibitory effect on cancers is of particular importance. One of the first clinically described activities of mushroom polysaccharides in the treatment of cancer comes from 1957 and was analyzed by Byerrum et al. (1957). The mechanism of this action on the immune system, confirmed in subsequent studies, involves the stimulation of immune system cells, including T lymphocytes and cytotoxic T lymphocytes (CTL), B lymphocytes, granulocytes (eosinophils and neutrophils), NK cells, or macrophages (Zhang et al. 2007; Roupas et al. 2012; Meng et al. 2016; Singdevsachan et al. 2016). This mechanism of action is particularly characteristic of β-1, 3-glucans. Numerous studies also suggest that β-glucans may enhance a specific cellular response by enhancing the secretion of IL-6, IL-8, IL-12 and IFN-Ɣ from neutrophils, macrophages and NK cells (Meng et al. 2016; Singdevsachan et al. 2016). In addition, β-glucans found in fruiting bodies of edible mushrooms can be factors that stimulate new effectors cells that contribute to the formation of, inter alia, antibodies against cancer antigens, which is less popular than the classic cytotoxic effect triggered by chemotherapy (Singdevsachan et al. 2016).

For the anticancer effect, the ability to bind other molecules (proteins, steroids) is also an important issue, which results in an increased anticancer activity. Most of the polysaccharides used are glucans, which when combined with other protein molecules can be converted into glycoproteins, glycopeptides or proteoglycans.

The conformation of the polysaccharide chain is also key issue to the therapeutic effect. The most active polysaccharides are most often complexes with proteins with a molecular weight of 10,000 kDa. Human macrophages possess a polysaccharide receptor that is highly specific for glucose and mannose molecules, from which the anticancer effect of polysaccharides possessing these groups may result (Zhang et al. 2007; Patel and Goyal 2012; Roupas et al. 2012; Meng et al. 2016; Tian et al. 2016).

Among the edible species of mushrooms rich in polysaccharides with the structure described above, the anticancer effect is exhibited, inter alia, by species like *Cantharellus cibarius* (chanterelle), *Armillaria mellea* (honey fungus), *Flammulina velutipes* (golden needle mushroom), *Macrolepiota procera* (parasol mushroom), *Lentinula edodes*, *Tremella fuciformis* (snow fungus), *Hericium erinaceus* (lion's mane mushroom), *Agaricus bisporus* (portobello mushroom), *Agaricus blazei* (almond mushroom), *Agaricus campestris* (field mushroom), *Pleurotus ostreatus* (oyster mushroom), *Sparassis crispa* (cauliflower fungus), *Grifola frondosa* (Maitake Mushroom), *Boletus edulis* (penny bun), *Imleria badia* (bay bolete) and *Lactarius deliciosus* (saffron milk cap) (Zhang et al. 2007; Patel and Goyal 2012; Meng et al. 2016; Singdevsachan et al. 2016; Tian et al. 2016).

All over the world, modern medicine, and in particular Eastern one, uses *L. edodes* as an agent enhancing the body's strength, and the extract of this species as an anticancer substance. The **therapeutic effect** is mainly due to the polysaccharide –lentinan, which action reduces the size of the tumor by up to 90% (Zhang et al. 2007; Patel and Goyal 2012; Roupas et al. 2012; Kosanić et al. 2016; Meng et al. 2016; Tian et al. 2016).

Lentinan in terms of its chemical structure is β (1 → 3) glucan with β (1 → 6) branching with a molecular weight from 400 to 800 kDa (Fig. 6.1). This polysaccharide is most commonly used to treat solid tumors of the stomach, lungs, breasts, large intestine and malignant leukemia. It acts by activating the immune system and affects the restoration of the body" correct defense response. It induces a humoral immune response of the body, which consists in restoring the activity of helper T cells in host cells occupied by the cancer. Lentinan has no **cytotoxic** activity, practically showing no side effects. However, local irritation after injection, sporadic fever and vomiting are possible. The use of this substance has been shown to increase the average survival time of patients (Patel and Goyal 2012; Roupas et al. 2012; Meng et al. 2016; Singdevsachan et al. 2016; Muszyńska et al. 2017a, b, c).

Studies on the mechanism of lentinan action prove that it is dependent on the thymus and consists in enhancing the response of T helper precursors and macrophages, and thus some cytokines, produced by lymphocytes, after the diagnosis of cancer cells. The induction of interferon (IFN-Ɣ) is also important for this activity. Lentinan is most often used in the treatment of gastric cancer as an aid during conventional treatment, including surgical tumor removal, chemotherapy or radiotherapy. This is a kind of synergistic effect, improving the general condition of the patient. Effective attempts have also been made to treat hepatocellular carcinoma, colorectal and pancreatic cancers while limiting side effects and thus improving the quality of life. Lentinan as an inhibitor of lymphocyte proliferation is also effective

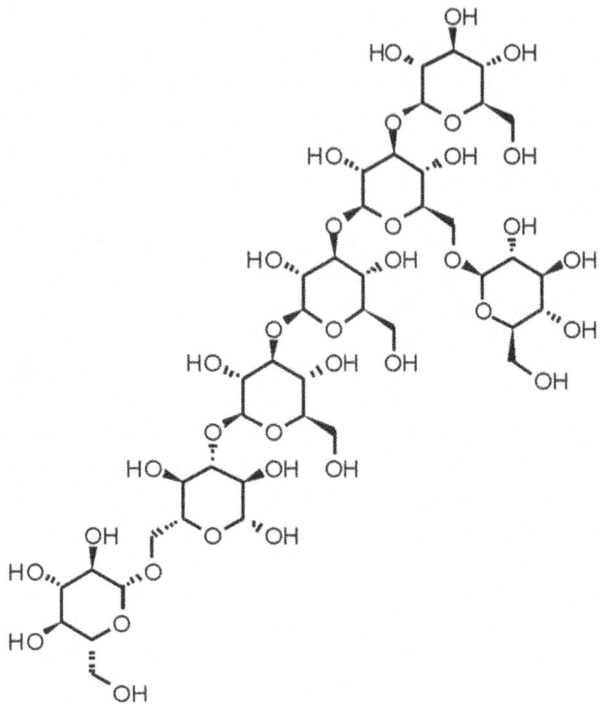

Fig. 6.1 Lentinan

in the treatment of leukemia. It has a selective **anti-proliferative** effect on the cancer-transformed skin cells (CH72), while not affecting healthy keratinocytes (C50). The use of lentinan as an adjuvant can improve the quality of life in patients with cancer because, like other fungal polysaccharides, it eliminates the side effects of chemo-and radiotherapy (Mantovani et al. 2008; Patel and Goyal 2012; Roupas et al. 2012; Zong et al. 2012; Lindequist et al. 2005; Meng et al. 2016; Singdevsachan et al. 2016; Muszyńska et al. 2017b).

Another example of polysaccharide responsible for anticancer activity is β-glucan isolated from *Sparassis crispa* fruiting bodies. Clinical trials were conducted in which patients suffering from tumors were administered orally powdered fruiting bodies of this species in an amount of 300 mg/day. Compared to the control group, a significant improvement was observed in many patients in the group supplemented with fruiting bodies (Roupas et al. 2012).

Lactarius deliciosus and *M. procera* are edible mushroom species exhibiting similar anticancer potential in *in vitro* studies. This activity was tested on human epithelial carcinoma cells (HeLa), human colorectal cancer (LS174) and human lung cancer (A549). The IC_{50} values ranged from 19.01 to 74.01 μg/mL for the *L. deliciosus* extract, and from 25.55 to 68.49 μg/mL for the *M. procera* extract, depending on the type of cell lines. The stronger anticancer effect against A549 and

LS174 cell lines was demonstrated for *M. procera*, while *L. deliciosus* was more active against HeLa (Kosanić et al. 2016).

At the Centro de Investigaciones Biológicas in Madrid, two α-glucan polysaccharide fractions were isolated from *A. mellea* fruiting bodies. The main fraction consisted of linear chains of α-(1, 3)-and α-(1, 4)-glucan linked to the protein, while the other fraction contained α-(1, 3)-glucan. A fraction containing a combination of β-glucan and a peptide part was also obtained. This fraction showed an anticancer activity. During the examination of the chemical structure of peptide-glucan sugar part, the presence of glucose molecules connected by β-(1, 3) and β-(1, 6) bonds has been demonstrated (Muszyńska et al. 2011d).

High-β-glucan polysaccharide fractions derived from *A. blazei* species have been shown to be effective in treating prostate cancer, both dependent and independent of androgens. The induction of prostate cancer cell apoptosis was directly related to the activation of caspase-3 as a pro-apoptotic agent. The polysaccharide fraction obtained from this species is effective in the treatment of tumors in *in vivo* studies, with no cytotoxic effects. In addition, in therapy using 5-fluorouracil (5-FU), this species protected against leukopenia as a side effect of treatment with this substance. The properties of β-glucan obtained from this mushroom species have been improved by modifications involving the incorporation of sulfate groups that improve solubility (Mantovani et al. 2008; Patel and Goyal 2012; Roupas et al. 2012; Zong et al. 2012).

Also pleuran (β-1, 3-D-glucan), obtained from *P. ostreatus* species deserves an attention (Fig. 6.2). Its activity has been shown to slow down the formation of precancerous lesions of the 'Aberrant Crypt Foci' type (ACF) in the colon in rats of the Wistar strain. This action consisted in inhibiting the proliferation of cancer cells and inducing their apoptosis. In addition, a new polysaccharide (POPS-1), obtained from hot water extracts of this mushroom species showed significantly reduced toxicity in combination with commonly used 5-FU (Patel and Goyal 2012; Roupas et al. 2012; Meng et al. 2016).

Fig. 6.2 Pleuran

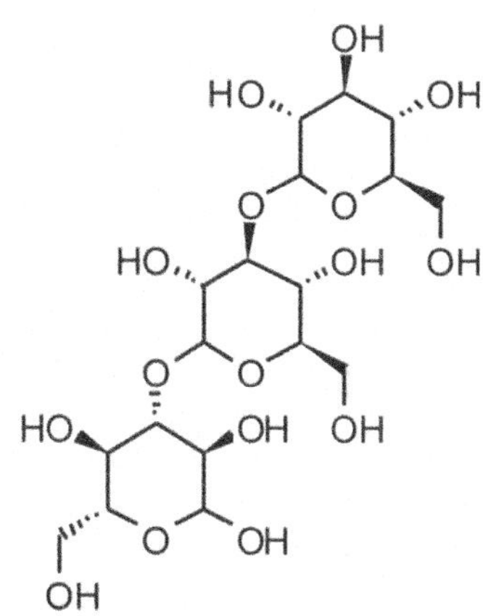

Diet rich in dried fruiting bodies of *P. ostreatus* reduced the toxicity in mice treated with cyclophosphamide and decreased pathological changes arising as a result of the appearance of dimethylhydrazine induced colon cancer in rats. The mechanism of action resulted from the strong antioxidant potential of these fruiting bodies. In turn, aqueous extracts obtained at high temperature from *H. erinaceus* fruiting bodies turned out to be rich in β-glucan, the administration of which contributed to the reduction of tumor mass in mice with induced colorectal cancer. The reduction in tumor mass was due to the induction of tumor necrosis factor and NK cell secretion, as well as macrophage activation and inhibition of angiogenesis (Patel and Goyal 2012; Roupas et al. 2012; Lindequist et al. 2005; Ruthes et al. 2015; Singdevsachan et al. 2016; Ment et al. 2016).

Grifola frondosa species is also a source of β-glucan, including the active fraction D, and MD fraction obtained by further purification. The MD fraction works by enhancing the activity inhibiting the spread of tumors, stimulating NK cells and reducing **nephrotoxicity** and **immunosuppression** induced by cisplatin treatment. MZF heteropolysaccharide with *in vivo* activity inhibiting tumor growth by stimulating cellular immunity was also isolated from this species. The use of whole powdered fruiting bodies with the addition of isolated MD fraction was used in patients with cancer diseases in the stages of tumor development II-IV. The use of such a combination resulted in an improvement in the general condition of patients, and even regression of the tumors themselves: 68.8% of patients diagnosed with breast cancer, 58.3% of patients with liver cancer and 62.5% of patients treated for lung cancer (Patel and Goyal 2012; Roupas et al. 2012; Zong et al. 2012; Cheung 2013; Lindequist et al. 2005).

Water insoluble polysaccharides such as, chemically sulfated polysaccharides (S-GAP-P) have also been classified as effective in the treatment of human gastric cancer (cancer cells SGC-7901). *Grifola frondosa* species proved to be effective not only in combination therapy alongside 5-FU, but also in S-GAP-P monotherapy, in which it induced the **apoptosis** of SGC-7901 cancer cells. The effect was dose-dependent. β-glucan from *G. frondosa* also showed a cytotoxic effect on human prostate cancer cells (PC-3). Induction of apoptosis has been confirmed in *in-vitro* studies of androgen-independent tumor (Patel and Goyal 2012; Roupas et al. 2012; Ruthes et al. 2015; Meng et al. 2016; Singdevsachan et al. 2016).

Interesting results were obtained in studies carried out using the species *F. velutipes* and *A. blazei*. The preventive effect on the incidence of cancer was observed especially in the group of farmers involved in the commercial cultivation of this edible mushroom–these species were a popular element of their daily diet. As a result of the consumption of mushrooms, the incidence of cancer was estimated at 40% lower compared to the general incidence in the population. To confirm the results obtained, tests were also carried out using mice. They were regularly fed the same species of mushrooms, and then cancer cells were introduced into the animal organisms. It was clearly observed that in the research group, in contrast to the control group, no cancer was developing, which confirms the strong preventive anticancer effect of polysaccharides from the species *F. velutipes* and *A. blazei* (Zhang et al. 2007).

6.1.2.3 Chitin

Mushroom cell walls contain a mixture of fibrous elements and matrix elements, rich in **chitin** (crosslinked N-acetylglucosamine polymer) and polysaccharides, such as β-D-glucans and mannans (Fig. 6.3). These elements of the cell wall are not digested in the human digestive tract, and thus constitute dietary fiber (Cheung 2013). Total fiber content varies on average from 2.7 to 4 g per 100 g of edible parts of fruiting bodies. However, the percentage content of chitin itself, for example, in *A. melenea* fruiting bodies calculated on the basis of glucosamine determinations in hydrolysates, calculated on the dry mass of the mushroom, is 6.4%. Physicochemical properties and molecular mass of chitosans isolated from mushrooms are identical to those isolated from chitin cuticle of crustaceans (Muszyńska et al. 2011b).

Chitosans reduce the concentration of LDL cholesterol (in the blood and liver) and serum triacylglycerides. In this way, they reduce the risk of cardiovascular diseases. They also affect the absorption of cholesterol from the gastrointestinal tract, causing a decrease in its total concentration in the blood and LDL fraction, without changing the concentration of HDL fraction (Rajewska and Bałasińska 2004; Cheung 2013; Muszyńska et al. 2013b). The mechanism of this phenomenon is not fully understood, but fiber found in mushrooms can be a good alternative to fiber from other foods. The hypocholesterolemic effect may be caused by a change in the absorption of nutrients from the gastrointestinal tract associated with the presence of fiber in the diet, effect on intestinal (or pancreatic) secretion, and indirectly on the metabolism of lipoproteins or bile acid. Considering the relatively high content of fiber and low fat content, edible mushrooms can become part of the diet, which is aimed at counteracting atherosclerosis (Fukushima et al. 2000; Siwulski et al. 2014).

Due to the presence of a large amount of dietary fiber in edible mushrooms, especially glucans (increasing the viscosity of the digestive content) and chitin, the excretion of bile acids and neutral steroids increases. In the acidic environment of the stomach, the amino groups present in the chitosan molecules assume a positive charge and combine with negatively charged bile acid residues. Low pH makes bile

Fig. 6.3 Chitin

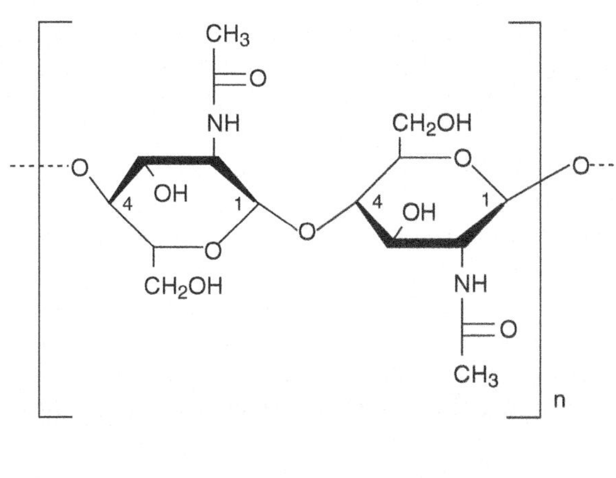

acid and chitosan complexes become insoluble and are excreted from the body. Chitin and its partially deacetylated form, i.e., chitosan, are widely used in the pharmaceutical industry. They are not only carriers for many drugs, but also are themselves components of slimming preparations. The slimming effect is related to the temporarily reduced absorption of lipids from various foods (Muszyńska 2012).

The effectiveness of chitosan has been proven through research focusing on two groups of volunteers who were on a low-calorie diet. Patients taking chitosan regularly reduced weight significantly faster within a month. In the research group, the weight loss was 7 kg, while in the control group this loss was only 3 kg (Rajewska and Bałasińska 2004). In addition to beneficial effects on fat metabolism and weight loss effect, studies have also shown an improvement in sugar metabolism. The consumption of *A. bisporus* species in this case resulted in a 24.7% reduction in blood glucose levels in rats that had deliberately induced type II diabetes. The desired effect was also obtained in rats with hypercholesterolemia, who after eating the mushroom, in addition to the decrease in total cholesterol, LDL and triglycerides also achieved an increase in the level of HDL fraction beneficial for the proper functioning of the organism (Jeong et al. 2010).

6.1.2.4 Lectins

Lectins are a kind of glycoproteins, necessary for the proper growth of mushroom fruiting bodies. They are also a protective factor for the mushrooms against the toxins from the environment, for example pesticides, as well as bacteria and even viruses (Varrot et al. 2013; Singh et al. 2015). Lectins found in edible mushrooms turned out to be the subject of research, due to the anticancer effect associated with the immunomodulatory potential, realized by stimulating the maturation of immune cells. A mixture of lectins (lectin A and lectin B) isolated from *A. bisporus* species (ABL) has **antiproliferative** activity on epithelial tumor cells, without inducing a direct cytotoxic effect. *In-vitro* studies using HT-29-human colon cancer cells have shown that lectins isolated from *A. bisporus* have the ability to induce apoptosis in these cells, bypassing the **cytotoxic** effect. This effect may be associated with an increase in **caspase-3** activity (Wang et al. 1998;Carrizo et al. 2005; Roupas et al. 2012; Singh et al. 2015). For AAL lectin obtained from the *Agrocybe aegerita* species, a precise relationship was established between the elements of the structure and the induced anticancer effect on both mouse and human cells. This is important because it may be the first step to design specific molecules that will later be used in therapy. AAL acts by inducing apoptosis in cancer cells. To achieve this activity, the dimeric organization of the lectin molecule is necessary. The presence of glucose and galactose in the carbohydrate functional group is also necessary (Ng 2004; Li et al. 2008; Singh et al. 2015).

In addition, lectins from edible mushroom species such as *G. frondosa, F. velutipes, Tricholoma mongilicum, Volvariella volvacea, Pleurotus citrinopileatus* and *P. ostreatus* have been found to possess immunomodulatory potential as well as anticancer, anti-proliferative and cytotoxic properties.

It was possible to isolate **heterodimeric** lectin from *V. volvacea* species, which has the potential to inhibit the growth and development of cancer cells. Importantly, this lectin lengthens the survival time of sarcoma 180 tumor-bearing mice. Elongation of the survival period is dose dependent. In contrast, the *G. frondosa* species has been tested for its cytotoxic effect on HeLa tumor cells. Lectin specific for N-acetylgalactosamine was responsible for this effect (Li et al. 2008; Singh et al. 2015). In turn, *P. citrinopileatus* is a species rich in homodimeric lectin that inhibits the growth of the murine sarcoma 180 cancer as early as 20 days in case of intra peritoneal, daily administration at a dose of 5 mg/kg body weight. The anti-cancer effect is similar to that exhibited by *P. ostreatus* (Li et al. 2008; Patel and Goyal 2012).

Lectin isolated from *P. ostreatus* species showed anticancer activity against mouse H-22 hepatoma and sarcoma S-180 tumors, reducing the number of cancer cells (Ng 2004; Li et al. 2008; Singh et al. 2015). This lectin works by inhibiting the growth of both these tumors by 75% and 88%, respectively. To achieve this effect, a dose of 1.5 mg lectin/kg body weight by intra peritoneal injection for 20 days was required. This treatment caused a decrease in the weight of mice in comparison to the control group, but the survival time significantly increased (Ng 2004).

Lectins derived from *T. mongolicum* species (TML1 and TML2) inhibit the development of sarcoma 180. They are compounds with a molecular weight of approximately 36–38 kDa. Both TML1 and TML2 inhibit the growth of these cells, with TML2 being more effective. The activity of both these lectins is based additionally on the stimulation of macrophages for the production of nitrite ions (NO^{2-}). In addition, they show anticancer activity by stimulating the immune system *in vivo*, while the **anti-proliferative** effect is only observed *in vitro* (Wang et al. 1996; Wang et al. 1998; Ng 2004; Singh et al. 2015).

Mushroom lectins have also been tested as compounds with therapeutic potential in the prevention and treatment of diabetes, inter alia, due to the induction of β-cell division of pancreatic islets. The study aimed to determine the rate of β cell regeneration in mice and to understand the mechanisms of their proliferation. The analysis was carried out on the basis of glucose concentration measurements and the degree of insulin secretion after the administration of fungal lectins. Comparing the results, the effect of lectins from *A. bisporus* on reducing the blood glucose level, increasing its tolerance and increase in β-cell mass of pancreatic is proved (Wang et al. 2012). Analyzing also the composition and properties of fatty acids obtained from *C. cibarius* species, it has been shown that as PPAR-γ agonists, they inhibit the development of insulin resistance and are therefore effective anti-diabetic agents (Hong et al. 2012). *Armillaria mellea* fruiting bodies have demonstrated the presence of a therapeutically important peptide-prosomatostatin, which has anticancer activity especially in the treatment of pancreatic cancer (Muszyńska et al. 2013b).

The fruiting bodies and mycelia of *Antrodia camphorata* in traditional Eastern medicine are in turn used as an anticancer agent due to the presence of ubiquinone derivative. Importantly, this compound has selective activity, affecting only cancer-affected cells, omitting healthy cells (Hu et al. 2016).

6.1.2.5 Amino Acids and Proteins

Anti-inflammatory properties of mushrooms are associated with the free amino acids contained in them (both endogenous and exogenous ones) that can affect the metabolism of prostaglandins. The anti-inflammatory properties of *P. ostreatus* are partly explained by the content of amino acids such as leucine, isoleucine, tyrosine and phenylalanine (Jedinak et al. 2011).

Ergothioneine is an important amino acid exogenous to humans, i.e., must be supplied with food. A good source of this compound is fruiting bodies of edible mushrooms (Fig. 6.4). This amino acid assimilated from the diet is particularly important for cells, tissues and organs sensitive to oxidative stress, such as erythrocytes, the lens of the eye, semen or skin (Chen et al. 2012). *In vivo* studies in the mouse HR-1 model (hairless) that ergothioneine isolated from *C. comatus* significantly reduced the amount of DNA damage and inhibited inflammation caused by UV-B radiation (Asahi et al. 2016). Additionally, ergothioneine–a strong **antioxidant** isolated from *A. bisporus*, has a beneficial effect during cancer therapy due to its strong anti-mutagenic properties as well as chemo-and radio protective activity (Chen et al. 2012; Muszyńska et al. 2017b).

Arginine present in *A. bisporus* species is another substance used in the supplementation of cancer patients. It not only improves the condition of the body and has a beneficial effect on the immune system, but also prolongs the life expectancy of cancer patients. This effect is due to delayed growth of tumor and the appearance of metastases (Novaes et al. 2013).

Also noteworthy are the young, closed, edible fruiting bodies of *Coprinus comatus*. Studies have shown that they can also be effective in the treatment of estrogen-independent breast cancer. In addition, their inhibitory effect on the development of prostate cancer cells (LNCaP cells) has been demonstrated. In turn, an immunomodulatory protein (FIP-fve), which has anticancer activity involving an activation of T lymphocytes, was isolated from *F. velutipes* species. This effect was tested against the mouse model of hepatoma. The aqueous extract turned out to be rich in another anticancer substance–flammulin. The presence of stable hemagglutinin, which is responsible for inhibiting the proliferation of tumor cells in the case of leukemia (L1210 cells), was also an important issue. Aqueous extracts from both *C. comatus* and *F. velutipes* have been shown to be effective in the treatment of estrogen-dependent and estrogen-independent breast cancer *in vitro*. The observed anticancer effect (caused by the induction of apoptosis of altered cells) was dependent on the dose of both extracts.

Fig. 6.4 Ergothioneine

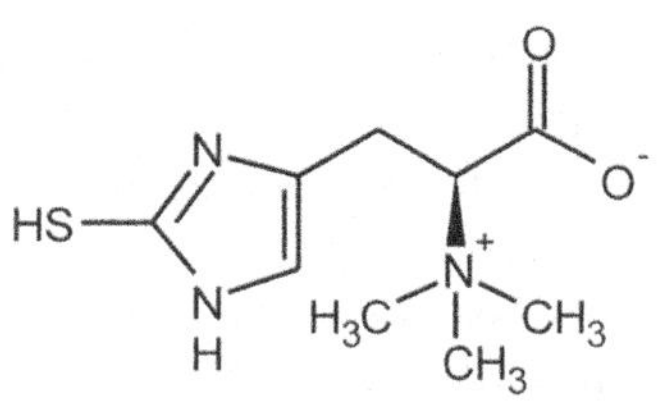

6.1.2.6 Fatty Acids

Fatty acids contained in mushrooms may support anti-inflammatory processes in the human body, due to the high content of unsaturated fatty acids (Ayaz et al. 2011; Öztürk et al. 2011). Polyunsaturated fatty acids (PUFA) are precursors of eicosanoids, signaling molecules necessary for proper regulation of cellular processes in muscles, blood vessels, nerve cells and in the immune system. Eicosanoids provide a balance between inflammatory and anti-inflammatory processes (Dennis and Norris 2015).

The polyunsaturated fatty acids (PUFAs) include the n-3, n-6 and n-9 acids. Maintaining proper proportions of fatty acids of the n-3 to n-6 series in the diet is crucial for preventing the development of cardiovascular diseases and cancers. α-Linolenic acid (ALA) is an essential ingredient in normal nutrition, a precursor to the long-chain PUFAs of then-3 series. It also has anti-inflammatory effects (Gdula-Argasińska et al. 2015).

Comparative studies on the composition of more than a dozen species of mushrooms showed the highest share of the following fatty acids: linoleic (C18:2 n-6), oleic (C18:1 n-9) and palmitic (C16:0) (Ayaz et al. 2011).

In experiments conducted by Grzywacz et al. (2016) on murine macrophages RAW 264.7 activated with LPS, anti-inflammatory effects of extracts from *I. badia* biomass have been demonstrated. It involved an inhibition of COX-2 expression, prostaglandin E2 synthase (cPGES), p50 and p65 NF-κB subunits, which can be explained by high content of unsaturated fatty acids. Biomass from *I. badia* was characterized by a 5% content of ALA, which did not appear in fruiting bodies (Grzywacz et al. 2016).

In the *C. cibarius* extracts, the presence of fatty acids with agonist activity toward peroxisome proliferator-activated receptors (**PPAR-γ**) has been demonstrated. PPAR receptors play a special role in the metabolism of lipids and carbohydrates, in the differentiation of adipocytes and in the regulation of inflammatory processes. Extracts of *C. cibarius* species therefore have a therapeutic effect in the case of inflammatory diseases and some cancers (Hong et al. 2012).

6.1.2.7 Phenolic Compounds

A particularly important group of secondary metabolites found in mushroom fruiting bodies with proven antioxidant and anti-inflammatory properties *in vitro* and *in vivo* are **phenolic compounds** (Czapski 2005; Elmastas et al. 2007; Ferreira et al. 2009). Species as *C. cibarius, A. bisporus, B. edulis, Calocybe gambosa, Hygrophorus marzuolus* and *L deliciosus* contain the high content (about 15 mg/g DM) of one of the most active antioxidants in mushrooms, i.e., caffeic acid (Reis et al. 2012; Muszyńska et al. 2013a).

Caffeic acid also demonstrates anti-inflammatory activity. In *in vitro* studies using human HUVEC vascular cells and TNF-α induced U937 monocytes, the potential role of caffeic acid in the treatment of inflammatory changes in cardiovascular disease has been demonstrated. Significant reduction of monocyte adhesion to HUVEC cells was observed in cultures supplemented with caffeic acid. In addition, caffeic acid inhibited MPC-1 expression in monocytes (chemoattractant of protein 1), IL-8 and translocation of NF-κB in the nucleus (Moon et al. 2009).

Anti-inflammatory properties associated with the activity of phenolic compounds derived from edible mushrooms: *A. bisporus, B. edulis, C. cibarius, H. marzuolus, L. deliciosus* and *P. ostreatus*, were tested *in vitro* using LPS activated RPS 264.7 macrophages. The results of the experiments showed the inhibitory effect of extracts on the expression of inflammatory markers such as IL-1β and IL-6 and the production of NO. The species of *A. bisporus, C. cibarius* and *L. deliciosus* showed the highest anti-inflammatory efficacy in the tested model (Palacios et al. 2011). In another experiment, the anti-inflammatory properties of the extract from *Elaphomyces granulatus* were confirmed. Inhibition of COX-2 activity in RAW 264.7 macrophages was observed and the antioxidant activity of the extracts was found. The effect obtained was probably related to the content of phenolic compounds, including syringic acid isolated from *E. granulates* (Stanikunaite et al. 2009).

6.1.2.8 Indole Compounds

Indole compounds have a particularly strong effect on the immune and nervous systems of animals. The *C. cibarius* species contain a large amount of serotonin −17.61 mg/100 g DM. and kynurenine sulfate −3.62 mg/100 g DM as well as lower amounts of melatonin, L-tryptophan, 5-hydroxytryptophan, 5-methyltryptophan, indole, indole-3-acetonitrile. Studies have shown that the content of indole compounds in biomass from *in vitro* cultures (mycelium) can be much higher than in fruiting bodies (Muszyńska et al. 2016a).

The content of non-hemolynogenic indole compounds and their release into artificial digestive juices was tested in the *A. bisporus* species. The following indole compounds have been identified and identified in the fruiting bodies: L-tryptophan, 5-hydroxy-L-tryptophan, melatonin, serotonin, tryptamine and 5-methyltryptamine, which indicates that this popular edible species is their good food source (Muszyńska et al. 2016b).

In the human body, L-tryptophan, one of the amino acids, can be transformed into other indole derivatives with high biological activity, for example serotonin, melatonin and niacin (vitamin B3) (Fukuwatari and Shibata 2013).

Melatonin, a compound with a broad spectrum of action in the body, is also an effective scavenger of free radicals, protecting cells from damage and the development of inflammatory reactions. Melatonin has been shown to regulate cytokine production by preventing translocation of NF-κB into the cell nucleus. Moreover, it can affect the reduction of damage associated with acute inflammation. It is related to the inhibition of adhesion molecules generation by leukocytes, thus affecting their migration process (Reiter et al. 2000). The positive effect of melatonin on the course of inflammatory neurodegenerative diseases, such as dementia, Alzheimer's and Parkinson's disease, or multiple sclerosis has also been demonstrated (Esposito and Cuzzocrea 2010). Scientific experiments indicate that serotonin (5-hydroxytryptamine) reduced allergic pneumonia in the C57BL/6 mouse model by 70–90%. In addition, a decrease in expression induced by transglutaminase 2 allergens was observed as well as a decrease in serotonylation of proteins in the pulmonary endothelium and a decrease in the migration of leukocytes and eosinophils. The addition of serotonin to cell cultures resulted in a reduction in the serotonylation of TNF-α induced protein. This indicates the important role of serotonin in the leukocyte migration process, which may be important in the treatment of allergic diseases (Abdala-Valencia et al. 2012).

Analysis of selected species of mushrooms indicates that even thermally processed (boiled) fruiting bodies of *C. cibarius, I. badia, L. deliciosus, M. procera, P. ostreatus* and *S. bovinus* are a valuable reservoir of compounds with **antidepressant** activity (Turner et al. 2006; Muszyńska et al. 2011c). The highest content of serotonin was demonstrated among the analyzed indole compounds in *C. cibarius* fruiting bodies. The species containing significant amounts of this neurotransmitter are also: *Leccinum scabrum, A. mellea, I. badia, B. edulis, L. deliciosus* and *P. ostreatus*.

Serotonin, a substance with multidirectional pharmacological activity that plays, inter alia, the role of the neurotransmitter in the central nervous system together with melatonin is regulated by the daily cycle (Fig. 6.5). Serotonin, generated endogenously in the brain, plays a very important role in regulating sleep, anxiety, aggression, body temperature, mood, the course of maturation, regeneration and inhibition of the aging process of cells, thus contributing to the general strengthening of the body's immune system. It is also one of the factors regulating the contraction and relaxation of blood vessels (Turner et al. 2006; Muszyńska et al. 2011c).

Serotonin, after ingestion, does not penetrate the blood-brain barrier into the central nervous system, but it can regulate the functioning of the gastrointestinal (intestinal) tract (Birdsall 1998). In patients with asthma, this substance causes bronchospasm. It is involved in the pathogenesis of migraine and vascular headaches.

The wide range of activities regulated by serotonin is explained by the existence of 7 types of serotonin receptors (5-HT), and several subtypes within them. Many classes of drugs work via serotonin receptors, being agonists or antagonists of these or by affecting the release of **serotonin**. These drugs include, above all, antidepressants, anxiolytics, antiemetics and antimigraine (Birdsall 1998).

Fig. 6.5 Serotonin

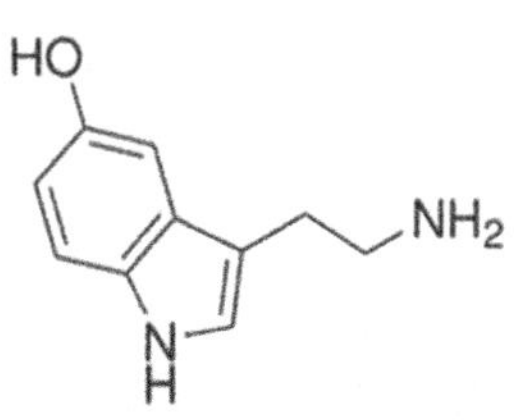

Information on the occurrence of indole compounds in *Basidiomycota* species also applies to L-tryptophan, which is a biogenetic precursor to all **indole compounds** (e.g., dopamine, melatonin, serotonin, and adrenaline) and vitamins (e.g., niacin) (Birdsall 1998; Turner et al. 2006; Muszyńska et al. 2011b). The determined L-tryptophan content ranged from 0.16 to 25.90/100 g DM (in extracts from *S. bovinus* fruiting bodies) (Muszyńska and Sułkowska-Ziaja 2012).

L-tryptophan is an amino acid that is exogenous to the human body and must therefore be supplied with food. Among the extracts from heat-treated fruiting bodies, the extract of the fruiting bodies of the *Suillus bovinus* contained the largest amount of L-tryptophan-17.71 mg/100 g DM (Muszyńska and Sułkowska-Ziaja 2012).

Processed edible mushrooms, especially *S. bovinus*, can be its source, which is why they are an alternative to animal foods. In the case of *B. edulis*, the amount of L-tryptophan was higher in the processed material than in the dried fruiting bodies. 5-Hydroxytryptophan (5-HTP), a direct precursor of serotonin and melatonin, was present in both thermally unprocessed and processed fruiting bodies. However, the highest amounts of this compound in the discussed species were found in uncooked mushrooms. The maximum amounts of this metabolite were found in fruiting bodies extracts of: *L. edodes* 24.83 mg/100 g DM, *M. procera* (22.94 mg/100 g DM), *S. bovinus* (15.83 mg/100 g DM). According to the latest research, 5-HTP is a potential drug in the treatment of Alzheimer's disease (Birdsall 1998; Muszyńska et al. 2011a, b; Muszyńska and Sułkowska-Ziaja 2012; Muszyńska et al. 2013c).

5-Methyltryptophan was only marked in *L. scabrum* fruiting bodies. This compound was also detected in cooked fruiting bodies of four species: *B. edulis, C. cibarius, L. deliciosus* and *P. ostreatus*, in amounts comparable to those found in *L. scabrum* (Muszyńska et al. 2011a, b; Muszyńska and Sułkowska-Ziaja 2012; Muszyńska et al. 2013c). The indole compound determined in the studied species was also melatonin. It was found in small amounts in extracts from the fruiting bodies of *B. edulis, C. cibarius, L. deliciosus, L. edodes* and *M. procera* (from 0.07 to 1.29 mg/100 g DM) (Muszyńska et al. 2011a, b; Muszyńska and Sułkowska-Ziaja 2012; Muszyńska et al. 2013c).

6.1.2.9 Enzymes

Lactase present in *Agrocybe cylindracea* species is an important enzyme with anticancer activity. Lactases also occur, inter alia, in the species of *H. erinaceus, A. blazei, L. edodes, P. ostreatus, C. cibarius*, or *Pleurotus eryngii*. However, their anticancer activity was not analyzed. Lactase is used to purify contaminated water as a biosensor or fabric dye. It is a lignin-degrading enzyme, participates in mushrooms morphogenesis, and further catalyzes oxidation reactions of organic compounds, including phenolic compounds (Zhang et al. 2010).

The studies carried out have shown that the lactase present in *A. calvacea* has anti-proliferative activity against MCF-7 breast cancer cells (IC_{50} = 6.5 μM) and HepG2 liver cancer cells (IC_{50} = 5.6 μM). The activity of lactase isolated from this species of mushrooms was the highest while maintaining the pH in the range of 3–4, while at the pH of about 9, its activity completely disappeared. In addition, raising the temperature to 50 °C had a positive effect on the enzyme activity; however, a further temperature increase adversely affected enzyme degradation (Elmastas et al. 2007; Endo et al. 2010).

The activity against MCF-7 and HepG2 tumor cells was also demonstrated for the lactase isolated from the edible species of *Clitocybe maxima*. The IC_{50} for these cancer cells was 3.0 μM and 12.3 μM, respectively (Zhang et al. 2010).

Another example of an enzyme with anticancer activity is tyrosinase extracted from *A. bisporus*, characterized by a high degree of similarity to that found in the human body. This mushroom species is a very good and cheap source of tyrosinase (Labus et al. 2011; Zaidi et al. 2014; Kampmann et al. 2015). Studies carried out on the enzyme isolated from *A. bisporus* have shown that it has a protective effect on human lymphoma cell lines, preventing them from the negative effects of damaging factors such as perhydrol (Shi et al. 2002). In addition, the genoprotective activity of this species, resulting from the presence of tyrosinase was investigated. The putative DNA-protective effect was associated with the pathway of tyrosine up to L-DOPY, followed by conversion of this metabolite to dopaquinone (Shi et al. 2002; Jani et al. 2016).

6.1.2.10 Terpenoids

The compounds from the terpenoid group with anti-inflammatory, anti-proliferative and anticancer potential were also isolated from the *Poria cocos* of the *Polyporaceae* family. The anti-inflammatory activity of extracts from this mushroom has been confirmed in an *in vitro* model using RAW 264.7 macrophages. A decrease in the generation of pro-inflammatory mediators was observed by inactivating the NF-κB signaling pathway (Jeong et al. 2014). In turn, the anticancer activity of mushroom terpenoids was tested on the U937 line. The concentration and time-dependent effects of anti-proliferative and pro apoptotic effects of the extracts studied were related to the release of cytochrome C to the cytosol, caspase-3,-8 and-9 activation, PARP degradation and loss of mitochondrial membrane potential. The obtained results indicate the potential of the *P. cocos* species in the treatment of leukemia (Choi 2015). A series of triterpenes with a similar inhibitory effect on the NF-κB signaling pathway were isolated from another species, *Inonotus obliquus* (Jiang et al. 2008). In turn, the *A. camphorata* extracts showed NO, TNF-α and IL-12 reducing properties (Rao et al. 2007).

6.1.2.11 Tocopherols and Steroids

Tocopherols, the most active of which is α-tocopherol, have anti-inflammatory, anticancer activity and also prevent peroxidation of cell membranes phospholipids (Jiang 2014).

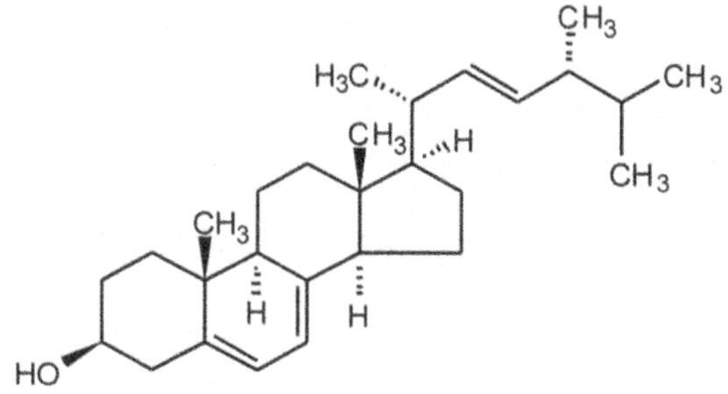

Fig. 6.6 Ergosterol

In many species of mushrooms, the presence of ergosterol (a precursor to vitamin D) (Fig. 6.6), ergocalciferol and other sterols (including ergosta-7,22-dienol, ergosta-5,7-dienol, ergosta-7-enol, cerevisterol, β-sitosterol, ergosterol peroxide, β-sitosterol, or 7-dehydrostigmasterol) has been detected. Among others, the fruiting bodies of *A. bisporus* are a rich source of ergosterol (about 61.5 mg/100 g).and ergocalciferol (Muszyńska et al. 2017a). A good source of vitamin D is also *C. cibarius*, in which the content of ergocalciferol even after a few years of storage of dried specimens is in the range of 0.12–6.3 μg/g DM (Phillips et al. 2011; Jiang 2014).

In addition, it turned out that subjecting the mushrooms to irradiation with UV-B and UV-C radiation allows them to increase the content of **vitamin D_2** (Strange et al. 2015; Drori et al. 2016). It is indicated that even more than half of the world's population suffers from a deficiency of this vitamin (Stepień et al. 2013). One of the mechanisms of the anti-inflammatory action of ergosterol and its derivatives is the inhibition of NF-κB translocation into the cell nucleus and thus the prevention of the expression of proinflammatory genes (Phillips et al. 2011).

The ergosterol present in the fruiting bodies of edible mushrooms, for example *I. badia* and *A. bisporus*, has anti-inflammatory and anticancer activity (Barros et al. 2008a). Anti-inflammatory properties in the mouse model were also showed for an extract of *L. edodes* enriched with ergosterol. The studies showed that in C57B1/6 mice with mitogen-induced (concanavalin A) liver inflammation, supplementation with Shiitake extract, enriched with vitamin D, caused a significant reduction in liver damage. The histopathological image of tissues was improved as well as the plasma level of transaminases and INF-γ decreased. In addition, the anti-inflammatory effect of vitamin D and mushroom extract was synergistic (Drori et al. 2016). In another experiment, it was observed that supplementation with powdered *A. bisporus* fruiting bodies enriched with vitamin D2, contributed to a significant reduction in the level of hsCRP protein after four weeks, which is a marker of inflammation in humans (Stepień et al. 2013).

The anticancer effect is also demonstrated by ergosterol (5, 7, 22-ergostatrien-3β-ol), present not only in *A. bisporus*, but also in other species from the Agaricales taxon. It occurs in most species of mushrooms from the Basidiomycota group. Ergosterol accounts for 83–89% of the total content of mushroom sterols. It inhibits the process of angiogenesis, which is associated with solid tumors, prevents tumor growth and prevents the migration and proliferation of cancer-affected cells.

This effect was determined by *in vitro* culturing of cancer cell lines as well as in *in-vivo* studies on rats (Yuan et al. 2008; Shao et al. 2010; Roupas et al. 2012; Novaes et al. 2013).

The content of ergosterol in species of cultivated mushrooms such as *A. bisporus, L. edodes*, or *P. ostreatus* is between 3.7–5.1 mg/g DM, while for wild growing species of *Cantharellus tubaeformis*, *C. cibarius* and *B. edulis*, these amounts are slightly lower, in the range of 1.4–4.0 mg/g DM. The protective effect of ergosterol on lymphocyte levels in patients undergoing chemotherapy has also been confirmed. This therapy is safe and well tolerated by patients. In addition, ergocalciferol (vitamin D2) present in edible mushrooms, resulting from ergosterol as a result of exposure to UV rays with a wavelength of 280–320 nm, is one of the preventive factors in cancer therapy (Shao et al. 2010; Roupas et al. 2012; Novaes et al. 2013).

In order to determine the antioxidant activity of mushroom extracts, lipids of rat liver microsomes were used in *in-vitro* tests, which were peroxidized according to the Fe^{2+}/ascorbate method. *A. mellea* and *Pleurotus cornucopiae* showed significant inhibition of **lipid peroxidation** (22.8% and 19.5%, respectively) at a concentration of 100 mg/mL, compared to the control group. Experiments have shown that the compound responsible for antioxidant activity is ergosterol peroxide (Krzyczkowski et al. 2009; Kampmann et al. 2015). As the concentration increased, this compound showed an increasing tendency to inhibit lipid peroxidation. The antioxidant activity of ergosterol peroxide was also compared with the effect of other antioxidants. It was confirmed that ergosterol peroxide showed a stronger effect than α-tocopherol and thiourea (by 19.2% and 21.5%, respectively) (Elmastas et al. 2007; Krzyczkowski et al. 2009).

Vitamin D_2, especially its hydroxylated form, is one of the potential anticancer drugs used in the treatment of, inter alia, melanoma. This effect is due to the inhibition of keratinocyte differentiation in *in vivo* studies and thus inducing protection against photodamage after topical administration. In the study of melanoma, ergosterol and dihydroergosterol also inhibited DNA synthesis due to the local metabolism of ergosterol, expressing cytochrome P-450scc by melanoma cells. Similarly, 17α-and 24-dihydroxyergosterol inhibited the proliferation of human epidermal keratinocytes cell lines. These results consistently confirm the antiproliferative effect and thus the anticancer effect of ergosterol metabolites on cells not only animal but also human (Słomiński et al. 2015).

Scientific research also indicates the existence of a connection between the appropriate level of vitamin D_2, also this of mushroom origin, and a reduced risk of prostate, ovarian, breast and large intestine cancer (Shao et al. 2010). Some species of mushrooms are a rich source of ergocalciferol, apart from *A. bisporus*, these include for example *G. frondosa, Morchella* spp., and *L. edodes* (Phillips et al. 2011).

Ergosterol derivative compounds, including ergosterol peroxide, also have antioxidant and anticancer activity, in this case based on cytotoxic activity against cancer cell lines and inhibition of their growth. It was possible to isolate ergosterol peroxide, or other sterols of this kind of action, inter alia, from the species of *Paecilomyces tenuipes* and *Cordyceps sinensis*. The *P. tenuipes* species also exhibited anticancer activity in *in vivo* studies (Lindequist et al. 2005; Hong et al. 2007).

In addition, ergosterol peroxide was first isolated from *H. erinaceus* species, which exhibits anticancer activity. The discussed substance was also obtained from other species of edible mushrooms considered to be medicinal, including *V. volvacea, I. badia, B. edulis, Suillus bovinus, Morchella esculenta*and *A. mellea* (Krzyczkowski et al. 2009).

Ergosterol peroxide isolated from the edible *Sarcodon aspratus* species has been shown to inhibit the growth of promyelocytic leukemia (HL60) cells. This effect was noticed for a dose above 10 μM of ergosterol peroxide (Takei et al. 2005).

Hypsizigus marmoreus species is also a source of ergosterol and ergosterol peroxide. It turned out that thanks to compounds from the group of sterols, fruiting bodies of this species can inhibit the TPA-induced (13-acetate-12-O-tetradecanoylphorbol) not only inflammatory ear swelling, but also tumor growth in mice during the two-stage carcinogenesis induced by TPA and DMBA (7 12-dimethylbenz[α]anthracene) (Yaoita et al. 2002).

Agaricus blazei species, in addition to anti-leukemia, also showed anticancer activity against KATO III stomach cancer cells and LU99 lung cancer cells. This activity was possible due to the presence of a steroid–blazein. The activity of six other steroids isolated in the form of acetone extracts from *A. blazei* species, which showed anticancer activity, was also examined. Based on the obtained research results, it was found that these compounds have anti-mutagenic potential (Lindequist et al. 2005; Endo et al. 2010; Patel and Goyal 2012; Roupas et al. 2012).

The bioactive steroid ergosta-4, 6, 8 (14), 22-tetra-3-one with antiproliferative and cytotoxic activity against **HepG2** cells was isolated from the edible species of *Russula cyanoxantha*. Under the influence of this compound, HepG2 cells undergo apoptosis as a result of inhibition of **cell cycle** (G2/M phase), chromatin condensation and cell nucleus fragmentation (Endo et al. 2010; Patel and Goyal 2012; Roupas et al. 2012).

6.1.2.12 Vitamins

The composition of five species of fungi from Uganda, *Termitomyces microcarpus, Termitomyces tyleranus, Termitomyces clypeatus, Polyporus tenucuilus* and *Volvariella speciosa* was investigated. The dried mushrooms contained folic acid, niacin, vitamin C and thiamine and small amounts of riboflavin, β-carotene (Fig. 6.7)

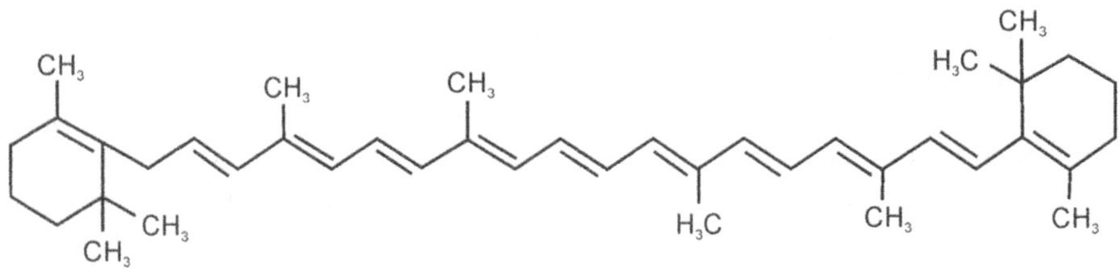

Fig. 6.7 β-carotene

and α-tocopherol. In these species, however, pantothenic acid, biotin and vitamin B_{12} have not been detected. The examined mushrooms contained a large amount of folic acid, the shortage of which is a frequent problem in the diet (Nakalembe et al. 2015).

The fruiting bodies of, inter alia, *A. bisporus* and *C. cibarius* are rich in B vitamins group, including thiamine, riboflavin, biotin and pyridoxine. These species are also a good source of tocopherols, carotenoids, and vitamin C (Barros et al. 2008b; Muszyńska et al. 2016a; Muszyńska et al. 2017a).

6.1.3 Other Compounds of Biological Activity

6.1.3.1 Statins

In 1971, while searching for new antibiotics produced by mushrooms, Akira Endo working for the pharmaceutical company Sankyo Co. discovered a class of compounds that reduced plasma cholesterol levels. Two years later, the research group isolated the compound responsible for the hypolipemic effect–mevastatin from *Penicillium citrinium*. Mevastatin served as the basis for further synthetic compounds from the statins group commonly used today (Wang et al. 1998; Carrizo et al. 2005; Li et al. 2008; Varrot et al. 2013; Singh et al. 2015).

Statins are inhibitors of 3-methyl-glutaryl coenzyme A reductase, the basic enzyme of endogenous cholesterol biosynthesis. They block the pathway of mevalonic acid, which causes lowering of cholesterol in the body. The **hypolipidemic** properties are used to treat: hypercholesterolemia, strokes and cardiovascular diseases. Other effects of statins proven *in vitro* and *in-vivo* include **cytotoxic** and **cytostatic** activity against various tumor cell lines. They result from such mechanisms of these compounds action as proapoptotic, anti-metastatic and anti-angiogenic activity. Natural statins include, inter alia, lovastatin, which in large quantities is observed in *P. ostreatus*, especially in its lamellar hymenophore. The addition of this species to the diet effectively reduces the accumulation of cholesterol in the blood and liver, reduces the production of VLDL and LDL for increasing HDL levels, attacks reduce the absorption of cholesterol and the activity of 3-hydroxy-3-methylglutaryl-coenzymeA reductase (HMG-CoA) in the liver. Another statin (eritadermin) having the above-mentioned properties is found in the Asian species *L. edodes* (Wang et al. 1998; Carrizo et al. 2005; Li et al. 2008; Varrot et al. 2013; Singh et al. 2015).

6.1.3.2 Theanine

The results of controlled trials in which edible mushrooms showed a synergistic, beneficial effect on reducing the incidence of cancer along with the consumption of green tea are also interesting. Theanine found in green tea also becomes a substance considered in cancer therapy. A popular species that becomes the source of theanine as a result of fermentation is *I. badia*.

Theanine shows a similar effect to synthetic anticancer substances such as irinotecan, doxorubicin or cisplatin, hence its effectiveness during therapy is assumed (Patel and Goyal 2012; Roupas et al. 2012).

Armillaria mellea fruiting bodies showed the presence of therapeutically important peptide-prosomatostatin, which has anticancer activity especially in the treatment of pancreatic cancer (Muszyńska et al. 2013b). The fruiting bodies and mycelia of *Antrodia camphorata* in traditional Eastern medicine are in turn used as an anticancer agent due to the presence of ubiquinone derivative. Importantly, this compound has selective activity, affecting only cancer cells, omitting the healthy cells (Hu et al. 2016).

6.1.3.3 Agaritine

Agaritine present in hot-obtained aqueous extracts was isolated from *A. blazei*. Agaritine was active against human lymphoma cells (U937), inducing their apoptosis *in vitro*. This activity also involved other leukemia cell lines, including HL60, MOLT4 and K562. The effect of this substance involved the damage to tumor cells DNA, through their fragmentation and to release cytochrome c from them. This mushroom species is used as an adjuvant during chemotherapy (Endo et al. 2010).

Currently, the consumption of mushrooms from Agaricales taxon containing agaritine is not only completely safe, but also beneficial for human health. *Agaricus bisporus* species has also been shown to be effective in inhibiting leukemia; among others HL60 cells by induction of their apoptosis (Lindequist et al. 2005; Endo et al. 2010; Roupas et al. 2012).

6.1.3.4 Bioelements

The bioelements with antioxidant and anti-inflammatory properties, accumulated by mushrooms, include, inter alia, zinc, copper, iron and selenium (Kalač 2010; Reczyński et al. 2013). The microelement with significant anti-inflammatory effects found in mushrooms is zinc. Its content in fruiting bodies of various species of edible mushrooms is in the range of 150–200 mg/kg DM (Reczyński et al. 2013; Muszyńska et al. 2015). The studies have also demonstrated good ability to accumulate zinc (II) ions by biomass from *in vitro* cultures from substrates enriched with this element (Reczyński et al. 2013; Muszyńska et al. 2015).

The mechanism of anti-inflammatory action of zinc is based on the induction of transcription of zinc-dependent transcription factors such as MTF-1 or A20 and inhibition of NF-κB activation, which is reflected in a decrease in the production of proinflammatory cytokines (Prasad 2014; Grzywacz et al. 2015). In turn, disorders of zinc homeostasis or its chronic deficiency contribute to the weakening of immunity, increase the production of pro-inflammatory cytokines and may affect the passage of inflammation into a chronic state (Cousins et al. 2006; Prasad 2014).

The studies conducted with the use of RAW 264.7 macrophages demonstrated the anti-inflammatory and anti-oxidative effect of *I. badia* biomass extracts obtained from substrates enriched with zinc (II) compounds. Cells incubated with extracts and activated with LPS showed a significant decrease in the **expression** of COX-2, cPGES, NF-κB p50 and NF-κB p65 proteins as well as the increase in GSTM1 expression, compared to the inflammatory cells induced by LPS (Ahmad et al. 2013). Zinc is a compound with antidepressant activity. The role of this element in the treatment of depression consists in modulating the sensitivity of NMDA-type glutamate receptors necessary for the proper action of antidepressants (Muszyńska et al. 2011a, c; Muszyńska and Sułkowska-Ziaja 2012; Muszyńska et al. 2013c).

Natural sources of zinc are fruiting bodies of edible mushrooms in which the content of this ingredient is from 25–200 mg/kg of dry matter. According to research, the best source of zinc is the fruiting bodies of *Lycoperdon perlatum*, in which its content is in the range of 150–200 mg/kg of dry matter. A good source of this bioelement is also the fruiting bodies of: *A. campestris, B. edulis, M. procera, I. badia* or *L. scabrum* (Kalač 2010; Reczyński et al. 2013).

Edible mushrooms are one of the best food sources of selenium (Falandysz 2008; Dogan et al. 2016). The content of selenium in fruiting bodies varies from about 0.5 to 20 mg/kg DM. Some species contain much more of it, for example selenium content in the fruiting bodies of *Albatrellus prescaprae* was 400 mg/kg DM, while in *A. sporporus* fruiting bodies this value reached the level of 0.150 mg/kg DM (Falandysz 2008; Maseko et al. 2014; Dogan et al. 2016). The element is essential for the proper functioning of the immune system.

Selenium with proteins forms seleno proteins, involved in the proper differentiation and proliferation and activation of cells of the immune system, thus affecting the innate and adaptive immune response. The immunoregulatory function of selenium is also manifested by effects on key leukocyte functions, such as adhesion and migration, **phagocytosis**, as well as cytokine secretion, which may be important in autoimmune diseases and chronic inflammation. Selenoproteins also play an important role in the cells in antioxidative processes. Selenium is an important factor in the fight against free radicals, which results, inter alia, from its presence in the structure of superoxide dismutase (**SOD**) or **glutathione peroxidase** (Huang et al. 2012; Maseko et al. 2014).

Magnesium is an element that guarantees proper physiological activity of the neuromuscular system. In edible mushroom species, the content of this element is in the range of 25–125 mg/kg DM, and the best is *Boletus edulis*, in which magnesium content reaches 75–125 mg/kg DM (Muszyńska et al. 2011a, c; Muszyńska and Sułkowska-Ziaja 2012; Muszyńska et al. 2013c).

Copper, like zinc, is a cofactor of many enzymes, including these of protective activity under oxidative stress. These enzymes include superoxide dismutase and ceruloplasmin-the main serum oxidase, which plays an important role in the transport of copper and iron homeostasis. An important role of copper in the regulation of inflammatory processes has been demonstrated (Moon et al. 2009). The high content of this element was demonstrated in such mushrooms species as *Lycoperdon perlatum* and *M. procera* (Kalač 2010).

6.1.4 Summary

It should be borne in mind that edible mushrooms contain a wealth of pharmacologically active compounds which properties we are only beginning to discover. Reports from recent years indicate that edible mushrooms extract show beneficial health and therapeutic effects, especially in relation to civilization diseases, such as those with cancer, inflammation, neurodegenerative or immune background. Certainly, some edible mushrooms can already be described as "superfood" and recommended as a valuable component of the daily diet.

References

Abdala-Valencia H, Berdnikovs S, McCary CA, Urick D, Mahadevia R, Marchese ME, Swartz K, Wright L, Mutlu GM, Cook-Mills JM (2012) Inhibition of allergic inflammation by supplementation with 5-hydroxytryptophan. Am J Physiol Lung Cell Mol Physiol 303:642–660

Ahmad MS, Ahmad S, Gautam B, Afzal M (2013) Antigenotoxic and anti-clastogenic potential of Agaricus bisporus against MMS induced toxicity in human lymphocyte cultures and in bone marrow cells of mice. Egypt J Med Hum Genet 14:395–402

Akramiene D, Kondrotas A, Didziapetriene J, Kevelaitis E (2007) Effects of beta-glucans on the immune system. Medicina (Kaunas) 43:597–606

Asahi T, Wu X, Shimoda H, Hisaka S, Harada E, Kanno T, Nakamura Y, Kato Y, Osawa T (2016) A mushroom-derived amino acid, ergothioneine, is a potential inhibitor of inflammation-related DNA halogenations. Biosci Biotechnol Biochem 80:313–317

Ayaz FA, Chuang LT, Torun H, Colak A, Seslin E, Presley J, Smith BR, Glew RH (2011) Fatty acid and amino acid compositions of selected wild-edible mushrooms consumed in Turkey. Int J Food Sci Nutr 62:328–335

Barros L, Cruz T, Baptista P, Estevinho LM, Ferreira IC (2008a) Wild and commercial mushrooms as source of nutrients and nutraceuticals. Food Chem Toxicol 46:2742–2747

Barros L, Venturini BA, Baptista P, Estevinho LM, Ferreira IC (2008b) Chemical composition and biological properties of portuguese wild mushrooms: a comprehensive study. J Agric Food Chem 56:3856–3862

Birdsall TC (1998) 5-Hydroxytryptophan: a clinically-effective serotonin precursor. Sci Rev Alternat Med 3:271–280

Brakhage AA, Schroeckh V (2011) Fungal secondary metabolites – strategies to activate silent gene clusters. Fungal Genet Biol 48:15–22

Brakhage AA (2013) Regulation of fungal secondary metabolism. Nat Rev Microbiol 11:21–32

Byerrum R, Clarke D, Lucas E, Ringler R, Stevens J, Stock CC (1957) Tumor inhibitors in Boletus edulis and other Holobasidiomycetes. Antibiot Chemother 7:1–4

Carrizo ME, Capaldi S, Perduca M, Irazoqui FJ, Nores GA, Monaco HL (2005) The anti-neoplastic lectin of the common edible mushroom (*Agaricus bisporus*) has two binding sites, each specific for a different configuration at a single epimeric hydroxyl. J Biol Chem 280:10614–10623

Chen SY, Ho KJ, Hsieh YJ, Wang LT, Mau JL (2012) Contents of lovastatin, g-aminobutyric acid and ergothioneine in mushroom fruiting bodies and mycelia. LWT - Food Sci Technol 47:274–278

Cheung PCK (2013) Mini-review on edible mushrooms as source of dietary fiber: preparation and health benefits. Food Sci Human Wellness 2:162–166

Choi YH (2015) Induction of apoptosis by an ethanol extract of *Poria cocos* Wolf. in human leukemia U937 cells. Oncol Rep 34:2533–2540

Cousins RJ, Liuzzi JP, Lichten LA (2006) Mammalian zinc transport, trafficking, and signals. J Biol Chem 281:24085–24089

Czapski J (2005) Antioxidant activity and phenolic content in some strains of mushrooms (*Agaricus bisporus*). Veg Crop Res 62:165–173

Dennis EA, Norris PC (2015) Eicosanoid storm in infection and inflammation. Nat Rev Immunol 15:511–523

Dogan H, Coteli E, Karatas F (2016) Determination of glutathione, selenium, and malondialdehyde in different edible mushroom species. Biol Trace Elem Res 174:459–463

Drori A, Shabat Y, Ben Ya'acov A, Danay O, Levanon D, Zolotarov L, Ilan Y (2016) Extracts from *Lentinula edodes* (Shiitake) edible mushrooms enriched with vitamin D exert an anti-inflammatory hepatoprotective effect. J Med Food 19:383–389

Echigo R, Shimohata N, Karatsu K, Yanom F, Kayasuga-Kariya Y, Fujisawa A, Ohto T, Kita Y, Nakamura M, Suzuki S, Mochizuki M, Shimizu T, Chung U, Sasaki N (2012) Trehalose treatment suppresses inflammation, oxidative stress, and vasospasm induced by experimental subarachnoid hemorrhage. J Transl Med 10:80

Elmastas M, Isildak O, Turkekul I, Temur N (2007) Determination of antioxidant activity and antioxidant compounds in wild edible mushrooms. J Food Compos Anal 20:337–345

Endo M, Beppu H, Akiyama H, Wakamatsu K, Ito S, Kawamoto Y, Shimpo K, Sumiya T, Koike T, Matsui T (2010) Agaritine purified from *Agaricus blazei* Murrill exerts anti-tumor activity against leukemic cells. Biochim Biophys Acta 1800:669–673

Esposito E, Cuzzocrea S (2010) Anti-inflammatory activity of melatonin in central nervous system. Curr Neuropharmacol 8:228–242

Falandysz J (2008) Selenium in edible mushrooms. J Environ Sci Health C 26:256–299

Ferreira ICFR, Barros L, Abreu (2009) Antioxidants in wild mushrooms. Curr Med Chem 16:1543–1560

Fox EM, Howlett BJ (2008) Secondary metabolism: regulation and role in fungal biology. Curr Opin Microbiol 11:481–487

Fukushima M, Nakano M, Morii Y, Ohashi T, Fujiwara Y, Sonoyama K (2000) Hepatic LDL receptor mRNA in rats is increased by dietary mushroom (*Agaricus bisporus*) fiber and sugar beet fiber. J Nutr 130:2151–2156

Fukuwatari T, Shibata K (2013) Nutritional aspect of tryptophan metabolism. Int J Tryptophan Res 6:3–8

Gdula-Argasińska J, Czepiel J, Woźniakiewicz A, Wojtoń K, Grzywacz A, Woźniakiewicz M, Jurczyszyn A, Perucki W, Librowski T (2015) N-3 fatty acids as resolvents of inflammation in the A549 cells. Pharmacol Rep 67:610–661

Grzywacz A, Gdula Argasińska J, Kała K, Opoka W, Muszyńska B (2016) Anti-inflammatory activity of biomass extracts of the bay mushroom, Imleria badia (*Agarico mycetes*), in RAW 264.7 cells. Int J Med Mushrooms 18:769–779

Grzywacz A, Gdula-Argasińska J, Muszyńska B, Tyszka-Czochara M, Librowski T, Opoka W (2015) Metal responsive transcription factor 1 (MTF-1) regulates zinc dependent cellular processes at the molecular level. Acta Biochim Pol 62:491–498

Hong IP, Nam SH, Sung GB, Chung IM, Hur H, Lee MW, Kim MK, Guo SX (2007) Chemical components of Paecilomyces tenuipes (Peck) Samson. Mycobiology 35:215–218

Hong SS, Lee JH, Jeong W, Kim N, Jin HZ, Hwang BY, Lee HJ, Lee SJ, Jang DS, Lee D (2012) Acetylenic acid analogues from the edible mushroom Chanterelle (*Cantharellus cibarius*) and their effects on the gene expression of peroxisome proliferator-activated receptor-gamma target genes. Bioorg Med Chem Lett 22:2347–2234

Hu YD, Zhang BB, Xu GR, Liao XR, Cheung PCK (2016) A mechanistic study on the biosynthetic regulation of bioactive metabolite Antroquinonol from edible and medicinal mushroom *Antrodia camphorata*. J Funct Foods 25:70–79

Huang Z, Aaron H, Rose AH, Hoffmann PR (2012) The role of selenium in inflammation and immunity: from molecular mechanisms to therapeutic opportunities. Antioxid Redox Signal 16:705–743

Jani AR, Nag S, Setty SR (2016) Visualization of intracellular Tyrosinase activity in vitro. Bio Protocol 6:e1794

Jedinak A, Dudhgaonkar S, Wu Q, Simon J, Sliva D (2011) Anti-inflammatory activity of edible oyster mushroom is mediated through the inhibition of NF-κB and AP-1 signaling. Nutr J 10:52–62

Jeong JW, Lee HH, Han MH, Kim GY, Hong SH, Park C, Choi YH (2014) Ethanol extract of Poria cocos reduces the production of inflammatory mediators by suppressing the NF-kappa B signaling pathway in lipopolysaccharide-stimulated RAW 264.7 macrophages. BMC Complement Altern Med 14:101

Jeong SC, Jeong YT, Yang BK, Islam R, Koyyalamudi SR, Pang G, Cho KY, Song CH (2010) White button mushroom (Agaricus bisporus) lowers blood glucose and cholesterol levels in diabetic and hypercholesterolemic rats. Nutr Res 30:49–56

Jiang J, Grieb B, Thyagarajan A, Sliva D (2008) Ganoderic acids suppress growth and invasive behavior of breast cancer cells by modulating AP-1 and NF-kappaB signaling. Int J Mol Med 21:577–584

Jiang Q (2014) Natural forms of vitamin E: metabolism, antioxidant, and anti-inflammatory activities and their role in disease prevention and therapy. Free Radic Biol Med 72:76–90

Kalač P (2010) Trace element contents in European species of wild growing edible mushrooms: a review for the period 2000–2009. Food Chem 122:2–15

Kampmann M, Hoffrichter AC, Stalinski D, Wichmann R (2015) Kinetic characterization of tyrosinase containing mushroom (*Agaricus bisporus*) cells immobilized in silica alginate. J Mol Catal B Enzym 116:124–133

Kosanić M, Ranković B, Rančić A, Stanojković T (2016) Evaluation of metal concentration and antioxidant, antimicrobial, and anticancer potentials of two edible mushrooms *Lactarius deliciosus* and *Macrolepiota procera*. J Food Drug Anal 24:477–484

Kozarski MS, Klaus AS, Nikšić MP, Van Griensven L, Vrvić MM, Jakovljević DM (2014) Antioxidative and immunomodulating activities of polysaccharide extracts of the medicinal mushrooms *Agaricus bisporus*, *Agaricus brasiliensis*, *Ganoderma lucidum* and *Phellinus linteus*. Chem Ind 68:305–320

Krzyczkowski W, Malinowska E, Suchocki P, Kleps J, Olejnik M, Herold F (2009) Isolation and quantitative determination of ergosterol peroxide in various edible mushroom species. Food Chem 113:351–355

Labus K, Turek J, Bryjak J (2011) Efficient Agaricus bisporus tyrosinase immobilization on cellulose-based carriers. Biochem Eng J 56:232–240

Li YR, Liu QH, Wang HX, Ng TB (2008) A novel lectin with potent antitumor, mitogenic and HIV-1 reverse transcriptase inhibitory activities from the edible mushroom *Pleurotus citrinopileatus*. Biochim Biophys Acta 1780:51–57

Lindequist U, Niedermeyer THJ, Jülich WD (2005) The pharmacological potential of mushrooms. J Evid Based Complementary Altern Med 2:285–299

Ma L, Chen H, Dong P, Lu X (2013) Anti-inflammatory and anticancer activities of extracts and compounds from the mushroom *Inonotus obliquus*. Food Chem 139:503–508

Mantovani MS, Bellini MF, Angeli JPF, Oliveira RJ, Silva AF, Ribeiro LR (2008) β-Glucans in promoting health: prevention against mutation and cancer. Mutat Res 658:154–161

Maseko T, Howell K, Dunshea FR, Ng K (2014) Selenium-enriched Agaricus bisporus increases expression and activity of glutathione peroxidase-1 and expression of glutathione peroxidase-2 in rat colon. Food Chem 146:327–333

Meimaroglou DM, Galanopoulou D, Markaki P (2009) Study of the effect of methyl jasmonate concentration on aflatoxin B 1 biosynthesis by *Aspergillus parasiticus* in yeast extract sucrose medium. Int J Microbiol:1–7

Meng X, Liang H, Luo L (2016) Antitumor polysaccharides from mushrooms: a review on the structural characteristics, antitumor mechanisms and immunomodulating activities. Carbohyd Research 424:30–41

Moon MK, Lee YJ, Kim JS, Kang DG, Lee HS (2009) Effect of Caffeic acid on tumor necrosis factor-alpha-induced vascular inflammation in human umbilical vein endothelial cells. Biol Pharm Bull 32:1371–1377

Muszyńska B, Maślanka A, Sułkowska-Ziaja K, Ekiert H (2011a) Analysis of indole compounds in *Armillaria mellea* fruiting bodies. Acta Pol Pharm 68:93–97
Muszyńska B, Sułkowska-Ziaja K, Ekiert H (2011b) Indole compounds in fruiting bodies of some edible Basidiomycota species. Food Chem 125:1306–1308
Muszyńska B, Sułkowska-Ziaja K, Ekiert H (2011c) Indole compounds in some culinary – medicinal higher Basidiomycetes from Poland. Int J Med Mushrooms 13:449–454
Muszyńska B, Sułkowska-Ziaja K, Ekiert H (2011d) Właściwości lecznicze i dietetyczne wybranych jadalnych grzybów wielkoowocnikowych. Farm Pol 67:551–560
Muszyńska B (2012) Jadalne gatunki grzybów źródłem substancji dietetycznych i leczniczych. Wydawnictwo ZOZ Ośrodka UMEA Shinoda-Kuracejo, Kraków, Poland
Muszyńska B, Sułkowska-Ziaja K (2012) Analysis of indole compounds in edible Basidiomycota species after thermal processing. Food Chem 132:455–459
Muszyńska B, Sułkowska-Ziaja K, Ekiert H (2013a) Phenolic acids in selected edible Basidiomycota species: *Armillaria mellea*, *Boletus badius*, *Boletus edulis*, *Cantharellus cibarius*, *Lactarius deliciosus* and *Pleurotus ostreatus*. Acta Sci Pol-Hortoru 12:107–116
Muszyńska B, Sułkowska-Ziaja K, Łojewski M, Opoka W, Zając M, Rojowski J (2013b) Edible mushrooms in prophylaxis and treatment of human diseases. Med Rev 101:170–183
Muszyńska B, Sułkowska-Ziaja K, Wójcik A (2013c) Levels of physiological active indole derivatives in the fruiting bodies of some edible mushrooms (Basidiomycota) before and after thermal processing. Mycoscience 54:321–332
Muszyńska B, Smalec A, Sułkowska-Ziaja K, Opoka W, Reczyński W, Baś B (2015) Culinary-medicinal *Agaricus bisporus* (white button mushroom) and its in vitro cultures as a source of selected biologically-active elements. J Food Sci Technol 52:7337–7344
Muszyńska B, Kała K, Firlej A, Sułkowska-Ziaja K (2016a) *Cantharellus cibarius* – culinary-medicinal mushroom content and biological activity. Acta Pol Pharm 73:589–598
Muszyńska B, Kała K, Sułkowska-Ziaja K, Krakowska A, Opoka W (2016b) *Agaricus bisporus* and its *in-vitro* culture as a source of indole compounds released into artificial digestive juices. Food Chem 199:509–515
Muszyńska B, Kała K, Rojowski J, Grzywacz A, Opoka W (2017a) Composition and biological properties of Agaricus bisporus fruiting bodies. Pol J Food Nutr Sci 67:1–10
Muszyńska B, Grzywacz-Kisielewska A, Kała K, Gdula-Argasińska J (2017b) Anti-inflammatory properties of edible mushrooms: a review. Food Chem 243:373–381
Muszyńska B, Pazdur P, Lazur J, Sułkowska-Ziaja K (2017c) Lentinula edodes (Shiitake) – biological activity. Med Rev 108:189–195
Muta T (2006) Molecular basis for invertebrate innate immune recognition of (1–>3)-beta-D-glucan as a pathogen-associated molecular pattern. Curr Pharm Des 12:4155–4161
Nakalembe I, Kabasa JD, Olila D (2015) Comparative nutrient composition of selected wild edible mushrooms from two agro-ecological zones, Uganda. Springerplus 4:433
Ng TB (2004) Peptides and protein from fungi. Peptides 25:1055–1073
Novaes MR, Valadares F, Reis MC, Gonçalves DR, Menezes MC (2013) The effects of dietary supplementation with *Agaricales* mushrooms and other medicinal fungi on breast cancer: evidence-based medicine. Clinics 66:2133–2139
Novak M, Vetvicka V (2009) Glucans as biological response modifiers. Endocr Metab Immune Disord Drug Targets 9:67–75
Opoka W, Kała K, Krężałek R, Sułkowska-Ziaja K, Maślanka A, Muszyńska B (2017) TLC-densitometry analysis of indole compounds in mycelial culture of *Imleria badia* and *Agaricus bisporus* enriched with precursors – serine or Anthranilic acid. Acta Chromatogr:1–8
Öztürk M, Duru ME, Kivrak S, Mercan-Doğan N, Türkoglu A, Özler MA (2011) In vitro antioxidant, anticholinesterase and antimicrobial activity studies on three Agaricus species with fatty acid compositions and iron contents: a comparative study on the three most edible mushrooms. Food Chem Toxicol 49:1353–1360
Palacios I, Lozano M, Moro C, D'Arrigo M, Rostagno MA, Martinez JA, Garcia-Lafuente A, Guillamon E, Villares A (2011) Antioxidant properties of phenolic compounds occurring in edible mushroom. Food Chem 128:674–678

Palazón J, Cusidó RM, Bonfill M, Mallol A, Moyano E, Morales C, Piñol MT (2003) Elicitation of different Panax ginseng transformed root phenotypes for an improved ginsenoside production. Plant Physiol Biochem 41:1019–1025

Patel S, Goyal A (2012) Recent developments in mushrooms as anti-cancer therapeutics: a review. 3 Biotech 2(1):1–15

Phillips KM, Ruggio DM, Horst RL, Minor B, Simon RR, Feeney MJ, Byrdwell WC, Haytowitz DB (2011) Vitamin D and sterol composition of 10 types of mushrooms from retail suppliers in the United States. J Agric Food Chem 59:7841–7853

Prasad AS (2014) Zinc is an antioxidant and anti-inflammatory agent: its role in human health. Front Nutr 1:1–10

Rajewska J, Bałasińska B (2004) Związki biologicznie aktywne zawarte w grzybach jadalnych i ich korzystny wpływ na zdrowie. Postepy Hig Med Dosw 58:352–357

Rao YK, Fang SH, Tzeng YM (2007) Evaluation of the anti-inflammatory and anti-proliferation tumoral cells activities of *Antrodia camphorata, Cordyceps sinensis*, and *Cinnamomum osmophloeum* bark extracts. J Ethnopharmacol 114:78–85

Reczyński W, Muszyńska B, Opoka W, Smalec A, Sułkowska-Ziaja K, Malec M (2013) Comparative study of metals accumulation in cultured in vitro mycelium and naturally grown fruiting bodies of *Boletus badius* and *Cantharellus cibarius*. Biol Trace Elem Res 153:355–362

Reis FS, Martins A, Barros L, Ferreira IC (2012) Antioxidant properties and phenolic profile of the most widely appreciated cultivated mushrooms: a comparative study between in vivo and in vitro samples. Food Chem Toxicol 50:201–207

Reiter RJ, Calvo JR, Karbownik M, Qi W, Tan DX, Ann NY (2000) Melatonin and its relation to the immune system and inflammation. Acad Sci 917:376–386

Roupas P, Keogh J, Noakes M, Margetts C, Taylor P (2012) The role of edible mushrooms in health: evaluation of the evidence. J Funct Foods 4:687–709

Ruthes AC, Smiderle FR, Iacomini M (2015) D-Glucans from edible mushrooms: a review on the extraction, purification and chemical characterization approaches. Carbohydr Polymers 117:753–761

Shao S, Hernandez M, Kramer JKG, Rinker DL, Tsao RTJ (2010) Ergosterol profiles, fatty acid composition, and antioxidant activities of button mushrooms as affected by tissue part and developmental stage. J Agric Food Chem 58:11616–11625

Shi YL, Benzie IF, Buswell J (2002) Role of tyrosinase in the genoprotective effect of the edible mushroom *Agaricus bisporus*. Life Sci 70:1595–1608

Singdevsachan SK, Auroshree P, Mishra J, Baliyarsingh B, Tayung K, Thatoi H (2016) Mushroom polysaccharides as potential prebiotics with their antitumor and immunomodulating properties: a review. Bioact Carbohydr Diet Fibre 7:1–14

Singh SS, Wang H, Chan YS, Pan W, Dan X, Yin CM, Akkouh O, Ng TB (2015) Lectins from edible mushrooms Molecules 20:446–469

Siwulski M, Sobieralski K, Sas-Golak I (2014) Wartość odżywcza i prozdrowotna grzybów. Zywn-Nauk Technol Ja 1:16–28

Słomiński A, Semak I, Zjawiony J, Wortsman J, Gandy MN, Li J, Zbytek B, Li W, Tuckey RC (2015) Enzymatic metabolism of ergosterol by cytochrome P450scc (CYP11A1) to biologically active 17α,24-dihydroxyergosterol. Chem Biol 12:931–939

Smiderle FR, Alquini G, Tadra-Sfeir MZ, Iacomini M, Wichers HJ, Van Griensven LJLD (2013) Agaricus bisporus and Agaricus brasiliensis (1 → 6)-ß-d-glucans show immunostimulatory activity on human THP-1 derived macrophages. Carbohydr Polym 94:91–99

Stanikunaite R, Khan SI, Trappe JM, Ross SA (2009) Cyclooxygenase-2 inhibitory and antioxidant compounds from the truffle *Elaphomyces granulatus*. Phytother Res 23:575–578

Stepień M, O'Mahony L, O'Sullivan A, Collier J, Fraser WD, Gibney MJ, Nugent AP, Brennan L (2013) Effect of supplementation with vitamin D_2-enhanced mushrooms on vitamin D status in healthy adults. J Nutr 2:e29

Strange RC, Shipman KE, Ramachandran S (2015) Metabolic syndrome: a review of the role of vitamin D in mediating susceptibility and outcome. World J Diabetes 6:896–911

Takei T, Yoshida M, Ohnishi-Kameyama M, Kobori M (2005) Ergosterol peroxide, an apoptosis-inducing component isolated from *Sarcodon aspratus* (Berk.). Biosci Biotechnol Biochem 69:212–215

Tian Y, Zhao Y, Huang J, Zeng H, Zheng B (2016) Effects of different drying methods on the product quality and volatile compounds of whole shiitake mushrooms. Food Chem 197:714–722

Turner EH, Loftis JM, Blackwell AD (2006) Serotonin a la carte: supplementation with the serotonin precursor 5-hydroxytryptophan. Pharmacol Ther 109:325–338

Varrot A, Basheer SM, Imberty A (2013) Fungal lectins: structure, function and potential applications. Curr Opin Struct Biol 23:678–685

Wang H, Ng TB, Ooi VEC (1998) Lectins from mushrooms. Mycol Res 102:897–906

Wang HX, Liu WK, Ng TB, Ooi VEC, Chang ST (1996) The immunomodulatory and antitumor activities of lectins from the mushroom *Tricholoma mongolicum*. Immunopharmacology 31:205–211

Wang Y, Liu Y, Wang H, Li C, Qi P, Bao J (2012) Agaricus bisporus lectins mediate sislet β cell proliferation through regulation of cell cycle proteins. Exp Biol Med 237:287–296

Wannet WJ, Op den Camp HJ, Wisselink HW, van der Drift C, Van Griensven LJ, Vogels GD (1998) Trehalose treatment suppresses inflammation, oxidative stress, and vasospasm induced by experimental subarachnoid hemorrhage. Biochim Biophys Acta 1425:177–188

Wu T, Zivanovic DFA, Sams CE (2004) Chitin and chitosan–value-added products from mushroom waste. J Agric Food Chem 52:7905–7910

Yaoita Y, Yoshihara Y, Kakuda R, Machida K, Kikuchi M (2002) New sterols from two edible mushrooms, *Pleurotus eryngii* and *Panellus serotinus*. Chem Pharm Bull 50:551–553

Yuan JP, Kuang HC, Wang JH, Liu X (2008) Evaluation of ergosterol and its esters in the pileus, gill, and stipe tissues of agaric fungi and their relative changes in the comminuted fungal tissues. Appl Microbiol Biotechnol 80:459–465

Zaidi KU, Ali AS, Ali SA (2014) Purification and characterization of melanogenic enzyme tyrosinase from button mushroom. Enzyme Res 2014:1–6

Zhang GQ, Wang YF, Zhang XQ, Ng TB, Wang HX (2010) Purification and characterization of a novel laccase from the edible mushroom *Clitocybe maxima*. Process Biochem 45:627–633

Zhang M, Cui SW, Cheung PCK, Wang Q (2007) Antitumor polysaccharides from mushrooms: a review on their isolation process, structural characteristics and antitumor activity. Trends Food Sci Technol 18:4–19

Zong A, Cao H, Wang F (2012) Anticancer polysaccharides from natural resources: a review of recent research. Carbohydr Polym 90:1395–1410

Chapter 7
Biotechnological Requirements for the Commercial Cultivation of Macrofungi: Substrate and Casing Layer

Jaime Carrasco, Maria L. Tello, Margarita Perez, and Gail Preston

7.1 Mushroom Growth and Development

The consumption of mushrooms dates back to antiquity. Its **organoleptic** and medicinal properties have been appreciated since the time of ancient Greece and Rome, when one of the most appreciated species was baptized as Caesar's mushroom (*Amanita caesarea*), while some poisonous species have been used to dethrone emperors, popes and kings (Van Griensven (1988).

According to Chang and Miles (2004), mushrooms are defined as the fruiting body of macrofungi that appears below (hypogeous) or above (epigeous) the ground, which is large enough to be identified by the naked eye and picked by hand. Although most mushrooms are produced by **basidiomycetes** (mainly from the *Agaricomycotina* taxa), some **ascomycetes**, like the genera *Morchella* and *Tuber*, also generate mushrooms.

Some 100 species of mushrooms have been cultivated commercially, among them, around 20 have been exploited on an industrial scale. The most cultivated genera worldwide (accounting for 85% of the world's cultivation of edible mushrooms) are *Lentinula edodes* (shiitake), *Agaricus* (mainly *Agaricus bisporus*), *Pleurotus* spp. (5 or 6 species), *Auricularia* and *Flammulina*, to support an industry that has been valued at approximately $58 billion in 2013 (Royse et al. 2017). The challenge of mushroom production lies in the need to integrate the various activities that must be perfectly coordinated for the commercial exploitation of the product, as well as the management of waste (Pardo-Giménez et al. 2017).

J. Carrasco (✉) · G. Preston
Department of Plant Sciences, University of Oxford, Oxford, UK
e-mail: jaime.carrasco@plants.ox.ac.uk

M. L. Tello · M. Perez
Centro Tecnológico de Investigación del Champiñón de La Rioja (CTICH), Autol, Spain

B. P. Singh et al. (eds.), *Biology of Macrofungi*, Fungal Biology,
https://doi.org/10.1007/978-3-030-02622-6_7

As noted above, the majority of the cultivated species belong to the basidiomycetes division which generally presents a relatively simple life cycle. The basidiospores are the only known spores of this subdivision and none sclerotia or latency structures have been described. The cycle, therefore, consists of the germination of the basidiospores, under certain environmental conditions, to produce hyphae with usually a single nucleus (monokaryon) that fuse with hyphae in the vicinity to develop hyphal compartments with two nuclei (dikaryon) that comprise the mushroom mycelium (vegetative tissue) (Schardl and Craven 2003). A sharp and controlled environmental change drives the morphological shift from the vegetative to the reproductive tissue while inducing fructification and promoting the development of basidiomes (Noble et al. 2009). Mushrooms are, therefore, multicellular structures generated by the differentiation of the cells from the vegetative mycelium (Eastwood and Burton 2002).

7.2 Commercial Mycelium Spawn

The mycelium or inoculum of the crop, known colloquially as "spawn" is commercially produced in specialized laboratories under sterile environment. Initially the mycelium can be obtained by plating tissue from a wild mushroom on nutritive growth medium or by reproducing a mycelium through germination of single or multiple spores (Chang and Miles 2004). The commercial inoculum is obtained by enhancing the growth of the mycelium (from a collection of selected strains) on the surface of different carriers, including organic carriers such as cereal grains (rye, millet or sorghum) or sawdust, or synthetic speed spawn (Speed Spawn™ from L. F. Lambert Spawn Company, Coatesville, USA or Fusion™ from Amycel Spawn Mate Inc., San Juan Bautista, USA). In addition, liquid spawn is available for the grower. The process consists of hydrating the cereal grains/sawdust with hot water (35–45% w/w), and mixing with a combination of gypsum and calcium carbonate to prevent caking and provide an adequate pH. The grain is then sterilized at 121 °C for at least 2 h. After cooling, the grain is inoculated with the mycelium of the selected strain under axenic conditions. Next, the grain is transferred in commercial formats to an incubation room (temperature will be set up depending on the species), where the mycelium develops, invading the grains (Fig. 7.1). The mycelium

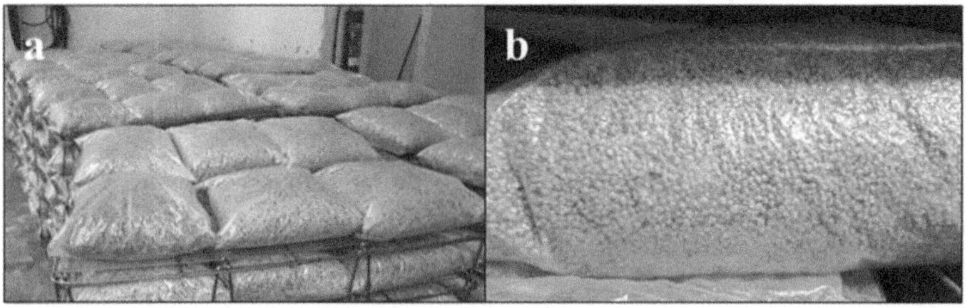

Fig. 7.1 Commercial spawn grain ready for inoculation: (**a**) Bags of commercial spawn; (**b**) Millet grains overgrowth with mushroom mycelium

thus grown under conditions of strict hygiene constitutes the "spawn" to inoculate the selective growing substrate (Carrasco 2016).

Liquid mycelium can be generated by resuspending fungal spores in nutritive broth, minimal medium or simply water plus some drops of an anionic surfactant such as Tween 80 in order to avoid clusters of spores. The suspension of germinated/non-germinated spores may be used to inoculate the selective substrate.

7.3 Nutritional Requirements for Cultivated Mushrooms

Mushrooms are **heterotrophic** organisms which require external nutrients. The vegetative mycelium supplies nutrients for the growth of basidiomes. Mushrooms produce a number of enzymes including **lignin-degrading** enzymes (laccases, lignin peroxidases, manganese peroxidases, arylalcohol oxidase, aryl-alcohol dehydrogenases or quinone reductases), and hemicellulose and **cellulose-degrading** enzymes (xylanase, cellulases or cellobiose dehydrogenase), that generate extracellular radicals to facilitate the degradation of several lignocellulosic substrates (Sánchez 2009; Kabel et al. 2017; Vos et al. 2017a) (Fig. 7.2).

Due to their metabolic requirements, cultivation of edible mushrooms combines the production of nutritional-rich food (mushrooms in general contain approximately 90% water, between 27% and 48% of protein in dry matter, d.m., less than

Fig. 7.2 Modern compost yard: (**a**) Watering and fermentation of wheat straw. (**b**) Phase I compost. (**c**) Phase II fermenting in camera. (**d**) Camera for the germination of compost: Phase III

60% of carbohydrates in d.m. and between 2–8% of lipids in d.m. (Sánchez 2004) with the reduction of waste, since most of the substrates employed are designed from agricultural by-products (Sánchez 2010).

Mushrooms require oxygen and a specific pH in order to grow properly. C and N are the two main components of fungi for structural and energy requirements; P, K and Mg are considered also macronutrients for mushrooms while other trace elements such as Fe, Zn, Mn, Cu, Se or Mo are considered necessary for diverse functions (Chang and Miles 2004). Noteworthy, some edible mushrooms have high selenium content, a vital element for human health which is commonly deficient in most human diets (Falandysz 2008).

7.3.1 Substrate Employed in Mushroom Cultivation

The substrate designed for mushroom cultivation is a selective media for the growth of different species which provides nutrients and support for the development of the mycelium and the fructification of the basidiomes. The preparation of substrates suitable for mushroom cultivation includes composted (through fermentation and pasteurization) and non-composted (mix and water + sterilization) mixtures with agricultural or lignin-based forestry by-products as the main ingredients (Fig. 7.3).

There is not standard formula fixed for growing each species of mushroom. Depending on the country or the area where the mushroom farm is located, a huge diversity of raw materials can be employed. Rice straw, for instance, is widely used for the production of substrates in Asia. On the other hand, in Europe, wheat straw is more commonly used. Therefore, the substrate recipe for growing fungi is adapted to the most abundant agricultural wastes produced locally and continuously throughout the year with sufficient quality and quantity. This fact contributes to create profitability while reducing transport costs and managing wastes efficiently (Pardo et al. 2017). Table 7.1 summarizes the substrates designed to cultivate the most common edible mushrooms.

7.3.1.1 Preparation of Selective Substrate Through Fermentation

The substrates designed for the cultivation of some species are produced through a solid-state fermentation process along with the action of native **microbiota**, which matures the raw ingredients through different stages. At the end of the process a selective substrate, ready for the inoculation of the target mycelium, is achieved. *Agaricus* species and *Pleurotus* spp. are mostly cultivated using these substrates:

Agaricus bisporus **Agaricus bisporus** (Lange) Imbach, is the best example and the most detailed process to produce selective substrate through fermentation. A two-stage process takes place from agricultural waste materials. At the beginning of phase I, raw materials (wheat straw and horse or chicken manure) are mixed and wetted. Phase I takes around 6–7 days during which the compost mass reaches temperatures

Fig. 7.3 Commercial cultivation on substrates prepared by fermentation: (**a**) White button mushroom; (**b**) Portobello mushroom; (**c**) *Agaricus blazei*; (**d**) *Pleurotus ostreatus*

up to 80 °C due to microbial activity. Initially the mesophilic microbiota starts to develop (which turns carbohydrates and proteins into heat and ammonia). The **mesophilic** microbiota is naturally replaced by thermophilic microbiota when the temperature rises. These reactions soften the compost mass. Subsequently, phase II starts with an initial temperature of 50 °C in the compost, followed by a 2-day period at 60 °C (pasteurization) and a 3-day period at 45 °C (conditioning), at the end of the phase II the mass of compost cools down until 25 °C, when the substrate is ready for spawning. Phase III compost is produced by incubating the compost at 25 °C for 16–18 days until the vegetative mycelium fully colonizes the compost mass. Only carbohydrates that are more difficult to degrade remain in the compost at the start of Phase III when the lignin degrading machinery of *A. bisporus* starts to work (Kabel et al. 2017; Pardo et al. 2017; Vos et al. 2017a).

The **microbial community** plays a decisive role along the process. Several fungal and bacterial species have been identified along the composting cycle. Among them, for instance, the thermophilic fungus *Scytalidium thermophilum* removes ammonia during phase II and therefore suppresses competitors of *A. bisporus* such as *Trichoderma* spp. (Vos et al. 2017b). In addition, some authors remark on the importance of bacterial dynamics under composting conditions. According to Vieira

Table 7.1 Substrates recipe for the edible mushroom species most commonly cultivated

Species	Substrate récipe	Treatment	Casing	Commercial Format	References
Agaricus bisporus	In Europe, composted mixture of wheat straw (40 to 50% of the total dry weight), horse manure or stable bedding (20–25%), poultry manure (10–15%) and gypsum (5 to 10%)	Composting (2 phases): Fermentation and pasterurisation	Yes	>Phase II, Phase III or Phase IV > bulk or compo-blocks (15–25 kg / 0.25 m^2)	Kabel et al. 2017; Pardo et al. 2017)
Pleurotus ostreatus	Mixture of cottonseed hulls and wheat straw. The substrate is milled to a length of about 2 to 6 cm. Mix and water to the desired level	Pasteurization (60 °C for 1 to 2 h)	No	The substrate is spawned and filled (from 10 to 15 kg) into black perforated polyethylene bags	(Sánchez 2010)
Lentinula edodes	Non-composted (n-c) mix: sawdust, rice bran, wheat bran, corn bran and $CaCO_3$; moisture content (m.c.) to 60%	Sterilisation (121 °C, 1 h)	No	>Natural logs >Polypropylene bags	(Chang and Miles 2004)
Auricularia sp.	A n-c mixture of 90% sawdust, 9% rice bran and 1% CaCO3; m.c. to 65%	Sterilisation (121 °C, 1 h-80 min)	No	>Polyethylene bags	(Liang et al. 2016; Wu et al. 2017)
Flammulina velutipes	N-c mix, mostly sawdust (80%) and rice bran (20%). Mix and water to 58–60%	Sterilisation (121 °C-1 h/ 95 °C-4 h)	No	>Bag cultivation >Bottle cultivation	(Chang and Miles 2004; Harith et al. 2014; Zhou et al. 2017)
Pleurotus eryngii	N-c mix of sawdust with materials like straw, maize stalks or cotton seed hulls, enriched with nitrogen-rich additives. C/N ratio of 30–40 and m.c. around 70%	Sterilisation (121 °C, 90 min)	Yes/No	>Bag cultivation (1–4 kg) > Bottle cultivation	(Rodriguez-Estrada et al. 2009)
Agrocybe aegerita	N-c lignocellulosic materials: Wheat straw, cotton waste, sawdust, corn cobs and peanut shells; m.c. 65%	Sterilisation (121 °C, 1 h)	No	>Polyethylene bags	(Rodriguez-Estrada et al. 2009)
Ganoderma lucidum	N-c mix of 75–77% sawdust, 2% $CaCO_3$, 5–9% corn bran, 5–9% oak bran, 3.5–4% wheat bran; m. c. 65%	Sterilisation (121 °C, 1–2 h)	No	>Natural logs >Polypropylene bags >Bottle cultivation	Zhou 2017; Maszlavér 2008; Pérez-Clavijo et al. 2016)

Volvariella volvácea	N-c mix by soaking cotton wastes with particle size of 2–3 cm in water for 24 h; mixed with wheat bran (10%) and $CaCO_3$ (2%).	Sterilisation (121 °C, 2 h)	No	>Polyethylene bags	(Philippoussis et al. 2001)
Hypsizygus tessulatus	N-c mix of sawdust (18.2%), rice bran (8.8%), soybean shell (5.3%), corncob meal (4.7%) and $CaCO_3$ (0.5%); m.c. 63–70%	Sterilisation (121 °C, 1 h)	No	>Bag cultivation >Bottle cultivation	(Harada et al. 2004; Yamanaka 2017)
Pholiota nameko	N-c mix of hardwood sawdust (90%) + corn bran, wheat bran, and dried tofu refuse (10%). Mixed and watered to a 64–65%	Sterilisation (121 °C, 1 h)	No	>Polyethylene bags	Yamanaka (2017)
Grifola frondosa	N-c mix of sawdust powder with 15% rice bran and 5% wheat bran; m.c. 65%	Sterilisation (121 °C-2 h/ 110 °C-7 h)	No	>Polypropylene bags (2.5 kg)	Mayuzumi and Mizuno (1997)
Calocybe indica	N-c mix: Coconut coir, kash, paddy straw, maize stalks, rice straw, sorghum stalks, sugarcane bagasse, sugarcane leaf, wheat bran or waste cotton chopped to 3–5 cm; m.c. 65%	Sterilisation (121 °C, 1 h)	Yes	>Polypropylene bags	(Amin et al. 2010; Lakshmipathy et al. 2017)
Hericium erinaceus	N-c mix: Sugarcane bagasse, sawdust, cottonseed hulls, corncobs, and chopped up paddy straw - with rice or wheat bran, sucrose and gypsum	Sterilisation (121 °C, 1 h)	No	>Polyethylene bags	(Chang and Miles 2004)

and Pecchia (2017), *Bacillales* members showed potential to be a key group of bacteria during phase I composting in relation to the breakdown of raw materials and the selectivity of the ultimate substrate.

Several species from the genera *Agaricus* can be cultivated commercially using this compost recipe, including *Agaricus bisporus* (Lange) Imbach (Fig. 7.4a), *Agaricus bitorquis* (Quelet) Saccardo or *Agaricus subrufescens* Peck (Fig. 7.4b, c) (Gea et al. 2003; Pardo-Giménez et al. 2014; Pardo et al. 2017).

Pleurotus ostreatus The oyster mushroom, ***Pleurotus ostreatus*** (Jacq: Fries) (Fig. 7.4d), has been cultivated in different countries using a variety of lignocellulosic substrates. Wheat straw is the most common base material employed. In contrast to the compost employed in white button mushroom cultivation, the straw must be milled to a length of about 2 to 6 cm. One of the most common substrates used is a mixture of cottonseed hulls and wheat straw. Some additives such as calcium sulphate or urea can be incorporated within the mixture as nitrogen source. The materials are thoroughly mixed and watered and subsequently the substrate is subjected to

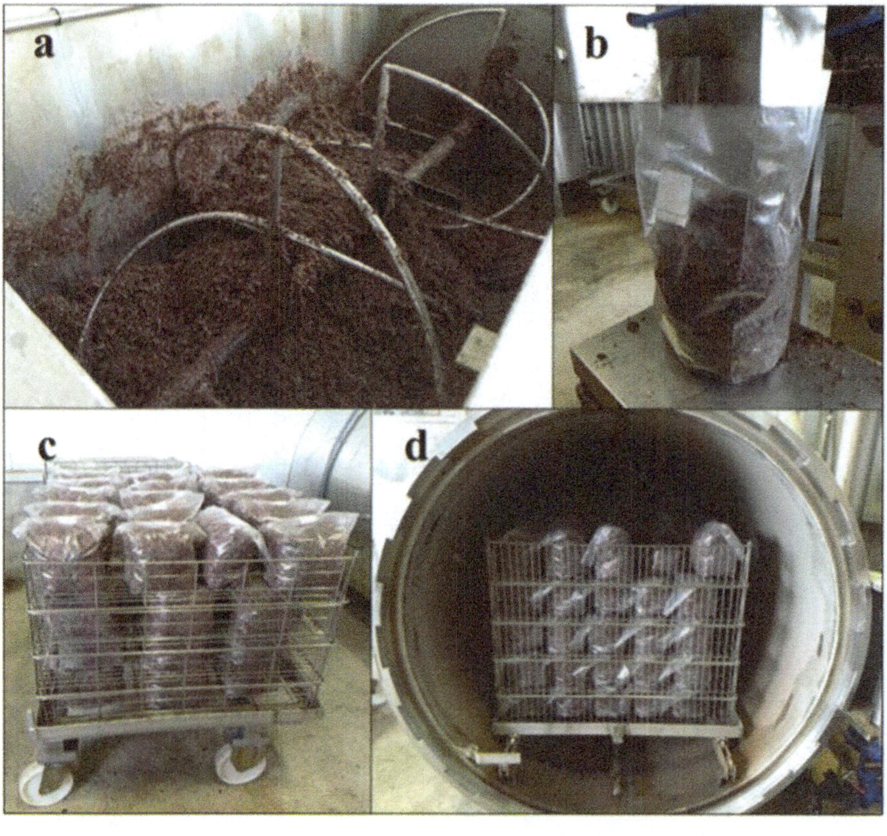

Fig. 7.4 Preparation of non-composted substrates: (**a**) Mixing and watering ingredients; (**b**) Filling polyethylene bags with mix substrate; (**c**) Bags of substrate ready for sterilisation; (**d**) Substrate bags in autoclave

fermentation/pasteurization heat processing (approximately 60–65 °C, 8 h) through injection of live steam. A progressive cooling of the substrate is conducted (approximately 15 h) until the mixture is ready for spawning (Pardo and Pardo-González 2009; Sánchez 2010). Beside *P. ostreatus*, other edible species from the genera *Pleurotus* are suitable for being cultivated with this kind of substrate, including *P. pulmonarius* (Fr.) Quél., *P. citrinopileatus* Sing. (Lentinaceae), *P. sajor-caju* (Fr.) singer and *P. cistidiosus* O.K. Mill (Chang and Miles 2004).

7.3.1.2 Non-composted Materials: Sterilization of Substrates

Non-composted substrates are based on lignicolous material, such as sawdust, with different additives, which undergoes a sterilization process for a determined period in order to eliminate mushroom competitors or parasites (Fig. 7.4). Details of some of the most common species cultivated by this methodology are summarized below:

Lentinula edodes Shiitake mushroom, ***Lentinula edodes*** (Berk.) Pegler (Fig. 7.5a, b), seems to have overtaken the white button mushroom as the most cultivated species

Fig. 7.5 Commercial cultivation on sterilised substrates: (**a, b**) Shiitake; (**c**), (**d**) *Auricularia* sp.; (**e**) *Hypsizygus tessulatus*; (**f**) *Flammulina velutipes*

worldwide (Royse et al. 2017). Although shiitake has been traditionally cultivated in natural logs, the fastest and the most efficient way of cultivation consists of cultivation in polypropylene bags, process so-called "bag log" cultivation (Sánchez 2010). The synthetic substrate is based on a mixture of several **lignocellulosic** materials such as sawdust, rice bran, wheat bran, corn bran and $CaCO_3$. The mixture is wetted to a final moisture content of 60%. The substrates are pressed in the form of small logs and they must be sterilized at 121 °C for 1 h before spawning at room temperature (Chang and Miles 2004).

***Auricularia* sp.** The Jew's ear mushrooms, essentially the species *Auricularia auricula-judae* (Bull.) J. Schröt. and *Auricularia polytricha* (Mont.) Sacc. (Fig. 7.5c, d), are produced in a synthetic medium designed from the mixture of different agronomic by-products (Fig. 7.5c, d). A mixture of 90% sawdust, 9% rice bran and 1% $CaCO_3$ can be considered as substrate formula (Liang et al. 2016), although some other materials such as sugarcane bagasse, rice straw, and rice husk can be also incorporated to the mixture at different ratios (Wu et al. 2017).

The water content of the mixture must be adjusted to approximately 65% (w/w). Then, the substrate is introduced in polyethylene bags and sterilized in autoclave at 121 °C for at least 1 h – 80 min. After sterilization the substrate is ready for the inoculation of the mushroom at room temperature.

Flammulina velutipes The golden needle mushroom, mainly ***Flammulina velutipes*** (Curtis) Singer (Fig. 7.5f), is mostly cultivated in China. The bag cultivation (Fig. 7.5e, f) and the bottle cultivation are the two methods for cultivation that are widely employed (Zhou et al. 2017). The substrate designed for the cultivation of *Flammulina velutipes* is mostly based on sawdust (80%) and rice bran (20%). In addition, some other materials have been studied as a potential component for the substrate employed on the golden needle mushroom cultivation, among them for instance rubber wood sawdust, paddy straw, palm empty fruit bunches or palm-pressed fiber (Harith et al. 2014). The ingredients are thoroughly mixed, water is added to achieve a moisture content of 58 to 60%, and the medium is mixed again (Zhou et al. 2017). Polypropylene bags or bottles are filled with the substrate and finally sterilized through autoclaving at 121 °C for 1 h or at 95 °C for 4 h (Chang and Miles 2004). Finally, mushroom spawn is inoculated on the substrate at room temperature.

Pleurotus eryngii The king oyster mushroom, *Pleurotus eryngii* (DC: Fr.) Quel. has been cultivated mostly on wheat straw-based substrates (Fig. 7.6). When the king oyster mushroom was cultivated on three different agricultural wastes, wheat straw (WS), cotton waste (CW) and peanut shells (PS), WS in non-composted substrates supported faster growth, faster colonization rate and favored earliness of the crop (Philippoussis et al. 2001). Besides, WS enhances yield and mushroom size with positive correlation for C/N ratio and mushroom yield (Philippoussis et al. 2001).

Although, the substrate composition designed for growing ***P. eryngii*** may vary, it consists basically of a mixture of sawdust and more aerated materials like straw, maize stalks or cotton seed hulls, enriched with nitrogen-rich additives. A C/N ratio

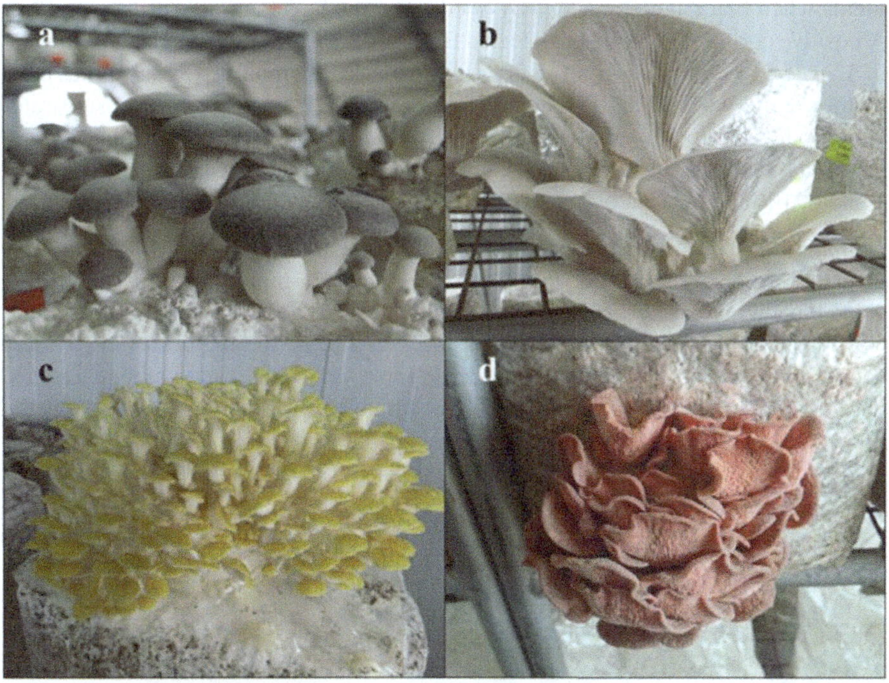

Fig. 7.6 Several species of *Pleurotus* cultivated on sterilised substrates: (**a**) *P. eryngii*; (**b**) *P. pulmonarius*; (**c**) *P. citrinopileatus*; (**d**) *P. djamor* var. *roseus*

of 30–40 and moisture content around 70% are optimal (Rodriguez-Estrada et al. 2009). Bottles are barely used in Western Europe, but artificial logs–substrates are produced in all sorts of bags and shapes. Substrate portions of 1 to 4 kg are filled into bags which are provided with different filter systems for gas exchange. A mixture of cottonseed hulls (56% in formula, 12% moisture), corn distiller's waste (4% in formula, 11% moisture), calcium sulfate (1%), ground soybean (12% in formula, 7% moisture) and aged Northern red oak (Q. rubra L.) sawdust (27% in formula, 18% moisture) was also successfully used for king oyster mushroom cultivation. The mixed ingredients were watered to reach 60% moisture content. Then, moistened substrate was packed into polypropylene (PP) bags and autoclaved at 121 °C for 90 min. Cold substrate was then inoculated with grain spawn (1.2% w/w) (Rodriguez-Estrada et al. 2009).

Agrocybe aegerita The black poplar mushroom, ***Agrocybe aegerita*** (V. Brig.) Singer (Fig. 7.7a), is a high quality mushroom which usually grows in the wild, forming clusters in poplar and other tree trunks. Non-composted materials based on wheat straw, cotton waste, sawdust, corn cobs and peanut shells has been reported as lignocellulosic substrates for the growth of *A. aegerita* (Kleofas et al. 2014). The culture on wheat straw supplemented with different residues showed the highest yields when black tea pomace, a by-product of the beverage industry, was added. A positive correlation has been detected for substrate C/N ratio and mushroom yield in *A. aegerita* (Philippoussis et al. 2001). The materials employed as substrates must be mixed and watered until 65% moisture is achieved, subsequently the substrates are introduced into polyethylene bags and sterilized in an autoclave at 121 °C for at least 1 h. Once the substrate is cold, the spawn is inoculated within the substrate at a rate of 10% (w/w) (Kleofas et al. 2014).

Ganoderma The "reishi", ***Ganoderma lucidum*** (Curtis) P. Karst. (Fig. 7.7b), is a genus of wood decaying polypore fungi of economic importance due to its high valued medicinal properties and mostly cultivated and consumed in Asia. This species has been traditionally cultivated using wood logs, the methods most widely adopted for commercial production are the short wood log, and synthetic sawdust bag and bottle procedures (Zhou 2017). Substitute cultivation of reishi is conducted

Fig. 7.7 Several mushrooms cultivated on sterilized substrates: (**a**) *Agrocybe aegerita*; (**b**) *Ganoderma sp.*; (**c**) *Pholiota nameko*; (**d**) *Calocybe indica*; (**e**) *Hericium erinaceus*; (**f**) *Hypsizygus marmoreus* (syn. *tessulatus*)

through the preparation of a non-composted substrate based on sawdust, cottonseed hull, rice straw, wheat straw, corncob powder, and other agricultural by-products as main source of carbon. Furthermore, wheat bran, rice bran, corn powder, ammonium sulfate, urea, and some other compounds containing nitrogen are incorporated to the mixture plus some additives to correct pH such $CaCO_3$ and gypsum (2%) in a ratio of 1:1 (w/w) and KH_2PO_4 2% (Zhou 2017). The mixture should be wetted to provide moisture content around 65%. Finally, the substrates are transferred into heat-resistant polypropylene bags (plugged with cotton plugs using PVC rings) or glass bottles (sealed with Kraft paper) are autoclaved at 121 °C for 1–2 h. A mixture of 75–77% sawdust, 2% $CaCO_3$, 5–9% corn bran, 5–9% oak bran, 3.5–4% wheat bran, has been successfully used to produce wild collected strains. However, according to our results it is important to note that different strains required a slightly different substrate formulation (Maszlavér 2008; Pérez-Clavijo et al. 2016).

Volvariella volvacea The paddy straw mushroom, ***Volvariella volvacea*** (Bull. Ex Fr.), is a commercial variety that requires high temperatures during its crop cycle. Certain waste materials have been used for cultivation of this species, including: paddy straw, water hyacinth, oil palm bunch, oil palm pericarp waste, banana leaves and sawdust, cotton waste, sugarcane bagasse, composted mixtures of tropical wood wastes and pineapple skin waste, wood waste and some other substrates (Chang and Miles 2004) (Fig. 7.7c, d).

Non-composted cotton-based substrate showed higher colonization by *V. volvacea* with earlier fructification in comparison to wheat straw and peanut shell based substrates. In addition, cotton waste substrate enhanced yield and mushroom size. A positive relation was observed between cellulose content of the substrates and the mushroom yield (Philippoussis et al. 2001). Non-composted substrate suitable for *V. volvacea* cultivation can be prepared by soaking cotton wastes including stem-leaf residues and gin trash with particle size of 2–3 cm in water for 24 h to subsequently drain. The cotton wastes are mixed with wheat bran (10%) and $CaCO_3$ (2%). Afterward, the substrates are disposed in heat resistant bags and sterilized at 120 °C for at least 2 h, prior to inoculation with the mushroom spawn (Philippoussis et al. 2001).

Hypsizygus marmoreus **(syn.** *tessulatus***)** The beech mushroom, ***Hypsizygus marmoreus*** (Peck) Bigel. (Fig. 7.5e, f), a species originally cultivated in Japan, is now widely cultivated throughout Japan and China. The commercial substrate employed for commercial production is based on sawdust or corncobs.

Some substrates for high productivity of the mushroom consist of a sawdust base, for instance: a) birch (*Betula ermanii* Cham) sawdust (18.2%), rice bran (8.8%), soybean shell (5.3%), corncob meal (4.7%) and $CaCO_3$ (0.5%); b) birch sawdust (22.8%), rice bran (14.2%), the commercial supplement Oruga K-1 (Katsuragi Sangyo corp.) (2.5%). Moisture content of the substrate was adjusted to 63%, based on the fresh weight of the mixture of solid materials (Harada et al. 2004). Another formula employed in Japan consists of 3.9% (wet weight) of corncobs, 11.1 of rice bran, 3.9% of wheat bran, 3.9% of soybean hull, 2% of cottonseed hulls, 2% of dry *tofu* refuse and 74.5% sawdust. The mixture is wetted to raise the moisture content up to 65–75% (w/w). After **sterilization** (121 °C for at least 1 h) and cooling the substrate, the mass is inoculated on the substrate surface at a rate of 2.5–2.9% of sawdust spawn (Yamanaka 2017). The substrate is introduced in bottles or heat resistant plastic bags and sterilized in autoclave at 121 °C for at least 1 h prior to inoculation with sawdust spawn at room temperature.

Pholiota nameko The production of ***Pholiota nameko*** (T. Ito) S. Ito & S. Imai (Fig. 7.7c) began with log cultivation in Japan. Currently it is mostly cultivated in China and Japan using small bags or bottles filled with sterile substrate. The non-composted substrate employed for the cultivation of this species is based on hardwood sawdust (90% of total fresh substrate weight) with different supplements such as corn bran, wheat bran, and dried tofu refuse (10% of total fresh substrate weight). The mixed substrate is adjusted to 64–65% moisture (Yamanaka 2017).

Grifola frondosa The "maitake" mushroom, ***Grifola frondosa*** (Dicks.) Gray, is a mushroom appreciated both for its culinary and medicinal properties. It was first cultivated 35 years ago and nowadays is mostly cultivated in Japan and China. Three different methods have been described for the cultivation of "maitake": (1) bottle culture in small bottles (800–1000 ml); (2) bag culture: sawdust is mixed with rice and wheat brans. After the moisture of the mixture is adjusted, the mixture is packed in a polypropylene bag (this is the most used method); (3) outdoor bed culture: requires about 6 months from inoculation to fruiting body formation (Mayuzumi and Mizuno 1997). The substrate is based on a non-composted mixture based on sawdust powder obtained from deciduous broadleaf trees with 15% rice bran and 5% wheat bran. After mixing thoroughly, the moisture is adjusted to 65%. 2.5 kg polypropylene bags are filled with the substrates and autoclaved at 120 °C for at least 2 h or 110 °C for 7 h. When cold, the substrate is inoculated with a spore suspension or mycelium from spawn manufacturers (Mayuzumi and Mizuno 1997).

Calocybe indica The milky mushroom, ***Calocybe indica*** Purkayastha & A. Chandra (Fig. 7.7d), is an edible species mostly cultivated in India. The substrates employed for milky mushroom cultivation varied from different **lignocellulosic** materials from agricultural wastes. The base materials, including coconut coir, kash, paddy straw, maize stalks, rice straw, sorghum stalks, sugarcane bagasse, sugarcane leaf, wheat bran or waste cotton are chopped to 3–5 cm bits to facilitate humectation (up to 65% moisture level). The substrate is introduced in polypropylene bags prior to autoclave at 121 °C at 15 psi for 1 h. Alternatively, heat treatments using steam for 30 minutes at 85 °C in a closed chamber can be applied. Once cooled down, the substrate is ready for spawning (Amin et al. 2010; Lakshmipathy et al. 2017).

Hericium erinaceus The "houtou" mushroom, ***Hericium erinaceus*** (Bull. Ex Fr.) (Fig. 7.7e), was firstly cultivated in China. It is now grown in plastic bag cultures on a variety of substrates: sugarcane bagasse, sawdust, cottonseed hulls, corncobs, and chopped up paddy straw - each supplemented with rice or wheat bran, sucrose, gypsum, and sometimes additional ingredients (Chang and Miles 2004).

7.4 Casing Materials

The phase III compost fully colonized by the *Agaricus* mycelium and the substrates employed to grow other species such as *Calocybe indica* is not sufficient to produce mushrooms. To obtain a commercially viable production it is necessary to apply a casing layer which requires certain physical, chemical and biological characteristics to reproduce the appropriate environment in which the morphological shift from the **vegetative mycelium** to the reproductive one takes place (Pani 2012; Pardo-Giménez et al. 2017). Besides, the yield and biological efficiency (BE) of some species such as *Pleurotus eryngii* var. *eryngii* have been increased when a casing overlay was applied in the substrates employed for the cultivation (Rodriguez-Estrada et al. 2009).

Casing layers consist of mixtures of different natural materials. Among them, peat based materials amended with sugar beet lime to correct **pH** are the most commonly used for the growth of *A. bisporus* (Noble et al. 2009). Noteworthy, some of the microbial casing inhabitants are directly involved in mushroom fructification. Axenic casing does not support mushroom production so the presence of living organisms in the casing material employed as top layer over the colonized compost is required (Noble et al. 2009).

Since peat is a natural resource, the exploitation of natural peatlands is limited due to the environmental impact and an increasingly restrictive legislation. Many other materials have been evaluated as alternatives to peat including: soil, coconut fiber, paddy straw, charcoal, spent mushroom compost or pine bark (Pardo et al. 2004; Noble et al. 2009; Colauto et al. 2010). However, to date none of them offer the benefits achieved with peat-based materials in terms of physical, chemical and biological properties. Therefore, the search for alternative materials is one of the challenges for the mushroom industry.

7.5 Conclusions and Perspectivas

The cultivation of mushrooms is a very efficient agricultural activity which uses by-products from other vegetable crops or forestry to generate high quality food. Besides, while cultivated under indoor-controlled conditions, mushroom crops are not susceptible of damage due to adverse weather conditions, therefore limiting crop losses. In addition, mushrooms present a nutritional value that can be compared to those of eggs, milk, and meat, while the water content is an average of 90% (Sánchez 2010). To sum up, it is highly advisable to introduce mushrooms in your diet with the aim of reducing the protein dependence with meat origin.

Acknowledgement This article has been funded by the European Union's Horizon 2020 research and innovation programme under the Marie Sklodowska-Curie grant agreement No 742966. Pictures attached in the text have been kindly loaned by CTICH (Maria L Tello).

References

Amin R, Khair A, Alam N, Lee TS (2010) Effect of different substrates and casing materials on the growth and yield of *Calocybe indica*. Mycobiology 38(2):97–101

Carrasco J (2016) Study of cobweb disease caused by Cladobotryum mycophilum in Spanish mushroom crops. PhD thesis. Universidad de Castilla-La Mancha. https://ruidera.uclm.es/xmlui/handle/10578/9752

Chang S, Miles PG (2004) Mushrooms: cultivation, nutritional value, medicinal effect and environmental impact, 2nd edn. CRC Press, Boca Raton

Colauto NB, da Silveira AR, da Eira AF, Linde GA (2010) Alternative to peat for *Agaricus brasiliensis* yield. Bioresour Technol 101(2):712–716

Eastwood D, Burton K (2002) Mushrooms-a matter of choice and spoiling oneself. Microbiol Today 29:18–23

Falandysz J (2008) Selenium in edible mushrooms. J Environ Sci Health C Environ Carcinog Ecotoxicol Rev 26(3):256–299

Gea FJ, Tello JC, Navarro MJ (2003) Occurrence of *Verticillium fungicola* var. *fungicola* on *Agaricus bitorquis* mushroom crops in Spain. J Phytopathol 151(2):98–100

Harada A, Gisusi S, Yoneyama S, Aoyama M (2004) Effects of strain and cultivation medium on the chemical composition of the taste components in fruit-body of *Hypsizygus marmoreus*. Food Chem 84(2):265–270

Harith N, Abdullah N, Sabaratnam V (2014) Cultivation of *Flammulina velutipes* mushroom using various agro-residues as a fruiting substrate. Pesq Agrop Brasileira 49(3):181–188

Kabel MA, Jurak E, Mäkelä MR, de Vries RP (2017) Occurrence and function of enzymes for lignocellulose degradation in commercial *Agaricus bisporus* cultivation. Appl Microbiol Biotechnol 101(11):4363–4369

Kleofas V, Sommer L, Fraatz MA, Zorn H, Rühl M (2014) Fruiting body production and aroma profile analysis of *Agrocybe aegerita* cultivated on different substrates. Nat Res 5:233–240

Lakshmipathy R, Harikrishna P, Naidu B, Rao DB (2017) Feasibility of maize stalks for milky mushroom cultivation. Int J Curr Microbiol App Sci 6(2):1294–1299

Liang CH, Wu CY, Lu PL, Kuo YC, Liang ZC (2016) Biological efficiency and nutritional value of the culinary-medicinal mushroom *Auricularia* cultivated on a sawdust basal substrate supplement with different proportions of grass plants. Saudi J Biol Sci. https://doi.org/10.1016/j.sjbs.2016.10.017

Maszlavér P (2008) Cultivation possibilities for production of reishi *Ganoderma lucidum* (curt.: fr.) Karst in Hungary. Doctoral dissertation, Ph. D Thesis, Corvinus University of Budapest

Mayuzumi Y, Mizuno T (1997) Cultivation methods of maitake (*Grifola frondosa*). Food Rev Intl 13(3):357–364

Noble R, Dobrovin-Pennington A, Hobbs PJ, Pederby J, Rodger A (2009) Volatile C8 compounds and *Pseudomonads* influence primordium formation of *Agaricus bisporus*. Mycologia 101(5):583–591

Pani BK (2012) Sporophore production of milky mushroom (*Calocybe indica*) as influenced by depth and time of casing. Int J Advanced Bot Res 2(1):168–170

Pardo A, Pardo-González JE (2009) Elaboration of new substrates for cultivating *Pleurotus ostreatus* (Jacq..) R Kumm. Based on degraded substrates from edible fungi cultivation. ITEA 105(2):89–98

Pardo A, De Juan AJ, Pardo J (2004) Assessment of different casing materials for use as peat alternatives in mushroom cultivation. Evaluation of quantitative and qualitative production parameters. Span J Agric Res 2(2):267–272

Pardo JE, Zied DC, Alvarez-Ortí M, Peñaranda JA, Gómez-Cantó C, Pardo-Giménez A (2017) Application of hazard analysis and critical control points (HACCP) to the processing of compost used in the cultivation of button mushroom. Int J Recycl Org Waste Agric 6(2):179–188

Pardo-Giménez A, Pardo JE, Carrasco J, Álvarez-Ortí M, Zied DC (2014) Use of Phase II mushroom compost in *Agaricus subrufescens* production. In; Proceedings of 8th International Conference on Mushroom Biology and Mushroom Products (ICMBMP8), New Delhi, India

Pardo-Giménez A, Pardo JE, Zied DC (2017) Casing materials and techniques in *Agaricus bisporus* cultivation. In: Diego CZ, Pardo-Giménez A (eds) Edible and medicinal mushrooms: technology and applications. Wiley-Blackwell, Chichester

Pérez-Clavijo M, Tello-Martín ML, Roncero-Ramos I, Grifoll-García V (2016) Optimization of *Ganoderma lucidum* cultivation and comparison of bioactive compounds in fruiting bodies, spores and mycelium. In: Proceedings of the 19th International congress on the science. Ed. Baars JJP and Sonnenberg ASM

Philippoussis A, Zervakis G, Diamantopoulou P (2001) Bioconversion of agricultural lignocellulosic wastes through the cultivation of the edible mushrooms *Agrocybe aegerita*, *Volvariella volvacea* and *Pleurotus* spp. World J Microbiol Biotechnol 17(2):191–200

Rodriguez-Estrada AE, Jimenez-Gasco MM, Royse DJ (2009) Improvement of yield of *Pleurotus eryngii* var. *eryngii* by substrate supplementation and use of a casing overlay. Bioresour Technol 100(21):5270–5276

Royse DJ, Baars J, Tan Q (2017) Current overview of mushroom production in the world. In: Diego CZ, Pardo-Giménez A (eds) Edible and medicinal mushrooms: technology and applications. Wiley-Blackwell, Chichester

Sánchez C (2004) Modern aspects of mushroom culture technology. Appl Microbiol Biotechnol 64(6):756–762

Sánchez C (2009) Lignocellulosic residues: biodegradation and bioconversion by fungi. Biotechnol Adv 27(2):185–194

Sánchez C (2010) Cultivation of *Pleurotus ostreatus* and other edible mushrooms. Appl Microbiol Biotechnol 85(5):1321–1337

Schardl CL, Craven KD (2003) Interspecific hybridization in plant-associated fungi and oomycetes: a review. Mol Ecol 12(11):2861–2873

Van Griensven L (1988) The cultivation of mushrooms. Darlington Mushroom Laboratories, Rustington

Vieira FR, Pecchia JA (2017) An exploration into the bacterial community under different pasteurization conditions during substrate preparation (Composting–Phase II) for *Agaricus bisporus* Cultivation. Microb Ecol:1–13

Vos AM, Jurak E, Pelkmans JF, Herman K, Pels G, Baars JJ, Hendriz E, Kabel MA, Lugones LG, Wösten HA (2017a) H_2O_2 as a candidate bottleneck for MnP activity during cultivation of *Agaricus bisporus* in compost. AMB Expr 7(1):124

Vos AM, Heijboer A, Boschker HT, Bonnet B, Lugones LG, Wösten HA (2017b) Microbial biomass in compost during colonization of *Agaricus bisporus*. AMB Expr 7(1):12

Wu CY, Liang CH, Wu KJ, Shih HD, Liang ZC (2017) Effect of different proportions of agrowaste on cultivation yield and nutritional composition of the culinary-medicinal jelly mushroom *Auricularia polytricha* (higher basidiomycetes). Int J Med Mushrooms 19(4):377–385

Yamanaka K (2017) Cultivation of mushroom in plastic bottles and small bags. In: Zied DC, Pardo-Giménez A (eds) Edible and medicinal mushrooms: technology and applications. Willey, Hoboken

Zhou XW (2017) Cultivation of *Ganoderma lucidum*. In: Diego CZ, Pardo-Giménez A (eds) Edible and medicinal mushrooms: technology and applications. Wiley-Blackwell, Chichester

Zhou F, Li Q, Song C, Li Z, Tan Q, Shang X, Li Y (2017) High-yield cultivating techniques of *Flammulina velutipes*. Agric Sci Technol 18(6):1010

Chapter 8
Role of Mushroom Fungi in Decolourization of Industrial Dyes and Degradation of Agrochemicals

Sachin Gupta, Sudheer K. Annepu, Baby Summuna, Moni Gupta, and Sunil A. Nair

8.1 Introduction

White-rot fungi are known to have the abilities to degrade diverse agricultural pollutants (Tortella et al. 2013). For the cleanup of contaminants, application of fungal technology has shown promise since 1985 when the white rot species *Phanerochaete chrysosporium* was found to metabolize a number of important environmental pollutants and the ability is generally attributed to the lignin degrading enzymatic system of the fungus. Similar degrading capacity was later described for another white rot fungal species (Sasek and Cajthaml 2005). White rot fungi possess a number of advantages that can be exploited in bioremediation systems. As the key components of their lignin-degrading system are extracellular, these fungi can degrade insoluble chemicals such as lignin or an extremely diverse range of very persistent or toxic environmental pollutants (Barr and Aust 1994).

The inherent growth habits of the vegetative mycelium for rapid colonization of the substrates and extension of hyphal structures enables the penetration of soil

S. Gupta (✉)
Division of Plant Pathology, Sher-e-Kashmir University of Agricultural Sciences and Technology of Jammu, Jammu, India

S. K. Annepu
ICAR-Directorate of Mushroom Research, Solan, India

B. Summuna
Sher-e-Kashmir University of Agricultural Sciences and Technology of Kashmir, Wadoora, India

M. Gupta
Division of Biochemistry, Sher-e-Kashmir University of Agricultural Sciences and Technology of Jammu, Jammu, India

S. A. Nair
Dr. Y.S. Parmar University of Horticulture and Forestry, Solan, India

B. P. Singh et al. (eds.), *Biology of Macrofungi*, Fungal Biology,
https://doi.org/10.1007/978-3-030-02622-6_8

reaching the pollutants which is not possible for other organisms (Reddy and Mathew 2001). This process can maximise physical, mechanical and enzymatic contact with the targeted environment (Maloney 2001). In addition, cheaply and abundantly available **lignocellulosic** materials are used as a nutrient source for the growth of the fungi. They can tolerate a wide range of environmental conditions, such as temperature, pH and moisture levels (Maloney 2001) and do not require pre-conditioning to a particular pollutant, because their degradative system is induced by nutrient deprivation (Barr and Aust 1994).

Edible mushrooms and other white rot fungi have strong non-specific, extra cellular lignolytic enzymes that have the ability to breakdown the lignin and cellulose, which are composed of long chains of carbon and hydrogen (Arun et al. 2008). Recently, greater emphasis has been paid to the discharge of effluents containing organic pollutants such as polycyclic aromatic hydrocarbons (**PAH**), polychlorinated biphenyl (**PCBs**), dioxins, pesticides, explosives, dyes, solvents, etc. which are structurally similar to lignin and cellulose system of the substrates. Wood rotting fungi an inherent capacity to biodegrade these xenobiotic compounds which otherwise take quite long time for their complete mineralization. A brief review on various aspects of mushrooms, associated fungi, white rot fungi, spent mushroom substrate (SMS) and their utilization in removal of synthetic dyes and degradation of various agro-chemicals is discussed in this chapter.

8.2 Removal of Textile and Industrial Dyes

Synthetic dyes are widely used in many industries such as textiles, paper, plastics, leather, food and pharmaceutical (Saratale et al. 2013). The effluent waste water released after utilization of synthetic dyes from different industries resulting in polluting the environment. It was estimated that about 280,000 tons of textile dyes are discharged every year in industrial effluents (Jin et al. 2007). In textile industry, to dye 1 kg of cotton requires 0.6–0.8 kg NaCl, 30–60 g dye stuff and 70–150 L water. After dying process, 20–30% of unfixed reactive dyes released through the waste water with an average concentration of 2000 ppm, high salt content and dyeing auxiliaries (Babu et al. 2007). Among several synthetic dyes used in the textile industry, azo compounds are occupying the maximum share up to 70% with a broad range of colours and different structures (Lang et al. 2013).

Chemically, the **azo dyes** belong to the class of aromatic and heterocyclic compounds with azo bond (-N=N-). These compounds are recalcitrant in nature and reported to posses the carcinogenic properties (Saratale et al. 2011). Due to the inefficient discharge of the effluent waste water, these compounds accumulate in the water bodies and shows negative effects in terms of Biological Oxygen Demand (BOD), Chemical Oxygen Demand (**COD**), dissolved oxygen (DO), colour, etc. (Saratale et al. 2013). Release of these effluents causes serious environmental problems and the toxic effects caused by these industrial dyes created attention among the researchers to find out suitable methods for their removal (Ayed et al. 2011).

The energy intensive conventional effluent management strategies were found ineffective in removal of the toxic compounds. They also resulted in formation of hazardous compounds due to incomplete removal of the dyes. These limitations compelled to find out the alternate strategies using the biological and biotechnological techniques for remediation of the dyes. Recent studies focussed on biodegradation of azo dyes by fungi and bacteria due to their higher enzymatic activities and strong adaptability (Dos Santos et al. 2007). Despite of wider application of bacteria, their decolourization ability is often inhibited by the aromatic amines (Qu et al. 2010). In contrast, fungi can degrade complex organic compounds through catalysis with extracellular ligninolytic enzymes including laccase, manganese peroxidase and lignin peroxidase (Gomi et al. 2011).

During the decolourization process many intermediates such as phenolics, aromatic amines are generated. These intermediates can be highly toxic and having lower biodegradability than the dyes itself and causes the secondary pollution. Interestingly, several fungal species have shown high efficiency in the removal of these aromatic compounds. Moreover, large surface area of fungi offers an efficient system of degradation (Mishra and Malik 2013). Awasthi et al. 2014 reported multiple mechanisms for degradation of organic and inorganic contaminates by the fungi. However, studies are largely confined to single pollutant degradation often using the pure cultures of fungi (Singh and Singh 2014).

The capacity of an individual strain for uptake of specific dye varies greatly. It is an established fact that the industrial effluents contain a cocktail of various organic contaminates and heavy metals Yadav et al. 2010). For example, in textile industry heavy metals are used as mordant and will be released into the waste water along with the residual dyes. Similarly, tanning industry, paper industries also generating effluents contaminated with wide ranges f metals and dyes. Considering these factors, Mahapatra et al. (2014) proposed that remediation of such effluents by microbial strains in a consortium. Undoubtedly, microbial consortia have a clear advantage for remediation with a rich metabolic network. In a consortium each microbial strain exhibits a specialized role in uptake of a particular contaminates. With this approach several contaminants can be successfully removed from the targeted sites.

The rate of pollutant degradation in the waste water system declines in a long run. This is due to the fact that, microbial biomass being washed and reduces its concentration. Moreno-Garrido (2008) suggested **immobilization** of fungi as a strategy for maintenance degrading biomass. Aerobic white rot fungi such as *Phanerochaete chrysosporium* and *Trametes versicolor* can degrade wide variety of textile dyes (Yang et al. 2013). These fungi have efficient and non selective lignolytic systems capable to degrade azo, heterocyclic and other polymeric dyes (Solıs et al. 2012).

8.2.1 *Fungal Decolouration Mechanisms*

8.2.1.1 Biosorption

Biosorption mechanism which involves the physic-chemical interactions such as adsorption, deposition, ion-exchange, etc. plays a major role in dyes decolouration y the fungi. Singh (2006); Kaushik and Malik (2009) presented a detailed review on ability of fungal mass on decolouriztion of dyes from the waste affluent water. The fungal biomass from *Aspergillus niger* which contains the functional groups such as carboxyl and amino phosphate fractions showed the decolouration of four different dyes viz., Basic blue 9, Acid Blue 29, Congo Red and Disperse Red-I by Biosorpion (Fu and Viraraghavan 2002). Carboxyl and amino groups combinedly decolourized the Basic Blue 9 dye. Whereas, for amino group alone was acted as binding side for Acid Blue 29. Electro static was reported to be involved between the binding site and the dye compounds. For decolourization of Congo Red dye, functional groups such as carboxyl acid, amino and phosphate groups, lipid fractions together acted as binding sites. Lipid fractions and amino groups acted as binding sites for Disperse Red-I dye. Along with the electrostatic attraction, physical and chemical adsorption mechanisms were also played important role in degradation of both Congo Red and Disperse Red-I.

The efficiency of adsorption capacity by **fungal biomass** can be further enhanced with suitable pre-treatment process. Treating the biomass with selected organic and inorganic chemicals such as formaldehyde, H_2SO_4, $NaHCO_3$, NaOH, $CaCl_2$, etc. and physical methods by exposing the biomass to high temperature by autoclaving are some of the pre-treatment methods suggested by several researchers. Autoclaving and chemical treatment with 0.1 N NaOH, 0.1 M HCl, 0.1 M H_2SO_4 were together used for living biomass of *A.niger* (Fu and Viraraghavan 2000). It was reported that, the rate of the biosorption of Basic Blue 9 dye increased from 1.17 mg/g to 18.54 mg/g by autoclaving the biomass. While, chemical pretreatment of biomass with 0.1 M H_2SO_4 resulted in enhanced biosorption of Acid Blue 29 dye from 6.63 mg/g to 13.83 mg/g. The mechanism involved in physical treatment of autoclaving is, fungal vegetative structure of fungal biomass could be changed to positive charged surface by the acid pre treatment and increases the affinity of **anionic dye** (such as Acid Blue 29) to bind with fungal biomass. Arica and Bayramoglu (2007) observed the enhancement in biosorption capacity o the *Lentinus sajor-caju* to remove Reactive Red 120 by autoclaving for 10 min and 100 °C.

8.2.1.2 Decolourization of Dyes by Enzymatic Action

Fungi degrade the lignocellulosic substrate material which is composed of lignin, cellulose and hemicelluoses with the help of various lignoclytic enzymes such as laccase, **manganese peroxidase** (MnP) and **lignin peroxidase** (LiP). The relative contribution of individual enzyme to the decolorization of dyes may be different for each fungus. Recently, there has been an increased interest in determining the exact mechanism of how these organic pollutants are broken down. Depending on the enzyme involved, different pathways have been reported in the literature and reviewed below.

Degradation of Azo Dyes by Azo Reductases

The role of azo reductases on degradation of azo dyes and azo compounds are widely reported in the literature. Under anerobic conditions, microbes possessing the reductase enzymes efficiently degraded the azo compounds (Chen et al. 2005). The enzymatic degradation activity hastened under the presence of reducing catalysts such as nicotinamide adenine dinucleotide (**NADþ**), nicotinamide adenine dinucleotide phosphate (**NADPþ**), etc. Due to the cytosolic nature of the reductase enzymes, it is assumed that the azo compounds have to be transported in to the cell before they are degraded by the anerobic microbes. However, for certain dyes with high ionic charge and high molecular weight, alternate mechanisms were suggested in which the dye compounds may pass though the membrane of the cell, instead of transporting into it (Robinson et al. 2001).

It was further hypothesized that, the reduction activity of the dye, may not be dependent on the intracellular uptake of the macromolecules. This alternate hypothesis suggested the role of the reduction of these dyes in the extracellular environment of the microbes. To enable this function, the fungi should link with the intracellular electron transport system and the molecules o dye compounds. This mechanism requires the direct contact between the cell surface, redox mediators and the dye. This linkage needs the presence of electron transport component in the outer membrane of the fungal cells. The catalyzing agents which can increase the degradation of dye molecules by extra cellular enzymes can be generated by the internal fungal metabolism or may be supplemented externally. Aerobic conditions retards the reduction process by oxidizing the redox mediators, hence these reactions must be carried out in the absence of oxygen.

Degradation of Azo Dyes by Laccase Enzyme

Laccase are the copper containing oxidase enzymes commonly found in higher fungi and other micro organisms (Pointing 2001). Fungi utilize laccases for degradation of different lignocellulosic substrates through oxidation activity (Majeau et al. 2010). By utilizing the oxygen as a secondary substrate, the fungal laccases oxidizes several aromatic and non aromatic compounds structurally similar to the dye compounds (Tuor et al. 1995). During this reduction process the protons will be abstracted and radical are generated. These radicals further regulate the reactions such as polymerization, hydration or proton abstraction. Similarly, the laccase concentration from the mushroom mycelium was increased when it was exposed to the heavy metals such as cadmium, lead, copper, zinc and mercury. These metals were absorbed subsequently and decolourized by the laccase enzymes (Hitivani and Mecs 2003). Degradation of phenols and several other aromatics in effluents either in the free or immobilized form have been explored by using various enzymes (Table 8.1).

Table 8.1 Degradation of Industrial dyes by some white-rot fungi

Fungal species	Conditions	Decolouration process	Reference
Pleurotus ostreatus	pH 3.5–5.2, 30 °C,	Enzyme degradation (Lac, LiP)	Tapia-Tussell et al. (2011)
Pleurotus sajarcaju	pH 4.5, 25 °C	Adsorption and degradation	Munari et al. (2008)
Pleurotus florida	pH 4.5, 25 °C	Enzyme degradation (Lac)	Sathiya Moorthi et al. (2007)
Pleurotus pulmonaris	pH 4.5, 25 °C	Biodegradation, adsorption and laccase	Tychanowicz et al. (2004)
Agaricus bisporus	pH 1.0	Biosorption	Akar et al. (2009)
Trametes versicolor	pH 4.5, 25 °C	Degradation	Cano et al. (2012)
Ganoderma lucidum	pH 4.5, 25 °C	Adsorption and degradation	Baccar et al. (2011)
Phanerochaete chrysosporium	pH 4.5, 30 °C	Degradation	Sen et al. (2012)
Phanerochaete chrysosporium	pH 4–7, 24–34 °C	Degradation	Sharma et al. (2009)
Ganoderma sp.,	pH 5.5, 28 °C	Enzyme degradation (Lac)	Anastasi et al. (2011)
Schizophyllum commune	pH 2.0; 30 °C	Bioaccumulation	Renganathan et al. (2006)

Role of Peroxidises in Dyes Degradation

Yousefi and Kariminia (2010) studied the decoulrization of Acid Orange 7 by the fungal biomass of *Coprinus cinereus*. The role of **peroxidises** enzyme produced from the fungal biomass resulted in decolouration of the above dye. Further it was also reported that the time required for degradation of the time can be shortened by maintaining the pH, temperature and dye concentration at optimum range. The fungal peroxidises including lignin peroxidase (LiP) and manganese peroxidase (MnP) is similar to the plant peroxidases such as horseradish peroxidase (HRP).

8.3 Degradation of Pesticides and Other Agro Chemicals

Efficient crop production practices depends upon judicious management of biotic factors such as insect pests, diseases, weeds etc., which otherwise are imminent threat to crop yield and quality. Indiscriminate use of pesticides for management of these biotic stress factors not only contributes to environmental pollution but also releases exogenous chemical compounds known as **xenobiotics**. These chemicals remain un-degraded by microorganisms and remains in the soil persistently. According to Agnihotri (1999) amongst South Asian countries, India ranks first in the usage of pesticides. As per reports of the Kookana et al. (1998) and Nawaz et al. (2011), only 5% of the applied pesticides are effective in managing the target insect pests. The remaining part of the **pesticides** (water soluble ones) act as off target entering surface and ground water bodies (Casara et al. 2012) often polluting the

Table 8.2 Degradation of agricultural pollutants by different fungi

Fungal species	Type of pollutant	Reference
Phanerochaete chrysosporium	Lindane, DDT, Atrazine, BTEX	Bumpus et al. (1985)
		Hickey et al. (1994)
		Yavad and Reddy (1993)
Pleurotus florida	Lindane	Bumpus et al. (1985)
Pleurotus pulmonarius	Atrazine	Masaphy et al. (1996)

water bodies affecting human health by interfering in the food chain (Strandberg et al. 1998). On the other side some pesticides are retained by soil and organic matter (Xiao et al. 2011) and these are called hydrophobic pesticides. **Biodegradation** of these toxic substances depends upon the presence of the suitable microorganisms (Frazar 2000) (Table 8.2). The other reasons for failure of biodegradation are due to lack of nutrients for growth of microorganisms or recalcitrant nature of the toxic compound (Field et al. 1993). However there are possibilities of survival of some micro organisms undergoing metabolic processes wherein release of energy makes them efficient converters of toxic pesticide into harmless products (Kirk and Farrell 1987; Hatakka 2001).

The United States Environmental Agency (USEPA) has identified several classes of chemicals viz., polycyclic aromatic hydrocarbons, pentachlorophenols, polychlorinated biphenyls, 1,1,1- trichloro - 2,2-bis (4-chlorophenyl) ethane, benzene, toluene, ethylbenzene xylene and trinitrotoluene for their role as priority pollutants causing health hazards. As per the workers (Lau et al. 2003; Verdin et al. 2004) the burning of fossil fuels, coal mining, oil drilling and wood burning leads to generation of Polycyclic aromatic hydrocarbons (**PAH**) a primary recalcitrant environmental contaminants. As piling up of these toxic compounds causes' serious health hazards and sustainability of our ecosystem is already on the brink of precipice, the need of the hour is to evolve ourselves and reduce and annihilate the environmental damage caused due to pollutants and wastes.

Besides unbalancing and causing disturbance to the equilibrium of the ecosystem, environmental contamination affects the flora, fauna and human being too. Hamman (2004) has thus advised a rapid cost-effective method for cleanup of the contaminated ecosystem. Though incineration or disposal of wastes by burning is the common practice, it is costly in terms of energy consumption. Thus the only environment friendly and less hazardous way of disposal and mopping up of the contaminated sites is by employing compatible micro-organisms (Finley et al. 2010).

8.3.1 Mechanisms Involved in Degradation of Agro Chemicals

The use of fungi either natural inhabitant or externally introduced to degrade the pollutants involves enzymatic mineralization, chelation, biosorption and precipitation (Sturman et al. 1995; Zhang and Chiao 2002).

8.3.1.1 Enzymatic Mineralization

Degradation of pollutants by lignolytic fungi was widely studied by several mycologists and environmental researchers. However, the precise role of enzymes in pesticides degradation is not established clearly, However, few evidences which suggest that complex of lignin degrading enzymes (lignin peroxidase, manganese peroxidase and laccase) is responsible for pesticides degradation by these fungi. All lignolytic fungi did not produce these enzymes in a similar fashion. Even if the enzymes have been detected, it has often been difficult to find out a clear correlation between the targeted pollutant for degradation and the enzyme involved in its degradation.

Biodegradation of xenobiotic compounds (which are structurally similar to lignin) by laccase and other lignolytic enzymes has attracted the interest among the researchers and its degradation properties and mechanisms involved are widely studied (Trejo-Hernadez et al. 2001). Textile effluent degradation by *Trametes versicolor* and azo dyes degradation by *Pyricularia oryzae* are attributed to the laccase activity released by these fungi. Duran and Eposoto (Duran and Esposito 2000) have studied the role of laccase from the *Correna unicolor* in removal of 2, 4 DCP in contaminated soil colloids (Table 8.2).

Recently, Demir (2004) studied the biodegradation of benzene and toulene by *T. versicolor* and found that toluene is completely removed within 4 h with initial concentration of 50 mg/litre, while it takes 36 h when the concentration of 300 mg/litre. In case of benzene, by the end of the 4th hour with an initial concentration of 50 mg/litre, the biodegradation was completed. While at 300 mg/litre initial concentration it continued for 42 h. However, addition of verataryl alcohol (a laccase inducer) to the medium stimulated the enzymatic system and biodegradation was completed in a relatively shorter period (Demir 2004).

Degradation of phenanthrene by *T. versicolor* was studied by Han et al. (2004) and purified its laccase after 36 h. of incubation under shaken and static fungal growth conditions and during this period about 46 and 65% of initial phenanthrene concentration (100 mg/litre) in the media, respectively was removed under two different growing conditions. Although highest phenanthrene removal takes place (76.7%) at lower phenanthrene concentration (10 mg/litre), but the transformation rate becomes highest (0.82 mg/litre) at a higher concentration (100 mg/litre). However, no transformation of the compound occurs with the purified laccase of *T. versicolor*. Mineralization of dioxin 2, 7-dichlorobenzene-P-dioxin by *P. chrysosporium* was studied by Valli et al. (1992). The results showed that purified lignin peroxidase and manganese peroxidase are capable to mineralize in a multi-step pathway. Esposite et al. (1998) also reported the ability of different **actinomycetes** to degrade diuron in soil using manganese peroxidase. In a study carried out at National Research Centre for Mushroom, Solan, India, the SMS fungi (*Trichoderma* sp. and *Aspergillus* sp.) with superior lignolytic activity, were also found to have higher pesticides biodegradation capacity (Ahlawat et al. 2006).

Besides having so many reports on enzyme mediated degradation of xenobiotics compounds, some researchers have reported contradictory evidences. Degradation of TNT by non-lignolytic strains of *P.chrysosporium* was reported by Jackson et al. (1999). Bending et al. (2002) also reported 86% degradation of atrazine and terbutylazine by white rot fungi in liquid culture and found no relationship between degradation rate and their lignolytic activity. Under liquid culture conditions, *P.chrysosporium* has been found to bio-transform lindane independently of its production of **lignolytic** enzymes (Mousin et al. 1996).

8.3.1.2 Metal Leaching

Chelation of metals occurs with organic acids produced by fungi such as *Aspergillus* sp. And *Penicillium* sp. (Schinner and Burgstaller 1989) extracted Zinc from the industrial waste using the citric acid produced by a *Penicillium* sp. Microbes' secret extracellular polymeric material (capsules, slimes etc.) of anionic character (Ferris and Beveridge 1989; Sly et al. 1990) and these polymers bind metal ions by salt bridging, aggregation, metal hydrolysis and colloidal binding.

8.3.1.3 Metal Bio-sorption

"Bio-sorption" is passive, non-directed metabolism, mediated by physico-chemical interactions between microbial surfaces and heavy metal ions (Shumate and Strandberg 1985). Usually this process occurs due to the combinations of several processes such as electrostatic interactions, formation of ionic bonds, complexation, ion exchange, nucleation etc. The complexity of microbial surfaces and physical/chemical properties of metal ions facilitates this interaction (Kapoor and Viraraghavan 1995). Biosorption of cadmium by *Penicillium digitatum* and *Pseudomonas aeruginosa* (3.5 mg/g, 57.4 mg/g) has also been reported by and Chang et al. (1997). It has been further reported by Kuo and Regan (1998) that SMS acts as an adsorption medium for the removal of **pesticides** (Carbaryl, carbofuran) in the concentration range of 30 mg/litre and SMS adsorbs carbamate pesticides from aqueous solution successfully.

8.3.1.4 Microbial Catalyzed Metal Transformation

Microorganisms carry out oxidation, reduction, methylation or demethylation of metals (Bhide et al. 1996) species. Enzymatic biotransformation to less soluble precipitating species has also been reported for metals like Pb^{+2}, Se^{+4}, Se^{+6} and U^{+6}. Bio-reduction of selenium in the **microbial biomass** and removal of chromium from METEX have been reported by Pumpel and Paknikar (2001). Microbial volatilization degrades toxic metals from wastewater, such as volatilization of Hg^{+2} as dimethylmercury and of selenium as dimethyl selenide (Ron and Royse 1992; Brisley 1990).

8.3.1.5 Precipitation

The organic matter will be oxidized by Sulfate reducing bacteria e.g. *Desulfo vibrio, Desulfoto maculum* (SRB) using sulfate as an electron acceptor. Hydrogen sulfide and a by product of the metabolism react with soluble metal ions to form metal sulfide precipitates (Pumpel and Paknikar 2001). Enzymatic cleavage of glycerol-phosphate by *Citrobacter* species generates HPO_4^{2-} which in-turn forms insoluble precipitates and these precipitates remain bound to the cells (Yong et al. 1987). Under a favorable chemical environment i.e. high pH, high local concentration of metals and anions with in a **biofilm** and in the presence of nucleation sites, metals can precipitate in various chemical forms (Pumpel and Paknikar 2001).

8.4 Limitations of Using Fungi as Tools of Bioremediation

The role of various fungal strains in biodegradation and **bioremediation** of xenobiotics and other organic, inorganic pollutants has been studied. However, certain drawbacks are restricting their wider application. Fungal degradation of pesticides was found relatively slower and incomplete process resulting in incomplete removal of contaminates from the target sites. This incomplete removal of pollutants many times results in accumulation of secondary metabolites which might be harmful (Badawi et al. 2009). Moreover, the chosen fungal strain needs to be adapted to the prevailing conditions before exerting its degradation mechanism against the pollutants (Kulshreshtha et al. 2014). Physico-chemical properties of the soil and prevailing climatic conditions limit the bulk transfer of the fungal strains under the field conditions (Boopathy 2000). Hence the bioremediation process must be designed in such a way that it can easily be adapted to the given contaminated site which increases the accessibility and bioavailability of the pollutants. The mechanisms involved in these process needs to warrant with further investigations.

Another important limitation in fungal degradation of dyes and organic pollutants is availability of capable fungal strains. To mitigate this problem fungal and bacterial consortium can be tested on degradation of different pollutants rather than individual strains. Identification of genes that can confer the degradation ability of microorganisms can be identified and addition of these genes to the indigenous micro-organisms is another strategy (Singh 2008).

References

Agnihotri NP (1999) Pesticide: safety evaluation and monitoring. All India coordinated Project (AICRP) on Pesticide Residues. Indian Agricultural Research Institute, New Delhi, pp 132–142

Ahlawat OP, Gupta P, Raj D, Vijay B (2006) Dye decolorization potential of spent substrates from *Agaricus bisporus* and *Pleurotus* sp.- a laboratory study. Mushroom Res 15:75–82

Akar ST, Gorgulu A, Kaynak Z, Anilan B, Akar T (2009) Biosorption of reactive Blue 49 dye under batch and continuous mode using a mixed biosorbent of macrofungus *Agaricus bisporus and Thuja orientalis cones*. Chem Eng J 148(1):26–34

Anastasi A, Parato B, Spina F, Tigini V, Prigione V, Varese GC (2011) Decolourisation and detoxification in the fungal treatment of textile wastewaters from dyeing processes. New Biotechnol 29:38–45

Arica M, Bayramoglu G (2007) Biosorption of Reactive Red 120 dye from aqueous solution by native and modified fungus biomass preparations of *Lentinus sajorcaju*. J Hazard Mater 149:499–507

Arun A, Prevee Raja P, Arthi R, Ananthi M, Sathish Kumar K, Eyini M (2008) Polycyclic aromatic hydrocarbons (PAHs) biodegradation by basidiomycetes fungi, *Pseudomonas* isolate and their co cultures: comparative *in vivo* and silico approach. Appl Biochem Biotechnol 151:132–142

Awasthi MK, Pandey AK, Bundela PS, Wong JW, Selvam A (2014) Evaluation of thermophilic fungal consortium for organic municipal solid waste composting. Bioresour Technol 168:214–221

Ayed L, Mahdhi A, Cheref A, Bakhrouf A (2011) Decolorization and degradation of azo dye Methyl Red by an isolated *Sphingomonas paucimobilis*: biotoxicity and metabolites characterization. Desalination 274:272–277

Babu BR, Parande AK, Raghu S, Kumar TP (2007) Cotton textile processing: waste generation and effluent treatment. J Cotton Sci 11:141–153

Baccar R, Blanquez P, Bouzid J, Feki M, Attiya H, Sarra M (2011) Decolorization of a tannery dye: from fungal screening to bioreactor application. Biochem Eng J 56:184–189

Badawi N, Ronhede S, Olsson S, Kragelund BB, Johnsen AH, Jacobsen OS, Aamand J (2009) Metabolites of the phenylurea herbicides chlorotoluron, diuron, isoproturon and linuron produced by the soil fungus *Mortierella* sp. Environ Pollut 57:2806–2812

Barr D, Aust S (1994) Mechanisms white rot fungi use to degrade pollutants. Environ Sci Technol 28:78–87

Bending G, Friloux M, Walker A (2002) Degradation of contrasting pesticides by white rot fungi and its relationship with lignolytic potential. FEMS Microbiol Lett 212:59–63

Bhide JV (1996) Microbiological processes for the removal of hexavalent chromium from chromate bearing cooling tower effluent. Biotechnol Lett 18:667–672

Boopathy R (2000) Factors limiting bioremediation technologies. Bioresour Technol 74:63–67

Brisley CL (1990) Bioremediation of metal-contaminated surface and ground waters. J Geom 8:204–223

Bumpus J, Tien M, Wright D, Aust S (1985) Oxidation of persistent environmental pollutants by white rot fungi. Science 228:1434–1436

Cano M, Solis M, Solis A, Loera O, HI P, Teutl MMM (2012) Decoloracion de CD2 (cafe directo 2) por enzimas intra-celulares y extra celulares de *Trametes versicolor*. Interciencia 37:294–298

Casara KP, Vecchiato AB, Lourencetti C, Pinto AA, Dores EF (2012) Environmental dynamics of pesticides in the drainage area of the Sao Lourenco River headwaters, Mato Grosso state, Brazil. J Braz Chem Soc 23:1719–1731

Chang JS, Law R, Chang CC (1997) Biosorption of lead, copper and cadmium by biomass of *Pseudomonas aeruginosa* PU21. Water Res 31:1651–1658

Chen H, Hopper SL, Cerniglia CE (2005) Biochemical and molecular characterization of an azoreductase from *Staphylococcus aureus*, a tetrameric NADPH-dependent flavoprotein. Microbiology 151:1433–1441

Demir G (2004) Degradation of toluene and benzene by *Trametes versioclor*. J Environ Biol 25(1):19–25

Dos Santos AB, Cervantes FJ, Van Lier JB (2007) Review paper on current technologies for decolourisation of textile waste-waters: perspectives for anaerobic biotechnology. Bioresour Technol 98:2369–2385

Duran N, Esposito E (2000) Potential applications of oxidative enzymes and phenoloxidase-like compounds in wastewater and soil treatment: a review. Appl Catal B Environ 28:83–99

Esposite E, Paulillo SM, Manfio GP (1998) Biodegradation of the herbicide diuron in soil by indigenous actinomycetes. Chemosphere 37:541–548

Ferris FG, Beveridge TJ (1989) Metal interaction in microbial biofilms in acidic and neutral pH environments. Appl Environ Microbiol 55:1249–1257
Field JA, De Jong E, Feijoo-Costa G, De Bont JAM (1993) Screening for ligninolytic fungi applicable to the biodegradation of xenobiotics. Trends Biotechnol 11:44–48
Finley SD, Broadbelt LJ, Hatzimanikatis V (2010) In Silico Feasibility of Novel Biodegradation Pathways for 1, 2, 4 - Trichlorobenzene. BMC Syst Biol 4(7):4–14
Frazar C (2000) The bioremediation and phytoremediation of pesticide contaminated sites, national network of environmental studies (NNEMS) fellow, Washington, DC
Fu Y, Viraraghavan T (2000) Removal of a dye from an aqueous solution by fungus *Aspergillus niger*. Water Qual Res J Can 35:95–111
Fu Y, Viraraghavan T (2002) Removal of Congo Red from an aqueous solution by fungus *Aspergillus niger*. Adv Environ Res 7:239–247
Gomi N, Yoshida S, Matsumoto K, Okudomi M, Konno H, Hisabori T, Sugano Y (2011) Degradation of the synthetic dye amaranth by the fungus Bjerkandera adusta Dec 1: inference of the degradation pathway from an analysis of decolorized products. Biodegradation 22:1239–1245
Hamman S (2004) Bioremediation capabilities of white- rot fungi. Biodegradation 52:1–5
Han M, Choi H, Song H (2004) Degradation of phenanthrene by *Trametes versicolor* and its laccase. J Microbiol 42:94–98
Hatakka A (2001) Biodegradation of lignin. In: Steinbüchel A (ed) Biopolymers. Vol 1: Hofrichter M., Steinbuchel A. (eds.) Lignin, Humic Substances and Coal. Wiley- VCH, Weinheim, pp 129–180
Hickey W, Fuster D, Lamar R (1994) Transformation of atrazine in soil by *Phanerochaete chrysosporium*. Soil Biol Biochem 26:1665–1671
Hitivani N, Mecs L (2003) Effects of certain heavy metals, on the growth, dye decolouration and enzyme activity of *Lentinula edodes*. Ectoxicol Environ Safety 55(2):199–203
Jackson M, Houl L, Banerjee H, Sridhar R, Dutta S (1999) Disappearance of 2, 4-dinitrotoulene and 2-amino, 4, 4-dinitrotoulene by *Phanerochaete chrysosporium* under non-lignolytic conditions. Bull Environ Contam Toxicol 62:390–396
Jin XC, Liu GQ, Xu ZH, Tao WY (2007) Decolorization of a dye industry effluent by Aspergillus fumigatus XC6. Appl Microbiol Biotechnol 74:239–243
Kapoor A, Viraraghavan T (1995) Fungal biosorption- an alternative treatment option for heavy metal bearing waters: a review. Bioresour Technol 53:195–206
Kaushik P, Malik A (2009) Fungal dye decolorization: recent advances and future potential. Environ Int 35:127–141
Kirk TK, Farrell RL (1987) Enzymatic combustion: the microbial degradation of lignin. Ann Rev Microbiol 41:465–505
Kookana RS, Baskaran S, Naidu R (1998) Pesticide fate and behavior in Australian soils in relation to contamination and management of soil and water: a review. Aust J Soil Res 36:715–764
Kulshreshtha S, Mathur N, Bhatnagar P (2014) Mushroom as a product and their role in mycoremediation. AMB Express 4:29
Kuo W, Regan R (1998) Aerobic carbamate bioremediation aided by compost residuals from the mushroom industry, laboratory studies. Compost Sci Util 6:19–29
Lang W, Sirisansaneeyakul S, Ngiwsara L, Mendes S, Martins LO, Okuyama M, Kimura A (2013) Characterization of a new oxygen-insensitive azo reductase from *Brevibacillus laterosporus* TISTR1911: toward dye decolorization using a packed-bed metal affinity reactor. Bioresour Technol 150:298
Lau KL, Tsang YY, Chiu SW (2003) Use of spent mushroom compost to bioremediate PAH- contaminated samples. Chemosphere 52:1539–1546
Mahapatra DM, Chanakya HN, Ramachandra TV (2014) Bioremediation and lipid synthesis through mixotrophic algal consortia in municipal wastewater. Bioresour Technol 168:142–150
Majeau JA, Brar SK, Tyagi RD (2010) Laccases for removal of recalcitrant and emerging pollutants. Bioresour Technol 101:2331–2350

Maloney S (2001) Pesticide degradation. In: Gadd G (ed) Fungi in bioremediation. Cambridge University Press, Cambridge, London

Masaphy S, Henis Y, Levanon D (1996) Manganese enhanced biotransformation of atrazine by the white rot fungus *Pleurotus pulmonarius* and its correlation with oxidation activity. Appl Environ Microbiol 62:3587–3593

Mishra A, Malik A (2013) Recent advances in microbial metal bioaccumulation. Crit Rev Environ Sci Technol 43:1162–1222

Moreno-Garrido I (2008) Microalgae immobilization: current techniques and uses. Bioresour Technol 99:3949–3964

Mousin CP, Ericuad C, Malosse C, Laugero C, Asther M (1996) Biotransformation of the insecticide lindane by the white rot basidiomycetes *Phanerochaete chrysosporium*. Pestic Sci 47:51–59

Munari FM, Gaio TA, Calloni R, Dillon AJP (2008) Decolorization of textile dyes by enzymatic extract and submerged cultures of *Pleurotus sajor-caju*. World J Microbiol Biotechnol 24:1383–1392

Nawaz K, Hussain K, Choudary N, Majeed A, Ilyas U, Ghani A, Lin F, Ali K, Afghan S, Raza G, Lashari MI (2011) Eco-friendly role of biodegradation against agricultural pesticides hazards. Afr J Microbiol Res 5:177–183

Pointing SB (2001) Feasibility of bioremediation by white-rot fungi. Appl Microbiol Biotechnol 57:20–33

Pumpel T, Paknikar KM (2001) Bioremediation technologies for metal containing waste waters using metabolically active microorganism. Adv Appl Microbiol 48:135–169

Qu Y, Shi S, Ma F, Yan B (2010) Decolorization of reactive dark blue K-R by the synergism of fungus and bacterium using response surface methodology. Bioresour Technol 101:8016–8023

Reddy C, Mathew Z (2001) Bioremediation potential of white rot fungi. In: Gadd G (ed) Fungi in bioremediation. Cambridge University Press, Cambridge, London

Renganathan S, Thilagaraj WR, Miranda LR, Gautam P, Velan M (2006) Accumulation of Acid Orange 7 Acid Red 18 *and Reactive Black 5 by growing Schizophyllum commune*. Bioresour Technol 97:2189–2193

Robinson T, McMullan G, Marchant R, Nigam P (2001) Remediation of dyes in textile effluent: a critical review on current treatment technologies with a proposed alternative. Bioresour Technol 77:247–255

Ron EZ, Royse DJ (1992) Interactions of bacteria with cadmium. Biodegradation 3:161–170

Saratale RG, Saratale GD, Chang JS, Govindwar SP (2011) Bacterial decolorization and degradation of azo dyes: a review. J Taiwan Inst Chem Eng 42:138–157

Saratale RG, Gandhi SS, Purankar MV, Kurade MB, Govindwar SP, Oh SE, Saratale GD (2013) Decolorization and detoxification of sulfonated azo dye C.I. Remazol Red and textile effluent by isolated *Lysinibacillus* sp. RGS. J Biosci Bioeng 115:658–667

Sasek V, Cajthaml T (2005) Mycoremediation. Current state and perspectives. Int J Med Mushrooms 7:360–361

Sathiya Moorthi P, Munuswamy D, Sellamuthu PS, Kandasamy M, Thangavelu KP (2007) Biosorption of textile dyes and effluents by *Pleurotus florida* and *Trametes hirsuta* with evaluation of their laccase activity. Iran J Biotechnol 5:114–118

Schinner F, Burgstaller W (1989) Extraction of zinc from industrial waste by a P*enicillium* sp. Appl Environ Microbiol 55:1153–1156

Sen K, Pakshirajan K, Santra SB (2012) Modelling the biomass growth and enzyme secretion by the white rot fungus. Appl Biochem Biotechnol 167(4):705–713

Sharma P, Singh L, Dilbaghi N (2009) Biodegradation of Orange *dye by Phanerochaete chrysosporium* in simulated wastewater. J Sci Ind Res 68:157–161

Shumate SE, Strandberg GW (1985) Accumulation of metals by microbial cells. In: Moo YM (ed) Comprehensive biotechnology, vol 4. Progamon Press, New York

Singh H (2006) Mycoremediation: Fungal Bioremediation. John Wiley and Sons Inc., New York

Singh DK (2008) Biodegradation and bioremediation of pesticide in soil: concept, method and recent developments. Indian J Microbiol 48:35–40

Singh M, Singh DK (2014) Biodegradation of endosulfan in broth medium and in soil microcosm by Klebsiella sp. M3. Bull Environ Contam Toxicol 92:237–242

Sly LI, Arunpairojana V, Dixon DR (1990) Binding of colloidal MnO_2 by extracellular polysaccharides of *Pedomicrobium manganicum*. Appl Environ Microbiol 56:2791–2794

Solıs M, Solıs A, Perez HI, Manjarrez N, Flores M (2012) Microbial decolouration of azo dyes: a review. Process Biochem 47:1723–1748

Strandberg B, Strandberg L, Bergqvist P, Falandysz J, Rappe C (1998) Concentrations and biomagnification of 17 chlordane compounds and other organochlorines in harbor porpoise (*Phocoena phocoena*) and herring from the southern Baltic Sea. Chemosphere 37:2513–2523

Sturman PJ, Stewart PS, Cunningham AB, Bouwer EJ, Wolfram JH (1995) Engineering scale-up *in situ* bioremediation processes: a review. J Contam Hydrol 19:171–203

Tapia-Tussell R, Perez-Brito D, Rojas-Herrera R, Cortes-Velazquez A, Rivera-Muenoz G, Solis-Pereira S (2011) New laccase-producing fungi isolates with biotechnological potential in dye decolorization. Afr J Biotechnol 10:10134–10142

Tortella G, Durán N, Rubilar O, Parada M, Diez MC (2013) Are white-rot fungi a real biotechnological option for the improvement of environmental health? Crit Rev Biotechnol 15:1350–1365

Trejo-Hernadez M, Lopez- Mungia A, Ramirez R (2001) Residual compost of *Agaricus bisporus* as a source of crude laccase for enzymatic oxidation of phenolic compounds. Process Biochem 36:635–639

Tuor U, Winterchalter K, Fietcher A (1995) Enzymes of white rot fungi involved in lignin degradation and ecological determinants for wood decay. J Biotechnol 41:65–74

Tychanowicz GK, Zilly A, deSouza CGM, Peralta RM (2004) Decolorization of industrial dyes by solid-state cultures of *Pleurotus pulmonarius*. Process Biochem 39:855e859

Valli K, Wariish H, Gold M (1992) Degradation of 2, 7-dicholrodibenzo-p-dioxin by the lignin degrading basidiomycetes *Phanerochaete chrysosporium*. J Bacteriol 174:2131–2137

Verdin AA, Sahraoui LHR, Durand R (2004) Degradation of benzo(a)pyrene by mitosporic fungi and extracellular oxidative enzymes. Int Biodeterior Biodegrad 53:65–70

Xiao P, Mori T, Kamei I, Kiyota H, Takagi K, Kondo R (2011) Novel metabolic pathways of organochlorine pesticides dieldrin and aldrin by the white rot fungi of the genus *Phlebia*. Chemosphere 85:218–224

Yadav AK, Kumar N, Sreekrishnan TR, Satya S, Bishnoi NR (2010) Removal of chromium and nickel from aqueous solution in constructed wetland: mass balance, adsorption-desorption and FTIR study. Chem Eng J 160:122–128

Yang S, Hai FI, Nghiem LD, Price W, Roddick F, Moreira MT, Magram SF (2013) Understanding the factors controlling the removal of trace organic contaminants by white-rot fungi and their lignin modifying enzymes: a critical review. Bioresour Technol 141:97–108

Yavad J, Reddy C (1993) Degradation of benzene, toluene, ethylbenzene and xylene (BTEX) by the lignin degradation basidiomycetes *Phanerochaete chrysosporium*. Appl Environ Microbiol 59:756–762

Yong C, Mcaskie LD, Dean ALR, Chetham AK, Jakeman RJB, Skarnulis AJ (1987) Cadmium accumulation by *Citrobacter* spp.: the chemical nature of the accumulated metal precipitate and its location in bacterial cells. J Gen Microbiol 133:539

Yousefi V, Kariminia HR (2010) Statistical analysis for enzymatic decolorization of acid orange 7 by *Coprinus cinereus* peroxidase. Int Biodetr Biodegr 64:245–252

Zhang J, Chiao C (2002) Novel approaches for remediation of pesticide pollutants. Int J Environ Pollut 18:423–433

Chapter 9
Mushrooms: Isolation and Purification of Exopolysaccharides

Yuxiao Wang, Xiaojun Huang, and Shaoping Nie

9.1 Introduction

Mushrooms are abundant sources of nutritional value. Mushrooms refer to the fleshy fruit body of the fungi with roots of forests or broad-leaf trees above the ground. Underground mycelium of mushrooms' fruit body is the main part of the fructification process (Pavel 2009). Mushrooms are heavily consumed especially in Japan, Korea and China, due to their special texture, flavor and aroma. The number of natural mushrooms on Earth is estimated at 140,000, but only about 10% are declared (Ghosh 2015). Meanwhile, among the approximately 14,000 named species, about 2000 are proved to meet rules of safe edibility for mankind (Wasser 2002). As a commercial industry in North America and Europe, *Agaricus bisporus* (Fig. 9.1a) are the most popular cultivated mushrooms, such as brown mushroom, white mushroom or Portobello. *Lentinula edodes* (also named shiitake) (Fig. 9.1b), *Dictyophora indusiata* (also named Kinugasatake) (Fig. 9.1c), *Ganoderma lucidu* (also named Lingzhi or Reishi), *Auricularia auricular-judae* ('wood-ear' mushroom) (Fig. 9.1d), *Pleurotus ostreatus* (or oyster mushroom) (Fig. 9.1e) and *Flammulina velutipes* (golden needle mushroom or enokitake) (Fig. 9.1f) are consumed and largely grown in China, Taiwan, Korea and Japan (Drori et al. 2016; Deng et al. 2016; Han 2010; Castro-Alves et al. 2017; Zhao et al. 2017; Bao et al. 2016).

Every year, approximately 650 species of mushroom are consumed at a cost of 5–6 billion because of their nutritional properties. Research has indicated that mushrooms not only contain low energy value and low fat level, but also contain essential amino acids and high amounts of proteins during the past two decades (Manzi et al. 2004). The mushroom-derived polysaccharides play a major part

Y. Wang · X. Huang · S. Nie (✉)
State Key Laboratory of Food Science and Technology, Nanchang University, Nanchang, Jiangxi Province, China
e-mail: spnie@ncu.edu.cn

B. P. Singh et al. (eds.), *Biology of Macrofungi*, Fungal Biology,
https://doi.org/10.1007/978-3-030-02622-6_9

Fig. 9.1 Some traditional mushrooms

among components with all kinds of bioactivities (Wang et al. 2017). This chapter reviews the main methods for mushroom polysaccharides (just referring to exopolysaccharides) isolation and purification, in order to facilitate the application of mushroom-derived polysaccharide on nutrition, medicine and nutraceuticals.

9.2 Extraction of Exopolysaccharides

The active polysaccharides contained in fruiting bodies, the cultured broth and mycelium biomass of mushrooms (Wu et al. 2006) are the subject of intense research. They are applied in diverse industry due to their high potential properties. Therefore, several methods are available for the purpose of improving the yield and purity. The extraction yield of mushroom-derived polysaccharides is affected by mushroom type and extraction conditions like temperature, pH, pressure and fineness of solid particle grinding and ionic strength of the solvents (Zhang et al. 2004).

Several steps are involved in isolation and purification of polysaccharides. First step, crude polysaccharides are extracted from raw material by subsequent extractions (Lei 2016). Second step, preparative chromatography, chemical reagents and specific enzymes are used to purify the crude polysaccharides. Finally, polysaccharide fractions can be separated according to their different solubility, sensitivity to chemical or enzymatic destruction, affinity to specific molecules and complexation with metal cations (Castro-Alves 2016).

Aqueous solvents, almost hot/cold water are used to dissolve crude polysaccharide samples before purification. Generally, after water extraction, the crude samples are freeze-dried for better storage, and then ground to a fine powder to shorten the dissolving time in subsequently purification process. Based on their molecular weight distributions, solubility in different solvents, and their different ionic properties, polysaccharides are purified with the employment of different resolvent like hot water, alkali and acidic solutions at different temperatures. Several methods for mushroom polysaccharides extraction are listed in Table 9.4.

9.2.1 Aqueous Extraction

Aqueous extraction is one most conventional method commonly used in extracting polysaccharide (Abdullah et al. 2017). Before water extraction, mushroom samples are pretreated with organic solvent, such as ethanol, methanolic, acetone or a mixture of chloroform and methanol, to remove some apolar compounds like terpenes, phenols and lipids (Ullah et al. 2018). The crude polysaccharide is extracted reputedly with hot water. Centrifugation is used to separate supernatant from residue. Supernatants from each repeating unit are collected (Ruthes et al. 2015; Manna et al. 2017).

9.2.2 Alkaline Extraction

After aqueous extraction, the remainder residues can be extracted with alkaline basic solutions, generally NaOH or KOH under different extraction conditions (Khatua and Acharya 2017). Siu et al. (2016) collected the residue and subjected it to the combination of 0.5 M NaOH and 0.05 M $NaBH_4$ at 4 °C about 6 h for three times. Liu et al. dipped the obtained residue into 5% alkali solution for 24 h three times (Liu et al. 2013). The extraction solution was neutralized by acetic acid until pH = 7, and vacuum evaporated usually below 45 °C. Bhunia et al. boiled the residue with 4% NaOH for further extraction (Bhunia et al. 2011). Commonly, $NaBH_4$ was added to the alkaline solutions to protect reducing end-units in polysaccharide chains from the degradation (Smiderle et al. 2006; Smiderle et al. 2008a).

9.2.3 Acidic Extraction

Aqueous solutions cannot work availably enough for extracting water-insoluble polysaccharides. Consequently, acidic solvents are used to separate the water insoluble part. Kao et al. (2012) used 2 M HCl to extract the waste residue, and further degraded at 50 °C for 4 h to provide insoluble residue. Acid hydrolysate fractions of different concentration (45% and 55%, successively) were obtained. Szwengiel et al. optimized extraction conditions to 3.8% HCl at 30 °C (Szwengiel and Stachowiak 2016). Further modification of water-insoluble polysaccharides, such as sulfated modification, chlorosulfonic acid-pyridine in dimethyl formamide, were studied for the purpose of improving water-solubility, chemical properties and bioactivities of polysaccharide (Liu et al. 2010).

9.2.4 *Pressurized Liquid Extraction*

Pressurized liquid extraction (PLE) is a sophisticated technique that liquid water is used as extraction solution at temperatures above 100 °C (0.1 MPa), but below 374 °C (22.1 MPa) (Plaza and Turner 2015). The entirely extraction procedure of PLE is suffering on high pressure and high temperature, in order to fully extracting the polysaccharide from fungus. The former requires the solvent below its boiling point, while the latter is emerged in the high solubility and diffusion rate. Under such conditions, a much higher extraction rate can be achieved with less volume extraction solvents and shorter extraction time, making this procedure more efficiently. PLE has been used as an effectual strategy for polysaccharides extracting in recent years (Table 9.1). Compared with other extraction methods, the selectivity, recovery and yields are all improved. In addition, with the application of PLE, the extraction process can avoid over-time exposure to light and oxygen. For instance, Palanisamy et al. extracted β-glucans from *Agaricus bisporus, Pleurotus ostreatus* and *Lentinula edodes* with the optimized extraction conditions: 5 cycles of 5 min at 10.3 MPa, 200 °C (Marimuthu et al. 2014). Pressure, temperature and time are the three critical conditions for PLE-aided extraction. The polysaccharide yield increased from 80.6% to 88.6% with the extraction time increased from 10 to 70 min. But continued increases in extraction time didn't make the polysaccharide yield higher. As for the extraction pressure, the higher pressure can make higher extraction recovery; the highest pressure can reach up to 15.2 MPa. Pressures over this special point didn't bring any increase in polysaccharide yield. It is reported an optimal conditions of 70 min at 10.1 MPa under 28 °C is likely to achieve the theoretically maximum polysaccharide recovery (Tiffany et al. 2007). However, because of the differences in polysaccharide structure and mushroom source, the optimized extraction conditions may vary within a certain range (Santoyo et al. 2010). Based on these advantages, PLE has been used more frequently.

Table 9.1 Compilation of works found in the literature on PLE of Polysaccharides from natural mushroom sources

Source	Temp. (°C)	Pressure (MPa)	Time(min)	Static/ Dynamic	Ref.
Chlorella vulgaris	150	10.3	20	Static	Santoyo et al. (2010)
Dunaliella salina	160	10.3	15	Static	Santoyo et al. (2012)
Golden oyster mushroom	200	0.002 to 5.0	20	Static	Jo et al. (2013)
Haematococcus pluvialis	100	10.3	20	Static	Santoyo et al. (2012)
Himanthalia elongata	100	10.3	20	Static	Santoyo et al. (2011)

9.2.5 Ultrasound-Assisted Extraction

Ultrasonic-assisted extraction (UAE) is another efficient and expeditious technique to improve the polysaccharide extraction rate (Table 9.2). The sound waves produce alternating low-pressure (rarefaction) cycles and high-pressure (compression). During the low-pressure (rarefaction) cycle, vacuum bubbles are produced by the high-intensity ultrasonic waves. When they accumulate enough volumes, they stop absorbing energy and collapse violently during the high-pressure cycle for extraction and fundamental interaction. Over the past two decades, UAE has been broadly used in the fractionation of vegetables materials (Riera et al. 2010). Conditions including extraction temperature, extraction time, ultrasonic power and ratio of water to sample are optimized in order to improve the polysaccharides yield.

For example, with the operating optimum condition of 350 W, 35 min at 90 °C and a ratio of 5 times water, polysaccharides were efficiently separated from Black Fungus. From this experiment, Riera et al. drew the conclusion that the significant parameters affecting the separation process were ultrasonic power and ratio of water to sample (Komura et al. 2010). The optimal conditions for *Agaricus bisporus* polysaccharides (ABP) separation were as followed: ultrasonic power 230 W, at 70 °C for 62 min, and a ratio of water volume to raw material weight of 30. Under this condition, the recovery of ABP can reach 6.02% (Tian et al. 2012). The yield of polysaccharides from *Cordyceps Sinensis* can also increase significantly with the ultrasound intensity. Both the increased extraction temperature and reduced the par-

Table 9.2 Compilation of works found in the literature on UAE of Polysaccharides from natural mushroom sources

Materia	Ultrasonic power(w)	Solid-liquid ratio(g/mL)	Temp. (°C)	Time(min)	yield(%)	Ref.
Agaricus Bisporus	230	1:30	70	62	6.02	Tian et al. (2012)
	400	1:10	25	15	4,70	Aguiló-Aguayo et al. (2017)
	400	– [a]	60	20	–	Lespinard et al. (2015)
Black Fungus	350	1:5	90	35	–	Riera et al. (2010)
Ganoderma lucidum	671	1:12.5	28	45	15.00	Chen et al. (2014)
Paecilomyces hepiali	300	1:160	25	11	9.48	Yu et al. (2011)
Thelephora ganbajun	500	1:70	40	11	–	Xu et al.(2016)
Tricholoma matsutake	365	1:53.5	–	2.67	7.97	You et al. (2013); You et al.(2014)
Tuber huidongense	99.65	1:24.65	70.1	40.39	7.17	Chen et al. (2016)
Yunzhi mushroom	30	1:40	40	480	5.95	Pan et al. (2010a, b)

Note: – [a]: Information didn't be provided

ticle size can lead to a higher extraction rate and yield (Cheung 2014). Cheung and Wu analyzed the experimental data that the increasing solid-to-liquid ratio and/or decreasing the liquid volume would increase the polysaccharide yield and extraction rate (Cheung and Wu 2013).

9.2.6 Microwave-Assisted Extraction

Microwave-assisted extraction (MAE) is a substituted technique to improve the extraction rate. It is broadly applied into the separation of bioactive compounds with lower quantities of solvents and less pre-preparation (Table 9.3). Many researchers discovered that microwaves can heat the molecules up through dipole rotation and ionic conduction. But ion migration is resisted by the extraction medium, heating the slight microscopic traces of moisture. The tremendous pressure produced in microwave treatment can break the cell wall and therefore relieve the active compounds contained in materials. MAE has been engaged in the separation of polysaccharides from mushroom in recent several years. The extracting conditions of *Cordyceps militaris* polysaccharides were a ratio of water to material 31.1 mL/g and a microwave power of 744.8 (Song et al. 2009). Zhang et al. optimized extraction conditions of polysaccharides from *Agaricus blazei* Murrill to a 400 W microwave power and a ratio of solution to solid 32.7 under temperature of 74.64 °C for time 29.37 min. Under this condition, the recovery can reach up to 12.35% (Zhang et al. 2012c). Likewise, microwave technique combined with ultrasound-assisted extraction was employed in *Lycoris aurea* to shorten the extraction time and improve extraction efficiency (Zhang et al. 2009).

The ultrasonic and microwave combined assisted extraction (UMAE, for short) is considered more efficient for polysaccharides extraction. UMAE was also used in

Table 9.3 Compilation of works found in the literature on MAE of Polysaccharides from natural mushroom sources

Materia	Microwave power(w)	Solid-liquid ratio(g/mL)	Temp. (°C)	Time(min)	yield(%)	Ref.
Agaricus blazei Murrill	400	1: 32.7	74.64	29.37	12.35	Zhang et al. (2012c)
Cordyceps militaris	744.8	1: 31.1	65	4.2	5.94	Song et al. (2009)
Ganoderma lucidum,	284	1: 11.6	–[a]	11.68	–	Huang and Ning (2010)
Hericium Erinaceum	100	–	140	5	–	Ookushi et al. (2006)
Fomitopsis ulmaria	400	1:40	–	2.5	8.36	Zhao et al. (2015)
Inonotus obliquus	90	1:20	85	19	3.25	Chen et al. (2010)

Note: –[a]: Information didn't be provided

the process of Lycium barbarum at the following optimized operating conditions: microwave power of 500 W for 10 min and ultrasonication for 30 min at 50 °C (Dong et al. 2011). Similarly, the optimal extraction conditions of the polysaccharides from *Ganoderma lucidum* were microwave power of 284 W, ultrasonic power of 50 W, solid/ water ratio of 1:11.6 and extraction time of 701 s (Huang and Ning 2010). Under these conditions, the yield was increased over 27.7%.

9.2.7 Supercritical Fluid Extraction

Supercritical fluid extraction (SFE) is commonly used for mostly active compounds extraction (Pereira and Meireles 2010). SFE was reacting through a corresponding solvent at certain pressure and temperature above its supercritical point. When approaching to the critical point, small changes in pressure or temperature will change the final yield. Carbon dioxide is a kind of frequently-used solvent due to the characters of environment-friendly, low viscosity, high liquid-like density and diffusivity. According to those traits, a particular solvent with the characteristics of penetrating solid matrices and dissolving materials should be provided. Supercritical carbon dioxide (SC-CO_2) extraction requires low temperature, but no organic solvent. Bioactive components can be protected from decomposition.

In addition, in the dynamic procedure, the fresh supercritical CO_2 flows through materials continuously and the mass transfer is intensive, providing an excellent performance. Conditions including time, temperature, CO_2 flow rate and operating pressure need farther optimization. Some researchers used this technique to extract polysaccharides from *Ganoderma lucidum* (Fu et al. 2009). The results suggested that polysaccharide yields were improved with the CO_2 flow rate from 4 to 10 kg/h. Nevertheless, when flow rate exceeded the upper limit, the yield didn't increase obviously. Meanwhile, operating pressure was one of the indispensable factors. In the range of 27–35 MPa, higher the operating pressure can bring higher yield increased, similar to the temperature parameter. With regard to extraction time, the yield would improve along with time increased. Therefore, when the conditions were time 4 h, temperature 25 °C, pressure 35 MPa and CO_2 flow rate 10 kg/h, the recovery of polysaccharides from *Ganoderma lucidum* implemented promptly with a maximum effectiveness (Fu et al. 2009).

9.3 Purification of Exopolysaccharides

Separation and purification of polysaccharides are one of the most important steps before structure analysis. Proteins and pigment are impurities require to be removed. Based on their different dissolution properties from polysaccharide, organic solvent can be used to remove these impurities. Organic solvents like ethanol and methanol are used to eliminate amino acids, monosaccharides, phenols and other compounds

(Palacios et al. 2011). Proteins are removed by using enzyme protease at conditions of 1 h at 40 °C or precipitated with trifluoroacetic acid (TFA). Then, ethanol is added to precipitate polysaccharides from the solution. Adding concentrated NaCl properly is beneficial to precipitation. And ethanol or acetone is used consequently to wash the precipitant.

Extracted polysaccharides, with a broad dispersity and different molecular sizes, usually require further purification to separate different polysaccharide fractionation. Ethanol precipitation, acidic precipitation with acetic acid, gel filtration, ion-exchange and affinity chromatography are several commonly used methods for fractionation, which are summerized in Table 9.4. Some key factors, such as molecular weights, branching degree and pattern of branches, especially the constituent of raw materials; can affect fractionation and purification during this process.

9.3.1 Freeze–Thawing

Freeze–thawing is an efficient and simple process to acquire pure polysaccharides. The concentrated crude water extracts are frozen and then slowly thawed at room temperature, and this process repeats for several times. The supernatant and precipitation achieved by centrifugation are frozen. Then it thawed slightly until precipitation producing in supernatant is eliminated completely. Pure fractions could be commonly obtained in precipitation. Freeze-thawing method is appropriate for different polysaccharide with diverse branch degrees. The molecules presenting fewer branches or a linear structure are induced to create precipitation (Ruthes et al. 2013b).

9.3.2 Treatment with Solvent

Ethanol is a commonly used solvent in purifying polysaccharide (Dong et al. 2012). It can dehydrate the polysaccharides and separate the high molecular weight precipitate from the low ones. In addition, polysaccharides can be dissolved some apolar solvents, such as DMSO and $(NH_4)_2SO_4$ (Smiderle et al. 2013a), therefore, separate polysaccharide from other compounds (Lehtovaara and Gu 2011).

9.3.3 Treatment with Fehling Solution

Some polysaccharides can separate from other composition by copper salts. Cu and other metal ions in **Fehling solution** can precipitate with some active groups (OH, NH_2 and COOH), therefore separate different fractions. Generally, after freeze-thawing treatment, the rest parts are combined with Fehling solution. There is a

Table 9.4 Isolation and purification of polysaccharides from natural mushroom sources

Source	Extraction	Purification Methods	M_w(Da)	Branching Position	Ref.
Agaricus bisporus	Cold water extract	Ethanol precipitation (3:1, v/v); Dialysis; Freeze-thawing; Fehling solution treatment; Closed dialysis (100 kDa)	$3.70 \times 10^5$and 4.40×10^5	O-2	Ruthes et al. (2013c)
Agaricus brasiliensis	Cold water extract	Ethanol precipitation (3:1, v/v); Dialysis; Freeze-thawing; Fehling solution treatment; Ultrafiltration (300 kDa)	1.90×10^4	O-2	Komura et al. (2010)
	Ethanol extract (75 °C); Water extract (75 °C)	Ethanol precipitation (4:1, v/v); Proteinase and Sevag treatment; Dialysis; Ethanol precipitation (4:1, v/v); DEAE-Sepharose column	3.9×10^5	O-2	Liu et al. (2011)
Amanita muscaria	A combination of hot water and hot water extract	Ethanol precipitation (3:1, v/v); Dialysis; Freeze-thawing; Fehling solution treatment; Closed dialysis (100 kDa)	2.60×10^4	O-2	Ruthes et al. (2013a)
Armillaria mellea	Ethanol extract (75 °C); Water extract (75 °C)	Ethanol precipitation (1:5, v/v); Sevag; Gradual ethanol precipitation; DEAE-Sephadex A-25; Sepharose CL-6B; Sephadex G-25	7.80×10^3	O-2	Sun et al. (2009)
Astraeus hygrometricus	6% NaOH extract (60 °C)	Ethanol precipitation (1:5, v/v); Dialysis; Ethanol re-precipitation; Sepharose-6B	1.60×10^5	O-6	Maiti et al. (2008)
Beauveria bassiana	0.1 N NaOH with 0.5 M $NaBH_4$ extract (22 °C)	96% ethanol precipitation (1:1, v/v); Dialysis; Re-suspensionin 50% ethanol; Sepharose CL-6B	2.00×10^5	O-4	Bernabé et al. (2011)
Boletus edulis	Ethanol extract (100 °C); Water extract (100 °C)	Ethanol precipitation (1:4, v/v); Sevag; Dialysis; Sepharose CL-6B	1.10×10^5	O-2	Wang et al. (2014a)
	Water extract (100 °C)	Gradual ethanol precipitation (30% and 60%); DEAE-Sepharose column; High-resolution Sephacryl S-300	1.08×10^4	–[a]	Zhang et al. (2014a)
Calocybe indica	Water extract (100 °C)	Sepharose-6B	2.00×10^5	O-6	Mandal et al. (2011)

(continued)

Table 9.4 (continued)

Source	Extraction	Purification Methods	M_w(Da)	Branching Position	Ref.
Chroogomphis rutilus	95% ethanol extract; Hot water extract	95% ethanol precipitation; Proteinase and Sevag treatment; Dialysis; 95% ethanol precipitation; DEAE Sepharose fast flow and Sepharose-6 Fast flow; Ethanol precipitation; Dialysis	3.20×10^4	O-2	Sun et al. (2010)
Coprinus comatus	95% ethanol extract; Hot water extract	Ethanol precipitation (3:1, v/v); DEAE-Sepharose CL-6B; Sepharose CL-6B column; Sepharose CL-4B column	–	O-2	Li et al. (2013)
Cordyceps militaris	$CHCl_3$-MeOH extract; Hot water extract; 5% KOH extract (100 °C)	Neutralization; Dialysis; Freeze-thawing; Me_2SO_4 treatment (50 °C); Dialysis (100 kDa)	2.30×10^4	O-6	Smiderle et al. (2013b)
Flammulina velutipes	Hot water extract; 25% KOH extract (100 °C)	Neutralization; Dialysis; Freeze-thawing; Me_2SO_4 treatment (40 °C)	3.10×10^5	O-4	Smiderle et al. (2006)
	Hot water extract	Ethanol precipitation (3:1, v/v); Dialysis; Freeze-thawing; Fehling solution treatment; Ultrafiltration (300 kDa and 30 kDa)	–	O-2	Smiderle et al. (2008a)
	Hot water extract	Sephacryl S-300 and Sephacryl S-400	1.30×10^4	O-2	Zhang et al. (2012d)
Fomitella fraxinea	0.9% sodium chloride extract	Ethanol precipitation; DEAE-cellulose	1.50×10^4	O-2	Cho et al. (2011)
Ganoderma atrum	Hot water extract	Ethanol precipitation (1:1, v/v); Sevag treatment; Dialysis; Ethanol precipitation (4:1, v/v); Superdex-200	6.90×10^4	O-2	Zhang et al. (2014b)
	–	Superdex G-200 column	2.00×10^5	–	Zhang et al. (2012b)

Ganoderma lucidum	Hot water extract	Ultrafiltration (10–100 kDa); DEAE-Sepharose fast flow; Sephacryl S-300	1.20×10^4	O-2	Ye et al. (2008)
	Ethanol extract (100 °C); Water extract (100 °C)	Ultrafiltration (10–100 kDa); DEAE-Sepharos fast flow column; Sephacryl S-300	1.20×10^4	O-2	Ye et al. (2008)
	Ethanol extract (100 °C); Water extract (100 °C)	Ultrafiltration (10–100 kDa); DEAE-Sepharos fast flow column; Sephacryl S-300	7.00×10^3	–	Ye et al.(2009)
	Ethanol extract (100 °C); Water extract (100 °C)	2 M NH_4OH treatment; Neutralization; Dialysis; Sephadex G-75; DEAE-52 column	7.80×10^4	O-2 / O-3	Pan et al. (2012)
Ganoderma sinense	Water extract (100 °C)	Ethanol precipitation (4:1, v/v); Sevag treatment; Sepharose CL-6B; Gel-permeation chromatography S-400 HR; Sepharcyl S-300	8.30×10^5	–	Han et al. (2012)
Grifola frondosa	Hot water extract	Deproteinated by trichloroacetic acid; Dialysis (3.5 kDa); Ethanol precipitation (3:1, v/v); DEAE-cellulose; Sephacryl S-300 high resolution	1.50×10^5	O-2	Wang et al. (2014b)
Hericium erinaceus	Ethanol extract; Hot water extract	Ethanol precipitation; DEAE-Sepharose fast flow; Sephacryl S-400 high resolution	2.00×10^5	O-2	Zhang et al. (2012a)
Inonotus levis	5% NaOH extract with $NaBH_4$	Proteinase treatment; Dialysis; Anion-exchange chromatography	–	–	Vinogradov and Wasser (2005)
Lactarius camphoratum	Water extract (100 °C)	Dialysis; Ethanol precipitation (3:1, v/v); Sevag treatment; DEAE cellulose; Sephadex G-200	9.30×10^3	O-6	Wang et al. (2013)
Lactarius rufus	Cold water extract	Ethanol precipitation (3:1, v/v); Dialysis; Freeze-thawing; Fehling solution treatment; Closed dialysis (100 kDa)	1.40×10^4	O-2	Ruthes et al. (2012)
Laetiporus sulphureus	Hot water extract	Ethanol precipitation (3:1, v/v); Dialysis; Freeze-thawing; Fehling solution treatment; Ultrafiltration (30 kDa)	2.80×10^4	O-2	Alquini et al. (2004)

(continued)

Table 9.4 (continued)

Source	Extraction	Purification Methods	M_w(Da)	Branching Position	Ref.
Lecanicillium muscarium	0.1 N NaOH with 0.5 M $NaBH_4$ extract (22 °C)	Ethanol precipitation (1:1, v/v); Dialysis; Re-suspensionin 50% ethanol; Sepharose CL-6B	2.00×10^5	O-2	Bernabé et al. (2011)
Lentinus edodes	Cold water extract	Ethanol precipitation (3:1, v/v); Dialysis; Freeze-thawing; Fehling solution treatment; Ultrafiltration (300 KDa)	1.60×10^4	O-2	Carbonero et al. (2008)
	Water extract (100 °C)	70% ethanol precipitation; Sevag treatment; DEAE cellulose; Sepharose CL-6B	1.80×10^4	–	Jeff et al.(2013)
Lentinus squarrosulus	Water extract (100 °C)	Ethanol precipitation; Sepharose 6B	1.96×10^5		Bhunia et al. (2010)
Lepista sordida	Ethanol extract (85 °C); Water extract (80 °C)	Ethanol precipitation (3:1, v/v); Freeze-thawing, Proteinase and Sevag treatments; DEAE cellulose; Sepharose CL-6B	4.00×10^4	O-2	Luo et al. (2012)
Lineolata rhizophorae	0.1 N NaOH extract (22 °C)	Re-suspensionin 50% ethanol; Sepharose CL-6B	–	O-3	Gimenez-Abian et al. (2007)
Phellinus baumii	Ethanol extract; Water extract (100 °C)	Gradual ethanol precipitation (30% and 60%); DEAE-Sepharose fast flow column	2.00×10^6	–	Ge et al. (2009)
Pleurotus eryngii	Hot water extract	Ethanol precipitation; DEAE-Sepharose fast flow; Sephacryl S-300 high resolution and Sephacryl S-100 high resolution	1.9×10^4	O-2	Zhang et al. (2013a)
Pleurotus geesteranus	95% ethanol extract; Hot water extract (100 °C)	Ultrafiltration; DEAE-Sepharose fast flow; Sephacryl S-300 high resolution	1.30×10^4	O-2	Zhang et al. (2013b)
Pleurotus florida	Hot water extract	Gradual ethanol precipitation (80%); Dialysis; Ethanol precipitation (5:1, v/v); Gel-permeation chromatography	4.80×10^4	O-2	Rout et al. (2006)
	Cold water extract	Ethanol precipitation (5:1, v/v); Dialysis; Sepharose-6B	1.70×10^5	O-2	Maity et al. (2014)

Leurotus florida and Calocybe indica var. APK2	Water extract (100 °C)	Ethanol precipitation (1:5, v/v); Wash with ethanol: Acetone mixture; Dialysis; Sepharose-6B	2.01×10^5	O-4	Dey et al. (2013)
	Water extract (100 °C)	Ethanol precipitation; Dialysis; Sepharose-6B	2.25×10^5	O-2 and O-4	(Maity et al.(2011a)
Pleurotus ostreatus	95% ethanol extract; Hot water extract (100 °C)	Sevag treatment; Dialysis; Ethanol precipitation (3:1, v/v); DEAE-sepharose CL-6B; Sephadex G-25	2.40×10^4	O-2	Sun and Liu (2009)
	Water extract (100 °C)	Gel-permeation chromatography	–	–	Maity et al. (2011b)
Pleurotus pulmonarius	CH_3Cl:MeOH extract(3:1, v/v); cold water extract	Ethanol precipitation (3:1, v/v); Dialysis; Freeze-thawing; Fehling solution treatment; Ultrafiltration (500 kDa and 30 kDa)	2.40×10^4	O-2	Smiderle et al. (2008b)
Pleurotus sajor-caju	Water extract (100 °C)	Ethanol precipitation (1:5, v/v); Sepharose-6B	3.50×10^4	O-2	Pramanik et al. (2005)
Russula albonigra	Water extract (100 °C)	Sepharose-6B	1.45×10^5	–	Nandi et al. (2013)
Tricholoma matsutake	Soak ethanol treatment; Water extract (100 °C)	Dialysis; ethanol precipitation (3:1, v/v); Sevag treatment; 30% H_2O_2treatment; DEAE cellulose; Sephadex G-100	8.90×10^4	O-6	Ding et al. (2010)
Termitomyces robustus	Water extract (100 °C)	Ethanol precipitation (1:5, v/v); 4% NaOH treatment; Re-precipitation with ethanol; Dialysis; Sepharose-6B	2.00×10^6	O-6	Mondal et al. (2008)
Volvariella bombycina	Water extract (100 °C)	Ethanol precipitation (1:5, v/v); Sepharose-6B	1.60×10^5	O-6	Das et al. (2008)
Volvariella diplasia	Water extract (100 °C)	Ethanol precipitation (1:5, v/v); 1% acetic acid treatment; Re-precipitation with ethanol (1:5, v/v); Dialysis; Sepharose-6B	1.76×10^5	–	Ghosh et al. (2008)
Volvariella volvacea	Water extract (100 °C)	Ethanol precipitation (1:5, v/v); Sepharose-6B	1.32×10^5	O-2	Nandan et al. (2011)
	Water extract (100 °C)	Ethanol precipitation; Sepharose-6B	1.95×10^5	O-6	Patra et al. (2011)

combination of two solutions named A and B in the Fehling solution, and equal volumes of them are prepared to be used (Jones and Stoodley 1965). Over the past years, the modified solution was widely used. It is reported that the A solution consisted of Copper (II) sulfate. B solution is a mixture of $KNaC_4H_4O_6 \cdot 4H_2O$ and KOH. Firstly, we should dissolve the polysaccharide parts in B solution, and mix with the equal volume of a solution. Then, samples are involved into B Solution. The diverse polysaccharide fractions are solubilized in lowest volume. When the Fehling solution is added, consistent stirring for more than 12 h is necessary to keep it dissolving. Then the mixture is preserved at 4 °C overnight. After centrifugation, soluble section and insoluble precipitate are separated. Finally, after neutralization and 48 h dialysis, they must be deionized with ion exchange resins before freeze dried.

9.3.4 Closed Dialysis and Ultrafiltration

Closed dialysis and **ultrafiltration** is more commonly used than Fehling solution method in polysaccharides purification. The concentration used for dialysis is usually 10 mg/ml. Membranes are regenerated cellulose with different refractive index elution profiles (5-1000 kDa; Sepctra/Por®; Milipore).

These methodologies have the advantages of little polysaccharide loss (Carbonero et al. 2008), however, it is difficult in practical application. This is because MWCO membrane is most commonly used in protein purification, however immaturely applied in separating different polysaccharide fractions (Komura et al. 2010). Unlike protein purification, even if dialysis or ultrafiltration membrane is shown effective in polysaccharide fractionation in theory, it is hard to realize large-scale fractionation.

9.3.5 Column Fractionation

Column fractionation is one of the most effective methods on purification of polysaccharides. Crude extracts may include some fractions with diverse molecular sizes (Zhang et al. 1998). Size exclusion chromatography (SEC) is a popular method that separates polysaccharides in accordance with different molecular size, which has great significance for laboratory separation (Zeng et al. 2012). Cross-linked agarose, dextran and polyacrylamide are three kinds of dominant column packing materials. The SEC column is appropriate to separate neutral polysaccharides, while a DEAE-cellulose column is always used to separate β-glucan. Therefore, ion exchange chromatography is an effective technique to separate neutral polysaccharide from charged polysaccharide.

References

Abdullah N, Abdulghani R, Ismail SM, Abidin M (2017) Immune-stimulatory potential of hot water extracts of selected edible mushrooms. Food Agr Immunol 28(3):374–387

Aguiló-Aguayo I, Walton J, As IV, Tiwari BK (2017) Ultrasound assisted extraction of polysaccharides from mushroom by-products. Food Sci Tech Brazil 77:92–99

Alquini G, Carbonero ER, Rosado FR, Cosentino C, Iacomini M (2004) Polysaccharides from the fruit bodies of the basidiomycete *Laetiporus sulphureus* (bull.: Fr.) murr. Fems Microbiol Lett 230(1):47–52

Bao H, You SG, Cao L, Zhou R, Wang Q, Cui SW (2016) Chemical and rheological properties of polysaccharides from fruit body of *Auricularia auricular*-judae. Food Hydrocoll 57:30–37

Bernabé M, Salvachúa D, Jiménezbarbero J, Leal JA, Prieto A (2011) Structures of wall hetero galactomannans isolated from three genera of entomopathogenic fungi. Fungal Biol UK 115(9):862–870

Bhunia SK, Dey B, Maity KK, Patra S, Mandal S, Maiti S, Maiti TK, Sikdar SR, Islam SS (2011) Isolation and characterization of an immunoenhancing glucan from alkaline extract of an edible mushroom, *Lentinus squarrosulus* (Mont.) singer. Carbohydr Res 346(13):2039–2044

Bhunia SK, Dey B, Maity KK, Patra S, Mandal S, Maiti S, Maiti TK, Sikdar SR, Islam SS (2010) Structural characterization of an immunoenhancing heteroglycan isolated from an aqueous extract of an edible mushroom, *Lentinus squarrosulus* (Mont.) singer. Carbohydr Res 345(17):2542–2549

Carbonero ER, Gracher AHP, Komura DL, Marcon R, Freitas CS, Baggio CH, Santos ARS, Torri G, Gorin PAJ, Lacomini M (2008) *Lentinus edodes* heterogalactan: Antinociceptive and anti-inflammatory effects. Food Chem 111(3):531–537

Castro-Alves V, Gomes D, Jr N, Sforça M (2017) Characterization and immunomodulatory effects of glucans from *Pleurotus albidus*, a promising species of mushroom for farming and biomass production. Int J Biol Macromol 95:215–223

Castro-Alves VC (2016) Speeding up the extraction of mushroom polysaccharides. Food Anal Method 9(9):2429–2433

Chen G, Zhang S, Ran C, Wang L, Kan J (2016) Extraction, characterization and antioxidant activity of water-soluble polysaccharides from tuber huidongense. Int J Biol Macromol 91:431–442

Chen T, Wu Y, Wu J, Ma L, Dong Z, Wu J (2014) Efficient extraction technology of antioxidant crude polysaccharides from *Ganoderma lucidum* (Lingzhi), ultrasonic-circulating extraction integrating with superfine-pulverization. J Taiwan Inst Chem Eng 45(1):57–62

Chen Y, Gu X, Huang S, Li J, Wang X, Tang J (2010) Optimization of ultrasonic/microwave assisted extraction (UMAE) of polysaccharides from Inonotus obliquus and evaluation of its anti-tumor activities. Int J Biol Macromol 46(4):429–435

Cheung Y, Wu J (2013) Kinetic models and process parameters for ultrasound-assisted extraction of water-soluble components and polysaccharides from a medicinal fungus. Biochem Eng J 79:214–220

Cheung YC (2014) High-intensity ultrasound for extraction and controlled degradation of high molecular weight polysaccharides from medicinal mushrooms: Process characteristics and product properties: Hong Kong Polytechnic University

Cho S, Yun B, Yoo I, Koshino H (2011) Structure of fomitellan a, a mannofucogalactan from the fruiting bodies of *Fomitella fraxinea*. Bioorg Med Chem Lett 2011 21(1):204–206

Das D, Maiti D, Chandra K, Mondal S, Ojha AK, Roy SK, Ghosh K, Islam SS (2008) NMR and Maldi-TOFMS analysis of a heteroglycan isolated from hot water extract of edible mushroom, *Volvariella bombycina*. Carbohydr Res 343(13):2258–2265

Deng C, Shang J, Fu H, Chen J, Liu H, Chen J (2016) Mechanism of the immunostimulatory activity by a polysaccharide from *Dictyophora indusiata*. Int J Biol Macromol 91:752–759

Dey B, Bhunia SK, Maity KK, Patra S, Mandal S, Behera B, Maiti TK, Sikdar SR, Islam SS (2013) Structure and immunology of a heteroglycan from somatic hybrid mushroom of *Pleurotus florida* and *Calocybe indica* var. APK2. Int J Biol Macromol 52(1):312–318

Ding X, Feng S, Cao M, Li M, Tang J, Guo C, Zhang J, Sun Q, Yang Z, Zhao J (2010) Structure characterization of polysaccharide isolated from the fruiting bodies of *Tricholoma matsutake*. Carbohyd Polym 81(4):942–947

Dong JZ, Wang ZC, Wang Y (2011) Rapid extraction of polysaccharides from fruits of *Lycium barbarum*, L. J Food Biochem 35(4):1047–1057

Dong Q, Wang Y, Shi L, Yao J, Li J, Ma F, Ding K (2012) A novel water-soluble β-D-glucan isolated from the spores of *Ganoderma lucidum*. Carbohydr Res 353:100–105

Drori A, Shabat Y, Ben YA, Danay O, Levanon D, Zolotarov L, Ilan Y (2016) Extracts from *Lentinula edodes* (shiitake) edible mushrooms enriched with vitamin D exert an anti-inflammatory hepatoprotective effect. J Med Food 19(4):383–389

Fu Y, Liu W, Zu Y, Shi X, Liu Z, Schwarz G, Efferth T (2009) Breaking the spores of the fungus *Ganoderma lucidum* by supercritical CO_2. Food Chem 112(1):71–76

Ge Q, Zhang AQ, Sun PL (2009) Structural investigation of a novel water-soluble heteropolysaccharide from the fruiting bodies of *Phellinus baumii* Pilát. Food Chem 114(2):391–395

Ghosh K, Chandra K, Roy SK, Mondal S, Maiti D, Das D, Ojha AK, Islam SS (2008) Structural investigation of a polysaccharide (Fr. I) isolated from the aqueous extract of an edible mushroom, *Volvariella diplasia*. Carbohydr Res 343(6):1071–1078

Ghosh K (2015) A review mushrooms: A source of immunomodulating and antitumor polysaccharides. J Phys Sci 20:239–252

Gimenez-Abian MI, Bernabe M, Leal JA (2007) Structure of a galactomannan isolated from the cell wall of the fungus *Lineolata* rhizophorae. Carbohydr Res 342(17):2599–2603

Han C (2010) Antinociceptive activity of agaricoglycerides extracted from mycelium of Ling Zhi or Reishi medicinal mushroom, *Ganoderma lucidum* (W. Curt.: Fr.) P. karst. (Aphyllophoromycetideae). Int J Med Mushrooms 12(3):273–278

Han XQ, Chan BC, Yu H, Yang YH, Hu SQ (2012) Structural characterization and immunomodulating activities of a polysaccharide from *Ganoderma sinense*. Int J Biol Macromol 51(4):597–603

Huang SQ, Ning ZX (2010) Extraction of polysaccharide from *Ganoderma lucidum* and its immune enhancement activity. Int J Biol Macromol 47(3):336–341

Jeff IB, Li S, Peng X, Kassim RM, Liu B (2013) Purification, structural elucidation and antitumor activity of a novel mannogalactoglucan from the fruiting bodies of *Lentinus edodes*. Fitoterapia 84(3):338–346

Jo EK, Heo DJ, Kim JH, Lee YH, Ju YC, Lee SC (2013) The effects of subcritical water treatment on antioxidant activity of golden oyster mushroom. Food Bioprocess Tech 6(9):2555–2561

Jones JKN, Stoodley RJ (1965) Fractionation using copper complexes. Methods Carbohyd Chem 5:36–38

Kao P, Wang S, Hung W, Liao Y, Lin C, Yang W (2012) Structural characterization and antioxidative activity of low-molecular-weights beta-1,3-glucan from the residue of extracted *Ganoderma lucidum* fruiting bodies. J Biomed Biotechnol 2012:1):1–1):8

Khatua S, Acharya K (2017) Alkaline extractive crude polysaccharide from *Russula senecis* possesses antioxidant potential and stimulates innate immunity response. J Pharm Pharmcol 69(12):1817–1828

Komura DL, Carbonero ER, Gracher AHP, Baggio CH, Freitas CS, Marcon R, Santos ARS, Gorin PAJ, Iacomini M (2010) Structure of *Agaricus* spp. fucogalactans and their anti-inflammatory and antinociceptive properties. Bioresour Technol 101(15):6192–6199

Lehtovaara BC, Gu FX (2011) Pharmacological, structural, and drug delivery properties and applications of 1, 3-β-glucans. J Agric Food Chem 59(13):6813–6828

Lei S (2016) Bioactivities, isolation and purification methods of polysaccharides from natural products: a review. Int J Biol Macromol 92:37–48

Lespinard AR, Bon J, Cárcel JA, Benedito J, Mascheroni RH (2015) Effect of ultrasonic-assisted blanching on size variation, heat transfer, and quality parameters of mushrooms. Food Bioprocess Technol 8(1):41–53

Li B, Dobruchowska JM, Gerwig GJ, Dijkhuizen L, Kamerling JP (2013) Structural investigation of water-soluble polysaccharides extracted from the fruit bodies of *Coprinus comatus*. Carbohyd Polym 91(1):314–321

Liu W, Wang H, Yao W, Gao X, Yu LL (2010) Effects of sulfation on the physicochemical and functional properties of a water-insoluble polysaccharide preparation from *Ganoderma lucidum*. J Agr Food Chem 58(6):3336–3341

Liu J, Zhang C, Wang Y, Yu H, Liu H (2011) Structural elucidation of a heteroglycan from the fruiting bodies of *Agaricus blazei* Murill. Int J Biol Macromol 49(4):716–720

Liu X, Wang L, Zhang C, Wang H, Zhang X (2013) Structure characterization and antitumor activity of a polysaccharide from the alkaline extract of king oyster mushroom. Carbohyd Polym 118:101–106

Luo Q, Sun Q, Wu L, Yang Z (2012) Structural characterization of an immunoregulatory polysaccharide from the fruiting bodies of *Lepista sordida*. Carbohyd Polym 88(3):820–824

Maiti D, Chandra K, Mondal S, Ojha AK, Das D (2008) Isolation and characterization of a heteroglycan from the fruits of *Astraeus hygrometricus*. Carbohydr Res 343(4):817–824

Maity K, Kar Mandal E, Maity S, Gantait SK, Das D, Maiti S, Maiti TK, Sikdar SR, Islam SS (2011a) Structural characterization and study of immunoenhancing and antioxidant property of a novel polysaccharide isolated from the aqueous extract of a somatic hybrid mushroom of *Pleurotus florida* and *Calocybe indica* variety APK2. Int J Biol Macromol 48(2):304–310

Maity KK, Patra S, Dey B, Bhunia SK, Mandal S, Das D, Majumdar DK, Maiti S, Maiti TK, Islam SS (2011b) A heteropolysaccharide from aqueous extract of an edible mushroom, *Pleurotus ostreatus* cultivar: structural and biological studies. Carbohydr Res 346(2):366–337

Maity P, Pattanayak M, Maity S, Nandi AK, Sen IK, Behera B, Maiti TK, Mallick P, Sikdar SR, Islam SS (2014) A partially methylated mannogalactan from hybrid mushroom. Pfle 1p: purification, structural characterization, and study of immunoactivation. Carbohydr Res 395(33):1–8

Mandal EK, Maity K, Maity S, Gantait SK, Maiti S, Maiti TK, Sikdar SR, Islam SS (2011) Structural characterization of an immunoenhancing cytotoxic heteroglycan isolated from an edible mushroom *Calocybe indica* var. APK2. Carbohydr Res 346(14):2237–2243

Manna DK, Maity P, Nandi AK, Pattanayak M, Panda BC, Mandal AK, Tripathy S, Acharya K, Sahoo AK, Gupta N, Roy S, Islam SS (2017) Structural elucidation and immunostimulating property of a novel polysaccharide extracted from an edible mushroom *Lentinus fusipes*. Carbohyd Polym 157:1657–1665

Manzi P, Marconi S, Aguzzi A, Pizzoferrato L (2004) Commercial mushrooms: nutritional quality and effect of cooking. Food Chem 84(2):201–206

Marimuthu P, Laila A, Alicia G, Alejandro R, Francisco RM, Guillermo R, Soler-Rivas C (2014) Repositorio institucional de la universidad autónoma de madrid. Biotechnol Prog 30(2):391–400

Mondal S, Chandra K, Maiti D, Ojha AK, Das D, Roy SK, Ghosh K, Chakraborty I, Islam SS (2008) Chemical analysis of a new fucoglucan isolated from an edible mushroom, *Termitomyces robustus*. Carbohydr Res 343(6):1062–1070

Nandan CK, Sarkar R, Bhanja SK, Sikdar SR, Islam SS (2011) Isolation and characterization of polysaccharides of a hybrid mushroom (backcross mating between PfloVv12 and *Volvariella volvacea*). Carbohydr Res 346(15):2451–2456

Nandi AK, Samanta S, Sen IK, Devi KSP, Maiti TK, Acharya K, Islam SS (2013) Structural elucidation of an immunoenhancing heteroglycan isolated from *Russula albonigra* (Krombh.) Fr. Carbohyd Polym 94(2):918–926

Ookushi Y, Sakamoto M, Azuma J (2006) Optimization of microwave-assisted extraction of polysaccharides from the fruiting body of mushrooms. J Applied Glycosci 53(4):267–272

Palacios I, Lozano M, Moro C, D M, Arrigo RMA (2011) Antioxidant properties of phenolic compounds occurring in edible mushrooms. Food Chem 128(3):674–678

Pan D, Wang L, Chen C, Teng B, Wang C, Xu Z, Hu B, Zhou P (2012) Structure characterization of a novel neutral polysaccharide isolated from *Ganoderma lucidum* fruiting bodies. Food Chem 135(3):1097–1103

Pan Y, Dong SW, Hao Y, Zhou Y, Ren XJ (2010a) Ultrasonic-assisted extraction process of crude polysaccharides from Yunzhi mushroom and its effect on hydroxyproline and glycosaminoglycan levels. Carbohyd Polym 81(1):93–96

Pan Y, Hao Y, Chu T, Li C, Zhang Z, Zhou Y (2010b) Ultrasonic-assisted extraction, chemical characterization of polysaccharides from Yunzhi mushroom and its effect on osteoblast cells. Carbohyd Polym 80(3):922–926

Patra S, Maity KK, Bhunia SK, Dey B, Mandal S, Maiti TK, Sikdar SR, Islam SS (2011) Structural characterization and study of immunoenhancing properties of heteroglycan isolated from a somatic hybrid mushroom (PfloVv1aFB) of *Pleurotus florida* and *Volvariella volvacea*. Carbohydr Res 346(13):1967–1972

Pavel K (2009) Chemical composition and nutritional value of European species of wild growing mushrooms: A review. Food Chem 113(1):9–16

Pereira CG, Meireles MAA (2010) Supercritical fluid extraction of bioactive compounds from botanic matrices: experimental data, process parameters. Food Bioprocess Tech 3(3):340–372

Plaza M, Turner C (2015) Pressurized hot water extraction of bioactives. TrAC-Trend Anal Chem 71:39–54

Pramanik M, Mondal S, Chakraborty I, Rout D, Islam SS (2005) Structural investigation of a polysaccharide (Fr. II) isolated from the aqueous extract of an edible mushroom, *Pleurotus sajor-caju*. Carbohydr Res 340(4):629–636

Riera E, Blanco A, García J, Benedito J, Mulet A, Gallego-Juárez JA, Blasco M (2010) High-power ultrasonic system for the enhancement of mass transfer in supercritical CO_2 extraction processes. Phys Procedia 3(1):141–146

Rout D, Mondal S, Chakraborty I, Islam SS (2006) The structure of a polysaccharide from fraction-II of an edible mushroom, *Pleurotus florida*. Carbohydr Res 341(8):995–1002

Ruthes AC, Rattmann YD, Carbonero ER, Gorin PAJ, Iacomini M (2012) Structural characterization and protective effect against murine sepsis of fucogalactans from *Agaricus bisporus* and *Lactarius rufus*. Carbohyd Polym 87(2):1620–1627

Ruthes AC, Carbonero ER, Córdova MM, Baggio CH, Santos ARS, Sassaki GL, Cipriani TR, Gorin PAJ, Iacomini M (2013a) *Lactarius rufus* (1→3), (1→6)-β-d-glucans: structure, antinociceptive and anti-inflammatory effects. Carbohyd Polym 94(1):129–136

Ruthes AC, Carbonero ER, Córdova MM, Baggio CH, Sassaki GL, Gorin PAJ, Santos ARS, Iacomini M (2013b) Fucomannogalactan and glucan from mushroom *Amanita muscaria*: structure and inflammatory pain inhibition. Carbohyd Polym 98(1):761–769

Ruthes AC, Rattmann YD, Malquevicz-Paiva SM, Carbonero ER, Córdova MM, Baggio CH et al (2013c) *Agaricus bisporus* fucogalactan: structural characterization and pharmacological approaches. Carbohyd Polym 92(1):184–191

Ruthes AC, Smiderle FR, Iacomini M (2015) D-glucans from edible mushrooms: a review on the extraction, purification and chemical characterization approaches. Carbohyd Polym 117(117):753–761

Santoyo S, Plaza M, Jaime L, Ez EI, Reglero G (2010) Pressurized liquid extraction as an alternative process to obtain antiviral agents from the edible microalga *Chlorella vulgaris*. Food Chem 58(15):8522–8527

Santoyo S, Plaza M, Jaime L, Ez EI, Reglero G (2011) Pressurized liquids as an alternative green process to extract antiviral agents from the edible seaweed *Himanthalia elongata*. J Appl Phycol 23(5):909–917

Santoyo S, Jaime L, Plaza M, Herrero M, Rodriguez-Meizoso I (2012) Antiviral compounds obtained from microalgae commonly used as carotenoid sources. J Appl Phycol 24(4):731–741

Siu K, Xu L, Chen X, Wu J (2016) Molecular properties and antioxidant activities of polysaccharides isolated from alkaline extract of wild *Armillaria ostoyae* mushrooms. Carbohyd Polym 137:739–746

Smiderle FR, Carbonero ER, Mellinger CG, Sassaki GL, Gorin PAJ, Iacomini M (2006) Structural characterization of a polysaccharide and a β-glucan isolated from the edible mushroom *Flammulina velutipes*. Phytochemistry 67(19):2189–2196

Smiderle FR, Carbonero ER, Sassaki GL, Gorin PAJ, Iacomini M (2008a) Characterization of a heterogalactan: some nutritional values of the edible mushroom *Flammulina velutipes*. Food Chem 108(1):329–333

Smiderle FR, Olsen LM, Carbonero ER, Marcon R, Baggio CH, Freitas CS, Santos AR, Torri G, Gorin PA, Iacomini M (2008b) A 3-O-methylated mannogalactan from *Pleurotus pulmonarius*: Structure and antinociceptive effect. Phytochemistry 69 (15):2731–2736

Smiderle FR, Alquini G, Tadra-Sfeir MZ, Iacomini M, Wichers HJ, Van Griensven LJLD (2013a) *Agaricus bisporus* and *Agaricus brasiliensis* (1→6)-β-D-glucans show immunostimulatory activity on human THP-1 derived macrophages. Carbohyd Polym 94(1):91–99

Smiderle FR, Sassaki GL, Van Griensven LJ, Iacomini M (2013b) Isolation and chemical characterization of a glucogalactomannan of the medicinal mushroom *Cordyceps militaris*. Carbohyd Polym 97(1):74–80

Song J, Lia D, Liu C (2009) Response surface analysis of microwave-assisted extraction of polysaccharides from cultured *Cordyceps milit*. J Chem Tech Biotechnol 84(11):1669–1673

Sun Y, Liang H, Zhang X, Tong H, Liu J (2009) Structural elucidation and immunological activity of a polysaccharide from the fruiting body of *Armillaria mellea*. Bioresour Technol 100(5):1860–1863

Sun Y, Li X, Yang J, Liu J, Kennedy JF (2010) Water-soluble polysaccharide from the fruiting bodies of *Chroogomphis rutilus* (Schaeff.: Fr.) O. K. Miller: isolation, structural features and its scavenging effect on hydroxyl radical. Carbohyd Polym 80(3):720–724

Sun Y, Liu J (2009) Purification, structure and immunobiological activity of a water-soluble polysaccharide from the fruiting body of *Pleurotus ostreatus*. Bioresour Technol 100(2):983–986

Szwengiel A, Stachowiak B (2016) Deproteinization of water-soluble β-glucan during acid extraction from fruiting bodies of *Pleurotus ostreatus* mushrooms. Carbohyd Polym 146:310–319

Tian Y, Zeng H, Xu Z, Zheng B, Lin Y, Gan C, Lo YM (2012) Ultrasonic-assisted extraction and antioxidant activity of polysaccharides recovered from white button mushroom (*Agaricus bisporus*). Carbohyd Polym 88(2):522–529

Tiffany CTL, Hsin HT, Ai YW, Chang CA (2007) Pressurized water extraction of polysaccharides as secondary metabolites from *Lentinula edodes*. J Agr Food Chem 55(10):4196–4201

Ullah MI, Akhtar M, Awais MM, Anwar MI, Khaliq K (2018) Evaluation of immunostimulatory and immunotherapeutic effects of tropical mushroom (*Lentinus edodes*) against eimeriasis in chicken. Trop Anim Health Prod 50(1):97–104

Vinogradov E, Wasser SP (2005) The structure of a polysaccharide isolated from *Inonotus levis* P. karst. Mushroom (Heterobasidiomycetes). Carbohydr Res 340(18):2821–2825

Wang F, Hou Y, Ding X, Hou W, Song B, Wang T, Li J. Zeng Y (2013) Structure elucidation and antioxidant effect of a polysaccharide from *Lactarius camphoratum* (Bull.) Fr. Int J Biol Macromol 62(11):131–136

Wang D, Sun S, Wu W, Yang S, Tan J (2014a) Characterization of a water-soluble polysaccharide from *Boletus edulis* and its antitumor and immunomodulatory activities on renal cancer in mice. Carbohyd Polym 105(1):127–134

Wang Y, Shen X, Liao W, Fang J, Chen X, Dong Q, Ding K (2014b) A heteropolysaccharide, l-fuco-d-manno-1,6-α-D-galactan extracted from *Grifola frondosa* and antiangiogenic activity of its sulfated derivative. Carbohyd Polym 101(1):631-641

Wang Q, Wang F, Xu Z, Ding Z (2017) Bioactive mushroom polysaccharides: A review on monosaccharide composition, biosynthesis and regulation. Molecules 22(6):955–967

Wasser S (2002) Medicinal mushrooms as a source of antitumor and immunomodulating polysaccharides. Appl Microbiol Biot 60(3):258–274

Wu Y, Sun C, Pan Y (2006) Studies on isolation and structural features of a polysaccharide from the mycelium of a Chinese edible fungus (*Cordyceps sinensis*). Carbohyd Polym 63(2):123–128

Xu D, Zheng J, Zhou Y, Li Y, Li S, Li H (2016) Extraction of natural antioxidants from the *Thelephora ganbajun* mushroom by an ultrasound-assisted extraction technique and evaluation of antiproliferative activity of the extract against human cancer cells. Int J Mol Sci 17(10):1664-1678

Ye L, Zhang J, Ye X, Tang Q, Liu Y, Gong C, Du X, Pan Y (2008) Structural elucidation of the polysaccharide moiety of a glycopeptide (GLPCW-II) from *Ganoderma lucidum* fruiting bodies. Carbohydr Res 343(4):746–752

Ye LB, Zhang JS, Yang Y, Zhou S, Liu YF (2009) Structural characterisation of a heteropolysaccharide by NMR spectra. Food Chem 112(4):962–966

You Q, Yin X, Zhao Y (2013) Ultrasound-assisted extraction of polysaccharides from the fruiting bodies of *Tricholoma matsutake* using response surface methodology. J Food Agric Environ 11(3):1969–1974

You Q, Yin X, Zhao Y, Jiang Z (2014) Extraction and purification of polysaccharides from pine medicinal mushroom, *Tricholoma matsutake* (higher basidiomycetes) fruit bodies. Int J Med Mushrooms 16(2):149–160

Yu S, Zhang Y, Li C, Zhang Q, Ma Z, Fan M (2011) Optimization of ultrasonic extraction of mycelial polysaccharides from *Paecilomyces hepiali* using response surface method. African J Biotechnol 10(75):17241–17250

Zeng WC, Zhang Z, Gao H, Jia LR, Chen WY (2012) Characterization of antioxidant polysaccharides from *Auricularia auricular* using microwave-assisted extraction. Carbohyd Polym 89(2):694–700

Zhang L, Zhou J, Yang G, Chen J (1998) Preparative fractionation of polysaccharides by columns packed with regenerated cellulose gels. J Chromatogr A 816(2):131–136

Zhang M, Zhang L, Cheung PCK, Ooi VEC (2004) Molecular weight and anti-tumor activity of the water-soluble polysaccharides isolated by hot water and ultrasonic treatment from the sclerotia and mycelia of *Pleurotus tuber-regium*. Carbohyd Polym 56(2):123–128

Zhang QM, Zhang LR, Chen JD, Pei AM, Zheng HL (2009) Microwave-assisted extraction and identification of polysaccharide from *Lycoris aurea*. Chem Nat Compd 45(4):474–477

Zhang AQ, Fu L, Xu M, Sun PL, Zhang JS (2012a) Structure of a water-soluble heteropolysaccharide from fruiting bodies of *Hericium erinaceus*. Carbohyd Polym 88(2):558–561

Zhang H, Li WJ, Nie SP, Chen Y, Wang YX (2012b) Structural characterisation of a novel bioactive polysaccharide from *Ganoderma atrum*. Carbohyd Polym 88(3):1047–1054

Zhang Z, Lv G, Pan H, Shi L, Fan L (2012c) Optimization of the microwave-assisted extraction process for polysaccharides in *Himematsutake*. Food Sci Tech Int Tokyo 17(6):461–470

Zhang AQ, Xiao NN, Deng YL, He PF, Sun PL (2012d) Purification and structural investigation of a water-soluble polysaccharide from *Flammulina velutipes*. Carbohyd Polym 87(3):2279–2283

Zhang AQ, Xu M, Fu L, Sun PL (2013a) Structural elucidation of a novel mannogalactan isolated from the fruiting bodies of *Pleurotus geesteranus*. Carbohyd Polym 92(1):236–240

Zhang AQ, Zhang Y, Yang JH, Sun PL (2013b) Structural elucidation of a novel heteropolysaccharide from the fruiting bodies of *Pleurotus eryngii*. Carbohyd Polym 92(2):2239–2244

Zhang AQ, Liu Y, Xiao NN, Zhang Y, Sun PL (2014a) Structural investigation of a novel heteropolysaccharide from the fruiting bodies of *Boletus edulis*. Food Chem 146(1):334–338

Zhang H, Nie SP, Yin JY, Wang YX, Xie MY (2014b) Structural characterization of a heterogalactan purified from fruiting bodies of *Ganoderma atrum*. Food Hydrocolloid 36(5):339–347

Zhao S, Li B, Chen G, Hu Q, Zhao L (2017) Preparation, characterization, and anti-inflammatory effect of the chelate of *Flammulina velutipes* polysaccharide with Zn. Food Agr Immunol 28(1):162–177

Zhao YT, Tang YQ, Liu KZ, Zhang Y (2015) A study about microwave-assisted extraction of polysaccharides form medicinal mushrooms *Fomitopsis ulmaria* (sor.: for.) bond. Et sing. Adv Mater Res 1073-1076:1837–1840

Chapter 10
Novel Prospective of Wild Mushroom Polysaccharides as Potential Prebiotics

Yuxiao Wang, Xiaojun Huang, and Shaoping Nie

10.1 Introduction

Prebiotics can advance the abundant lactic acid bacteria existed in human colon and assist antagonism on *Escherichia coli.* and *Salmonella sp.*, by discouraging their proliferation (Gibson et al. 2004). Polysaccharides and oligosaccharides are recently considered as prebiotics and added in functional foods. Mushrooms have been conventionally consumed to prevent and treat many disorders as dietary supplements due to its bioactive substance polysaccharide (Rathore et al. 2017). Recent years, researches have proved the bioactivities of mushrooms polysaccharide from extensive evidence (Ferrão et al. 2017). Polysaccharides from mushrooms can improve host immunity and active the host immune system, by exhibiting prominent antitumor, antimicrobial, antiviral activities (Roncero-Ramos and Delgado-Andrade 2017) and Immunoceuticals (Zhang et al. 2017). During the past, fibreis regarded as prebiotics, however, recent studies suggest that fibre compounds show the same behaviour as prebiotic (Campelo-Felix et al. 2017). Valued carbohydrates prebiotic usually consist of glucans or galactans, both of which can be fermented by anaerobic bacteria in human's large intestine (Nowak et al. 2018). Prebiotics glucans are mostly comprised by α/β-glucans (Khan et al. 2017), involving in antitumor, antioxidative, immunomodulatory, anti-inflammatory, anti-bacterial and antinociceptive activities.

Y. Wang · X. Huang · S. Nie (✉)
State Key Laboratory of Food Science and Technology, Nanchang University, Nanchang, Jiangxi Province, China
e-mail: spnie@ncu.edu.cn

B. P. Singh et al. (eds.), *Biology of Macrofungi*, Fungal Biology,
https://doi.org/10.1007/978-3-030-02622-6_10

10.2 Prebiotic

10.2.1 Criteria for Prebiotic

In 1995, Roberfroid and Gibson gave the term prebiotic and then the definition of prebiotic gradually entered people's field of vision (Gibson and Roberfroid 1995). The concept of prebiotic was put forward by certain criteria against gastric acidity or hydrolyzed by gastrointestinal absorption and mammalian enzymes. It can be fermentated by intestinal microflora, and activate the connection between intestinal bacteria and host (Gibson et al. 2004).

Afterwards, Wang et al. showed in Fig. 10.1 had pointed out the important criteria for prebiotics (Wang 2009). Firstly, prebiotics could not be digested by human's upper gut tract, suggesting prebiotics could restrain digestion process before they entered into the colon. Prebiotics can effectively stimulate beneficial bacteria like *Lactobacilli* and *Bifidobacterial*, while suppress the growth of *Clostridia* and *Bacteroides* (Macfarlane et al. 2008). This criterion allows the food ingredients to be prebiotics and fermented selectivity by potential bacteria in colon (Gibson et al. 2004). The selective fermentation process would further influence the reflection and changes on short-chain fatty acids (SCFAs), reductive enzymes, fecal moisture as well as immune system modulation and active carriers with mineral absorption (Douglas and Sanders 2008). Then, activity of intestinal bacteria, selective stimulation of the growth and healthy body is a quite essential standard for prebiotics (Gibson et al. 2004). Furthermore, there was a parameter that some prebiotics did not have selective fermentation. Duncan et al. and Langlands et al. studied that the ability of fermentation by prebiotic-inulin had induced a slight increment of gut bacteria named *Roseburia*, *Ruminococcus* and *Eubacterium*, respectively (Duncan et al. 2003; Langlands et al. 2004). Nobody owns the same bacterial species, which the gut microbiota has been in a dynamic state, even age, antibiotic, drugs, illness, diet may influence them (Macfarlane et al. 2006). With the increasing attention to gastrointestinal disorder and healthy problems, an attempt to improve the properties of prebiotics can be carried out to keep in touch without being degraded or chemically altered (Tuohy et al. 2003; Huebner et al. 2008).

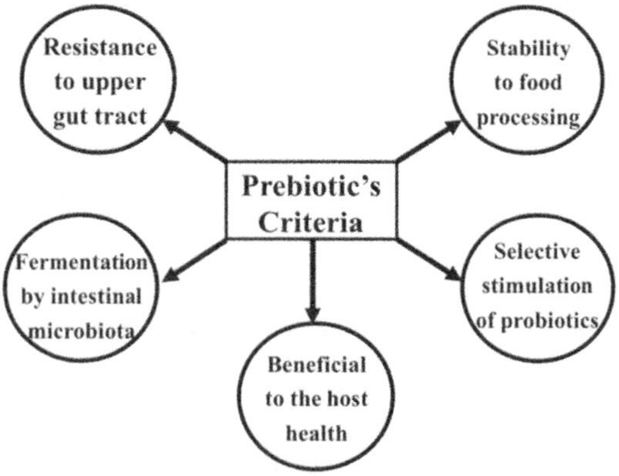

Fig. 10.1 The main prebiotics' criteria

10.2.2 Mechanism Action

Polysaccharides are considered as significant prebiotics, but the bioactive mechanism is different from oligosaccharides. In 2007, Stowell (2007) started to categorize the existing prebiotics. In his opinion, one of the established prebiotics groups, sorted of galactooligosaccharides, fructooligosaccharides, lactulose, inulin and polydextrose, while the other group of emerging prebiotics, sorted of lactitol, xylo-oligosaccahrides and isomaltooligosaccharides. Mannitol, maltodextrin, sorbitol and raffinose also belonged to prebiotics with proven biological activities (Mandal et al. 2009; Vamanu and Vamanu 2010). The starch-rich grains had natural prebiotic properties and some healthy benefits. Prebiotics are considered to improve colonic integrity, enhance immune function, reduce long chain fatty acids (LCFAs) in bowels and incidence and duration of intestinal infections as well as increase the quantities of SCFAs (Tuohy et al. 2003; Huebner et al. 2008) (Fig. 10.2).

However, these effects cannot happen to the consequences of prebiotic ingestion immediately. Prebiotics works in an indirect way. When changes are happened in the gastrointestinal microbiota compositions, prebiotics shows its positive effects. *Lactobacilli* and *Bifidobacteria* were beneficial bacteria. One of the crucial positive effects reflect the increasing amount of probiotics like *Lactobacilli* and *Bifidobacteria*, whereas restraint *Histolyticum sp.*'s growth (Palframan et al. 2003). Some researchers reported that the increasing number of *Lactobacilli* and *Bifidobacteriain* the gut was considered as prebiotic action. Gibson et al. (1995) found that *Bifidobacteria*

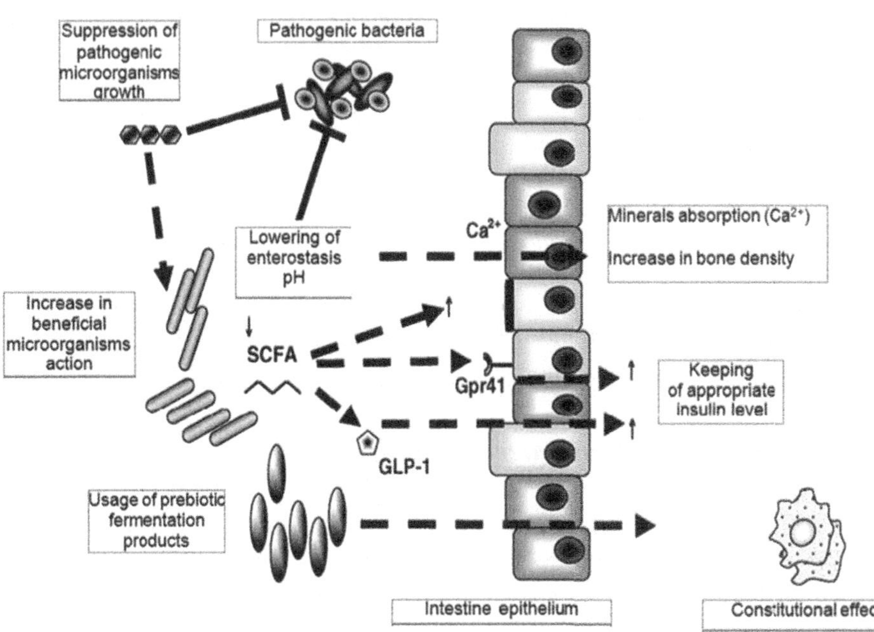

Fig. 10.2 Mechanism action of prebiotics

produced vitamin B, stimulated the immune system and inhibited pathogen growth. *Lactobacilli* was able to decrease infantile diarrhea and constipation.

Potential mechanisms are drawed for their beneficial effects listed below (Manning and Gibson 2004) (Fig. 10.2):

1. Effect of prebiotic on those components of the gut microbiota obviously through competitive substrate availability;
2. Regulation of dynamic process of carbohydrates utilized through certain microbiota composition;
3. Production of vitamins and the available modulation of gut microbiota;
4. Influence of probiotics bacteria on gut microbiota possibly through the aim resistance;
5. Improvement in competing ability for binding sites of pathogens;
6. Increment in epithelial cells' barrier function;
7. Anti-inflammatory by stimulating human innate immune response.

10.2.3 Mushroom-Derived Polysaccharide as Prebiotics

It is very important for human being to keep intestinal flora balanced. As an expectant edible industry, prebiotics have the advantages over the regulation of properties and intestinal flora's amount. Human cannot digest the components in mushroom owning prebiotic function. β-glucans are the main structures of polysaccharides occurring in *Pleurotus ostreatus* and *Lentinus edodes*, which are treated as prebiotics in recent reports. These two kinds of polysaccharides have the anti-inflammatory capability to prevent intestinal mucosa and restrain the numbers of intestinal ulcers in rats (Nosál'Ová et al. 2001).

Besides, lentinan with potential prebiotic ability can react on human gut health through selectively regulation the increase of probiotic bacteria including lactic acid bacteria and *Bifidobacteria* (Wong et al. 2005). Synytsya et al. (2009) found that glucans isolated from *P. ostreatus* and *P. eryngii* had capability of stimulating the development of *Bifidobacterium sp.*, *Enterococcus faecium* and probiotics-*Lactobacillus sp.* Reports showed that glucans obtained from *P. Eryngi* much more assistant of *Lactobacillus* strains than *P. ostreatus*.

10.2.4 Biological Properties as Prebiotic

Modern researches have demonstrated that food derived from *Ascomycetes* and *Basidiomycetes* benefits health. The polysaccharides extracted from mushroom have been consumed as functional dietary supplements. Experiments on mushroom polysaccharides suggested that they exhibited some biological properties as prebiotic (Hardy 2008). Therefore, details of the biological activity are discussed in the following part.

10.2.5 Gut Bacteria Protective Effects

Mushroom polysaccharides increase the resistance of gut bacteria, as well as enhance the survival of probiotics to reflect on host. Chou et al. (2013) found that the polysaccharides from three kinds of mushroom with comparatively low concentration can improve the survival rate of *Lactobacillus acidophilus* and *Bifidobacterium longum subsp.* in colder condition. This result also revealed that mushroom-derived polysaccharides had synergistic actions with the amino acids and peptides. Additionally, mushroom polysaccharides were exposed to artificial gastric juice in order to explore the digestive ability. While the researchers concluded that polysaccharides could stimulate the development of *Lactobacillus* strains more than inulin or fructo oligosaccharides. Besides, those polysaccharides survived in artificial human gastric juice more than 90%. Consequently, mushroom polysaccharides can be undigested in human stomach, then reach the colon and stimulate the development of some healthy bacteria (Nowak et al. 2018).

10.2.6 Antiobesity Effects

It is well known that prebiotics are regarded as fibers and a kind of non-digestible polysaccharides (mostly β-glucans), chancing the gastrointestinal microbiota and improving salubrious host. Compared with higher molecular weight (*Mw*) polysaccharides, *Bifidobacteria* and *Lactobacilli* are likely to ferment prebiotics with low-molecular-mass oligosaccharides more sensitively. β-glucans occuring in fungi cannot be digested and fermented by SCFAs, which offers new ideas and solution to prebiotics. Results showed the extracts from *Ganoderma lucidum* can modulate microbiota availably in obesity and sequentially low the risk of obesity *in vitro* (Chang et al. 2015; Delzenne and Bindels 2015).

10.3 Gut Health Improvement

Mushroom polysaccharides could exert a therapeutic influence on regulation of gut microbiota. Polysaccharides extracted from Ganoderma lucidum, Poria cocos were used to administer mice's daily for 15 days, and 16 Samplicon sequencing and ERIC-PCR studied gut microbiota of mice. Results showed that polysaccharides decreased OTUs varieties and reconstructed gut microbiota constitutes. Some researchers have been reported that when the experimental mice were dieted with polysaccharides, the bacteria were obviously increased, that resulted in a decrease of obesity and an increase of SCFA and lactic acid. According to above, results revealed that polysaccharides from *G. lucidum* and *P. cocos* are working as prebiotics by modulating gut microbiota compositions and then make human bodies in heathy balance (Khan et al. 2018). Furthermore, fruiting bodies of *G. lucidum* significantly improve the amount of lactic acid bacterial species from genera *Bifid bacterium* and *Lactobacillus* (Khan et al. 2017).

10.4 Immunomodulatory Activity

It is reported that mushroom polysaccharides have immunomodulatory activity rather than arresting the cancer cells directly. In other words, mushroom polysaccharides can assist the host adapting to diverse biological stresses and play a nonspecific role in the host. Mushroom polysaccharides achieve their effects and work best by biological response modifiers (BRM) through supporting body's major systems. Xu et al. (2015) studied the impacts of a kind of polysaccharide obtained from the fruit body of *L. edodes* on the immune body response of aged mice. Through contrasting the fecal microbiota in adult, old and polysaccharide-treated old mice, researchers found that the group of polysaccharide-treated old mice not only could improve cytokine levels in peripheral blood and recover the age-attenuated immune responses, but also have a partly reversal effect on the gut microbiota (Table 10.1).

Polysaccharides extracted from *Maitake* and *G. frondosa* can decrease the nephrotoxicity and immune suppression of cisplatin-induced in mice (Masuda et al. 2009). But under normal condition, mushroom polysaccharides produced little effect (Wasser 2011). There are two crucial functional units, adaptive immune system and innate immune system in the human body, which involve into the action of macromolecule (De Silva et al. 2012).

Yamanaka et al. (2012) found that polysaccharides reacted on innate immune system by activating Toll-like receptors. Dramatically reduction caused by Dentin-1-deficient mice would affect dendritic cells (Yamanaka et al. 2012). On the one hand, lentinan could strengthen the absolute phagocytosis and numbers of macrophage, resulting in the reduction in cancer cell size and the growth. On the other hand, it can induce a cascade of cytokine release, including factor-alpha (TNF-α), tumor necrosis and interleukins (IL), applying differentiation and proliferation of immune competent cells to the host protective mechanisms (Wasser 2002). The special components extracted from *Antrodia camphorata* were investigated to improve the concentration of Th1-type and Th2-type cytokines (Liu et al. 2013). Meanwhile, the complement and neutrophil system could be activated by mushroom polysaccharide, not only increasing the cytokine emerged by monocyte-macrophages, but also playing an important part in innate immunity (Wang et al. 1997) (Fig. 10.3).

10.5 Potential Applications and Future Perspectives

Generally, polysaccharides are considered as a kind of polymeric macromolecule, consist of monosaccharide in different linkage pattern. Compared with nuclear acids and proteins, polysaccharides can attach higher value to bioactivity research

Table 10.1 Some antitumor and immunomodulatory polysaccharides from mushrooms and their source, type, molecular weight and experiment

Fungi source	Polysaccharide source	Type	Molecular weight(Da)	Biological activity	Experiment	Ref.
Agaricus blazei	Fruiting body; Mycelium	Glucan, heteroglycan, glucan protein, glucomannan-protein complex	–[a]	Antitumor activity	*in vivo*	Mizuno et al. (1990), Mizuno (1995), Mizuna (1999)
Agaricus bisporus	Fruiting body	α-glucans and β-glucans	–	Immunomodulating and antioxidative activities	*in vitro*	Kozarski et al. (2011)
Armillaria mellea	Fruiting body	Heteropolysaccharide with a heterogeneous main chain (Glc*p* and Gal*p*)	7.80×10^3	Immunomodulating and inflammatory activities	*in vivo*	Sun et al. (2009)
Armillariella tabescens	Mycelium	Heteroglycan	–	Antitumor activity	*in vitro*	Kiho et al. (1992)
Auricularia auricula	Fruiting body	Glucan	–	Antitumor, immunomodulating, antiflammatory, hyperglycemia and antiradiative activities	*in vivo*	Ukai et al. (1983), Ukai et al. (2008)
Auricularia auricula-judae	Fruiting body	(1 → 4)-linked D-glucopyranosyl main chain with (1 → 6)-linked D-glucopyranosyl branch at O-6	3.40×10^4–2.88×10^5	Antitumor activity	*in vitro*	Ma et al. (2008), Ma et al. (2010)
	Fruiting body	(1 → 3)-β-D-glucan main chain with two (1 → 6)-β-D-glucosyl residues for every three glucose residues	2.07×10^6–2.15×10^6	Induced apoptosis of cancer cell	*in vitro*	Xu et al. (2012)
Boletus edulis	Fruiting body	Heteropolysaccharide with a heterogeneous main chain (Glc*p*, Gal*p* and Rha*p*)	1.10×10^5	Immunomodulatory activity	*in vivo*	Wang et al. (2014)

(continued)

Table 10.1 (continued)

Fungi source	Polysaccharide source	Type	Molecular weight(Da)	Biological activity	Experiment	Ref.
Bulgaria inquinans	Fruiting body	Heteromannan	7.40×10^3	Immunomodulatory, antioxidant and anti-malarial activities	*in vivo and in vitro*	Bi et al. (2011), Bi et al. (2013)
Calocybe indica	Fruiting body	Heteropolysaccharide with a heterogeneous main chain (Gal*p* and Glc*p*)	2.00×10^5	Immunomodulatory andcytotoxic activities	*in vivo and in vitro*	Mandal et al. (2011)
Clitopilus caespitosus	Fruiting body	Glucan	–	Antitumor activity	*in vivo*	Liang et al. (1996)
Cordyceps militaris	Fruiting body	(1 → 6)-β-D-mannopyranosyl backbone with (1 → 4)-α-D-glucopyranosyl and (1 → 6)-β-D-galactopyranosyl side branches at O-3, and terminated with β-D-galactopyranosyl residues and α-D-glucopyranosyl residues	2.60×10^4–5.00×10^4	Induced apoptosis of cancer cell	*in vitro*	Yu et al. (2007), Lee and Park (2010)
Dictyophora indusiata	Fruiting body	Heteroglycan, mannan, glucan	1.13×10^6	Antitumor and hyperlipidemia activity	*in vitro*	Hara et al. (1991), Liao et al. (2015)
Flammulina velutipes	Fruiting body	(1 → 3)-β-D-glucan	2.00×10^5	Increased the expression of cytokines	*in vivo*	Leung et al. (1997)
Ganoderma applanatum	Fruiting body	Glucan	–	Antitumor activity	*in vivo*	Nakashima et al. (2013)
Ganoderma lucidum	Fruiting body	hetero-β-D-glycans (glucurono-β-D-glucan, arabinoxylo-β-D-glucan, xylo-β-D-glucan, manno-β-D-glucan and xylomanno-β-D-glucan)	8.00×10^4–2.00×10^5	Induced cell-cycle arrest and apoptosis	*in vivo*	Zhang et al. (2010)

Grifola frondosa	Fruiting body	β-1,6-linked glucose residues with branches of β-1,3-linked glucose	–	Antitumor activity	*in vitro*	Nanba et al. (1987)
Hericium erinaceus	Fruiting body	glucoxylan with (1 → 3)-β and (1 → 6)-glucosidic linkages; galactoxyloglucan-protein complex included (1 → 3)-βand (1 → 6)-glucosidic linkages	–	Induced the secretion of cytochrome	*in vitro*	Mizuno et al. (1992)
Inonotus obliquus	Fruiting body; Mycelium	Glucan	–	Antitumor and immunomodulating activities	*in vitro*	Kim et al. (2005)
	Fruiting body	Xylogalactoglucan consisted of glucose, mannose, galactose, xylose, arabinose, and fucose	1.00×10^6	Induced apoptosis	*in vitro*	Kim et al. (2006)
Lactarius delicius	Fruiting body	Xylomannan	1.10×10^4	Antitumor activity	*in vivo*	Ding et al. (2012)
Lentinus edodes	Fruiting body	β-D-glucan	–	Antitumor activity	*in vivo*	Sasaki and Takasuka (1976)
	Fruiting body	Fucomannogalactan	1.60×10^4	Antinociceptive and anti-inflammatory activities	*in vivo*	Carbonero et al. (2008)
	Fruiting body	D-glucopyranose	2.03×10^5	Immunomudulatory activity	*in vitro*	Zheng et al. (2005)
Lentinus squarrosulus	Fruiting body	Heteropolysaccharide with a heterogeneous main chain (Glc*p* and Gal*p*)	1.96×10^5	Immunomudulatory activity	*in vitro*	Bhunia et al. (2010)
Lepista sordida	Fruiting body	Galactoglucan	4.00×10^4	Immunomodulatory activity	*in vitro*	Luo et al. (2012)
Morchella esculenta	Fruiting body	Heteroglycan	1.00×10^7	Antitumor and hyperglycemia activities	*in vitro*	Duncan et al. (2002)

(continued)

Table 10.1 (continued)

Fungi source	Polysaccharide source	Type	Molecular weight(Da)	Biological activity	Experiment	Ref.
Phellinus linteus	Fruiting body	Glucan	1.50×10^4	Antitumor activity	*in vivo*	Kim et al. (2004)
Pleurotus citrinopileatus	Mycelium	Galactomannan	$> 1.00 \times 10^5$	Antitumor activity	*in vivo*	Wang et al. (2005)
Pleurotus florida	Fruiting body	Glucan	–	Immunomodulatory activity	*in vitro*	Rout et al. (2004)
	Fruiting body	Glucan	1.26×10^5	Antitumor and immunomodulatory activities	*in vitro*	Ojha et al. (2010)
Pleurotus ostreatus cultivar	Fruiting body	Heteroglucan	1.87×10^5	Immunomodulatory activity	*in vitro*	Maity et al. (2011)
Polypours umbellatus	Mycelium	Glucan	–	Antitumor and immunomodulatory activities	*in vivo*	Yang et al. (2004)
Russula albonigra	Fruiting body	Heteropolysaccharide with a heterogeneous main chain (Glc*p*, Gal*p* and Fuc*p*)	PS-I: 1.88×10^5; PS-II: 1.32×10^5; PS-III: 0.92×10^5	Immunomodulatory activity	*in vitro*	Nandan et al. (2011)
Schizophyllum commune	Fruiting body	Schizophyllan (sizofiran or SPG/SCH)	1.02×10^5–1.04×10^5	Increased the immune responses, the expression of cytokines and NK cells' activity	*in vivo*	Yoneda et al. (1991)
Tremella fuciformis	Fruiting body; Mycelium; Culture broth	Heteroglycan	–	Antitumor, immunomodulatory, hyperlipidemia and hyperglycemia Activities	*in vitro*	Chen and Huang (2001)
Tricholoma crassum	Fruiting body	Heteroglucan	2.00×10^2	Immunomodulatory activity	*in vitro*	Patra et al. (2012)

Note:–[a] Information did not be provided

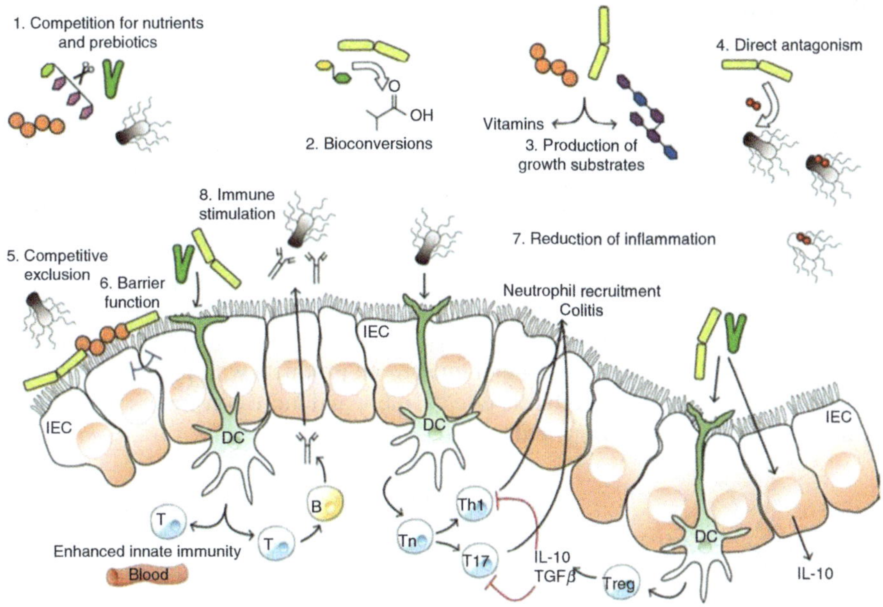

Fig. 10.3 Potential immunomodulatory mechanism of prebiotics

due to their availability and variability. In the last couple of years, discovery in new polysaccharide resources and the application of mushroom polysaccharides in food and medicine industrial have attracted more and more attention. The structure-activity relationship study of mushroom polysaccharides is also a hot spot in future study. Based on those publications mentioned above, mushroom production is the subject of exploration in developing and developed countries, most are as consume a large population of fungi as food or medicine. In fact, some of these biopolymers have already been made to the market as functional food or food supplementary based on their diverse biological activities in their ways.

However, more work has been down in investigating the properties and function of mushroom polysaccharide in the past decades. There are still some questions reminded to be solved. The biological mechanism is still unclear, and much more work is required to broaden its application in numerous industries. In addition, the research direction of hybrid mushrooms would be amplified by the development of techniques improvement with variability caused by the consequently modification and hybridization. The immense potentiality of those active components makes mushroom products quickly into the market. However, functions and properties of polysaccharide from certain hybrid mushrooms remains to be confirmed. In conclusion, studies on new biological properties and researches on their industrial application require further analysis in the future.

10.6 Conclusions

In the last few decades, mushroom has been demonstrated that it contains abundant bioactive polysaccharides, considered as medicinal compounds, therapeutic adjuvant, and health food supplements. The polysaccharides, such as galactans, mannans, even polysaccharide–protein complexes from mushrooms with biological activities are regarded as prebiotics. And these functional compounds from mushroom can promote human health, such as reducing the danger of adiposis, hypersensitivity, humor and degenerative ailments. The expecting property of fungi prebiotic polysaccharides is antitumor and immunomodulation activity, but the mechanism on these bioactivities are still unknown. Consequently, more efforts should be spared to produce beneficial mushroom products with prebiotic function. With the advantages of the prebiotic effects, more and more consumers will acquire health benefits by diet therapy in the future.

References

Bhunia SK, Dey B, Maity KK, Patra S, Mandal S, Maiti S, Maiti TK, Sikdar SR, Islam SS (2010) Structural characterization of an immunoenhancing heteroglycan isolated from an aqueous extract of an edible mushroom, *Lentinus squarrosulus* (Mont.) Singer. Carbohydr Res 345(17):2542–2549

Bi H, Gao T, Li Z, Ji L, Yang W, Jeff Iteku B, Liu E, Zhou Y (2013) Structural elucidation and antioxidant activity of a water-soluble polysaccharide from the fruit bodies of *Bulgaria inquinans* (Fries). Food Chem 138(2–3):1470–1475

Bi H, Han H, Li Z, Ni W, Chen Y, Zhu J, Gao T, Hao M, Zhou Y (2011) A Water-Soluble Polysaccharide from the fruit bodies of *Bulgaria inquinans* (Fries) and its anti-malarial activity. Evid Based Complement Alternat Med 2011:1–12

Campelo-Felix PH, Souza HJB, Figueiredo JDA, Fernandes RVDB, Botrel DA, de Oliveira CR, Yoshida MI Borges SV (2017) Prebiotic carbohydrates: effect on reconstitution, storage, release, and antioxidant properties of lime essential oil microparticles. J Agric Food Chem 65(2):445–453

Carbonero ER, Gracher AHP, Komura DL, Marcon R, Freitas CS, Baggio CH, Santos ARS, Torri G, Gorin PAJ, Lacomini M (2008) *Lentinus edodes* heterogalactan: antinociceptive and anti-inflammatory effects. Food Chem 111(3):531–537

Chang CJ, Lin CS, Lu CC, Martel J, Ko YF, Ojcius DM, Tseng SF, Wu TR, Chen YYM, Young JD, Lai HC (2015) *Ganoderma lucidum* reduces obesity in mice by modulating the composition of the gut microbiota. Nat Commun 6:7489–7506

Chen AW, Huang NL (2001) Production of the medicinal mushroom *Tremella fuciformis Berk* by mixed-culture cultivation on synthetic logs. Int J Med Mushrooms 32(2–3):1

Chou WT, Sheih IC, Fang TJ (2013) The applications of polysaccharides from various mushroom wastes as prebiotics in different systems. J Food Sci 78(7):1041–1048

De Silva DD, Rapior S, Fons F, Bahkali AH, Hyde KD (2012) Medicinal mushrooms in supportive cancer therapies: an approach to anti-cancer effects and putative mechanisms of action. Fungal Divers 55(1):1–35

Delzenne NM, Bindels LB (2015) Gut microbiota: *Ganoderma lucidum*, a new prebiotic agent to treat obesity? Nat Rev Gastroenterol Hepatol 12(10):553–554

Ding X, Hou Y, Hou W (2012) Structure elucidation and antioxidant activity of a novel polysaccharide isolated from *Boletus speciosus* forst. Int J Biol Macromol 50(3):613–618

Douglas LC, Sanders ME (2008) Probiotics and prebiotics in dietetics practice. J Am Diet Assoc 108(3):510–521
Duncan CJG, Pugh N, Pasco DS, Ross SA (2002) Isolation of a galactomannan that enhances macrophage activation from the edible fungus *Morchella esculenta*. J Agric Food Chem 50(20):5683–5685
Duncan SH, Scott KP, Ramsay AG, Harmsen HJ, Welling GW (2003) Effects of alternative dietary substrates on competition between human colonic bacteria in an anaerobic fermentor system. Appl Environ Microbiol 69(2):1136–1142
Ferrão J, Bell V, Calabrese V, Pimentel L, Pintado M, Fernandes TH (2017) Impact of mushroom nutrition on microbiota and potential for preventative health. J Food Nutr Res 5(4):226–233
Gibson GR (2004) From probiotics to prebiotics and a healthy digestive system. J Food Sci 69(5):141–143
Gibson GR, Beatty ER, Wang X, Cummings JH (1995) Selective stimulation of bifidobacteria in the human colon by oligofructose and inulin. Gastroenterology 108(4):975–982
Gibson GR, Probert HM, Loo JV, Rastall RA, Roberfroid MB (2004) Dietary modulation of the human colonic microbiota: updating the concept of prebiotics. Nutr Res Rev 17(2):259–275
Gibson GR, Roberfroid MB (1995) Dietary modulation of the human colonic microbiota: introducing the concept of prebiotics. J Nutr 125(6):1401–1412
Hara C, Kumazawa Y, Inagaki K, Kaneko M, Kiho T, Ukai S (1991) Mitogenic and colony-stimulating factor-inducing activities of polysaccharide fractions from the fruit bodies of *Dictyophora indusiata* fisch. Chem Pharm Bull 39(6):1615–1616
Hardy ML (2008) Dietary supplement use in cancer care: help or harm. Hematol Oncol Clin North Am 22(4):581–617
Huebner J, Wehling RL, Parkhurst A, Hutkins RW (2008) Effect of processing conditions on the prebiotic activity of commercial prebiotics. Int Dairy J 18(3):287–293
Khan I, Huang G, Li X, Leong W, Xia W, Hsiao WLW (2017) Polysaccharides from two medicinal mushrooms *Ganoderma lucidum* and *Poria cocos* reveal prebiotic effects in mice. The Functional Food Center's 22nd International Conference, Boston
Khan I, Huang G, Li X, Leong W, Xia W, Hsiao WLW (2018) Mushroom polysaccharides from *Ganoderma lucidum* and *Poria cocos* reveal prebiotic functions. J Funct Foods 41:191–201
Kiho T, Shiose Y, Nagai K, Ukai S (1992) Polysaccharides in fungi. xxx. antitumor and immunomodulating activities of two polysaccharides from the fruiting bodies of *Armillariella tabescens*. Chem Pharma Bulletin 40(8):2110–2114
Kim GY, Choi GS, Lee SH, Park YM (2004) Acidic polysaccharide isolated from *Phellinus linteus*, enhances through the up-regulation of nitric oxide and tumor necrosis factor-α from peritoneal macrophages. J Ethnopharmacol 95(1):69–76
Kim YO, Han SB, Lee HW, Ahn HJ, Yoon YD (2005) Immuno-stimulating effect of the endo-polysaccharide produced by submerged culture of Inonotus obliquus. Life Sci 77(19):2438–2456
Kim YO, Park HW, Kim JH, Lee JY, Moon SH (2006) Anti-cancer effect and structural characterization of endo-polysaccharide from cultivated mycelia of *Inonotus obliquus*. Life Sci 79:72–80
Kozarski M, Klaus A, Niksic M, Jakovljevic D, Jpfg H, Ljld VG (2011) Antioxidative and immunomodulating activities of polysaccharide extracts of the medicinal mushrooms *Agaricus bisporus*. Food Chem 129(4):1667–1675
Langlands SJ, Hopskin MJ, Coleman N, Cummings JH (2004) Prebiotic carbohydrates modify the mucosa-associated microflora of the human large bowel. Gut 53(11):1610–1616
Lee JW, Park JW (2010) Changes of fatty acid composition and oxidation stability of edible oils with frying number of french fried potatoes. J Korean Soc Food Sci Nutr 39(7):1011–1017
Leung MY, Fung KP, Choy YM (1997) The isolation and characterization of an immunomodulatory and anti-tumor polysaccharide preparation from *flammulina velutipes*. Immunopharmacol 35(3), 255–263.
Liang Z, Miao C, Zhang Y (1996) Influence of chemical-modified structures on antitumour activity of polysaccharides from *Clitopilus caespitosus*. Chin Pharm J 31(10):613–615

Liao W, Yu Z, Lin Z, Lei Z, Ning Z, Regenstein JM, Yang J, Ren J (2015) Biofunctionalization of selenium nanoparticle with *Dictyophora indusiata* polysaccharide and its antiproliferative activity through death-receptor and mitochondria-mediated apoptotic pathways. Sci Rep 5:18629

Liu Y, Sun J, Rao S, Su Y, Li J, Li C, Xu S, Yang Y (2013) Antidiabetic activity of mycelia selenium-polysaccharide from *Catathelasma ventricosum* in STZ-induced diabetic mice. Food Chem Toxicol 62:285–291

Luo Q, Sun Q, Wu L, Yang Z (2012) Structural characterization of an immunoregulatory polysaccharide from the fruiting bodies of Lepista sordida. Carbohyd Polym 88(3):820–824

Ma Z, Wang J, Zhang L (2008) Structure and chain Conformation of β-Glucan Isolated from *Auricularia auricula-judae*. Biopolymers 87(7):614–622

Ma Z, Wang J, Zhang L, Zhang Y, Ding K (2010) Evaluation of water soluble β-d-glucan from *Auricularia auricular-judae* as potential anti-tumor agent. Carbohydr Polym 80(3):977–983

Macfarlane GT, Steed H, Macfarlane S (2008) Bacterial metabolism and health-related effects of galacto-oligosaccharides and other prebiotics. J Appl Microbiol 104(2):305–344

Macfarlane S, Macfarlane GT, Cummings JH (2006) Review article: prebiotics in the gastrointestinal tract. Aliment Pharm Ther 24(5):701–714

Maity KK, Patra S, Dey B, Bhunia SK, Mandal S, Das D, Majumdar DK, Maiti S, Maiti TK, Islam SS (2011) A heteropolysaccharide from aqueous extract of an edible mushroom, *Pleurotus ostreatus* cultivar: structural and biological studies. Carbohydr Res 346(2):366–372

Mandal EK, Maity K, Maity S, Gantait SK, Maiti S, Maiti TK, Sikdar SR, Islam SS (2011) Structural characterization of an immunoenhancing cytotoxic heteroglycan isolated from an edible mushroom *Calocybe indicavar. APK2*. Carbohydr Res 346(14):2237–2243

Mandal V, Sen SK, Mandal NC (2009) Effect of prebiotics on bacteriocin production and cholesterol lowering activity of *Pediococcus acidilactici* LAB 5. World J Microbiol Biotechnol 25(10):1837–1847

Manning T, Gibson G (2004) Microbial-gut interactions in health and disease Prebiotics. Best Pract Res Clin Gastroenterol 18(2):287–298

Masuda Y, Inoue M, Miyata A, Mizuno S, Nanba H (2009) *Maitake* beta-glucan enhances therapeutic effect and reduces myelosupression and nephrotoxicity of cisplatin in mice. Int Immunopharmacol 9(4):620–629

Mizuno T (1995) *Yamabushitake*: bioactive substances and medicinal utilization. Food Rev Int 11(1):173–178

Mizuna T (1999) Bioactive substances in *Hericium erinaceus* (bull.: fr.) pers. (yamabushitake), and its medicinal utilization. Int J Med Mushrooms 1(2):105–119

Mizuno T, Inagaki R, Kanao T, Hagiwara T, Nakamura T, Ito H, Shimura K, Sumiya T, Asakura A (1990) Antitumor activity and some properties of water-insoluble hetero-glycans from "*Himematsutake*", the fruiting body of *Agaricus blazei Murill*. Agric Biol Chem 54(11):2897–2905

Mizuno T, Wasa T, Ito H, Suzuki C, Ukai N (1992) Antitumor-active polysaccharides isolated from the fruiting body of *Hericium erinaceum*, an edible and medicinal mushroom called *yamabushitake* or houtou. Biosci Biotechnol Biochem 56(2):347–348

Nakashima S, Umeda Y, Kanada T (2013) Effect of polysaccharides from *ganoderma applanatum* on immune responses. i. enhancing effect on the induction of delayed hypersensitivity in mice. Microbiol Immunol 23(6):501–513

Nanba H, Hamaguchi A, Kuroda H (1987) The chemical structure of an antitumor polysaccharide in fruit bodies of *Grifola frondosa* (*maitake*). Chem Pharm Bull 35(3):1162–1168

Nandan CK, Sarkar R, Bhanja SK, Sikdar SR, Islam SS (2011) Isolation and characterization of polysaccharides of a hybrid mushroom (backcross mating between PfloVv12 and *Volvariella volvacea*). Carbohydr Res 346(15):2451–2456

Nosál'Ová V, Bobek P, Cerná S, Galbavý S, Stvrtina S (2001) Effects of pleuran (*β*-Glucan isolated from *Pleurotus ostreatus*) on experimental colitis in rats. Physiol Res 50(6):575–581

Nowak R, Nowacka-Jechalke N, Juda M, Malm A (2018) The preliminary study of prebiotic potential of polish wild mushroom polysaccharides: the stimulation effect on *Lactobacillus* strains growth. Eur J Nutr 57:1511–1521

Ojha AK, Chandra K, Ghosh K, Islam SS (2010) Glucans from the alkaline extract of an edible mushroom, *Pleurotus florida*, Assam Florida: isolation, purification, and characterization. Carbohydr Res 345(15):2157–2163

Palframan R, Gibson GR, Rastall RA (2003) Development of a quantitative tool for the comparison of the prebiotic effect of dietary oligosaccharides. Lett Appl Microbiol 37(44):281–284

Patra P, Bhanja SK, Sen IK, Nandi AK, Samanta S, Das D, Devi KS, Maiti TK, Acharya K, Islam SS (2012) Structural and immunological studies of hetero polysaccharide isolated from the alkaline extract of *Tricholoma crassum (Berk.) Sacc*. Carbohydr Res 362:1–7

Rathore H, Prasad S, Sharma S (2017) Mushroom nutraceuticals for improved nutrition and better human health: a review. Pharma Nutr 5(2):35–46

Roncero-Ramos I, Delgado-Andrade C (2017) The beneficial role of edible mushrooms in human health. Curr Opin Food Sci 14:122–128

Rout D, Mondal S, Chakraborty I, Pramanik M, Islam SS (2004) Structural characterisation of an immunomodulating polysaccharide isolated from aqueous extract of *Pleurotus florida*, fruit-bodies. Med Chem Res 13(6–7):509–517

Sasaki T, Takasuka N (1976) Further study of the structure of lentinan, an anti-tumor polysaccharide from *Lentinus edodes*. Carbohydr Res 47(1):99–104

Stowell J (2007) Sweeteners and sugar alternatives in food technology. In: Mitchell H (ed) Calorie Control and Weight Management, pp 54–64. https://doi.org/10.1002/9780470996003.ch4

Sun YX, Liang HT, Zhang XT, Tong HB, Liu JC (2009) Structural elucidation and immunological activity of a polysaccharide from the fruiting body of *Armillaria mellea*. Bioresour Technol 100(5):1860–1863

Synytsya A, Míčková K, Synytsya A, Jablonský I, Spěváček J, Erban V, Kovarikova E, Copikova J (2009) Glucans from fruit bodies of cultivated mushrooms *Pleurotus ostreatus* and *Pleurotus eryngii*: structure and potential prebiotic activity. Carbohydr Polym 76(4):548–556

Tuohy KM, Probert HM, Smejkal CW, Gibson GR (2003) Using probiotics and prebiotics to improve gut health. Drug Discov Today 8(15):692–700

Ukai S, Kiho T, Hara C, Morita M, Goto A, Imaizumi N, Hasegawa Y (1983) Polysaccharides in fungi. xiii. antitumor activity of various polysaccharides isolated from *Dictyophora indusiata*, *Ganoderma japonicum*, *Cordyceps cicadae*, *Auricularia auricula-judae* and *Auricularia* species. Chem Pharm Bull 31(2):741–744

Ukai S, Morisaki S, Goto M, Kiho T, Hara C, Hirose K (2008) Polysaccharides in fungi. vii. acidic heteroglycans from the fruit bodies of *Auricularia auricula-judae* quel. Chem Pharm Bulletin 30(2):635–643

Vamanu E, Vamanu A (2010) The influence of prebiotics on bacteriocin synthesis using the strain *Lactobacillus paracasei* CMGB16. Afr J Microbiol Res 4(7):534–537

Wang D, Sun S, Wu W, Yang S, Tan J (2014) Characterization of a water-soluble polysaccharide from *Boletus edulis* and its antitumor and immunomodulatory activities on renal cancer in mice. Carbohydr Polym 105:127–134

Wang J, Hu S, Liang Z, Yeh C (2005) Optimization for the production of water-soluble polysaccharide from *Pleurotus citrinopileatus* in submerged culture and its antitumor effect. Appl Microbiol Biotechnol 67(6):759–766

Wang SY, Hsu ML, Hsu HC, Tzeng CH, Lee SS (1997) The anti-tumor effect of *Ganoderma lucidum* is mediated by cytokines released from activated macrophages and T lymphocytes. Int J Cancer 70(6):699–705

Wang Y (2009) Prebiotics: present and future in food science and technology. Food Res Int 42(1):8–12

Wasser S (2002) Medicinal mushrooms as a source of antitumor and immunomodulating polysaccharides. Appl Microbiol Biotechnol 60(3):258–274

Wasser SP (2011) Current findings, future trends, and unsolved problems in studies of medicinal mushrooms. Appl Microbiol Biotechnol 89(5):1323–1332

Wong KH, Wong KY, Kwan HS, Cheung PC (2005) Dietary fibers from mushroom sclerotia: 3. *in vitro* fermentability using human fecal microflora. J Agric Food Chem 53(24):9407–9412

Xu S, Xu X, Zhang L (2012) Branching structure and chain conformation of water-soluble glucan extracted from *Auricularia auricula-judae*. J Agric Food Chem 60(13):3498–3506

Xu X, Yang J, Ning Z, Zhang X (2015) Lentinula edodes-derived polysaccharide rejuvenates mice in terms of immune responses and gut microbiota. Food Funct 6(8):2653–2663

Yamanaka D, Tada R, Adachi Y, Ishibashi K, Motoi M, Iwakura Y, Ohno N (2012) *Agaricus brasiliensis*-derived β-glucans exert immunoenhancing effects via a dectin-1-dependent pathway. Int Immunopharmacol 14(3):311–319

Yan LL, Han JC (2008) Effect of *Agarcus blazei murill* polysaccharides on lactic acid bacteria. China Dairy Ind 36(12):7–10

Yang LJ, Wang RT, Liu JS, Tong H, Deng YQ, Li QH (2004) The effect of polyporus umbellatus polysaccharide on the immunosuppression property of culture supernatant of S180 cells. J Cell Mol Immunol 20(2):234–237

Yoneda K, Ueta E, Yamamoto T, Osaki T (1991) Immunoregulatory effects of Sizofiran (SPG) on lymphocytes and polymorphonuclear leukocytes. Clin Exp Immunol 86(2):229–235

Yu R, Yang W, Song L, Yan C, Zhang Z, Zhao Y (2007) Structural characterization and antioxidant activity of a polysaccharide from the fruiting bodies of cultured *Cordyceps militaris*. Carbohydr Polym 70(4):430–436

Zhang J, Tang Q, Zhou C, Jia W, Da Silva L, Nguyen LD, Reutter W, Fan H (2010) GLIS, a bioactive proteoglycan fraction from *Ganoderma lucidum*, displays anti-tumour activity by increasing both humoral and cellular immune response. Life Sci 87(19–22):628–637

Zhang L, Li CG, Liang H, Reddy N (2017) Bioactive mushroom polysaccharides: immunoceuticals to anticancer agents. J Nutraceuticals Food Sci 2(6(2)):1–6

Zheng R, Su J, Dai H, Wu M (2005) Characterization and immunomodulating activities of polysaccharide from *Lentinus edodes*. Int Immunopharmacol 5(5):811–820

Zhou BL, Liang Q, Zou Y, Ming J, Zhao GH (2011) Research progress in prebiotic properties of edible mushroom. Food Sci 32(15):303–307

Chapter 11
Pharmaceutic Prodigy of Ergosterol and Protein Profile of *Ganoderma lucidum*

Anna Goyal and Anu Kalia

11.1 Introduction

Basidiomycetous fungi, 'mushrooms' are quite substantially untapped source of substances of nutraceutical importance (Hawksworth 2001). However, few mushroom genera such as *Ganoderma lucidum* (Leyss.: Fr.) Karst, has been popularly considered for use as health food as well as medicine in China. The taxonomic hierarchy of *Ganoderma lucidum* places it in class Basidiomycetes, order Aphyllophorales and family Polyporaceae. Commonly, it is a white rot causing phytopathogenic fungus which affects different types of woody plants including trees. It's fruit body have been consumed as a dietary supplement, not only in Orient (China and Japan) for more than 2000 years, but also in other parts of the world (Chang and Miles 2004). It is acronymed as '*Ling Zhi*', '*Ling Chu*', '*Ling Chih*' in Chinese, '*Reishi*', '*Sachitake*' or '*Mannetake*' in Japanese and '*Youngzhi*' in Korean folklore. *Ganoderma* is also called "Miraculous *Zhi*" as it is anticipated to happiness, health and longetivity (Wasser 2005). Now it is artificially cultivated in more than 10 countries with China being leader in its production, followed by many South-east Asian countries including Korea, Indonesia, Japan, Malaysia, Vietnam, and Taiwan, besides USA (Figlas and Curvetto 2010).

Ganoderma lucidum secretes diverse metabolites such as triterpenes (>119 types) and polysaccharides. It has been utilized to treat a variety of disease conditions in different organs including liver (chronic hepatitis, hepatopathy), kidney (nephritis), general physical and mental health (anorexia, hypertension, insomnia),

A. Goyal (✉)
Department of Microbiology, College of Basic Sciences and Humanities, Punjab Agricultural University, Ludhiana, Punjab, India

A. Kalia
Electron Microscopy and Nanoscience Laboratory, Department of Soil Science, College of Agriculture, Punjab Agricultural University, Ludhiana, Punjab, India

B. P. Singh et al. (eds.), *Biology of Macrofungi*, Fungal Biology,
https://doi.org/10.1007/978-3-030-02622-6_11

skeletal system (arthritis), respiratory (asthma, bronchitis), alimentary (gastric ulcer), vascular (arteriosclerosis) disorders, and diseases(cancer, diabetes) involving multiple organ or tissue systems (Mizuno et al. 1995). The fruit body as a whole or its constituent parts possesses potent cytotoxic activity and thus has been utilized to prevent and/or cure diseases besides improving the health (Paterson 2006).

11.2 *Ganoderma*: Cultivation and Ergosterol

The sub-tropical prominence of *Ganoderma lucidum* is observed as its growth on a wide variety of deciduous trees, especially oak, maple, elm and willow (Chang and Miles 2004). The first report of *in vitro* cultivation of *Ganoderma* on solid substrate was performed by Naoi's in 1970's by spore separation method (Mizuno 1997). Later, different solid substrates such as wood logs, grain, sawdust, and cork residues have been tried for its cultivation. World production of this mushroom has been estimated to be around 6000 tonnes, with China being the largest contributor. In India, Rai (2003) has reported the first successful attempt for cultivation of this mushroom.

The sub-tropical occurrence and asymmetric distribution of *Ganoderma* in wild indicates its preference to adapt and grow under high temperature and moisture conditions. The demand for *G. lucidum* is increasing which has escalated research interventions for cultivation of mushroom under specific optimum conditions (Chang and Buswell 2008). Different members of the genus *Ganoderma* exhibit specific nutritional requirements besides peculiar growth and cultivation conditions (Mayzumi et al. 1997). Moreover, there exists differential preference for specific type of *Ganoderma* among the populations inhabiting different territories such as South Chinese black and Japanese red *Ganoderma.*

Currently, its commercial production has been practiced on the wood log, stumps of trees, bags filled with sawdust and bottle utilizing sawdust and agricultural wastes as the main media components (Wasser 2005; Peksen and Yakupoglu 2009). This mushroom exhibits variable growth and yield on same substrate derived from different sources or origins. The maximum yield has been reported on oak saw dust (BE 17.48%) and wheat bran (BE 18.63%) substrates when cultivated on different sawdust (beech, poplar and oak) and three types of bran obtained from corn, rice and wheat (Erkel 2009). It has been successfully grown in Punjab at Punjab Agricultural University, Ludhiana on even wheat and rice straw with maximum BE (31.23%, GL-I) obtained on wheat straw (Goyal et al. 2016). Artificial cultivation of *Ganoderma* involves occurrence of five distinct growth stages i.e. spawn run, primordia (Antler) formation, primordia (Young Conk) formation, and formation of fruiting body and cropping cycle (Chen 1999). Approximately, three to four months (90 to 120 days) are required for obtaining the final cropping from initial spawn run mycelia of this on artificial cultivation. However, this duration greatly varies according to the method of the cultivation used.

A great amount of literature suggests the immunomodulatory and therapeutic benefits of different metabolites and other compounds derived from *Ganoderma*. Similarly, research interventions for development of different growth substrates for easy and cost-effective cultivation of this mushroom of medicinal importance have also been expedited. However, several challenges due to interplay of different factors such as supplementation to production quality of this fungus need to be effectively addressed.

11.3 Ergosterol: Molecular Structure

Eumycotic fungi (molds) and unicellular yeast cells contain large amounts of Ergosterol ((22E)-ergosta-5, 7, 22-trien-3-beta-ol) as their primary sterol compound. Likewise, these microbes also contain zymosterol (5-alpha-cholesta-8, 24-dien-3-beta-ol), one of sterol unlikely present in higher plants (Axelsson et al. 1995). Ergosterol was first isolated from fungal genus *Claviceps* commonly termed as 'ergot fungus' (Czub and Baginski 2006). Ergosterol is a cell membrane constituent having a role similar to its counterpart in animal cells, 'cholesterol'. It also shares the synthesis pathway of cholesterol which involves lanosterol unlike plant sterol synthesis pathways (Christie and Han 2010). It exist in fungal cells both as free and esterified forms. Though their relative amounts vary among different species of the ergot fungi (Sio et al. 2000) (Fig. 11.1).

11.3.1 Biosynthesis of Ergosterol

The information on ergosterol synthesis has been derived from several mutants defective in ergosterol biosynthesis. These mutation studies helped in deciphering the role of ergosterol in maintenance of fungal and yeast cell membrane structure and integrity. As molds have multicellular characteristics, ergosterol biosynthetic pathway and biocatalysts involved in its synthesis have been elucidated in common **yeast** (*Saccharomyces cerevisiae*) (Japelt and Jette Jakobsen 2013). The synthesis pathway

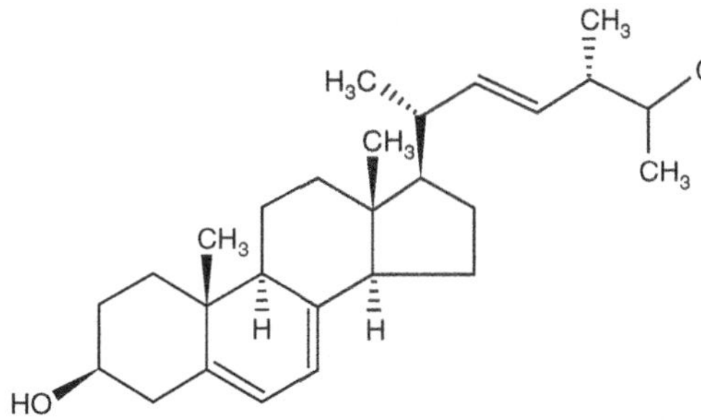

Fig. 11.1 Structure of ergosterol

is shared till zymosterol for both cholesterol and ergosterol (Lees et al. 1995). Likewise, both share the functions performed by the two molecules. Similar to cholesterol, ergosterol stabilizes the liquid-ordered phase, and can form lipid rafts along with sphingolipids. It also plays multiple key roles in regulation of growth in yeast (Grille et al. 2010). Moreover, this fungal specific sterol is the precursor for vitamin D_2 (ergocalciferol). Exposure of ergosterol with electromagnetic radiation in the ultra violet range leads to its photochemical conversion to ergocalciferol (Fig. 11.2).

The ergosterol biosynthesis is complex involving ~20 enzymes which generate squalene from mevalonate (Alcazar-Fuoli et al. 2008). It is synthesized via the lanosterol pathway though it is a smaller molecule formed by combining two molecules of a 15-C terpenoid, farnesyl pyrophosphate, into C-30 lanosterol. Later, removal of two methyl groups results in formation of ergosterol. However, unlike animal sterols, fungal sterols exhibit presence of methyl group at carbon 24. In *S. cerevisiae* the side chain alkylation is catalyzed by *S*-adenosyl methionine sterol methyl-transferase (ERG6) that converts zymosterol to fecosterol (Bach and Benveniste 1997). Mostly, plants cannot synthesize ergosterol, and any form of vitamin D2 presence is due to the endophytic fungi or a fungal infection (Japelt and Jette Jakobsen 2013).

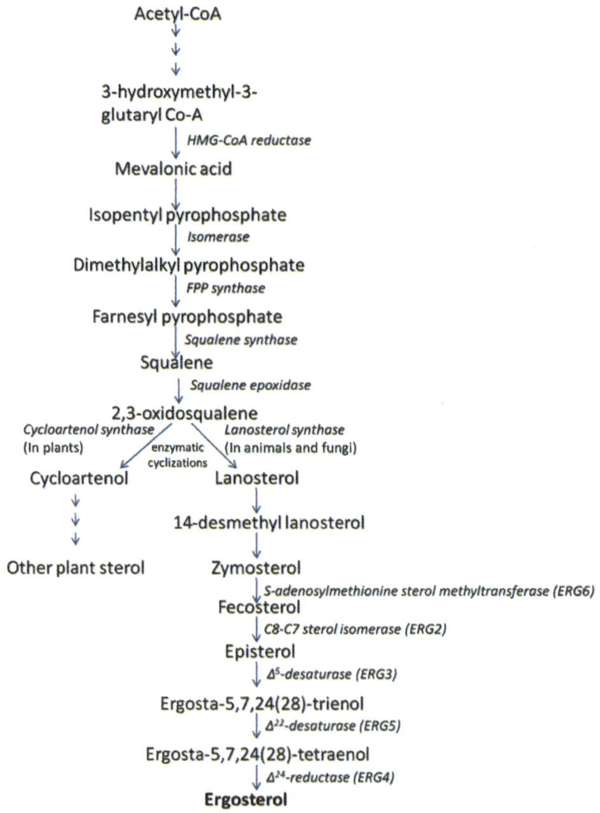

Fig. 11.2 Biosynthetic pathway for ergosterol synthesis

11.3.2 Determination of Ergosterol: Conventional and Advanced Techniques

Determination of the amount of ergosterol is widely used to estimate of fungal biomass in different types of environments as the ergosterol content and fungal dry mass has a strong correlation (Pasanen et al. 1999). Ergosterol regulates fluidity and asymmetry of the membrane and thus maintains the fungal cell integrity. It is used to diagnose the fungal invasion in grains and other substrates like wood, foliage, soil, mycorrhizal roots (Aswad et al. 2011). Ergosterol can be estimated through ergosterol assay. It is not produced by all fungi and even varies among same species depending on the physiological state of the fungus. The ergosterol assay is the most prominent method (Bermingham et al. 1995) as it indicates live fungal mass with relative accuracy compared to alternative methods as it readily gets rancid on cell death (Parkinson and Coleman 1991).

The original procedure for extraction of ergosterol was worked out by Seitz et al. (1977). It involves extraction of ergosterol in an alcoholic base followed by subsequent partitioning in a non-polar solvent such as hexane. Upon the evaporation of the solvent, the precipitates are redissolved in a known volume of fresh solvent. Final separation and quantification of ergosterol from total lipids is done by high-performance liquid chromatography (HPLC) (Gessner and Schmitt 1996). Others methods are based on UV absorption at specific wavelength maxima at 282 nm because of double bond at positions 5 and 7 (Lau et al. 2006). Currently, HPLC methods are regularly practiced, though a spectro-photometric method is a sensitive for determination of ergosterol content (Varga et al. 2006). *Ganoderma* biomass and ergosterol concentration was found to be positively correlated with *in vitro* HPLC with diode array detection (Aswad et al. 2011) and detection of white rot disease treatments may be made earlier to take remedial action.

The content of ergosterol can be used to assess the quality of *Ganoderma* spore and *Ganoderma* spore lipid (GSL) products. The contents of free and esterified ergosterol levels in *G. lucidum* lie between 0.8 and 1.6 mg/g (Yuan et al. 2007) Similarly, Shao et al. (2010) have reported that only free ergosterol were present in white and brown button mushrooms in the range of 2.04 to 4.82 mg/g dry matter (DM).

The ratio of free ergosterol to esterified forms varies in different parts of *G. lucidum* as the spore-producing cells (spores, tubes, hymenophore) have maximum percentage of ergosteryl esters (41.9 and 39.7% of total ergosterol) in comparison to the pileus and stipe (3.6 and 6.2%) (Yuan et al. 2007). Therefore, the amount of ergosterol in *Ganoderma* varied among the various strains and at different developmental stages, namely, the vegetative mycelium, spawns run, pinhead, and fruiting body phases. Ergosterol content from the fruit body of *Ganoderma* was observed to be maximum (7009 μg/g) (Goyal et al. 2016).

11.3.3 Ergosterol Applications

Ganoderma lucidum (Curt.: Fr.) Karst. serves as potential and long established medicine for preventing and curing numerous diseases (Wasser and Weis 1999). Triterpene derivatives such as oxygenated lanostane and other fungal steroids including ergosterol have been extracted out from *G. lucidum* having prospective therapeutic properties (Rosecke and Konig 2000).

11.3.3.1 Ergosterol: Precursor of Vitamin D

Windaus (1928) had pioneered the formation of vitamin D_2 from ergosterol after exposure to ultra violet radiations. The provitamin of vitamin D_2 [9,10-Secoergosta-5,7,10(19),22-tetraen-3-ol,(3β)- from Ergosta-5,7,22-trien-3-ol,(3β)-] is ergosterol, therefore, irradiating the fungal biomass with UV radiations ($\lambda = 280$ to 320 nm) results in formation of vitamin D_2 via conversion of ergosterol into viosterol followed by provitamin D_2 which undergoes thermal rearrangement to form Vitamin D_2 or ergocalciferol. It is one among the scarce Vitamin D rich available dietary sources of vegetarian origin (Fig. 11.3).

11.3.3.2 Antimycotic Drug Development

Ergosterol being fungal cell membrane specific sterol is an important target for antimycotic drug development (Arthington-Skaggs et al. 2000) and some antifungal drugs target the ergosterol biosynthetic pathway. Among other eukaryotes, ergosterol is located in the protists's cell membranes like trypanosomes on which antifungal agents can act for healing West African sleeping sickness (Roberts et al. 2003). An antifungal drug Amphotericin B attaches physically to ergosterol in the membrane and forms a pore in membranes which leads to exudation of ions and other particles from of the cell that results in cell death (Ellis 2002). Other antifungal drugs such as miconazole, itraconazole, and clotrimazole operate in a distinct way and block the conversion of lanosterol to ergosterol.

Ergosterol biosynthetic pathway is being intensively explored as antifungal drugs will inhibit important enzymatic conversions. The azole antifungal drugs inhibit the lanosterol 14-alpha demethylase, a cytochrome P450 family protein like Erg11 protein in fungal cell walls. Inhibition of 14α-demethylase leads to accumulation of ergosterol precursors like 14α-methylated sterols (lanosterol, 4, 14-dimethylzymosterol, and 24-methylenedihydrolanosterol), with exhaustion of ergosterol that results in the formation of altered plasma membrane. Other antifungal drugs such as fluconazole and voriconazole (a triazole in development) acts to inhibit the cytochrome P-450-dependent 14a-sterol demethylase activity.

Allylamine drugs, for eg. Terbinafine, inhibits the activity of squalene epoxidase (Valachovic et al. 2002). The sterols also regulate the antifungal activity of the

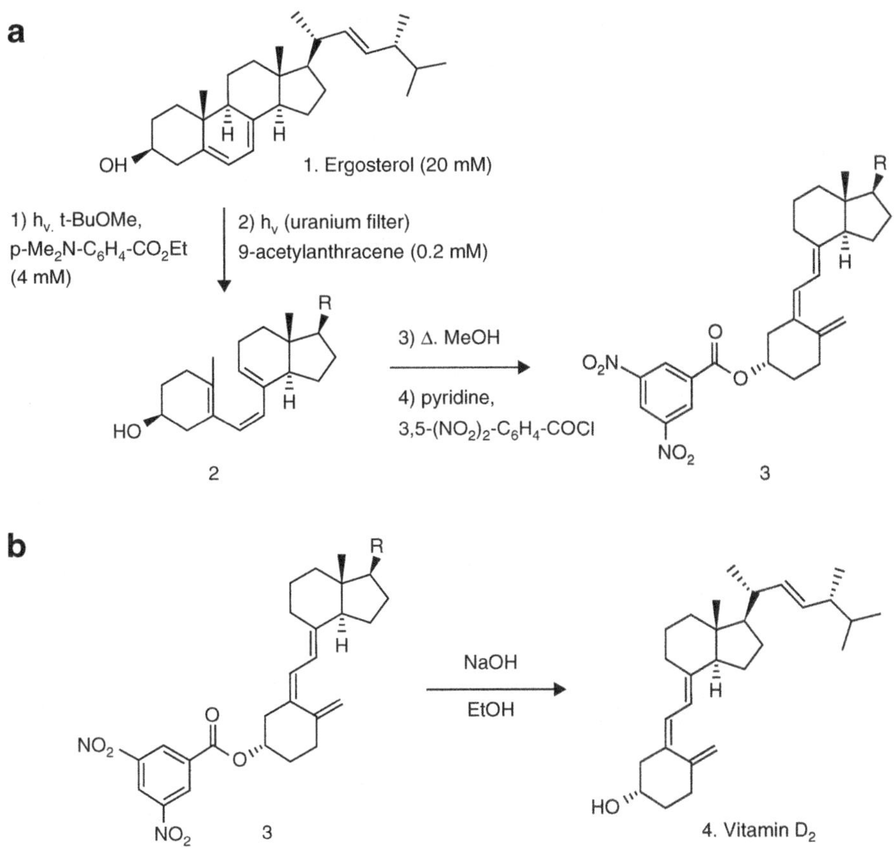

Fig. 11.3 Formation of vitamin D_2 from ergosterol

lipodepsipeptide syringomycin E from *Pseudomonas syringae*. In an *in vitro* study, the membrane sterol composition from *Saccharomyces cerevisiae* auxotroph strain FY-14, was amended to examine the sensitivity to syringomycin E and the authors reported higher sensitivity of the cells solely containing β-sitosterol or stigmasterol (Wangspa and Takemoto 1998).

11.3.3.3 Elicitor of Plant Immunity

Ergosterol acts to trigger the intrinsic plant immunity by elevating the activity of NADPH-oxidase and superoxide dismutase (SOD) enzymes. Both of these enzymes have been reported to impact the structural associations of plasma membrane in plants and hence, promote the formation of lipid rafts (Rossard et al. 2011). Ergosterol triggers the defense genes expression and enhancing the plant resistance to various pathogens (Lochman and Mikes 2006).

11.3.3.4 Biomedical Applications: Health Benefits

Ergosterol and its peroxide derivatives add up to enhance the pharmacological activities, like reducing inflammation and cardiovascular disease incidences, inhibiting the activity of cyclooxygenase enzyme and improving the antioxidant, antimicrobial, anti-complementary, and antitumor activities (Zaidman et al. 2005). Sterols and its oxygenated triterpenes derivatives also inhibit cholesterol synthesis *in vitro*. Ergosterol compounds may have a role in inhibition of blood vessel proliferation i.e. angiogenesis, which is desirable for curbing tumor growth and metastasis (Stamets and Yao 2002). Therefore, ergosterol acts by directly inhibiting the angiogenesis induced by solid tumors that results in its anti-tumor activity (Subbiah and Abplanalp 2003) and the same is also observed against KB cells and human PLC/PRF/5 cells *in vitro* (Lin et al. 2003). Ergosterol peroxide (EP) and dehydroergosterol peroxide (9(11)-DHEP) from *Ganoderma lucidum* can instigate the caspase-dependent cell apoptosis in susceptible cancer cells via the mitochondria-mediated pathway. These two fungal steroids have the potential to be used as natural chemopreventive agents as reported by its *in vitro* and *in vivo* studies (Zheng et al. 2009). Ergosterol peroxide can overcome the drug-resistance conferred by miR-378 and it can prove to be a likely a positive treatment to overcome the drug-resistance in tumor cells (Wu et al. 2012).

Steroidal components of *G. applanatum* possess the antibacterial activities and other broad range spectrum effects (Zhao et al. 2005). Moreover, two anti-aging compounds *Ganoderma* sides C and D, having a 4,6,8, (14),22-tetraene-3-one hydroxylation at C-9, significantly enhances the life period of K6001 yeast strain by regulating UTH1 gene expression to extend the replicative life cycle of yeast cells (Weng et al. 2011). Thus, ergosterol based anti-ageing substances could be developed to treat age related diseases.

11.3.3.5 Gano-Protein Province: Diversity and Functions

Therapeutic/Pharmaceutic Proteins

Several compounds with potential immune regulating mechanism have been extracted from *Ganoderma* such as polysaccharides (glucans), proteins and triterpenoids (ergosterol). Important immune regulating effects include induction of mitosis and immune effector cells that leads to the formation of cytokines, interleukins (ILs), tumor necrosis factor (TNF)-α, and interferons (Gao et al. 2003). The GMI, recombinant fungal immunodulatory protein, inhibits the EGF-induced phosphorylation and EGFR activation and AKT pathway kinases during the conditions of lung cancer (Lin et al. 2010). Ooi and Liu (2002) also observed that hot water extracted of *G. lucidum* polysaccharide suppressed of growth of solid tumour in mice with enhanced increase in immunomodulatory cytokines expression levels.

Protein bound polysaccharides from *G. lucidum*, have been known to exhibit superoxide and hydroxyl radical scavenging activity (Lee et al. 2001). A hot water extract of *G. lucidum* exhibited antioxidative effect on mouse liver and kidney lipid peroxidation (Shieh et al. 2001). Moreover, the extracts have the potential to reduce strand breakage in DNA caused by UV induced photolysis of hydrogen peroxide (Lee et al. 2001). Supplementation with antioxidants could represent an important therapeutic potential to minimize the damage.

Gradually, now focus is towards the establishment of natural antioxidants that are efficient and safe for use for which medicinal mushrooms are prominent source of such bioactive antioxidant compounds (Lee and Yun 2007). Selenium (Se) is an essential microelement living organisms and important component of many antioxidant enzymes like glutathione peroxidase (GSH-Px) which is associated with normal metabolism and health effect by its concentration in animals and humans. Benefits of selenium includes reducing the incidence and mortality of cancer, specifically in liver, prostate, colo-rectal and lung cancers (Rayman 2005). Zhao et al. (2004) have reported that *Ganoderma lucidum* can incorporate inorganic selenite present in the substrate into organic forms like seleno-proteins (56–61%) and seleno-polysaccharides (11–18%) and other components. Percentage absorption of selenium from selenium enriched medium in *Ganoderma lucidum* was reported to be around 7.2–9.9% and were absorbed as cytosolic embodies in form of selenoproteins (Goyal et al. 2015). The selenium supplementation can help to enhance the medicinal properties.

Ganoderma Specific Proteins: Physiological and Developmental Role

Basidiomycetes life cycle in dikaryotic cells is comprised of two defined stages, vegetative stage and a morphogenic stage which is triggered by the specific environmental signals. Understanding the phenomenon behind the shift from vegetative growth to fruit body formation will result in the commercial mushroom production improvement (Kamada 2002). Therefore, identification of certain key proteins responsible for onset of specific developmental stage is critical and is required to be performed. In addition, physiological mechanisms and enzymatic profile regulating lignocellulose biotransformation can enhance the process of edible and medicinal mushroom production (Vikineswary et al. 2006).

The development of fruit body from vegetative phase passes through a systematic manner. Morphological changes occur before the primordial formation due to the environmental stress like cold shock that causes dikaryotic mycelia in *L. edodes* to develop small primordial which further modifies to form fruit bodies (Przybylowicz and Donoghue 1990). Similarly in *S. commune*, SC4 hydrophobin, lines up the air spaces in its fruit bodies and thus, prevents the collapse of air channels during the drying and wetting cycles (Wessel 1994). Stress-related genes such as Hsp12, Nam9 and DNA repair helicase RAD15 have been reported to play an important role in fructification. Signal transduction mechanisms are complex in cells that recognizes and respond to the outer surrounding signals, for eg, calmodulin, phosphatase 2C and RAB18 were found to be specifically expressing during fruit body formation (Lee et al. 2002). In *Coprinopsis cinerea* fruiting was regulated by biotic-abiotic factors such as nutrients, temperature, light, and humidity, physiological stages and genetic factors, for eg., A and B mating type genes, dst1 and dst2 light regulators genes) (Gonzalez et al. 2011).

Thus several fructification genes are explored in *Agaricus bisporus, Agrocybe aegerita, Coprinus cinereus, Flammulina velutipes, Lentinula edodes, Schizophyllum commune* and *Tuber borchi* (Lacourt et al. 2002). The genes for hydrophobin proteins are present in many copy numbers that are expressed during the fruit body formation in basidiomycetes, as reported by Penas et al. (1998) and such genes have been isolated from *A. bisporus*, *F. velutipes* and *L. edodes* (Ng et al. 2000). Other genes for ATPase have been isolated from *L. edodes* and *T. borchii* (Leung et al. 2000), PriA from *L. edodes*, *Agrocybe aegerita* and EST from *P. ostreatus* (Lee et al. 2002). Two proteins Pf1 and Pf3 had been identified in *Flammulina velutipes* (Sakamoto 2010) which is involved in fruit body induction and one pileus-specific hydrophobin-like protein (PSH) is reported to be involved in the pileus formation. During the primordial formation and fruiting in *Pleurotus ostreatus,* a cytolytic protein Ostreolysin, is specifically expressed (Berne et al. 2007).

Hydrophobins

Hydrophobins have been characterized as compact cysteine-rich (nearing 100 ± 25 amino acids), secretory proteins that exists in large copy numbers in cell walls of fungi and forms the outermost layer. These proteins form outer surface of hydrophobic layers by self-assembly of secreted monomers in response to the external stimuli and this plays a conformational role for the developmental processes. These proteins are required in several morphological processes like sporulation, fructification and infection tube formation.

Different types of hydrophobin proteins are expressed at different stages of development of fruit body in basadiocarps and play an important function in protection against hyphal desiccation, parasitic attack, hyphal attachment, lowering of surface tension surrounding medium to allow hyphal aerial growth. Various types of hydrophobins genes have been isolated from different types of basadiocarps whose expressions are regulated at developmental stages for eg., ABH1, ABH2 and ABH3 in *Agaricus bisporus,* CoH1 and CoH2 in *Coprinus cinereus*, and SC1, SC3 and SC3 in *S. commune* (Lugones et al. 1998). These genes were first explored as mRNAs copies transcribed at developmental stages like sporulation, fructification and fungal infection tube formation. Ng et al. (2000) has reported two genes for class I hydrophobins (Le.hyd1 and Le.hyd2) from primordial cDNA library of *Lentinula edodes* with base pair size of 760 and 738 nucleotides.

The FT-IR analysis of proteins made during different stages of cultivation that the fruiting strains have an ordered protein structure with hydrophobic amino acids. Non fruiting strains indicated the presence of aromatic amino acids with very low amount of acidic amino acids like aspartic acid and glutamic acid indicating a positive role of hydrophobic amino acids and hydrophobin proteins in mushroom fructification process (unpublished data).

It has also been reported that the phenolic compounds polymerized by the laccase enzymes also adds to hydrophobicity of fruiting body surface and aids in the cross-linking of hyphal cell walls during the fruit body formation (Pelkmans et al. 2016). The esterase and peroxidase activity significantly increased during the pinning of the *Ganoderma lucidum* cultures thus, indicating the positive role of these enzymes during the fructification (Unpublished data thesis). Other reports suggest that cytochrome P450 enzymes, lectins, haemolysins and expansins also have a pivotal role in fructification of the mushrooms. In *Flammulina velutipes*, the proteolytic activity in mycelium was enhanced during the fructification to allow reallocation of amino-nitrogen from mycelium to the fruit body (Chao and Gruen 1987).

11.4 Future Prospects

A growing research interest is driving towards the development of novel formulations of the medicinal mushroom derived bioactives of potent nutraceutical and pharmaceutic importance. Moreover, this will lead to manufacturing of food supplements from medicinal mushrooms. Improved augmenting or additive effects can be obtained if the bioactive compounds of the medicinal mushrooms can be complexed and formulated as nano-emulsions or can be utilized to encapsulate other drugs or chemoprotectants. The latter approach is anticipated to exhibit additive benefits such as improved water disponibility of water phobic curcumin besides enhanced biological effects when complexed and encapsulated by 1–3/1–6, β-glucan derived from *Ganoderma* (Mai Huong et al. 2016). Another research aspect pertains to using nanoparticles as vehicles for targeted delivery of bioactives of *Ganoderma* origin. The effective dose administered can be decreased to nM concentrations of ganoderic acid-DM, a triterpene derived from *Ganoderma* that ensures improved permeability in the targeted cancerous cells besides minimizing the cytotoxicity issues for the normal body cells (Bryant et al. 2017).

11.5 Conclusion

Extensive research studies are being carried on *G. lucidum* in various parts of the world to improve its cultivation and therapeutic substances. It possesses several positive health benefits derived from its anti-oxidant, anti-microbial, anti-proliferative and organ protective action. Therefore, both conventional and novel quality control strategies will ensure identification of the bioactive compound diversity and standardization of ergosterol and other putative bioactives in *G. lucidum* preparations for improved bioavailability and efficacy. These bioactives can be anticipated to accelerate the intrusion of compounds of nutraceutical/pharmaceutic importance as food supplements for enhancing the nutritional quality of the food.

References

Alcazar-Fuoli L, Mellado E, Garcia-Effron G, Lopez JF, Grimalt JO, Cuenca-Estrella JM (2008) Ergosterol biosynthesis pathway in *Aspergillus fumigatus*. Steroids 73(3):339–347

Arthington-Skaggs BA, Warnock DW, Morrison CJ (2000) Quantitation of *Candida albicans* ergosterol content improves the correlation between in vitro antifungal susceptibility test results and in vivo outcome after fluconazole treatment in a murine model of invasive candidiasis. Antimicrob Agents Chemother 44(8):2081–2085

Aswad AWM, Sariaha M, Paterson RRM, Abidinb MAZ, Limac N (2011) Ergosterol analyses of oil palm seedlings and plants infected with *Ganoderma*. Crop Prot 30:1438–1442

Axelsson BO, Saraf A, Larsson L (1995) Determination of ergosterol in organic dust by gas chromatography-mass spectrometry. J Chromatogr B 666:77–84

Bach TJ, Benveniste P (1997) Cloning of cDNAs or genes encoding enzymes of sterol biosynthesis from plants and other eukaryotes: heterologous expression and complementation analysis of mutations for functional characterization. Prog Lipid Res 36:197–226

Bermingham S, Maltby SL, Cooke RC (1995) A critical assessment of the validity of ergosterol as an indicator of fungal biomass. Mycol Res 99:479–484

Berne S, Pohleven J, Vidic I, Rebolj K, Pohleven F, Turk T, Macek P, Sonnenberg ASM, Sepcic K (2007) Ostreolysin enhances fruiting in the oyster mushroom (*Pleurotus ostreatus*). Mycol Res 111:1431–1436

Bryant JM, Bouchard M, Haque A (2017) Anticancer activity of Ganoderic acid DM: current status and future perspective. J Clin Cell Immunol 8:535

Chang ST, Buswell JA (2008) Safety, quality control and regulational aspects relating to mushroom nutriceuticals. Proc. 6th Intl. Conf. Mushroom Biology and Mushroom Products GAMU Gmbh, Krefeld. pp: 188–195

Chang ST, Miles PG (2004) Mushrooms: cultivation, nutritional value medicinal effect and environmental impact, second edn. CRC Press, Boca Raton

Chao EE, Gruen HE (1987) Intracellular activity of mycelial proteinases during fruit-body development in *Flammulina velutipes*. Can J Bot 65:518–525

Chen AW (1999) Cultivation of the medicinal mushroom *Ganoderma lucidum* (Curt.: Fr.) P. karst. (Reishi) in North America. Int J Med Mush 1:263–282

Christie WW, Han X (2010) Lipid analysis - isolation, separation, identification and lipidomic analysis. Oily Press, Bridgwater

Czub J, Baginski MJ (2006) Comparative molecular dynamics study of lipid membranes containing cholesterol and ergosterol. Biophys J 90:2368–2382

Ellis D (2002) Amphotericin B: spectrum and resistance. J Antimicrob Chemother 49(1):7–10

Erkel EI (2009) The effect of different substrate mediums on yield of *Ganoderma lucidum* (Fr.) karst. J Food Agri Environ 7:841–844

Figlas D, Curvetto N (2010) Medicinal mushroom reishi (*Ganoderma lucidum*). Main toxicity and allergenicity studies, dosage, posology and side effects. pp 1–13 http://hongoscomestiblesymedicinales.com/P/3%20G.pdf

Gao YH, Sai XH, Chen GL, Ye JX, Zhou SF (2003) A randomized, placebo-controlled, multicenter study of *Ganoderma lucidum* (W. Curt.: Fr.) Lloyd (Aphyllophoromycetideae) polysaccharides (Ganopoly) in patients with advanced lung cancer. Int J Med Mushrooms 5:368–381

Gessner MO, Schmitt AL (1996) Use of solid-phase extraction to determine ergosterol concentrations in plant tissue colonized by fungi. Appl Env Microbiol 62(2):415–419

Gonzalez MN, Arndt M, Zomorrodi M, Majcherczyk A, Kües U (2011) Regulation of fruiting body formation in Coprinopsis cinerea. Proceedings of the 7th international conference on mushroom biology and mushroom products (ICMBMP7)

Goyal A, Kalia A, Sodhi HS (2015) Selenium stress in *Ganoderma lucidum:* A scanning electron microscopy appraisal. Afr J Microbiol Res 9(12):855–862

Goyal A, Kalia A, Sodhi HS (2016) High-performance liquid chromatography studies to estimate ergosterol content at different developmental stages of the lingzhi or reishi medicinal mushroom, *Ganoderma lucidum* (Agaricomycetes). Int J Med Mushrooms 18(12):1115–1120

Grille S, Zaslawski A, Thiele S, Plat J, Warnecke D (2010) The functions of steryl glycosides come to those who wait: recent advances in plants, fungi, bacteria and animals. Prog Lipid Res 49:262–288

Hawksworth DL (2001) Mushrooms: the extent of the unexplored potential. Int J Med Mush 3:333–337

Japelt RB, Jette Jakobsen J (2013) Vitamin D in plants: a review of occurrence, analysis, and biosynthesis. Front Plant Sci 4:136

Kamada T (2002) Molecular genetics of sexual development in the mushroom *Coprinus cinereus*. Bioessay 24:449–459

Lacourt I, Duplessis S, Abba S, Bonfante P, Martin F (2002) Isolation and characterization of differentially expressed genes in the mycelium and fruit body of *Tuber borchii*. Appl Environ Microbiol 68:4574–4582

Lau APS, Lee AKY, Chan CK, Fang M (2006) Ergosterol as a biomarker for the quantification of the fungal biomass in atmospheric aerosols. Atmos Environ 40:249–259

Lee IK, Yun BS (2007) Highly oxygenated and unsaturated metabolites providing a diversity of hispidin class antioxidants in the medicinal mushrooms *Inonotus* and *Phellinus*. Bioorg Med Chem 15:3309–3314

Lee JM, Kwon H, Jeong H, Lee JW, Lee SY, Baek SJ, Surh YJ (2001) Inhibition of lipid peroxidation and oxidative DNA damage by *Ganoderma lucidum*. Phytother Res 15:245–249

Lee SH, Kim BG, Kim KJ, Lee JS, Yun DW, Hahn JH, Kim GH, Lee KH, Suh DS, Kwon ST, Lee CS, Yoo YB (2002) Comparative analysis of sequences expressed during the liquid-cultured mycelia and fruit body stages of *Pleurotus ostreatus*. Fungal Genet Biol 35:115–134

Lees N, Skaggs B, Kirsch D, Bard M (1995) Cloning of the late genes in the ergosterol biosynthetic pathway of *Saccharomyces cerevisiae* – a review. Lipids 30:221–226

Leung GS, Zhang M, Xie WJ, Kwan HS (2000) Identification by RNA fingerprinting of genes differentially expressed during the development of the basidiomycete *Lentinula edodes*. Mol Gen Genet 262:977–990

Lin CH, Sheu GT, Lin YW, Yeh CS, Huang YH, Lai YC, Chang JG, Ko JL (2010) A new immunomodulatory protein from *Ganoderma microsporum* inhibits epidermal growth factor mediated migration and invasion in A549 lung cancer cells. Process Biochem 45(9):1537–1542

Lin SB, Li CH, Lee SS, Kan LS (2003) Triterpene enriched extracts from *Ganoderma lucidum* inhibit growth of hepatoma cells via suppressing protein kinase C, activating mitogen-activated protein kinases and G2-phase cell cycle arrest. Life Sci 72:2381–2390

Lochman J, Mikes V (2006) Ergosterol treatment leads to the expression of a specific set of defence-related genes in tobacco. Plant Mol Biol 62(2):43–51

Lugones LG, Wosten HAB, Wessels JGH (1998) A hydrophobin (ABH3) specifically secreted by vegetatively growing hyphae of *Agaricus bisporus* (common white button mushroom). Microbiol 144:2345–2353

Mai Huong L, Do HD, Thi HHT, Dung LV, Nguyen HN, Thi HNT, Nguyen LD, Hoang CK, Le HC, Thi THL, Trinh HT, Ha PT (2016) The dual effect of curcumin nanoparticles encapsulated by 1-3/1-6 β-glucan from medicinal mushrooms *Hericium erinaceus* and *Ganoderma lucidum*. Adv Nat Sci Nanosci Nanotechnol 7:045019

Mayzumi F, Okamoto H, Mizuno T (1997) Cultivation of Reishi (*Ganoderma lucidum*). Food Rev Int 13:365–382

Mizuno T (1997) Studies on bioactive substances and medicinal effect of Reishi *Ganoderma lucidum* in Japan. In: A committee (ed) *Proc 1st Int Symp Ganoderma lucidum in Japan*. Toyo-Igaku-Sha Co Ltd, Tokyo, pp 121–127

Mizuno T, Wang G, Zhang J, Kawagishi H, Nishitoba T, Li J (1995) *Ganoderma lucidum* and *Ganoderma tsugae*: bioactive substances and medicinal effects. Food Rev Int 11:151–166

Ng WL, Ng TP, Kwan HS (2000) Cloning and characterization of two hydrophobin genes differentially expressed during fruit body development in *Lentinula edodes*. FEMS Microbiol Lett 185:139–145

Ooi VE, Liu F (2002) Immunomodulation and anti-cancer activity of polysaccharide-protein complexes. Curr Med Chem 7:715–729

Parkinson D, Coleman DC (1991) Methods for assessing microbial populations, activity and biomass. Microbial communities, activity and biomass. Agric Ecosyst Environ 34:3–33
Pasanen AL, Yli-Pietila K, Pasanen P, Kalliokoski P, Tarhanen J (1999) Ergosterol content in various fungal species and biocontaminated building materials. Appl Environ Microb 65:138–142
Paterson RRM (2006) *Ganoderma* – a therapeutic fungal biofactory. Phytochemistry 67:1985–2000
Peksen A, Yakupoglu G (2009) Tea waste as a supplement for the cultivation of *Ganoderma lucidum*. World J Microbiol Technol 5:611–618
Pelkmans JF, Lugones LGZ, HAB W (2016) 15 fruiting body formation in Basidiomycetes. In: Wendland J (ed) Growth, differentiation and sexuality. The Mycota (A Comprehensive Treatise on Fungi as Experimental Systems for Basic and Applied Research). Springer, Cham, pp 387–405
Penas MM, Ásgeirsdóttir SA, Lasa I, Macià FAC, Pisabarro AG, Wessels JGH, Ramírez L (1998) Identification, characterization, and *in situ* detection of a fruit-body-specific hydrophobin of *Pleurotus ostreatus*. App Environmen Microbiol 64:4028–4034
Przybylowicz P, Donoghue J (1990) Shiitake growers handbook, the art and science of mushroom cultivation. Kendall/Hunt Publishing Co, USA, pp 151–155
Rai RD (2003) Successful cultivation of the medicinal mushroom Reishi, *Ganoderma lucidum* in India. Mush Res 12:87–91
Rayman MP (2005) Selenium in cancer prevention: a review of the evidence and mechanism of action. Proc Nutr Soc 64:527–542
Roberts CW, McLeod R, Rice DW, Ginger M, Chance ML, Goad LJ (2003) Fatty acid and sterol metabolism: potential antimicrobial targets in apicomplexan and trypanosomatid parasitic protozoa. Mol Biochem Parasitol 126(2):129–142
Rosecke J, Konig WA (2000) Constituents of various wood-rotting basidiomycetes. Phytochemistry 54:603–610
Rossard S, Roblin G, Atanassova R (2011) Ergosterol triggers characteristic elicitation steps in *Beta vulgaris* leaf tissues. J Exp Bot 61(6:1807–1816
Sakamoto Y (2010) Protein expression during *Flammulina velutipes* fruiting body formation. Mycosci 51:163–169
Seitz LM, Mohr HE, Burroughs R, Sauer DB (1977) Ergosterol as an indicator of fungal invasion in grains. Cereal Chem 54:1207–1217
Shao S, Hernandez M, Kramer JKG, Rinker DL, Tsao R (2010) Ergosterol profiles, fatty acid composition, and antioxidant activities of button mushrooms as affected by tissue part and developmental stage. J Agric Food Chem 58:11616–11625
Shieh YH, Liu CF, Huang YK, Yang JY, Wu IL, Lin CH, Lin SC (2001) Evaluation of the hepatic and renal protective effects of *Ganoderma lucidum* in mice. Am J Chin Med 29:501–507
Sio DF, Laratta B, Giovane A, Quagliuolo L, Castaldo D, Servillo L (2000) Analysis of free and esterified ergosterol in tomato products. J Agric Food Chem 48:780–784
Stamets P, Yao D (2002) Myco Medicinals: an informational treatise on the medicinal properties of mushrooms. Olympia, Wash.: Mycomedia productions, Fungi Perfecti
Subbiah MTR, Abplanalp W (2003) Ergosterol (major sterol of baker's and brewer's yeast extracts) inhibits the growth of human breast cancer cells *in vitro* and the potential role of its oxidation products. Int J Vitam Nutr Res 73:19–23
Valachovic MC, Randall SK, Nickels JT, Bard M (2002) Protein-protein interactions among C-4 demethylation enzymes involved in yeast sterol biosynthesis. Proc Natl Acad Sci U S A 99(15):9739–9744
Varga M, Bartok T, Mesterhazy A (2006) Determination of ergosterol in fusarium-infected wheat by liquid chromatography atmospheric pressure photoionization mass spectrometry. J Chromatogr 1103:278–283
Vikineswary S, Abdullah N, Renuvathani M, Sekaran M, Pandey A, Jones EBG (2006) Productivity of laccase in solid substrate fermentation of selected agro-residues by *Pycnoporus sanguineus*. Bioresour Technol 97:171–177
Wangspa R, Takemoto JY (1998) Role of ergosterol in growth inhibition of *Saccharomyces cerevisiae* by syringomycin E. FEMS Microbiol Lett 167:215–220

Wasser SP (2005) Reishi or Lingzhi (*Ganoderma lucidum*). Encycl Dietary Suppl. https://doi.org/10.1081/E-EDS-120022119

Wasser SP, Weis AL (1999) Medicinal properties of substances occurring in higher basidiomycetes mushrooms: current perspectives. Int J Med Mush 1:31–62

Weng Y, Lu J, Xiang L, Matsuura A, Zhang Y, Huang Q, Qi J (2011) *Ganoderma* sides C and D, two new anti-aging ergosterol from spores of the medicinal mushroom *Ganoderma lucidum*. Biosci Biotechnol Biochem 75(4):800–803

Wessel JGH (1994) Developmental regulation of fungal cell wall formation. Annu Rev Phytopathol 32:413–437

Windaus A (1928) Constitution of sterols and their connection with other substances occurring in nature. Noble Lecture

Wu QP, Xie YZ, Deng Z, Li XM, Yang W, Jiao CW, Fang L, Li SZ, Pan HH, Yee AJ, Lee DY, Li C, Zhang Z, Guo J, Yang BB (2012) Ergosterol peroxide isolated from *Ganoderma lucidum* abolishes microrna mir-378-mediated tumor cells on chemoresistance. PLoS One 7(8):e44579

Yuan JP, Wang JH, Liu X (2007) Distribution of free and esterified ergosterol in the medicinal fungus *Ganoderma lucidum*. Appl Microbiol Biotechnol 77:159–165

Zaidman BZ, Yassin M, Mahajna J, Wasser SP (2005) Medicinal mushroom modulators of molecular targets as cancer therapeutics. Appl Microbiol Biotechnol 67:453–468

Zhao HB, Wang SZ, He QH, Yuan L, Chen AF, Lin ZB (2005) *Ganoderma* total sterol (GS) and GS1 protect rat cerebral cortical neurons from hypoxia/reoxygenation injury. Life Sci 76:1027–1037

Zhao L, Zhao G, Zhao Z, Chen P, Tong J, Hu X (2004) Selenium distribution in a se-enriched mushrooms species of the genus *Ganoderma*. J Agric Food Chem 52:3954–3959

Zheng L, Si J, Wong YS (2009) Ergosterol peroxide and 9,11-dehydroergosterol peroxide from ling zhi or reishi medicinal mushroom, *Ganoderma lucidum* (W. Curt.: Fr.) P. Karst. (Aphyllophoromycetideae) mycelia inhibit the growth of human breast adenocarcinoma MCF-7 cells. Int J Med Mush 11(3):248–257

Chapter 12
Application of Wild Macrofungi as Anticancer Therapeutics

Peter Chiew Hing Cheong, Chon Seng Tan, and Shin Yee Fung

12.1 The Word Is Out

The power of social networking and mass media; Facebook, Twitter, YouTube and TEDMED, in triggering excitement, hope and spreading ideas worth selling and following, is undeniable. Mycologist Paul Stamets got the message out to the world, far and wide, that mushrooms are more than just useful. At 84, his mother was diagnosed with stage 4 breast cancer that had spread to her sternum and liver and her oncologist told her she was too old for radiation therapy and no more than 3 months. After a year of turkey tail mushroom supplementing addition to the standard drugs Taxol (**paclitaxel**) and Herceptin (**trastuzumab**) however, the spreading cancer was no longer detected on her (CNN 2016). The message was a clear and powerful one, positioning mushrooms, in different light among the practitioners and giving hope to patients and families, that there are anticancer therapeutic properties in mushrooms worth evaluating and for further research.

P. C. H. Cheong
Medicinal Mushroom Research Group (MMRG), Department of Molecular Medicine, University of Malaya, Kuala Lumpur, Malaysia

C. S. Tan
Ligno Research Foundation, Selangor, Malaysia

S. Y. Fung (✉)
Medicinal Mushroom Research Group (MMRG), Department of Molecular Medicine, University of Malaya, Kuala Lumpur, Malaysia

Centre for Natural Products Research and Drug Discovery (CENAR), University of Malaya, Kuala Lumpur, Malaysia

University of Malaya Centre for Proteomics Research (UMCPR), University of Malaya, Kuala Lumpur, Malaysia
e-mail: syfung@um.edu.my

B. P. Singh et al. (eds.), *Biology of Macrofungi*, Fungal Biology,
https://doi.org/10.1007/978-3-030-02622-6_12

12.2 Macrofungi and their Uses

Typical macro fungi, belonging to taxonomy classes Basidiomycetes, Ascomycetes and a few Zygomycetesare noticeable higher fungi with fruiting bodies (consisting of caps and stems) which erect above grounds, rotting woods or termite mounds when the conditions are ideal for them to produce from their gills tiny, wind or animal dependant for dispersal of reproductive particles known as spores. The more exotic ones; (i) **bolete**, have pores instead of gills on the underside of the cap to produce spores inside the pore tunnels, (ii) the **puffballs**, produce spores inside ball-like structures and puff the spores like smoke through tiny openings when squeezed, (iii) the **stinkhorns**, produce spores on slimy and smelly stalk, to be carried away by animals attracted to its smell (iv) the **cup fungi**, have their spores on the inner surface of their cup-like bodies (v) the **jelly fungi**, fire spores like catapults (vi) only the **subterranean truffles**, have spores deposited inside the fleshy fruiting bodies (Barbarathiers 2013). Mature truffles are often sniffed out from underground using female pigs, dogs and goats (Knapton 2008) as they make exquisite and expensive cuisines in several parts of the world. Most terrestrial macro fungi are **saprophytes**, serve to keep trees alive, breaking dead organic matters down and returning them to the soil but some are pathogens of plants (Mueller et al. 2007). They are homes and food too. Edible mushrooms are always referred first in many literatures and cultures (Fig. 12.1).

Wild edible mushrooms of different varieties are still mostly collected from the forests. Forests with different tree species provide the vegetation that promotes the greatest diversity of mushrooms (Montoya et al. 2003; Garibay-Orijel et al. 2006). Hunters and collectors rely on the presence of certain tree species to indicate the presence of valuable wild mushrooms. Users believe mountain mushrooms are superior and have much better flavour because they grow where there are many trees (Montoya et al. 2003). The efforts spent on collection and cultural use determine the importance and value of certain mushrooms (Knapton 2008). In Mexico, Brazil

Fig. 12.1 Common edible mushrooms: *Agaricus blazei* (**a**) portobello and (**d**) button mushroom, *Pleurotus eryngii* (**b**) oyster mushroom, *Flammulina velutipes* (**c**) enokitake mushroom, *Lyophyllum shimeji* (**e**) G hon shimeji mushrooms, *Lentinula edodes* (**f**) shiitake mushroom

and many western countries, mushrooms were mainly used as food followed by secondary function for trade, handicraft, insecticides and medicine (Montoya et al. 2003; Garibay-Orijel et al. 2006; Dias et al. 2004). In some cultures, mushrooms are sought for ceremonial, recreation, religious and remedy against fear of death (Montoya et al. 2003; Garibay-Orijel et al. 2006; Guzmán 2008; Riede 2010). A special church was built to worship a specimen of mushroom *Ganoderma lobatum* (Schwein) displayed in a crystal box at the main altar (Guzmán 2008). *Fomes fomentarius*, *Inonotus obliquus*, *Laricifomes officinalis* and *Calvatia gigantean* were the only few mushroom species used therapeutically by the Europeans (Wasser and Weis 1999; Poucheret et al. 2006). Only countries in the far east like China, Japan and Korea where the mushrooms are not only sought after as food but also used frequently as medicine and have been for centuries (Patel and Goyal 2012; Lemieszek and Rzeski 2012). Many mushrooms used in traditional practices are from the genera *Auricularia, Flammulina, Ganoderma, Grifola, Hericium, Lentinus* (Lentinula), *Pleurotus, Trametes* (Coriolus), *Schizophyllum* and *Tremella* (Ooi and Liu 2000). The different types of medicinal mushrooms are usually blended, mixed and used as concoction for maximum benefit (Ferreira et al. 2010).

12.3 The Need for Anticancer Therapeutics

Cancerous cells ignore signals to undergo natural cell cycle and differentiation. Death defect and hard to eradicate, they replicate in mass without control, blocking and disrupting normal functions of the body severely and when spread and invade other systems, one dies. Death from cancer spiralled downward 25% since 1991 in the United States, prolonging the lives of 2.4 million, based on available data from 2014. The avert is attributed to the advances in early detection, treatment and change of lifestyle (Siegel et al. 2017). Cancer SEEK (a new blood test) is a sensitive simple blood test that could detect early stages instead of late stages of eight different kinds of cancer concurrently without raising false-positive results (Cohen et al. 2018). About 46% (Fig. 12.2) over all cancer deaths were due to lung, colorectal, prostate and breast cancers (Siegel et al. 2017). In men, 42% of all newly diagnosed cancers in 2017 were from prostate, lung, and colorectal cases. In women, breast, lung, and colorectal cancers were the most common. Leukaemia killed almost a third (29%) of all paediatric cancer patients, followed closely by brain and nervous system cancers (26%) (Siegel et al. 2017). The number of new cases is expected to rise by about 70% (22 million) over the next 2 decades (WHO Media Centre 2017).

Basic research and deeper understanding of cancer, creative strategies to manipulate signalling pathways that would influence host responses and improved medical techniques are all needed to keep the statistics low. Cancer treatment and cure currently revolve around chemotherapy, radiation, surgery, immunotherapy, stem cells, hyperthermia, photodynamic therapy, blood transfusion and donation, lasers, complementary and alternative therapies. When modern medicine could no longer aid, cancer relapses and the body responds badly to the treatment or when the cost

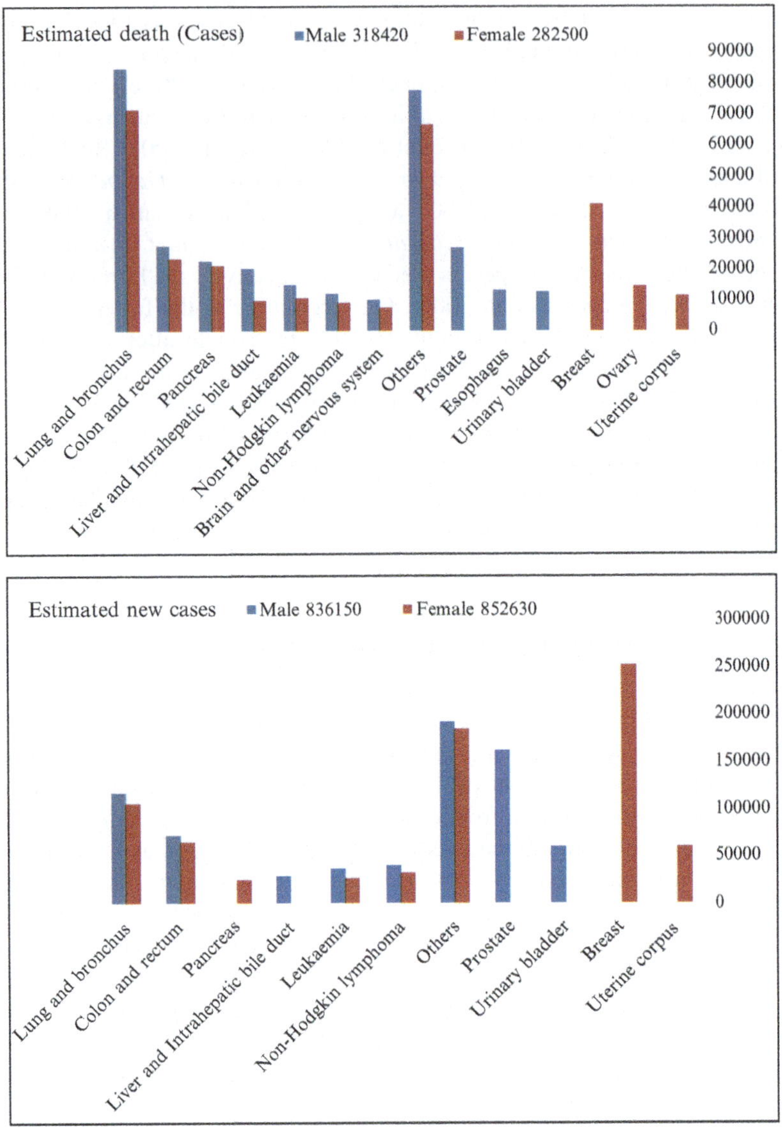

Fig. 12.2 Leading cancer types for estimated deaths and new cases in the United States Siegel et al. 2017

is deemed too much to bear, many cancer patients and families resort to alternative therapy which include Traditional Knowledge (TK), Traditional Chinese Medicine (TCM), Japanese traditional medicine (Kampo), classical Ayurvedic treatment, homeopathy and medicinal practices of other cultures. Word of caution is always given when it comes to alternative therapies. Most alternative therapies have not been through rigorous scientific testing and there is no to low scientific evidence that they would work. Some of the alternative therapies may be completely unsafe and could cause harmful side effects with chronic cases of patients hospitalised after consuming products tainted with toxic materials or heavy metals or herbs when used could interact adversely with modern drugs, causing serious side effects especially in ailing patients with certain medical conditions (UK CR 2016; Health NCFCAI 2016).

12.4 Medicinal Mushrooms

Fungal fruiting bodies, mycelium and culture broth are explored extensively for **anti-cancer** potentials (Wasser and Weis 1999). *Agaricus blazei* (*A. bisporus* and *A. subrufesce*), *Auricularia, Lentinula edodes, Grifola fondosa, Pleurotus eryngii, Lyophyllum shimeji* and *Flammulina velutipes* are among the edible mushrooms identified with anti-cancer properties (Owaid Mustafa et al. 2017). Other mushroom genus included in the literature are *Phellinus,Clitocybe, Antrodia, Cordyceps, Xerocomus, Calvatia, Schizophyllum, Suillus, Inocybe, Funlia, Lactarius, Albatrellus, Russula, Fomes* and *Lignosus* (Patel and Goyal 2012). Within the Basidiomycetes, there are approximately 200 mushroom species which could inhibit the growth of different cancers (Wasser and Weis 1999). Of the many medicinal mushrooms, only about 20 are cultivated in mass (Reddy 2015). *Ganoderma lucidum, Trametes versicolor* and *Inonotus obliquus* are often prepared as powder, brew, or extract, due to their hard texture and bitterness (Wasser 2002). *Lentinula edodes* leads the world production for edible mushrooms by 22% as of 2013, followed by *Pleurotus* (19%), *Auricularia* (18%), *Agaricus* (15%), *Flammulina* (11%), *Volvariella* (5%) and other mushrooms (10%) (Royse et al. 2017). *Ganoderma* (polypores) dominates the world production, with China being the biggest producer and exporter with capacity over 110,000 ton/year (Li et al. 2016).

In Japan, Korea, China and Taiwan, following favourable clinical trial results, *Agaricus subrufescens* (also known as *A. blazei* and *A. brasiliensis*) (Hetland et al. 2011), *Lentinula edodes* (Shiitake) and *Trametes versicolor* were developed and used as anti-cancer agents (Table 12.1) and alternative medicine (Friedman 2016). In Japan, polysaccharides from *T. Versicolor* are taken orally for almost 50 years by surgical patients (gastric, colorectal and small cell lung cancer) with almost no severe negative drug reactions (Katoh and Ooshiro 2007; Lemieszek and Rzeski 2012). In a reported animal study, 6 of 10 mice suffering from sarcoma had complete cancer regression when treated with Lentinan derived from *Lentinula edodes* (Shiitake mushroom) at 0.2 mg kg^{-1} and at slightly higher concentration of 1 mg kg^{-1}, all 10 mice showed complete cancer regression (Chihara et al. 1970). Triterpenes, polysaccharides and immuno-modulatory proteins of *Ganoderma lucidum* inhibit proliferation, growth, invasive behaviour and induce cell death of breast and prostate cancer (Paterson 2006). In addition, the compounds suppress angiogenesis and growth of prostate cancer cells. D-fraction of *G. fondosa* (Maitake) stimulates T-cells, Natural Killer cells (NK) and macrophages, increases interleukin-1

Table 12.1 Anti-cancer therapeutic potential of mushrooms against leading cancer types

Type of cancer	Mushroom used	References
Lung	*Ganoderma*	Sliva (2010), Kosanić et al. (2016), Kimura et al. (2004)
	Phellinus linteus	
	Lactarius deliciosus	
	Macrolepiota procera	
	Lignosus rhinocerus	
	Agaricus blazei	
	Fomes fomentarius	
Colorectal	*Trametes versicolor*	Torisu et al. (1990), Kawaguchi (2009)
	Lentinula edodes	
Prostate	*Trametes versicolor*	Kawaguchi (2009), Luk et al. (2011)
	Amanita phalloides	
Breast	*Trametes versicolor*	Riede (2012), Martin and Brophy (2010)
	Lignosus rhinoceros	
	Grifola frondosa	
Leukaemia	*Agaricus blazei*	Endo et al. (2010), Gao et al. (2007), Wan et al. (2010)
	Trametes versicolor	

(IL1) interleukin-2 (IL-2) and lymphokines production (Geng et al. 2017; Zhou et al. 2005; Kodama et al. 2002, 2003). Mushrooms also protect cells with its antioxidant properties and decrease the inflammatory factor Cyclooxygenase-2(COX2) enzyme which is commonly found in cancer physiology. Studies have also shown that mushrooms have anti-metastatic properties inhibiting the proliferation and spread of cancer. An inverse association of breast cancer risk was found in post-menopausal women consuming mushrooms on a daily and average frequent basis (Hong et al. 2008).

Selective cytotoxicity towards cancer cells without harming normal cells was demonstrated in mushroom properties from *Agaricus blazei* (Fujimiya et al. 1998), *Clitocybe nebularis* (Xu et al. 2011), *Coprinus comatus* (Zhang et al. 2017), *Ganoderma lucidum* (Liu and Zhong 2011) and *Lignosus rhinocerus* (Yap et al. 2015). *L. rhinocerus,* better known as the Tiger Milk Mushroom (TMM) (Fig. 12.3) has only recently been actively studied for its medicinal properties despite being used traditionally with records dating as early as 1664 (Evelyn and De la Bédoyère 2004). TMM is rare in the wild. It was harder to collect for its sclerotia underground, which would have shrunk in size by the time the fruiting bodies were found. The sclerotia are the more precious part of the mushroom as most of the medicinal properties are contained in the sclerotia (Lau et al. 2015; Lee et al. 2009, 2012). The high molecular weight fraction of the sclerotia is active against breast, lung and human cervical cancer cell lines (Lee et al. 2012; Pushparajah et al. 2016).

Psilocybin from various species of 'wild magic mushrooms' could provide an alternative solution to patients with advanced-stage cancer when it comes to managing anxiety and depression (Grob et al. 2011), improving moods and ameliorating

Fig. 12.3 Sclerotia and fruiting body of *Lignosus rhinocerus*, the Tiger Milk Mushroom

psychological distress. The mushrooms were found helping the body to recover thus enabling patients to live an improved quality of life. Over the years, numerous studies were carried out to determine the potential of several mushrooms on how they could ameliorate chemotherapy and radiation therapy associated side effects; loss of appetite, nausea, depression, bone marrow suppression, anaemia, lowered resistance and ability to carry out simple tasks (Patel and Goyal 2012; Zong et al. 2012; Kim et al. 2015).

Wild poisonous mushrooms have no use whatsoever as minute amount consumed would be detrimental and dangerous: damaging the heart, liver and kidneys (Montoya et al. 2003).The responsible toxins from a number of wild mushrooms covering molecular properties, clinical manifestation, mechanisms, potential toxicity were briefly reviewed (Jo et al. 2014). Mushroom collectors avoid these mushrooms by paying special attention to the morphological structures (lamellae), colours, location collected and the season (Montoya et al. 2003; Garibay-Orijel et al. 2006). Amatoxins (cyclopeptides), orellanus (*Cortinarius* species), gyromitrin (monomethyl hydrazine), muscarine, ibotenic acid, psilocybin and coprine were the seven categories of toxins (Vetter 1998; Jo et al. 2014). Amatoxins are heat and enzymatic resistant, could pass through plasma membrane and be absorbed into bloodstream quickly (Luo et al. 2014). Nonetheless, the scaffold of amatoxin could lead to new pharmaceuticals for cancer therapy (Nguyen et al. 1996; Luo et al. 2014; Riede 2017), inhibiting RNA polymerase II, replicating enzyme used extensively in cancer cells but not healthy normal somatic cells. This inhibits specifically cancer cell activity and not the immune response (Riede 2010, 2013).

Despite the practice in Asia and amount of research conducted, mushroom use for cancer treatment is still not widely practised in western medicine (Lemieszek and Rzeski 2012; Reis et al. 2017).

12.5 Mechanisms of Action of Mushrooms with Anti-Cancer Properties

Several mechanisms of action are proposed on how mushrooms could act against cancer cells (Table 12.2). They may act as reactive oxygen species (ROS) inducer, anti-mitotic, mitotic kinase inhibitors, topoisomerase inhibitors, RNA polymerase II inhibitors, apoptosis, angiogenesis, anti-metastatic, anti-cancer proliferators, immune-modulators: increase in T cell numbers; IL-12, Interferon-gamma (IFN-γ), Tumour Necrosis Factor-alpha (TNF-α), NK cell, phagocytes and the amelioration of a skewed Th1/Th2 balance and inflammation (Hetland et al. 2011).

Tremendous amount of work and effort were also dedicated to assess the anti-cancer efficacy of mushrooms by identifying the properties responsible, evaluating their toxicity, mode of action, aspects of host responses; up regulation or down regulation and the cell surface receptors involved (Table 12.3). Ferreira et al. (2010) compiled bioactive molecules of low and high molecular weight with anti-cancer potential. Xu et al. (2011), Singh et al. (2016) and Coulibaly and Youan (2017) compiled mushroom lectins with potential as anticancer diagnostic and therapy. Wasser (2002), Ramberg et al. (2010), Shah et al. (2011), Zong et al. (2012) and Singdevsachan et al. (2016) reviewed and highlighted bioactivities of the polysaccharides of mushrooms as sources of anticancer and immunomodulatory compounds. Nie et al. (2017) covered the specific receptors on the immune cells that could recognize the polysaccharide polymers acting as pattern recognition

Table 12.2 Anti-cancer mechanisms of mushrooms studied

Mechanism	Mushroom	References
ROS inducer	*Antrodia camphorata*	Hseu et al. (2017)
Anti-mitotic, mitotic kinase inhibition topoisomerase inhibition RNA polymerase II inhibition	*Amanita phalloides*	Riede (2012)
Apoptosis	*Agaricus blazei*	Endo et al. (2010)
	Fomes fomentarius	
Angiogenesis inhibitor	*Agaricus blazei*	Takaku et al. (2001)
Anti-metastatic	*Grifola fondosa*	Konno (2007), Nakamura et al. (2015), Kim et al. (2013)
	Cordyceps sinensis	
	Hericium erinaceus	
Immune-modulatory	*Agaricus blazei*	Guggenheim et al. (2014)
Increase in T cell numbers; IL-12, IFN-γ, TNF-α, NK cell, phagocytes	*Lentinula edodes*	
Th1/Th2 balance	*Pleurotus cornucopiae*	Tanaka et al. (2016)
	Agaricus blazei	

Table 12.3 Anticancer assessment and bioactive properties of medicinal mushrooms. Adapted from Ferreira et al. (2010); Mizuno et al. (1999); Coulibaly and Youan (2017); Itoh et al. (2008)

Mushroom	*In vitro* Cell type	*In vivo*	Human clinical	Compound/properties/(Selective cytotoxicity*)	Results	References
Agaricus blazei	Leukaemia	Ovarian cancer	Leukaemia	Low Molecular Weight	Agaritine towards: U937 (IC_{50} 2.7 μg/mL)	Wasser (2002), Li et al. (2016), Fujimiya et al. (1998), Mizuno et al. (1999), Endo et al. (2010), Gao et al. (2007), Takaku et al. (2001), Guggenheim et al. (2014), Itoh et al. (2008), Grube et al. (2001), Mizuno et al. (1998), Cristine da Silva de Souza et al. (2017)
Agaricus isporus	A549	Fibrosarcoma	Gynecological	Agaritine*		
Agaricus subrufescens	HT29	Lung cancer		Selenium	MOLT4 (IC_{50} 9.4 μg/mL)	
	MCF-7	Myeloma		Blazein	HL60 (IC_{50} 13.0 μg/mL)	
		Prostate cancer		Ergosterol	K562 (IC_{50} 16.0 μg/mL)	
		Sarcoma 180		Essential fatty acids	Normal lymphatic no effect >40 μg/mL	
				High Molecular Weight		
				Polysaccharide	MCF-7aro inhibition Blazein towards: HL-60, LU99, KATO III cell apoptosis inhibition of cancer size and neo-angiogenesis	
				β-(1 → 6)-; β-(1 → 3)-glucan, β-(1 → 6)-; α-1 → 4)-glucan, and acidic β-1 → 6)-; α-(1 → 3)-glucan		
				Proteoglucan	Hot water-soluble fraction: Increased Thy1.2-(pan T-cells)	
				Small protein 170 kDa*		
				Xyloglucan	L3 T4-(CD4, helper T-cells)	
				Mannan-protein complex	Lyt2-(CD8, cytotoxic T-cells)	
				Lectin	NK-cell activity	

(continued)

Table 12.3 (continued)

Mushroom	*In vitro* Cell type	*In vivo*	Human clinical	Compound/properties/(Selective cytotoxicity*)	Results	References
Agrocybe aegerita	HeLa SW480 SGC-7901 MGC80–3 BGC-823 HL-60 S-180	S-180 tumor in BALB/c mice	–	Lectin	Apoptosis and DNase activity.	Zhao et al. (2003), Jiang et al. (2012)
					Lectin towards: SW480 (IC_{50} 10 μg/mL)	
					SGC-7901(IC_{50} < 5 μg/mL)	
					HeLa (IC_{50} 10 μg/mL)	
					MGC80–3(IC_{50} 10 μg/mL)	
					BGC-823 (IC_{50} 10 μg/mL)	
					HL-60 (IC_{50} > 100 μg/mL)	
					S-180 (IC_{50} 50 μg/mL)	
Amanita phylloides	MKN45 MCF-7 HCT116 HT29 HeLa	MKN45 cancer-relapse mouse model human pancreatic cancer mouse skin xenograft model	Prostate	α-amanitin	RNA polymerase II inhibitor	Moldenhauer et al. (2012), Kume et al. (2016)
					Increased apoptosis and reduced cell proliferation	
					Restrains cancer relapse	

Antrodia camphorata	SKOV-3 A549 HepG2 urinary cancer, MCF-7 head and neck squamous cell carcinoma	SKOV-3 xenografted nude mice	–	CoQ0	Triggered ROS-mediated apoptosis and cytoprotective autophagy	Chung et al. (2014), Song et al. (2005), Peng et al. (2007), Yang et al. (2011), Lee et al. (2007), Su et al. (2017), Nakamura et al. (2004), Cao et al. (2011)
Antrodia cinnamomea	Nasopharyngeal carcinoma				Inhibit cancer cell growth	
					Adjuvants in radiotherapy and chemotherapy	
				Submerged culture filtrate ethyl acetate extract	Apoptosis	
				Pinicolol B		
				Maleic and succinic acid		
Boletus edulis	HT29	Sarcoma 180 mice	–	Selenium	DNA cytosine methyltransferase inhibitor	Lucas et al. (1957), Bovi et al. (2011)
	HepG2			Lectin BEL		
	MCF-7					
Calvatia gigantea	Sarcoma 180 mammary	Rabbits, guinea pigs, dog, rats, mice, monkeys	–	Calvacin	Anticancer in 14 of 24 lines of cancer	Roland et al. (1960), Eroğlu et al. (2016), Sternberg et al. (1963)
	Adenocarcinoma 755			Methanol extract	A549 (IC_{50} 500 μg/ml for 72 h) induces a wide spectrum of lesions in experimental mammals	
	L-1210 HeLa					
	A549					
Clitocybe clavipes	A431	Mice bearing ascitic A431 tumor	–	Clavilactones CB, CD and CA	Clavilactones: Tyrosine kinase inhibitor towards: A431, GROV-1, SKOV-3	Cassinelli et al. (2000)
	IGROV-1 SKOV-3			Clitocypin, a cysteine protease inhibitor		
Clitocybe nebularis	Mo-T			Lectin CNL*	CD and diacetyl-CA: Weak activity *in vivo*	
					Lectin CNL towards: Mo-T (IC_{50} 100 mg/L)	

(continued)

Table 12.3 (continued)

Mushroom	*In vitro* Cell type	*In vivo*	Human clinical	Compound/properties/(Selective cytotoxicity*)	Results	References
Cordyceps sinensis	B16 mouse melanoma (B16) and Lewis lung carcinoma (LLC)	Mouse model C57BL/6Cr mice Sprague-Dawley rats	–	Water extracts	Metastasis	Nakamura et al. (2015)
	PC-3			5,8-Epidioxy-24(*R*)-methylcholesta-6,22-dien-3β-D-glucopyranoside 5,6-Epoxy-24(*R*)-methylcholesta-7,22-dien-3β-ol		
Coprinus comatus	ES-2		–	Ethyl acetate extract	Apoptotic cells	Zhang et al. (2017), Rouhana-Toubi and Wasser (2015)
				Y3 protein*		
Flammulina velutipes	MCF-7	Lewis lung carcinoma	–	Genistein	Cdc2 kinase modulator	Ikekawa (2001)
	PC-3			Selenium	EA6-PII prolonged life span and post-treatment effective for cancer growth inhibition at 10 mg/kg	
				Protein flammulin		
				EA6 or EA6-P11 polysaccharides		
Fomitella fraxinea	RP9	–	–	Lectin	Mitogenic	Dalloul et al. (2006)
				Fomitellic acids A and B	DNA topoisomerase α and β inhibitors	
Fomes fomentarius	A549	–	–	Polysaccharide 12 kDa	Apoptosis	Kim et al. (2015)

Ganoderma lucidum	Several cancer cell lines	Sarcoma 180 mice	–	Lucidenic acid O	DNA polymerase α, β and RT inhibitors	Patel and Goyal (2012), Paterson (2006), Liu and Zhong (2011), Guggenheim et al. (2014)
					DNA polymerase α inhibitor	
					NF-κB and AP-1 inhibitors	
	MDA-MB-231 breast cancer			Lucidenic lactone	DNA topoisomerase inhibitor	
Ganoderma tsugae				Cerevisterol		
				Lucidumol A and B		
				Ganoderiol F		
				Ganodermanondiol		
				GanodermanontriolGanoderic acids A, F, H, W, X, Y, T, Mf, S		
				Polysaccharides Glycan-Glucan protein complex Bis-β-carboxyethylgermanium seqquioxide		
				Spore oil		
Gerronema	HL60		–	Gerronemins A-F	COX-2 enzyme inhibitors	Silberborth et al. (2002)
	U937					
	L1210					
	COS-7					
	HeLA					
Grifola frondosa	MCF-7	Oral and subcutaneous administered to Sarcoma-180 bearing mice	–	Ergosterol 1-Oleoyl-2-linoleoyl-3-palmitoylglycerol	Cyclooxygenase inhibitor	Martin and Brophy (2010)
				Ergosta-4,6,8,22-tetraen-3-one Polysaccharides β-glucans		
				β-(1 → 3)-; β-(1 → 6)-glucan, acidic β-D-glucan, β-(1 → 6)-; β-(1 → 3)-glucanAcidic xyloglucan		

(continued)

Table 12.3 (continued)

Mushroom	*In vitro* Cell type	*In vivo*	Human clinical	Compound/properties/(Selective cytotoxicity*)	Results	References
Hericium erinaceus	Huh-7	SCID mice	–	Galactoxyloglucan-protein complex	*H. erinaceus*extract: HepG2 (IC_{50} 2.50 mg/mL) Huh-7 (IC_{50} 1.50 mg/mL) HT-29 (IC_{50} 1.25 mg/mL) NCI-87 (IC_{50} 5.00 mg/ml)	Li et al. (2010), Kim et al. (2013), Li et al. (2014)
	HepG2					
Hericium erinaceum	HT-29			Glucoxylan-protein complex		
	NCI-87			Lectin		
Hericium caput-medusae					*H. erinaceum*extract: HepG2 (IC_{50} 56.1 μM)	
					MCF7 (IC_{50}76.5 μM)	
					No toxicity towards animals	
Hypsizygus marmoreus	–	Mice	–		21/36 control developed cancer	Ikekawa (2001)
					3/36 treated developed cancer	
Inonotus obliquus	–	Lewis lung carcinoma	–	Aqueous extract	Cancer agglomeration and inhibition of vascularization	Arata et al. (2016)
Lactarius deliciosus	HeLA	—		Methanol extract	HeLa (IC_{50} 19.01) A549 cells (IC_{50} 33.05) LS174 (IC_{50} 74.01)	Kosanić et al. (2016)
	A549					
	LS174					

Lentinula edodes (shiitake mushroom)	Human colon	Sarcoma 180 and various synergic and autochthonous cancersin mice	Gastric	Lentinan	272 advanced gastric cancer patients; prolong life, cancer regression, improvedimmune responses without toxic side effects.	Chihara et al. (1970), Hamuro and Chihara (1985), Ghoneum et al. (1995), Wasser and Weis (1999), Uno et al. (2000), Ikekawa (2001), Won (2002), Ng and Yap (2002), Cowawintaweewat et al. (2006), Kawaguchi (2009), Turner and Chaudhary (2009)
			Colorectal Breast	AHCC		
			Liver			
			Prostate			
			Bone marrow depression		80% reduction in cancer size, complete regression in animals	
					Increased IL-12, ITF-gamma, NK	
					Doubled binding capacity of NK cells. Reduced recurrence, increased survival rate after surgery in advanced cancer patients	
					Prevent bone marrow depression	
					Improved serologic response	
Lepiota Americana	Leukaemia	–	–	2-aminophenoxazin-3-one	Aromatase inhibitor	Didukh and Mahajna (2005)
				5,8-Epidioxy-24(*R*)-methylcholesta-6,22-dien-3β-ol	Sulfatase inhibitor	

(continued)

Table 12.3 (continued)

Mushroom	*In vitro* Cell type	*In vivo*	Human clinical	Compound/properties/(Selective cytotoxicity*)	Results	References
Lignosus rhinoceros	MCF7	Sprague Dawley (SD) rats	–	Serine proteases*, FIP, cold water extract	Cold water extract towards: MCF-7 (IC_{50} 96.7 μg/mL)	Lee et al. (2012), Hj Che Fauzi et al. (2013), Yap et al. (2015), Kong et al. (2016), Tan et al. (2016), Pushparajah et al. (2016), Lee et al. (2017)
	A549					
Lignosus tigris	HeLa				A549 (IC_{50} 466.7 μg/mL)	
Lignosus cameronensis	HCT 116				HCT116 (IC_{50} 600 μg/mL)	
					High molecular weight fraction of cold water extract towards: MCF-7 (IC_{50} 70.0 μg/mL)	
					A549 (IC_{50}76.7 μg/mL)	
					Cytotoxic serine protease fraction towards MCF-7 (IC_{50} 3.0 μg/mL)	
					No adverse effect found in test animals	
					FIP towards: MCF-7 (IC_{50} 0.34 μM) HeLa (IC_{50} 0.58 μM) A549 (IC_{50} 0.60 μM)	
Macrolepiota procera	HeLa	–	–	Methanol extract	HeLa (IC_{50} 29.39 μg/mL)	Kosanić et al. (2016)
	A549				A549 (IC_{50} 25.55 μg/mL)	
	LS174				LS174 (IC_{50} 68.49 μg/mL)	

Phellinus linteus	Macrophage	–	–	Hispolon	Increased production of NO, activation of PTK and PKC	Sliva (2010), Huang et al. (2011)
	Melanoma cancer			Phellifuropyranone A		
	Colon cancer			Meshimakobnol A and B	Inhibition of cell adhesion and invasion, inhibition of metastasis in mice	
	Sarcoma				Inhibition of proliferation and colony formation, cell cycle arrest at G2/M, decrease in cyclin B1; induction of apoptosis, Inhibition of cancer growth in mice	
	Splenocytes					
	NK-cells					
	Epidermoid					
	Breast cancer				Inhibition of cancer growth and pulmonary metastasis in mice	
	Bladder cancer				Induction of proliferation of β-cells	
	Lung cancer				Induction of production of IL-6, TNF-α, NO	
					Hispolon towards: Hep3B (IC_{50}35.91 μM)	
					J5 (IC_{50}54.51 μM)	
					HepG2 (IC_{50}87.61 μM)	
Pholiota spumosa	HepG2	–	–	Putrescin-1,4-dicinnamide Lectin (16kda)	Inducer of apoptosis and necrosis Lectin towards: HepG2 (IC_{50} 2.1 μM)	Zhang et al. (2009)
Pholiota adiposa	MCF7				MCF7 (IC_{50} 3.2 μM)	

(continued)

Table 12.3 (continued)

Mushroom	*In vitro* Cell type	*In vivo*	Human clinical	Compound/properties/(Selective cytotoxicity*)	Results	References
Piptoporus betulinus	Human colon adenocarcinoma human lung carcinoma and human breast cancer cell lines	–	–	(E)-2-(4-hydroxy-3-methyl-2-butenyl)-hydroquinone	MMPs inhibitor	Pleszczyńska et al. (2016)
				Polyporenic acid C		
Poria cocos	Human pancreatic cancer cell lines	–	–	Triterpenes	Anticancer and anti-invasive effects	Cheng et al. (2013)
Ramaria flava	HepG2	–	–	Aqueous, ethanol extract	MDA-MB-231(IC_{50}66.54 μg/mL)	Sadi et al. (2016), Liu et al. (2013)
	MDA-MB-231					
Russula lepida	HepG2	Sarcoma180	–		*Russula* extract towards	Zhang et al. (2010), Zhao et al. (2010)
Russula delica	MCF-7				HepG2 (IC_{50} 1.6 μM)	
					MCF7 (IC_{50} 0.9 μM)	
					67.6% reduction in the weight of cancer	
Schizophyllum commune	No direct growth inhibitory effect on cancer cell lines	Sarcoma 37, Erlich sarcoma, Yoshida sarcoma and Lewis lung	Recurrent and inoperable gastric cancer, stage 2 cervical cancer and advanced cervical carcinoma	Schizophyllan	Increased cellular immunity by restoring suppressed killer cell activity to normal levels.	Borchers et al. (1999)
					Restored mitosis of bone marrow cells suppressed by anticancer drugs and radiation.	

Trametes versicolor	Macrophages	Adenosarcoma, fibrosarcoma, mastocytoma, plasmacytoma, melanoma, sarcoma, carcinoma, mammary, colon, and lung	Breast cancer	PSK (krestin)	Orally safe at 3,6,9 g/day. Increased lymphocyte counts at 6 and 9 grams/day; (2) increased natural killer cell functional activity at 6 grams/day; (3) dose-related increases in CD8+ T cells and CD19+ B cells, but not CD4+ T cells or CD16 + 56+ NK cells.	Torisu et al. (1990), Zhang et al. (2003), Hattori et al. (2004), Katoh and Ooshiro (2007), Kinoshita et al. (2010), Wan et al. (2010), Torkelson et al. (2012)
					Acts directly on cancer cells and boost cellular immunity.	
					Effective against several cancers, prevent metastasis, but not alone.	
	Human Burkitt lymphoma		Colorectal cancer		Enhanced efficacy of anticancer drugs.	
	Human gastric					
	Human pancreatic cancer cells NOR-P1 HL-60					
Termitomyces clypeatus	U937	–	–	Extract	U937 (IC_{50} 25 ± 1.02 μg/mL)	Mondal et al. (2016)
					Reduced cancer volume, viable cancer cell count and improved haemoglobin content, RBC count, mean survival time, cancer inhibition and % increase life span.	

(continued)

Table 12.3 (continued)

Mushroom	*In vitro* Cell type	*In vivo*	Human clinical	Compound/properties/(Selective cytotoxicity*)	Results	References
Tricholoma lobayense		Animal models	–	Glycoprotein	Immunomodulatory	Ferreira et al. (2010)
Xylaria strain 45–93	HeLa	–	–	Cytotoxic Chaxine C	Hela (IC_{50} 2.24 µg/mL)	Liu et al. (2006), McCloskey et al. (2017)
	HT29			Lectin (28.8 kDa	HT29 (IC_{50} 2.51 µg/mL)	
Xylaria hypoxylon	HCT116				HCT116 (IC_{50} 3.50 µg/mL)	
	MCF7				MCF-7 (IC_{50} 3.77 µg/mL)	
	Vero				Vero (IC_{50}3.65 µg/mL)	
	M1				M1(IC_{50} 1.24 µM)	
	HepG2				HepG2 ((IC_{50} 0.74 µM)	

molecules (Markova et al. 2003; Bhardwaj et al. 2014). Several receptors were identified; Toll-like receptors (TLRs), mannose receptors, Dectin-1, Dectin-2, scavenger receptors (SR), and Type 3 Complement Receptors (CR3). Binding of the polysaccharides to these receptors activates the expression of pro-inflammatory cytokines and nitric oxide (Nie et al. 2017).

Low molecular weight triterpenoids such as Ganoderic A and Ganoderic B from medicinal mushroom *Ganoderma lucidum* have anticancer activities among other pharmacological activities (You et al. 2017; Shiao 2003). Till date, more than 130 triterpenoids with different structural features were identified from the fruiting bodies, cultured mycelium and spores of *G. lucidum* (Shiao 2003; Paterson 2006). Cordycepin isolated from *Ophiocordycep sinensis* has been found to exert antiangiogenic, antimetastatic and antiproliferative effects as well as inducing apoptosisin cancer cells (Hwang et al. 2017; Wang et al. 2017). Hispolon is aphenolic compound of *Phellinus igniarius*active against gastric cancer cells (Chen et al. 2008; Huang et al. 2011). Essential linoleic acid and conjugated linoleic acid in *A. Bisporus* decreased breast cancer cell proliferation but had no effect on noncancerous cell line. The effect was also shown in nude mice (Chen et al. 2006).

High molecular weight proteins with anticancer activities can be divided into two groups; proteins with direct anti-proliferative activities towards cancer cells and immunomodulatory proteins (Ivanova et al. 2014). Lectins can possess both mechanisms. Lectin CNL from *Clitocybe nebularis* in additional, is selectively cytotoxic towards human leukemic T cells (Pohleven et al. 2009). Over 27 mushroom lectins have shown anti-proliferative activities towards a range of cancer cells and 5 of these demonstrated 67–92% anti-proliferation activities in Sarcoma-180 cancer bearing animals (Singh et al. 2016). Another protein group with potent selective cytotoxicity reported recently is serine proteases of *L. rhinocerus* (Yap et al. 2015). The partially purified protein fraction, containing two serine proteases, demonstrated cytotoxicity activity towards MCF-7 breast cancer cells with IC_{50} 3.00 ± 1.01 μg/mL. The IC_{50} value on 184B5 cells were at 7.60 μg/mL. The proteolytic and cytotoxicity activities of the fraction were inhibited in the presence of phenylmethylsulfonyl fluoride, PMSF. A fungal immunomodulatory protein (FIP) from *L. rhinocerus* with highest homology (64%) to FIP-glu (LZ-8) from *Ganoderma lucidum* has also demonstrated anti-proliferative effect on HeLa, MCF-7 and A549 cancer cells (Pushparajah et al. 2016).

Most of the high molecular weight polysaccharides belong to the group of β-glucans and are chemically different in composition (Mizuno et al. 1999). The β-(1 → 3) linkages and 1 β-(1 → 6) branch points of the glucan are required for most anticancer activities. The β-glucans with mainly (1 → 6) linkages have lesser activity. Higher molecular weight glucans exert stronger pharmacological activities compared to lower molecular weight glucans (Mizuno et al. 1999). Different mushroom species, even strains within the same species could harbour different sets of polysaccharides with different chemical structures, molecular weight, branching rate and forms that could define their bioactivities (Poucheret et al. 2006; Lemieszek and Rzeski 2012; Wasser 2002).

Polysaccharides do not attack cancer cells directly, but activate different immune responses in the host to act against cancer cells (Wasser 2002). The anticancer mechanism requires an intact T-cell component with activity mediated through a thymus-dependent immune system (Borchers et al. 1999). The immune frontiers; the NK cells, T-cells, B-cells, and macrophages, can be stimulated using mushroom polysaccharides (Wasser and Weis 1999) providing alternatives for cancer patients to use mushroom polysaccharides as co-treatments with chemotherapy or post treatment.

Polysaccharide Lentinan from *Lentinula edodes*, polysaccharide Krestin (PSK) and polysaccharopeptide (PSP) derived from *Trametes versicolor* and *Schizophyllan* (SPG) from *Schizophyllum commune* demonstrated anticancer applications (Lemieszek and Rzeski 2012; Yamaguchi 2016; Daba and Ezeronye 2003).

In Japan, gastric cancer patients are treated with Lentinan. Clinical studies showed Lentinan when used together with chemotherapeutic agents could prolong the lifespan of patients with advanced gastric cancer and lowered recurrence (Yang et al. 2008). Lentinan restores T-cells and humoral immune responses (Ooi and Liu 2000) and prevents metastasis in animal models. *In vivo* studies showed it could enhance the cytotoxicity of anticancer drug with active ingredient trastuzumab. Complete cancer regression was observed when Sacrcoma 180 implanted mice were treated with lentinan-alpha at 1 mg kg^{-1} (Chihara et al. 1970).

Polysaccharide PSK (from Japan) and PSP (from China) were both developed as anticancer agents especially for colorectal and breast cancer treatment after extensive studies with both demonstrated enhancement of the cytotoxicity of chemotherapeutic drugs through *in vivo* studies (Katoh and Ooshiro 2007; Kinoshita et al. 2010) and significantly suppressed cancer growth through immunostimulatory activity, stop growth of various cancer cells through cell cycle arrest and apoptosis. PSP combined with chemotherapeutic agents lowered chemo-treatment side effects and increased chances and quality of survival and life in patients from clinical studies. PSP and PSK activities increased the production of interleukin-2 (IL-2), interleukin-6 (IL-6), interferon (IFN) on the humoral side, and T-cells proliferation on the cellular side (Cui and Chisti 2003; Ng 1998).

Schizophyllan (SPG) (Borchers et al. 1999; Spelman et al. 2017) derived from *Schizophyllum commune* Fr (Schizophyllaceae) is used clinically to treat cervical cancer patients in Japan. SPG stimulates immune system and activates NK cells, spleen cells, lymphoid cells as well as bone marrow cells and enhances the production of cytokines IL 1, 2 and 3. Clinical studies showed SPG prolonged life of patients with stage II cervical cancer when used in combination with treatment drugs.

Ganoderma mushrooms contain high amount of polysaccharides consisting of β-(1 → 3)-D-glucans with β-(1 → 6)-D-glucopyranosyl branches which displayed significant immunostimulatory properties (Borchers et al. 1999).The structural and functional similarities however translate differently in their effectiveness against specific cancers and in their abilities to stimulate different cellular responses; interferons (IFNs), interleukins (ILs) and others.

AHCC (Active Hexose Correlated Compound) is a low molecular weight approximately 5000 daltons mushroom-derived-alpha-glucan available in several species of mushrooms. AHCC from *Lentinula edodes* is widely used in Japan, after *A. blazei* supplements as alternative and complementary treatment of cancer due to its immune-enhancing functions (AHCC 2009; Shah et al. 2011). AHCC increased IL-12, ITF-γ and NKcell activities and doubled the binding capacity of NK cells to cancer targets in patients (Ghoneum et al. 1995; Uno et al. 2000). In advanced liver cancer patients, AHCC reduced recurrence and incidence of death post-surger (Uno et al. 2000; Cowawintaweewat et al. 2006).A castration-resistant prostate cancer patient showed serologic response improvement after treatment with AHCC (Turner and Chaudhary 2009). Bone marrow depression from chemotherapy could also be prevented using AHCC in a study on 12 different cancer patients (Won 2002).

The total amount of polysaccharides is higher in fruiting bodies compared to mycelia (Wasser 2002).The bioactivities exerted could be of different magnitude and nature. Of the 29 polysaccharide fractions obtained from *G. frondosa* fruiting bodies, 69% exhibited different levels of anticancer activity (Mizuno et al. 1986) while 86% of 28 polysaccharide fractions obtained from culture mycelium of the mushroom possessed anticancer activity (Zhang et al. 1994).

12.6 Concluding Remarks: The Way Forward and Challenges

Excluding the species from most of south-eastern Europe, Africa, western Asia or tropical eastern Asia, an estimated figure of 56, 679 species of macro fungi is proposed (Mueller et al. 2007). Of the 12,000 described wild species of mushrooms worldwide, only a couple hundreds of them are used for medicinal purposes including for use in cancer (Beulah et al. 2013; Rathore et al. 2017).

The exciting race is on for modern science to validate claims, safety for use, build new information and explore the unknown potential realms of medicinal macro fungi for cancer treatment. Using modern analytical instruments, assay systems, reliable statistical methods and elaborative collaborations worldwide, safety and efficacy data can be established of the potential drugs derived from medicinal macro fungi. Rare cancer patients would benefit as well having properties from mushroom as treatment options as current treatment information is hard to find.

Abbreviations

AHCC Active Hexose Correlated Compound
CNL *Clitocybe nebularis* lectin
COX2 Cyclooxygenase-2
CR Complement Receptors
FIP Fungal Immunomodulatory Protein
GLU *Ganoderma lucidum*

IFN	Interferon
IL	Interleukin
NK	Natural Killer
PSK	Polysaccharide Krestin
PSP	Polysaccharopeptide
RNA	Ribonucleic acid
ROS	Reactive Oxygen Species
SPG	Schizophyllan
TCM	Traditional Chinese Medicine
TK	Traditional Knowledge
TMM	Tiger Milk Mushroom
TNF	Tumour Necrosis Factor

Cell Lines

A431	Human epidermoid carcinoma
A549	Human lung adenocarcinoma
B16	Mouse melanoma cell line
BGC-823	Human gastric carcinoma cells
COS-7	African green monkey kidney fibroblast-like cell line
ES-2	Ovarian carcinoma
IGROV-1	Human ovarian carcinoma
HCT116	Human colon carcinoma cell line
HeLa	Human cervical cancer cell line
HepG2	Human hepatocellular carcinoma cell line
HL-60	Human leukaemia acute promyelocytic human leukaemia
HT29	Human colon colorectal adenocarcinoma cell line
Huh-7	Human hepatic cell model
L-1210	Murine lymphocytic leukaemia cell line
LLC	Lewis lung carcinoma
LS174	Colon carcinoma
M1	Leukaemia cell line
MCF-7	Human mammary adenocarcinoma cell line
MDA-MB-231	Human breast adenocarcinoma
MGC80–3	Human gastric adenocarcinoma cell line
MKN45	Gastric carcinoma
Mo-T	Human leukaemia T-cell
NCI-N87	Human gastric cancer cells
PC-3	Human prostate grade IV, adenocarcinoma
RP9	Avian cancer cell line
S-180	Sarcoma Cancer associated with hypercalcemia
SGC-7901	Gastric carcinoma cell line
SKOV-3	Human ovarian adenocarcinoma

A431	Human epidermoid carcinoma
SW480	Human colorectal cancer, epithelial
U937	Human lung lymphoblast
Vero	African green monkey kidney cell line

References

AHCC (2009) [Internet]. Wolters Kluwer Health. [cited 12.04.2018]. Available from: https://www.drugs.com/npp/ahcc.html
Arata S, Watanabe J, Maeda M, Yamamoto M, Matsuhashi H, Mochizuki M (2016) Continuous intake of the Chaga mushroom (*Inonotus obliquus*) aqueous extract suppresses cancer progression and maintains body temperature in mice. Heliyon 2(5):e00111
Barbarathiers (2013) What are macrofungi?
Beulah GH, Margret AA, Nelson J (2013). Marvelous medicinal mushrooms. https://www.semanticscholar.org/paper/Marvelous-Medicinal-Mushrooms-Beulah-Margret/01cf71c8a40786def4b2cd876485477404f5b7d2
Borchers AT, Stern JS, Hackman RM, Keen CL, Gershwin ME (1999) Mushrooms, tumors, and immunity. Proc Soc Exp Biol Med 221(4):281–293
Bovi M, Carrizo ME, Capaldi S, Perduca M, Chiarelli LR, Galliano M (2011) Structure of a lectin with antitumoral properties in king bolete (*Boletus edulis*) mushrooms. Glycobiology 21(8):1000–1009
Bhardwaj N, Katyal P, Sharma KA (2014) Suppression of inflammatory and allergic responses by pharmacologically potent fungus *Ganoderma lucidum*. Recent Pat Inflamm Allergy Drug Discov 8(2):104–117
Cao SM, Simons MJ, Qian CN (2011) The prevalence and prevention of nasopharyngeal carcinoma in China. Chin J Cancer 30(2):114–119
Cassinelli G, Lanzi C, Pensa T, Gambetta RA, Nasini G, Cuccuru G (2000) Clavilactones, a novel class of tyrosine kinase inhibitors of fungal origin. Biochem Pharmacol 59(12):1539–1547
Chen S, Oh S-R, Phung S, Hur G, Ye JJ, Kwok SL, Shrode GE, Belury M, Adams LS, Williams D (2006) Anti-aromatase activity of phytochemicals in white button mushrooms (*Agaricus bisporus*). Cancer Res 66(24):12026–12034
Chen W, Zhao Z, Li L, Wu B, Chen S-F, Zhou H, Wang Y, Li YQ (2008) Hispolon induces apoptosis in human gastric cancer cells through a ROS-mediated mitochondrial pathway. Free Radic Biol Med 45(1):60–72
Cheng S, Eliaz I, Lin J, Sliva D (2013) Triterpenes from *Poria cocos* suppress growth and invasiveness of pancreatic cancer cells through the downregulation of MMP-7. Int J Oncol 42(6):1869–1874
Chihara G, Hamuro J, Maeda YY, Arai Y, Fukuoka F (1970) Fractionation and purification of the polysaccharides with marked antitumor activity, especially lentinan, from *Lentinus edodes* (Berk.) Sing. (an edible mushroom). Cancer Res 30(11):2776–2781
Chung CH, Yeh SC, Chen CJ, Lee KT (2014) Coenzyme Q0 from *Antrodia cinnamomea* in submerged cultures induces reactive oxygen species-mediated apoptosis in A549 human lung cancer cells. Evid Based Complement Alternat Med 2014:10
CNN (2016) The 'forbidden fruit' of medicinal mushrooms [press release]
Cohen JD, Li L, Wang Y, Thoburn C, Afsari B, Danilova L et al (2018) Detection and localization of surgically resectable cancers with a multi-analyte blood test. Science 359:926–930. https://doi.org/10.1126/science.aar3247
Coulibaly FS, Youan BBC (2017) Current status of lectin-based cancer diagnosis and therapy. AIMS Mol Sci 4:1–27

Cowawintaweewat S, Manoromana S, Sriplung H, Khuhaprema T, Tongtawe P, Tapchaisri P, Chaicumpa W (2006) Prognostic improvement of patients with advanced liver cancer after active hexose correlated compound (AHCC) treatment. Asian Pac J Allergy Immunol 24(1):33–45
Cristine da Silva de Souza AC, Correa VG, de Goncalves GA, Soares AA, Bracht A, Peralta RM (2017) *Agaricus blazei* bioactive compounds and their effects on human health: benefits and controversies. Curr Pharm Des 23(19):2807–2834
Cui J, Chisti Y (2003) Polysaccharopeptides of *Coriolus versicolor*: physiological activity, uses, and production. Biotechnology advances, 21(2):109–122
Daba A, Ezeronye O (2003) Anti-cancer effect of polysaccharides isolated from higher basidiomycetes mushrooms. Afr J Biotechnol 2(12):672–678
Dalloul R, Lillehoj H, Lee J, Lee S, Chung K (2006) Immuno potentiating effect of a *Fomitella fraxinea*-derived lectin on chicken immunity and resistance to coccidiosis. Poult Sci 85(3):446–451
Dias ES, Abe C, Schwan RF (2004) Truths and myths about the mushroom *Agaricus blazei*. Sci Agric 61(5):545–549
Didukh MY, Mahajna JA (2005) Screening antitumor activity of low-molecular-weight compounds obtained from the fruit bodies of family *Agaricaceae Chevall.* (higher basidiomycetes). Int J Med Mushrooms 7(3):398–400
Endo M, Beppu H, Akiyama H, Wakamatsu K, Ito S, Kawamoto Y et al (2010) Agaritine purified from *Agaricus blazei* Murrill exerts anti-tumor activity against leukemic cells. Biochim Biophys Acta 1800(7):669–673
Eroğlu C, Seçme M, Atmaca P, Kaygusuz O, Gezer K, Bağcı G (2016) Extract of *Calvatia gigantea* inhibits proliferation of A549 human lung cancer cells. Cytotechnology 68(5):2075–2081
Evelyn J, De la Bédoyère G (2004) The diary of John Evelyn. Boydell Press, Woodbridge
Ferreira CFRI, Vaz JA, Vasconcelos MH, Martins A (2010) Compounds from wild mushrooms with antitumor potential. Anti Cancer Agents Med Chem 10(5):424–436
Friedman M (2016) Mushroom polysaccharides: chemistry and antiobesity, antidiabetes, anticancer, and antibiotic properties in cells, rodents, and humans. Foods 5(4):80
Fujimiya Y, Suzuki Y, Oshiman K-i, Kobori H, Moriguchi K, Nakashima H, Matumoto Y, Takahara S, Ebina T, Katakura R (1998) Selective tumoricidal effect of soluble proteoglucan extracted from the basidiomycete, *Agaricus blazei* Murill, mediated via natural killer cell activation and apoptosis. Cancer Immunol Immunother 46(3):147–159
Gao L, Sun Y, Chen C, Xi Y, Wang J, Wang Z (2007) Primary mechanism of apoptosis induction in a leukemia cell line by fraction FA-2-b-ss prepared from the mushroom *Agaricus blazei* Murill. Braz J Med Biol Res 40(11):1545–1555
Garibay-Orijel R, Cifuentes J, Estrada-Torres A, Caballero J (2006) People using macro-fungal diversity in Oaxaca, Mexico. Fungal Divers 21:41–67
Geng P, Siu KC, Wang Z, Wu JY (2017) Antifatigue functions and mechanisms of edible and medicinal mushrooms. BioMed Res Int 2017, 9648496. https://doi.org/10.1155/2017/9648496
Ghoneum M, Wimbley M, Salem F, McKlain A, Attallah N, Gill G (1995) Immunomodulatory and anticancer effects of active hemicellulose compound (AHCC). Int J Immunother 11(1):23–28
Grob CS, Danforth AL, Chopra GS, Hagerty M, McKay CR, Halberstadt AL, Greer GR (2011) Pilot study of psilocybin treatment for anxiety in patients with advanced-stage cancer. Arch Gen Psychiatry 68(1):71–78
Grube BJ, Eng ET, Kao Y-C, Kwon A, Chen S (2001) White button mushroom phytochemicals inhibit aromatase activity and breast cancer cell proliferation. J Nutr 131(12):3288–3293
Guggenheim AG, Wright KM, Zwickey HL (2014) Immune modulation from five major mushrooms: application to integrative oncology. J Integr Med 13(1):32
Guzmán G (2008) Hallucinogenic mushrooms in Mexico: an overview. Econ Bot 62(3):404–412
Hamuro J, Chihara G (1985) Immunopotentiation by the antitumor polysaccharide lentinan Pharmacology: Springer Boston MA 285–307
Hattori TS, Komatsu N, Shichijo S, Itoh K (2004) Protein-bound polysaccharide K induced apoptosis of the human Burkitt lymphoma cell line, Namalwa. Biomed Pharmacother 58(4):226–230

Health Ncfcai (2016) Complementary, alternative, or integrative health: what's in a name? https://nccih.nih.gov/health/integrative-health

Hetland G, Johnson E, Lyberg T, Kvalheim G (2011) The mushroom *Agaricus blazei* Murill elicits medicinal effects on tumor, infection, allergy and inflammation through its modulation of innate immunity and amelioration of Th1/Th2 imbalance and inflammation. Adv Pharmacol Sci 2011:157015

Hj Che Fauzi S, Y Farida Zuraina M, Mohd Nawi N, Leong L, Nurshahirah N, Florinsiah L, Rajab NF (2013) Antiproliferative effect of *Lignosus rhinocerotis*, the Tiger Milk mushroom on HCT 116 human colorectal cancer cells. Open Conf Proc J 4:65–70

Hong SA, Kim K, Nam SJ, Kong G, Kim MK (2008) A case–control study on the dietary intake of mushrooms and breast cancer risk among Korean women. Int J Cancer 122(4):919–923

Hseu Y-C, Tsai T-J, Korivi M, Liu J-Y, Chen H-J, Lin C-M (2017) Antitumor properties of Coenzyme Q(0) against human ovarian carcinoma cells via induction of ROS-mediated apoptosis and cytoprotective autophagy. Sci Rep 7:8062

Huang G-J, Deng J-S, Huang S-S, Hu M-L (2011) Hispolon induces apoptosis and cell cycle arrest of human hepatocellular carcinoma Hep3B cells by modulating ERK phosphorylation. J Agric Food Chem 59(13):7104–7113

Hwang JH, Park SJ, Ko WG, Kang S-M, Lee DB, Bang J et al (2017) Cordycepin induces human lung cancer cell apoptosis by inhibiting nitric oxide mediated ERK/slug signaling pathway. Am J Cancer Res 7(3):417–432

Ikekawa T (2001) Beneficial effects of edible and medicinal mushrooms on health care. Int J Med Mushrooms 3(4)

Itoh H, Ito H, Hibasami H (2008) Blazein of a new steroid isolated from *Agaricus blazei* Murrill (himematsutake) induces cell death and morphological change indicative of apoptotic chromatin condensation in human lung cancer LU99 and stomach cancer KATO III cells. Oncol Rep 20(6):1359–1361

Ivanova T, Krupodorova T, Barshteyn V, Artamonova A, Shlyakhovenko V (2014) Anticancer substances of mushroom origin. Exp Oncol 36:58–66

Jiang S, Chen Y, Wang M, Yin Y, Pan Y, Gu B (2012) A novel lectin from *Agrocybe aegerita* shows high binding selectivity for terminal N-acetylglucosamine. Biochem J 443(2):369–378

Jo WS, Hossain MA, Park SC (2014) Toxicological profiles of poisonous, edible, and medicinal mushrooms. Mycobiology 42(3):215–220

Katoh R, Ooshiro M (2007) Enhancement of antitumor effect of tegafur/uracil (UFT) plus leucovorin by combined treatment with protein-bound polysaccharide, PSK, in mouse models. Cell Mol Immunol 4:295–299

Kawaguchi Y (2009) Improved survival of patients with gastric cancer or colon cancer when treated with active hexose correlated compound (AHCC): effect of AHCC on digestive system cancer. Nat Med 1(1):1–6

Kim SP, Nam SH, Friedman M (2013) *Hericium erinaceus* (Lion's Mane) mushroom extracts inhibit metastasis of cancer cells to the lung in CT-26 colon cancer-tansplanted mice. J Agric Food Chem 61(20):4898–4904

Kim SH, Jakhar R, Kang SC (2015) Apoptotic properties of polysaccharide isolated from fruiting bodies of medicinal mushroom *Fomes fomentarius* in human lung carcinoma cell line. Saudi J Biol Sci 22(4):484–490

Kimura Y, Kido T, Takaku T, Sumiyoshi M, Baba K (2004) Isolation of an anti-angiogenic substance from *Agaricus blazei Murill*: its antitumor and antimetastatic actions. Cancer Sci 95(9):758–764

Kinoshita J, Fushida S, Harada S, Makino I, Nakamura K, Oyama K et al (2010) PSK enhances the efficacy of docetaxel in human gastric cancer cells through inhibition of nuclear factor-κB activation and survivin expression. Int J Oncol 36(3):593–600

Knapton S (2008) Truffles facts. The telegraph

Kodama N, Komuta K, Nanba H (2002) Can mistake MD-fraction aid cancer patients. Altern Med Rev 7(3):236–239

Kodama N, Komuta K, Nanba H (2003) Effect of Maitake (*Grifola frondosa*) D-fraction on the activation of NK cells in cancer patients. J Med Food 6(4):371–377

Kong BH, Tan NH, Fung SY, Pailoor J (2016) Sub-acute toxicity study of Tiger Milk mushroom *Lignosus tigris* Chon S. Tlan cultivar E sclerotium in Sprague Dawley rats. Front Pharmacol 7:246

Konno S (2007) Effect of various natural products on growth of bladder cancer cells: two promising mushroom extracts. Altern Med Rev 12(1):63–68

Kosanić M, Ranković B, Rančić A, Stanojković T (2016) Evaluation of metal concentration and antioxidant, antimicrobial, and anticancer potentials of two edible mushrooms *Lactarius deliciosus* and *Macrolepiota procera*. J Food Drug Anal 24(3):477–484

Kume K, Ikeda M, Miura S, Ito K, Sato KA, Ohmori Y (2016) α-amanitin restrains cancer relapse from drug-tolerant cell subpopulations via TAF15. Sci Rep 6:25895

Lau BF, Abdullah N, Aminudin N, Lee HB, Tan PJ (2015) Ethnomedicinal uses, pharmacological activities, and cultivation of *Lignosus* spp. (Tiger Milk mushrooms) in Malaysia–a review. J Ethnopharmacol 169:441–458

Lee TH, Lee CK, Tsou WL, Liu SY, Kuo MT, Wen WC (2007) A new cytotoxic agent from solid-state fermented mycelium of *Antrodia camphorata*. Planta Med 73(13):1412–1415

Lee SS, Chang YS, Noraswati MNR (2009) Utilization of macrofungi by some indigenous communities for food and medicine in Peninsular Malaysia. For Ecol Manag 257(10):2062–2065

Lee ML, Tan NH, Fung SY, Tan CS, Ng ST (2012) The Antiproliferative activity of *Sclerotia* of *Lignosus rhinocerus* (Tiger Milk mushroom). Evid Based Complement Alternat Med 2012:697603 http://dx.doi.org/10.1155/2012/697603

Lee SS, Tan NH, Pailoor J, Fung SY (2017) Safety evaluation of sclerotium from a medicinal mushroom, *Lignosus cameronensis* (cultivar): preclinical toxicology studies. Front Pharmacol 8:594

Lemieszek M, Rzeski W (2012) Anticancer properties of polysaccharides isolated from fungi of the Basidiomycetes class. Contemp Oncol 16(4):285–289

Li Y, Zhang G, Ng TB, Wang H (2010) A novel lectin with antiproliferative and HIV-1 reverse transcriptase inhibitory activities from dried fruiting bodies of the monkey head mushroom *Hericium erinaceum*. Biomed Res Int 2010

Li G, Yu K, Li F, Xu K, Li J, He S (2014) Anticancer potential of Hericium erinaceus extracts against human gastrointestinal cancers. J Ethnopharmacol 153(2):521–530

Li S, Dong C, Wen HA, Liu X (2016) Development of Ling-zhi industry in China – emanated from the artificial cultivation in the Institute of Microbiology, Chinese Academy of Sciences (IMCAS). Mycology 7(2):74–80

Liu RM, Zhong JJ (2011) Ganoderic acid Mf and S induce mitochondria mediated apoptosis in human cervical carcinoma HeLa cells. Phytomedicine 18(5):349–355

Liu Q, Wang H, Ng TB (2006) First report of a xylose-specific lectin with potent hemagglutinating, antiproliferative and anti-mitogenic activities from a wild ascomycete mushroom. Biochim Biophys Acta 1760(12):1914–1919

Liu K, Wang J, Zhao L, Wang Q (2013) Anticancer, antioxidant and antibiotic activities of mushroom *Ramaria flava*. Food Chem Toxicol 58:375–380

Lucas E, Ringler R, Byerrum R, Stevens J, Clarke DA, Stock CC (1957) Tumor inhibitors in *Boletus edulis* and other Holobasidiomycetes. Antibiot Chemother 7(1)

Luk S-U, Lee TK-W, Liu J, Lee DT-W, Chiu Y-T, Ma S (2011) Chemopreventive effect of PSP through targeting of prostate cancer stem cell-like population. PLoS One 6(5):e19804

Luo H, Hong S-Y, Sgambelluri RM, Angelos E, Li X, Walton Jonathan D (2014) Peptide macrocyclization catalyzed by a prolyl oligopeptidase involved in α-amanitin biosynthesis. Chem Biol 21(12):1610–1617

Markova, N., Kussovski, V., Drandarska, I., Nikolaeva, S., Georgieva, N., and Radoucheva, T. (2003). Protective activity of Lentinan in experimental tuberculosis. Int Immunopharma, 3(10-11), 1557–1562.

Martin KR, Brophy SK (2010) Commonly consumed and specialty dietary mushrooms reduce cellular proliferation in MCF-7 human breast cancer cells. Exp Bio Med 235(11):1306–1314

McCloskey S, Noppawan S, Mongkolthanaruk W, Suwannasai N, Senawong T, Prawat U (2017) A new cerebroside and the cytotoxic constituents isolated from *Xylaria allantoidea* SWUF76. Nat Prod Res 31(12):1422–1430
Mizuno T, Ohsawa K, Hagiwara N, Kuboyama R (1986) Fractionation and characterization of antitumor polysaccharides from Maitake, Grifola frondosa. Agric Biol Chem 50(7):1679–1688
Mizuno M, Morimoto M, Minato K-I, Tsuchida H (1998) Polysaccharides from *Agaricus blazei* stimulates lymphocyte T-cell subsets in mice. Biosci Biotechnol Biochem 62(3):434–437
Mizuno M, Minato KI, Ito H, Kawade M, Terai H, Tsuchida H (1999) Anti-tumor polysaccharide from the mycelium of liquid-cultured *Agaricus blazei* mill. IUBMB Life 47(4):707–714
Moldenhauer G, Salnikov AV, Lüttgau S, Herr I, Anderl J, Faulstich H (2012) Therapeutic potential of amanitin-conjugated anti-epithelial cell adhesion molecule monoclonal antibody against pancreatic carcinoma. J Nat Cancer Inst 104(8):622–634
Mondal A, Banerjee D, Majumder R, Maity TK, Khowala S (2016) Evaluation of in vitro antioxidant, anticancer and in vivo antitumour activity of *Termitomyces clypeatus* MTCC 5091. Pharm Biol 54(11):2536–2546
Montoya A, Hernández-Totomoch O, Estrada-Torres A, Kong A, Caballero J (2003) Traditional knowledge about mushrooms in a Nahua community in the state of Tlaxcala, México. Mycologia 95(5):793–806
Mueller GM, Schmit JP, Leacock PR, Buyck B, Cifuentes J, Desjardin DE, Halling RE, Hjortstam K, Iturriaga T, Larsson KH, Lodge DJ, May TW, Minter D, Rajchenberg M, Redhead SA (2007) Global diversity and distribution of macrofungi. Biodivers Conserv 16(1):37–48
Nakamura N, Hirakawa A, Gao JJ, Kakuda H, Shiro M, Komatsu Y (2004) Five new maleic and succinic acid derivatives from the mycelium of *Antrodia camphorata* and their cytotoxic effects on LLC tumor cell line. J Nat Prod 67(1):46–48
Nakamura K, Shinozuka K, Yoshikawa N (2015) Anticancer and antimetastatic effects of cordycepin, an active component of *Cordyceps sinensis*. J Pharmacol Sci 127(1):53–56
Ng ML, Yap AT (2002) Inhibition of human colon carcinoma development by lentinan from shiitake mushrooms (*Lentinus edodes*). J Altern Complement Med 8(5):581–589
Nguyen VT, Giannoni F, Dubois MF, Seo SJ, Vigneron M, Kédinger C, Bensaude O (1996) In vivo degradation of RNA polymerase II largest subunit triggered by alpha-amanitin. Nucleic Acids Res 24(15):2924–2929
Ng TB (1998) A review of research on the protein-bound polysaccharide (polysaccharopeptide, PSP) from the mushroom *Coriolus versicolor* (Basidiomycetes: Polyporaceae). Gen Pharmacol Vasc S 30(1):1–4
Nie S, Cui SW, Xie M (2017) Bioactive polysaccharides. Elsevier Science. https://doi.org/10.1016/C2015-0-04574-9. ISBN.978-0-12-809418-1
Ooi CVE, Liu F (2000) Immunomodulation and anti-cancer activity of polysaccharide-protein complexes. Curr Med Chem 7(7):715–729
Owaid Mustafa N, Barish A, Ali Shariati M (2017) Cultivation of *Agaricus bisporus* (button mushroom) and its usages in the biosynthesis of nanoparticles. Open Agric 2:537–543
Patel S, Goyal A (2012) Recent developments in mushrooms as anti-cancer therapeutics: a review. 3 Biotech 2(1):1–15
Paterson RRM (2006) Ganoderma – a therapeutic fungal biofactory. Phytochemistry 67(18): 1985–2001
Peng CC, Chen KC, Peng RY, Chyau CC, Su CH, Hsieh-Li HM (2007) *Antrodia camphorata* extract induces replicative senescence in superficial TCC, and inhibits the absolute migration capability in invasive bladder carcinoma cells. J Ethnopharmacol 109(1):93–103
Pleszczyńska M, Wiater A, Siwulski M, Lemieszek MK, Kunaszewska J, Kaczor J (2016) Cultivation and utility of *Piptoporus betulinus* fruiting bodies as a source of anticancer agents. World J Microbiol Biotechnol 32(9):151
Pohleven J, Obermajer N, Sabotič J, Anžlovar S, Sepčić K, Kos J, Kralj B, Strukelj B, Brzin J (2009) Purification, characterization and cloning of a ricin B-like lectin from mushroom *Clitocybe nebularis* with antiproliferative activity against human leukemic T cells. Biochim Biophys Acta 1790(3):173–181

Poucheret P, Fons F, Rapior S (2006) Biological and pharmacological activity of higher fungi: 20-year retrospective analysis. Cryptogam Mycol 27(4):311–333

Pushparajah V, Fatima A, Chong CH, Gambule TZ, Chan CJ, Ng ST, Tan CS, Fung SY, Lee SS, Tan NH, Lim RLH (2016) Characterisation of a new fungal immunomodulatory protein from Tiger Milk mushroom, *Lignosus rhinocerotis*. Sci Rep 6:30010

Ramberg JE, Nelson ED, Sinnott RA (2010) Immunomodulatory dietary polysaccharides: a systematic review of the literature. Nut J 9(1):54

Rathore H, Prasad S, Sharma S (2017) Mushroom nutraceuticals for improved nutrition and better human health: a review. Pharma Nutr 5(2):35–46

Reddy SM (2015) Diversity and applications of mushrooms. In: Bahadur B, Venkat Rajam M, Sahijram L, Krishnamurthy K (eds) Plant biology and biotechnology. Springer, New Delhi ISBN. 978-81-322-2285-9, pp 231–261

Reis FS, Martins A, Vasconcelos MH, Morales P, Ferreira ICFR (2017) Functional foods based on extracts or compounds derived from mushrooms. Trends Food Sci Technol 66:48–62

Riede I (2010) Tumor therapy with *Amanita phalloides* (death cap): stabilization of B-cell chronic lymphatic leukemia. J Altern Complement Med 16(10):1129–1132

Riede I (2012) Tumor therapy with *Amanita phalloides* (Death cap): long term stabilization of prostate-cancers. J Integr Oncol 1:101

Riede I (2013) Switch the tumor off: from genes to *Amanita* therapy. Am J Biomed Res 1:93–107

Riede I (2017) New therapy strategy for prostate cancer: *Amanita phalloides* treatment stabilizes best without pre-treatments (observational study pre-protocol). Br J Med Med Res 21:1–7

Roland JF, Chmielewicz ZF, Weiner BA, Gross AM, Boening OP, Luck JV, Bardos TJ, Reilly HC, Sugiura K, Stock CC, Lucas EH, Byerrum RU, Stevens JA (1960) Calvacin: a new antitumor agent. Science 132(3443):1897

Rouhana-Toubi A, Wasser SP (2015) The shaggy ink cap medicinal mushroom, *Coprinus comatus* (higher Basidiomycetes) extract induces apoptosis in ovarian cancer cells via extrinsic and intrinsic apoptotic pathways. Int J Med Mushrooms 17(12):1127–1136

Royse D, Baars JJP, Tan Q (2017) Current overview of mushroom production in the world: technology and applications. Zied DC, Pardo-GimÃ A, editors: John Wiley & Sons 5–13 p

Sadi G, Kaya A, Yalcin HA, Emsen B, Kocabas A, Kartal DI, Altay A (2016) Wild edible mushrooms from Turkey as possible anticancer agents on HepG2 cells together with their antioxidant and antimicrobial properties. Int J Med Mushrooms 18(1):83–95

Shah SK, Walker PA, Moore-Olufemi SD, Sundaresan A, Kulkarni AD, Andrassy RJ (2011) An evidence-based review of a *Lentinula edodes* mushroom extract as complementary therapy in the surgical oncology patient. J Parenter Enter Nutr 35(4):449–458

Shiao MS (2003) Natural products of the medicinal fungus *Ganoderma lucidum*: occurrence, biological activities, and pharmacological functions. Chem Rec 3(3):172–180

Siegel RL, Miller KD, Jemal A (2017) Cancer statistics. CA Cancer J Clin 67(1):7–30

Silberborth S, Stumpf A, Erkel G, Anke T, Sterner O (2002) Gerronemins A–F, cytotoxic biscatechols from a *Gerronema* species. Phytochemistry 59(6):643–648

Singdevsachan SK, Auroshree P, Mishra J, Baliyarsingh B, Tayung K, Thatoi H (2016) Mushroom polysaccharides as potential prebiotics with their antitumor and immunomodulating properties: A review. Bioact Carbohydr Dietary Fibre 7(1):1–14

Singh SR, Preet Kaur H, Rakesh Kanwar J (2016) Mushroom lectins as promising anticancer substances. Curr Protein Pept Sci 17(8):797–807

Sliva D (2010) Medicinal mushroom *Phellinus linteus* as an alternative cancer therapy. Exp Ther Med 1(3):407–411

Song TY, Hsu SL, Yeh CT, Yen GC (2005) Mycelia from *Antrodia camphorata* in submerged culture induce apoptosis of human hepatoma HepG2 cells possibly through regulation of FAS pathway. J Agric Food Chem 53(14):5559–5564

Spelman K, Sutherland E, Bagade A (2017) Neurological activity of Lion's Mane (Hericium erinaceus) J. Restorative Med 6(1):19–26

Sternberg SS, Philips FS, Cronin AP, Sodergren JE, Vidal PM (1963) Toxicological studies of calvacin. Cancer Res 23:1036–1044

Su YK, Shih PH, Lee WH, Bamodu OA, Wu ATH, Huang CC (2017) *Antrodia cinnamomea* sensitizes radio-/chemo-therapy of cancer stem-like cells by modulating microRNA expression. J Ethnopharmacol 207:47–56

Takaku T, Kimura Y, Okuda H (2001) Isolation of an antitumor compound from *Agaricus blazei* Murill and its mechanism of action. J Nutr 131(5):1409–1413

Tan NH, Fung SY, Pailoor J, Tan CS, Ng ST (2016) Nutritional composition, antioxidant properties and toxicology evaluation of the sclerotium of Tiger Milk mushroom *Lignosus tigris* cultivar E. Nutr Res 36(2):174–183

Tanaka A, Nishimura M, Sato Y, Sato H, Nishihira J (2016) Enhancement of the Th1-phenotype immune system by the intake of Oyster mushroom (Tamogitake) extract in a double-blind, placebo-controlled study. J Tradit Complement Med 6(4):424–430

Torisu M, Hayashi Y, Ishimitsu T, Fujimura T, Iwasaki K, Katano M (1990) Significant prolongation of disease-free period gained by oral polysaccharide K (PSK) administration after curative surgical operation of colorectal cancer. Cancer Immunol Immunother 31(5):261–268

Torkelson CJ, Sweet E, Martzen MR, Sasagawa M, Wenner CA, Gay J (2012) Phase 1 clinical trial of *Trametes versicolor* in women with breast cancer. ISRN Oncology 2012:7

Turner J, Chaudhary U (2009) Dramatic prostate-specific antigen response with activated hemicellulose compound in metastatic castration-resistant prostate cancer. Anti-Cancer Drugs 20(3):215–216

Uno K, Kosuna K, Sun B, Fujii H, Wakame K, Chikumaru S, Hosokawa G, Ueda Y (2000) Active hexose correlated compound (AHCC) improves immunological parameters and performance status of patients with solid tumors. Biotherapy 14(3):303–307

UK, C.R. (2016) Retrieved from http://www.cancerresearchuk.org

Vetter J (1998) Toxins of *Amanita phalloides*. Toxicon 36(1):13–24

Wan JM-F, Sit W-H, Yang X, Jiang P, Wong LL-Y (2010) Polysaccharopeptides derived from Coriolus versicolor potentiate the S-phase specific cytotoxicity of Camptothecin (CPT) on human leukemia HL-60 cells. Chin Med 5(1):16

Wang C-W, Hsu W-H, Tai C-J (2017) Antimetastatic effects of cordycepin mediated by the inhibition of mitochondrial activity and estrogen-related receptor α in human ovarian carcinoma cells. Oncotarget 8(2):3049–3058

Wasser S (2002) Medicinal mushrooms as a source of antitumor and immunomodulating polysaccharides. Appl Microbiol Biotechnol 60(3):258–274

Wasser SP, Weis AL (1999) Medicinal properties of substances occurring in higher basidiomycetes mushrooms: current perspectives. Int J Med Mushrooms 1:31–62. https://doi.org/10.1615/IntJMedMushrooms.v1.i1.30

WHO (2017) Cancer. Fact Sheet. WHO Media centre

Won JS (2002) The hematoimmunologic effect of AHCC for Korean patients with various cancers. Biotherapy 16(6):560–564

Xu X, Yan H, Chen J, Zhang X (2011) Bioactive proteins from mushrooms. Biotechnol Adv 29(6):667–674

Yamaguchi Y (2016) Immunotherapy of cancer: an innovative treatment comes of age. Springer, Tokyo

Yang P, Liang M, Zhang Y, Shen B (2008) Clinical application of a combination therapy of lentinan, multi-electrode RFA and TACE in HCC. Advan Ther 25(8):787–794

Yang HL, Kuo YH, Tsai CT, Huang YT, Chen SC, Chang HW (2011) Anti-metastatic activities of *Antrodia camphorata* against human breast cancer cells mediated through suppression of the MAPK signaling pathway. Food Chem Toxicol 49(1):290–298

Yap H-YY, Fung SY, Ng ST, Tan CS, Tan NH (2015) Shotgun proteomic analysis of tiger milk mushroom (*Lignosus rhinocerotis*) and the isolation of a cytotoxic fungal serine protease from its sclerotium. J Ethnopharmacol 174:437–451

You BJ, Tien N, Lee MH, Bao BY, Wu YS, Hu TC, Lee HZ (2017) Induction of apoptosis and ganoderic acid biosynthesis by cAMP signaling in *Ganoderma lucidum*. Sci Rep 7(1):318

Zhang J, Wang G, Li H, Zhuang C, Mizuno T, Ito H et al (1994) Antitumor polysaccharides from a Chinese mushroom,"yuhuangmo" the fruiting body of *Pleurotus citrinopileatus*. Biosci Biotechnol Biochem 58(7):1195–1201

Zhang H, Morisaki T, Nakahara C, Matsunaga H, Sato N, Nagumo F (2003) PSK-mediated NF-κB inhibition augments docetaxel-induced apoptosis in human pancreatic cancer cells NOR-P1. Oncogene 22(14):2088–2096

Zhang G, Sun J, Wang H, Ng T (2009) A novel lectin with antiproliferative activity from the medicinal mushroom *Pholiota adiposa*. Acta Biochim Pol 56(3):415–421

Zhang G, Sun J, Wang H, Ng TB (2010) First isolation and characterization of a novel lectin with potent antitumor activity from a Russula mushroom. Phytomedicine 17(10):775–781

Zhang P, Li K, Yang G, Xia C, Polston JE, Li G, Li S, Lin Z, Yang LJ, Bruner SD, Ding Y (2017) Cytotoxic protein from the mushroom *Coprinus comatus* possesses a unique mode for glycan binding and specificity. Proc Natl Acad Sci 14(34):8980–8985

Zhao C, Sun H, Tong X, Qi Y (2003) An antitumour lectin from the edible mushroom *Agrocybe aegerita*. Biochem J 374(2):321–327

Zhao S, Zhao Y, Li S, Zhao J, Zhang G, Wang H (2010) A novel lectin with highly potent antiproliferative and HIV-1 reverse transcriptase inhibitor activities from the edible wild mushroom *Russula delica*. Glycoconj J 27(2):259–265

Zong A, Cao H, Wang F (2012) Anticancer polysaccharides from natural resources: a review of recent research. Carbohydr Polym 90(4):1395–1410

Zhou S, Gao Y, Chan E (2005) Clinical trials for medicinal mushrooms: Experience with *Ganoderma lucidum* (W.Curt.:Fr.) Lloyd (Lingzhi Mushroom). Int J Med Mushrooms 7(1&2):111-118 https://doi.org/10.1615/IntJMedMushr.v7.i12.110

Chapter 13
Recent Advances in Cultivation of Edible Mushrooms

Meena Kapahi

13.1 Introduction

A mushroom is a characteristic fruit body of a macrofungus, generating large spores which can be seen with the naked eye (Chang and Miles 1992). Mushrooms are reported to be extremely diverse and plentiful worldwide. Of all the fungal species (1.5 million) on Earth (Hawksworth 2001), mushrooms comprise 16,000 species (Hawksworth 2012; Wasser 2010). However, theactual number of mushroom species on Earth far exceeds the estimated number. Approximately 30 genera contributing about 3000 species are considered to be edible; of which, around 30 species are commercially cultivated (Chang and Miles 2004). Majority of the cultivated mushrooms are saprophytes, do not possess chlorophyll and photosynthesis.

Mushroom cultivation is an age-old practice; dating back centuries. The Chinese and the Japanese have been reported to be amongst the first ones to grow mushrooms professionally. The most cultivated mushroom species are shiitake (***Lentinula edodes***), oyster mushroom (***Pleurotus*** spp.), white button mushroom (***Agaricus bisporus***), black fungus or wood-ear mushrooms (***Auricularia auricula***, *Auricularia polytricha*) and paddy straw mushroom (***Volvariella*** spp.) (Dhar 2017) as shown in Fig. 13.1. The cultivation of shiitake was started by Japanese approximately 2000 years ago (Ainsworth 1976); while, cultivation of button mushroom is quite recent, which is supposed to be one of the most widely grown mushrooms globally. The oyster mushrooms (*Pleurotus ostreatus*), initially gathered from forests in

M. Kapahi (✉)
Department of Chemistry, Manav Rachna University (MRU), Faridabad, India

Research Scholar, Department of Biotechnology, FET, Manav Rachna International Institute of Research and Studies (MRIIRS), Faridabad, India
e-mail: meenakapahi@mru.edu.in

B. P. Singh et al. (eds.), *Biology of Macrofungi*, Fungal Biology,
https://doi.org/10.1007/978-3-030-02622-6_13

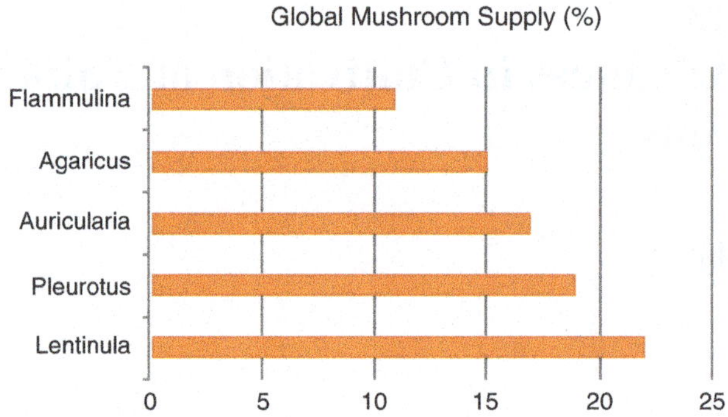

Fig. 13.1 Global mushroom supply of five major mushrooms

Fig. 13.2 An over view of market value of mushrooms (adapted from Royse et al. 2017)

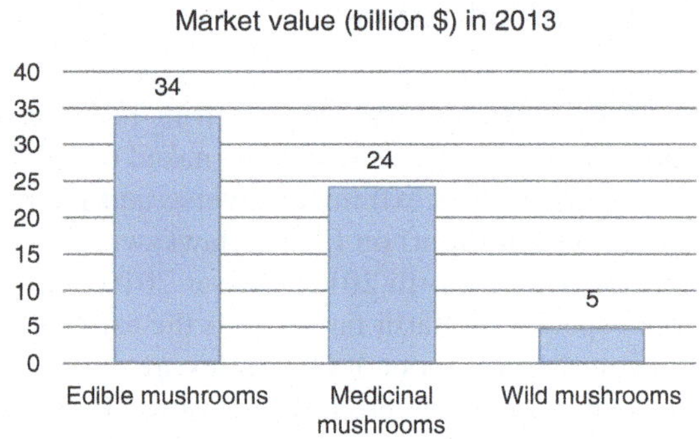

Florida, spread to all across the world (Chakravarty 2011). A tremendous increase in the mushroom cultivation can be attributed to an increase in awareness about their good taste, nutritional and therapeutic values.

13.2 Global Mushroom Production–An Overview

Global mushroom production, especially for edible and medicinal mushrooms, has increased more than 30 times as compared to 1.7 times increase in the global population over a period of 1978 to 2013, clearly indicating a whooping increase in the per capita consumption of mushrooms. Mushroom industry was placed at a market value of $63 billion in the year 2013 (Fig. 13.2) (Royse et al. 2017).

World mushroom production increased from 0.30 to 3.41 million tons from 1961 to 2010 along with the tremendous increase in the export of processed and fresh mushroom export as stated by Wakchaure (2011). China remained the major player in the production and consumption of cultivated edible mushrooms producing more than 30 billion kg (almost 87% of total production) as compared to about 1.3 billion kg produced by the rest of Asia (Royse et al. 2017); while, the contribution by EU, USA and rest of the sources remained at about 3.1 billion kg. Approximately, 95% of mushroom produced in China is consumed meeting domestic consumption with consumption per capita amounting to be over 10 kg/person/year (Wakchaure 2011). The speedy growth and market expansion of the mushroom business in China is a great example of rural development driven by bio-innovation and technological diffusion. Being one of the most important commercially grown crops, mushroom cultivation is also an excellent example of rural economic development and poverty alleviation as well as a typical recycle-economy and sustainable agriculture and forestry. China's mushroom cultivation is mostly practiced by small scale units run by families. Since 2004, China has been the global leader in the export of canned mushrooms. India's mushroom yield has also doubled. Countries like China and India are some of major exporters of mushrooms to the US market (canned) (Muhammad and Suleiman 2015) and the European countries (dried) with China dominating the export market. The European countries exhibit uneven pattern with Germany leading the dried mushroom import market followed by France, The Netherlands, Italy and the UK in 2015. Countries like Estonia, Slovenia and Poland have also shown promising market for mushroom imports (Muhammad and Suleiman 2015). The US represents one of the largest mushroom import markets and has comparatively lesser share in specialty mushroom export to countries like Japan and France (Muhammad and Suleiman 2015). In Latin America, Mexico, Chile and Brazil are the major mushroom producers growing *Agaricus* (95%) and *Pleurotus* spp. (5%) majorly. Africa hosts approximately 25% of global mushroom species but have very little contribution in the mushroom market. Shortage of technical and financial resources and promotional strategies, spawn availability and consumer preferences could be the possible reasons.

13.3 Mushrooms as Sources of Nutraceuticals and as Bioremediators

Mushrooms are the power houses of nutritional and therapeutic properties. They derive their nutrition from agro-wastes/lignocellulosic materials wastes by converting them in to a food rich in nutritional and therapeutic properties. However, the actual benefits depend upon the amount and frequency of mushroom intake in one's daily diet. Mushrooms contain carbohydrates (50–65%), proteins (19–35%) and fat (2–6%) with unsaturated fatty acids like palmitic, oleic and linoleic acids (Rathore et al. 2017). Mushrooms are reported to be the only vegetarian source for vitamin D containing no cholesterol (Rathore et al. 2017). They are reported to be having anti-oxidant (Roupas et al. 2012), anti-cancer (Kim et al. 2015)and other important medicinal and health promoting properties (Barros et al. 2007; Kim et al. 2007; Sarikurkcu et al. 2008). These functional characteristics are mainly due to their chemical composition (Kapahi and Sachdeva 2017). For their numerous nutraceutical benefits, mushrooms are being trusted and have become popular all over the globe.

Mushroom cultivation helps in environmental bioremediation by absorbing (biosorption) and accumulating pollutants in high concentrations in their bodies (Kalac and Svoboda 2000) and presents another influential role of mushrooms in the ecosystem. *Pleurotus*, *Armillaria*, *Agaricus*, *Boletus*, *Polyporus*, *Termitomyces* are among the most sought-after genre for the heavy metals bioremediation (Raj et al. 2011). High biosorption efficiency and shorter life cycle make them effective bioremediators. The by-product spent mushroom substrate, can be used in making compost, animal feed and in bioremediation; thus, representing a zero-waste model.

Apart from the multi-dimensional benefits of mushroom agro-industry, mushroom cultivation has a great potential for improving the socio-economic status by generating local employment and can serve as an entrepreneurial model for sensitive sections of the society.

13.4 Recent Trends

Mushroom cultivation is an art as well as science requiring both knowledge of scientific principles behind and the practical know-how of the cultivation techniques. It encompasses a number of steps which involve careful selection of a favourable culture of the mushroom; preparation of a quality spawn and a nutritious substrate (compost); inoculation of the compost; incubation and crop care during spawn run; and fruiting and harvesting under the right environmental conditions of temperature and humidity. The mushroom cultures were initially obtained from a high yielding mushroom by tissue/spore culture or breeding. The cultures are now available with recognized lab or centres encouraging strain purity. For spawn preparation, sterilized and healthy grains (wheat, rye etc.) are used to increase the pure mycelial cultures. Some of the mushroom species like *Pleurotus* and *Volvariella* can be cultivated using simple cultivation techniques; while, mushrooms like *A. bisporus*, *Flammulina velutipes* require special trainings and skill sets. A tremendous increase in the mushroom production can be attributed to the improvement in cultivation practices (spawn, substrate, casing, crop care, pest management etc.) and improved varieties. In the recent times, there has been an increase in awareness about their good taste, flavour and nutritional value. China has cultivated some of the well-known mushrooms like *A. auricula-judae*, *F. velutipes*, *L. edodes* and *V. volvacea* except *A. bisporus* (Chang and Miles 2004). The cultivation technique of *A. Bisporus* (popular as button mushroom) was introduced from France to other parts of the world. It is estimated that the mushroom cultivation began in China around 600 AD with *A. auricula* followed by *F. velutipes* and *L. edodes* cultivated around 800–900 AD, and *L. edodes*, first cultivated around 1000–1100 AD (Chang 2005). Early practices thrived on collecting mushrooms from natural habitats and using them for colonizing the substrates for mushroom cultivation. Five of the most popular mushrooms were cultivated before twentieth century. There was an unprecedented increase in the number of cultivated mushroom species brought around 1980s–1990s matching with the massive increase in global cultivated mushroom production (i.e. from 0.90 m tons in 1975 to 6.16 m tons in 1997). The amount of production depends on many factors like the technology involved, the costs involved in producing and bringing mushrooms from farms to markets and the associated market process, flavour, nutrition and texture. Improvement in the quality and in the production of oyster mushrooms has been the prime target of researchers in the area (Chakravarty 2011). Research in mushroom science is focussed on finding better cultivation techniques and developing new strains to give higher yields.

13.5 Cultivation Tends of Some of the Cultivated Edible Mushrooms

13.5.1 L. edodes

L. edodes (shiitake/xianggu/oakwood mushroom/black forest mushroom/black mushroom/golden oak mushroom) is one of the most important and the largest growing edible mushrooms in the world. This non-pathogenic fungus grows on dead and decaying broad leaf trees like shii trees, under natural conditions (Royse 2005). It is known for its unique taste and medicinal properties. It is the largest cultivated edible mushroom in China. China started with picking and drying of shiitake. Wu San Kang, a mushroom hunter, collected mushrooms in the forests in Zhejiang (Luo 2004). Wu San Kang started with its cultivation in around 1100 AD on wooden logs. He discovered that beaten logs would give better yield leading to the birth of the shocking technique. Its cultivation methods employ two main techniques - cultivation on wood logs and on synthetic substrate logs (Chang and Miles 2004; Stamets 2000). The primary way of *L. edodes* cultivation remained production on wood logs of shii tree. The cultivation methods on wood logs were further developed by Japanese researchers and scientists. In 1936, Kitajima used pure culture method and employed pure culture to produce spawn which provided independence to the growers for cultivation (Chang and Miles 2004). Its cultivation as a popular and a commercial venture started in the 1940s in Japan with adoption of new techniques. Of lately, use of artificial substrates of sawdust supplemented with nutrient sources has been a popular practice (Özçelik and Pekşen 2007).

In 1943, with the invention of Tanegoma spawn (wood peg spawn) by Kisaku Mori, Kyoto University, the mushroom industry got a major boom. As per the technique (spawning wood logs method), sterilized wood chips with mushroom culture were employed as inoculum by putting them into holes/cuts drilled into logs. With the ongoing research in mushroom farming and improved technology, China became the major producer of this speciality mushroom. China surpassed Japan using sawdust-based technique which reduced growing time and increased production efficiency (Royse 2014). In the 'cutting wood logs method', small cuts were made on the fresh wood logs to provide passage for the wind-borne spores. The cuts on wood would accelerate the fungus infestation on wood leading to mycelial spread and would yield fruiting bodies. Shiitake sawdust-based technique used glass

Table 13.1 A comparison of production (%) of *L. edodes* (1983–1997) in China and Japan

Country	1983	1985	1991	1992	1993	1994	1995	1996	1997
China	9.4	13.9	60.5	63.9	68.9	73.6	72.5	76.3	85.1
Japan	82.8	63.3	28.6	25.2	21.3	18.5	19.4	16.4	7.3

bottles as containers in the 1960s in China. In the 1980s, modelled sawdust bricks (brick or pressed cake method) were used. The Zhejiang Academy of Agricultural Sciences (ZAAS) was the first ever hub for exporting shiitake to Japan. Later on, innovative 'plastic bag method' enabled shorter production cycles and high production efficiency. In 1987, the use of the innovative 'synthetic log' method in China revitalized mushroom production in China and China overtook Japan as the world's major producer of xianggu.

In 1983, Japan was the world's largest producer of *L. edodes* with a share of 82.8% of total world production, and the production in China was approximately 9.4% (Table 13.1). China's production of xianggu mushrooms rose from 57% to 72.5% in the period from 1991 to 1995. The period from 1995 to 2000, witnessed an increase in production from about 0.5 million tons to more than two million tons; and to over four million tons in the year 2012. Shiitake production helped in upliftment of poor communities (Chang 2005). The year 2013 represented a 106.8% increase in production as compared to 2010. In the United States, shiitake production is done on sawdust substrates supplemented with nutrients.

The use of plastic bags containing different types of substrates under controlled environmental conditions is also a common practice these days. Shiitake mycelia can grow on natural and synthetic media; potato dextrose agar (PDA) is the most popular medium. It can be produced on both domestic and commercial scales. Temperature is one of the most important environmental conditions required for the cultivation and varies in different stages (for spore germination: 22–26 °C; mycelial run: 23–25 °C; and fruiting body: 10–20 °C). The optimum pH requirement employed for the preparation of mushroom bag/log is about 5.0–5.5 (Chang and Miles 2004; Stamets 2000). An important stage of shiitake cultivation is requirement of 'browning stage'turning mycelia from white to brown and forming a hard crust on the surface. Good quality strain, techniques like use of liquid spawn, heavy spawning, synthetic substrate and supplements can help in reducing spawn run duration and labour cost, microbial contamination and good yield of over 80%.

13.5.2 A. bisporus

A. bisporus (button mushroom, white/brown mushroom), a saprophyte, grows on dead plants. Its production has witnessed an ever-increasing trend since 1950s (Royse 2014). It is grown in more than 70 countries around the world. China became the major producer of *A. bisprous* in the year 1998. Its production increased approximately 11.7% during the last 10 years (2006–2010) (Royse et al. 2017; USDA 2015). Both the white and brown varieties saw tremendous growth patterns during this period. In the year 2013, China remained the major producer of *A. bisporus* producing about 54% of the global production (2013) followed by the USA (9%), Poland (7%), Netherlands (6%) and India (6%). Increasing consumer awareness, preferences towards vegetarian diet, advanced, innovative and cheap mushroom cultivation technology have been favouring the increasing consumption and production of the button mushroom. The market for button mushrooms (fresh/processed) is dependent on regions and their respective consumer demands. Depending on the type, the market is segmented into fresh and processed (frozen/canned/dried) mushrooms.

The technique employed for *A. bisporus* requires time, technical know-how and space. Since 1950s, the substrate or compost required for cultivating *A. bisporus* requires the two-step treatment - outdoors (first and initial phase) and indoors (second and conditioning phase under limited environment) (Miller 1993). The substrate mixture or ingredients are collected, moistened/pre-wetted and made into a heap/stack for 1–2 weeks. This first step involves breaking down of organic substances present in the mixture. The heap is subjected to turning after time intervals in order to homogenize the entire substrate. The compost becomes dark brown in colour and has moisture holding capacity. In the next step, i.e., indoor fermentation, the substrate is pasteurized for compost formation (Fermor and Grant 1985; Overtjins 1998) in order to increase the process of decomposition and to create right nutrient environment (Miller 1993). *A. bisporus* cultivation on non-composted raw materials has also been reported in the 1960s. Since then, more efforts have been laid on to lessen the time required to produce a good-quality mushroom compost. In order to tackle the odour and water pollution problems, different techniques of composting have been developed; for example, indoor composting in different countries like Italy, Australia and the Netherlands (Laborde et al. 1993). Mushrooms fruiting bodies are produced in flushes at an interval of one week; each subsequent flush producing lesser fruiting body due to decreasing amount of substrate nutrients. New and modified techniques have been developed to increase the yield. Methods like application of delayed release nutrients while spawning or casing or injecting nutrients during the culture have also been experimented (San Antonio 1966; Zied et al. 2011). The double cropping (Royse et al. 2008; Royse and Sanchez 2008 a, b; Royse and Chalupa 2009) method involves removing casing, supplementing with nutrients in between (Royse 2010; Royse et al. 2008) followed by re-casing of the substrate (Fig. 13.3).

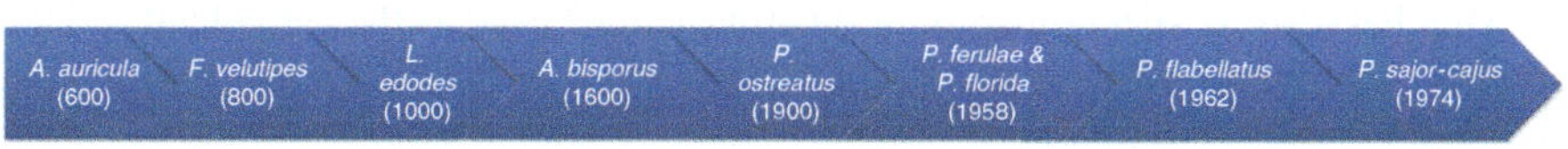

Fig. 13.3 Some of the cultivated edible mushrooms with their year of cultivation

13.5.3 Pleurotus spp.

Pleurotus spp. (oyster mushrooms) is one of the most popular mushrooms due to flavour, better shelf life and a good taste. The genus is characterised by white spores, a fleshy cap and an eccentric attachment of the stem. Its cultivation started in Germany (1917) utilizing tree stubs or wood logs (Upadhyay and Singh 2010). There has been a marked improvement in the techniques. Now, they are the easiest and the most economical mushrooms to cultivate; and proudly occupy the second place among the mushrooms with maximum production all over the globe. The species produces white mycelia and are popularly called 'white rot fungi'. Different varieties of oyster mushrooms display different colours like brown, gray, white, yellow and pink; each having its unique flavour. The species includes *P. ostreatus* (black oyster), *P. florida* (white oyster), *P. djamor* (pink oyster), *P. eryngii* (king oyster), *P. citrinopleatus* (golden oyster) and *P. sajor-caju* (Indian oyster) as reported by Knop et al. (2015). These varieties can be cultivated using different techniques with little variations.

The species can be easily grown on non-composted agro-wastes or lignocellulosic materials (Savoie et al. 2007) with a simple technology and minimal finances. Utilization of easily available lignocellulosic agro-waste (e.g. wheat and rice straw, sawdust) (Hussain et al. 2002) depends on the oxidation and hydrolytic enzyme action. Substrates like cotton waste (Tan 1981) amended with protein rich cereal bran (Kinugawa et al. 1994) improve the yield. Sawdust has been reported to be a suitable substrate (Shah et al. 2004); but requires nutrients (like nitrogen and potassium) and composting for better mycelial growth (Obodai et al. 2002). *P. eryngii* has also been reported to exhibit higher yield on sawdust as compared to rice straw (Moonmoon et al. 2010). The species can be grown in a variety of sterilized containers; for e.g., plastic bags, bottles, trays etc. The containers should be strong enough to hold the substrate during mycelial run (Mamiro et al. 2014). *P. ostreatus*, with a distinguished aroma and a soft texture, shows resistance to pests and diseases. It requires a temperature range of 18–22 °C (Sánchez 2010). *P. florida* (Fig. 13.4–13.6), characterized by decurrent gills running till the stipe base, also grows at a temperature range of 18–22 °C.

Better resistance to pests, higher biological efficiency, better yield and adaptation to environmental factors are some of the desirable characteristics of oyster mushrooms. Oyster mushrooms can help in alleviating problems of hunger in various developing nations and in rejuvenating local economies. There are various modifications of the technique which are quite simple and easy. The lignocellulosic waste is sterilised and mixed with the suitable spawn. The spawn run usually takes approximately 15 days.

After harvesting, spent mushroom substrate can be used as the cattle feed and the compost leaving no waste at all.

Fig. 13.4–13.6 Different stages of oyster mushroom (*Pleurotus florida*)

13.5.4 F. velutipes

F. velutipes (winter mushroom/golden needle mushroom/velvet stem/Enoki/Enokitake) occurs naturally on trees like elms, poplar, birch and plum during autumn to early winter. The mushroom is rich in flavour, taste and offers many therapeutic properties.

Morphologically, it has yellowish/orangish brown colour under natural conditions; while and cultivated ones (in the absence of sunlight) are white in colour. Initially cultivated in China (800 AD), the species now is cultivated in different parts of the world like Siberia, Europe, Africa, Japan and Australia. In 1928, it was cultivated using substrate of rice bran and sawdust in Japan (Nakamura 1981). In 1960s, Japan emerged as its major producer. In 1990s, China surpassed Japan in its production and produced more than 1.5 million tons in 2010 (Royse 2014). Its production is done on sawdust and other agro based substrates like sugarcane bagasse, corncobs etc. in bottles or bags.

13.6 Conclusion

Mushrooms are bountiful sources of nutrients and medicinal properties. Global production of edible mushrooms has increased manifold over the past few decades owing to an increasing awareness among people. World's mushroom demands are met majorly by five genera in the order of *L.edodes* > *Pleurotus* spp. > *A.auricula-judae* > *A. bisporus* > *F. velutipes*. China is the major producer and consumer of edible mushrooms all over the world. Mushroom production offers multiple benefits of tackling solid agro-wastes, converting it to a highly nutritious and therapeutic food and generating employment or entrepreneurial opportunies for the sensitive sections of the society.

References

Ainsworth GC (1976) Introduction to the History of Mycology. Cambridge University Press, Cambridge

Barros L, Baptista P, Estevinho LM, Ferreira ICFR (2007) Bioactive properties of the medicinal mushroom *Leucopaxillus giganteus* mycelium obtained in the presence of different nitrogen sources. Food Chem 105:179–186

Chakravarty B (2011) Trends in mushroom cultivation and breeding. Aust J Agric Eng 2(4):102–109

Chang ST (2005) Witnessing the development of the mushroom industry in China. Acta Edulis Fungi 12:3–19

Chang ST, Miles PG (1992) Mushroom biology—a new discipline. Mycologist 6:64–65

Chang ST, Miles PG (2004) Mushrooms: cultivation, nutritional value, medicinal effect, and environmental impact. CRC press, Florida

Dhar BL (2017) Mushroom and human civilization. In: Zied DC, Pardo-Gime'nez A (eds) Edible and medicinal mushrooms: technology and applications, 1st edn. Wiley

Fermor TR, Grant WD (1985) Degradation of fungal and actinomycete mycelia by *Agaricus bisporus*. J Gen Microbiol 129:15–22

Hawksworth DL (2001) Mushrooms: The extent of the unexplored potential. Int J Med Mushrooms 3:333–337

Hawksworth DL (2012) Global species numbers of fungi: Are tropical studies and molecular approaches contributing to more robust estimate? Biodivers Conserv 21:2425–2433

Hussain M, Khan SM, Khan SM, Chohan MA (2002) Effect of different sterilization methods on the production of oyster mushroom (*Pleurotus ostreatus*) on different substrates. In: Integrated

plant disease management. Proceeding of 3rd national conference of plant pathology, NARC, Islamabad: 1–3 Oct 2001, pp. 159–160
Kalac P, Svoboda L (2000) A review of trace element concentrations in edible mushrooms. Food Chem 69:273–281
Kapahi M, Sachdeva S (2017) Mycoremediation potential of Pleurotus species for heavy metals: a review. Bioresour Bioproces 4(1):32. https://doi.org/10.1186/s40643-017-0162-8
Kim HY, Yoon DH, Lee WH, Han SK, Shrestha B, Kim CH, Lim MH, Chang W, Lim S, Choi S, Song WO, Sung JM, Hwang KC, Kim TW (2007) *Phellinus linteus* inhibits inflammatory mediators by suppressing redox-based NF-jB and MAPKs activation in lipopolysaccharide-induced RAW 264.7 macrophage. J Ethnopharmacol 114:307–315
Kim SH, Jakhar R, Kang SC (2015) Apoptotic properties of polysaccharide isolated from fruiting bodies of medicinal mushroom *Fomes fomentarius* in human lung carcinoma cell line. Saudi J Biol Sci 22:484–490
Kinugawa K, Phusawang W, Chinbenjapho IS, Fukada S, Tanesaka E, Okada M, Tsutsui H (1994) Progress Report (1991–1993) of joint research program of Kinki and Chiang Mai Universities on the promotion of mushroom research. Mem Fac Agr Kinki Univ 27:93–113
Knop D, Yarden O, Hadar Y (2015) The ligninolytic peroxidases in the genus *Pleurotus*: divergence in activities, expression, and potential applications. Appl Microbiol Biotechnol 99:1025–1038
Laborde J, Lanzi G, Francescutti B, Giordani E (1993) Indoor composting: general principles and large scale development in Italy. In: Chang ST, Buswell JA, Chiu SW (eds) Mushroom Biology and Mushroom Products. The Chinese University Press, Hong Kong, pp 93–113
Luo XC (2004) Progress of xiang-gu (shiitake) cultivation in China. Mush Sci 15:307–311
Mamiro DP, Mamiro PS, Mwatawala MW (2014) Oyster mushroom (*Pleurotus* spp.) Cultivation technique using reusable substrate containers and comparison of mineral content with leafy vegetables. J Appl Biosci 80(1):7071–7080
Miller FC (1993) Conventional composting system. In: Tan Nair NG (ed) Proceedings of the second AMGA/ISMS international workshop-seminar on *Agaricus* compost. Sydney, pp 1–18
Moonmoon MM, Uddin N, Ahmed S, Shelly NJ, Khan MA (2010) Cultivation of different strains of king oyster mushroom (*Pleurotus eryngii*) on saw dust and rice straw in Bagladesh. Eur J Biol Sci 2(1):13–18
Muhammad BL, Suleiman B (2015) Global development of mushroom biotechnolgy. Int J Emerging Trends Sci Techno 12(6):2660–2669
Nakamura K (1981) Mushroom cultivation in Japan. Asaki Publication House, Japan
Obodai M, Sawyerr LCB, Johnson PNT (2002) Yield of seven strains of oyster mushrooms (*Pleurotus* spp.) grown on composted sawdust using various woods for effective cultivation of Oyster mushroom. Pak J Bot 44(1):399–402
Overtjins A (1998) The conventional phase II in trays or shelves. Mushroom J 584:15–21
Özçelik E, Pekşen A (2007) Hazelnut husk as a substrate for the cultivation of shiitake mushroom (*Lentinula edodes*). Bioresour Technol 98(14):2652–2658
Raj DD, Mohan B, Vidya Shetty BM (2011) Mushrooms in the remediation of heavy metals from soil. Int J Environ Pollut Control Manag 3(1):89–101
Rathore H, Prasad S, Sharma S (2017) Mushroom nutraceuticals for improved nutrition and better human health: A review. Pharma Nutr 5(2):35–46
Roupas P, Keogh J, Noakes M, Margetts C, Taylor P (2012) The role of edible mushrooms in health: evaluation of the evidence. J Funct Foods 4:687–709
Royse DJ (2005) Speciality mushrooms and their cultivation. Hortic Rev 19:59–97
Royse DJ (2010) Effects of fragmentation, supplementation and the addition of phase II compost to 2nd break compost on mushroom (*Agaricus bisporus*) yield. Bioresour Technol 101:188–192
Royse DJ (2014) A Global perspective on the high five: Agaricus, *Pleurotus, Lentinula, Auricularia & Flammulina*. Proceedings of the 8th international conference on mushroom biology and mushroom products (ICMBMP8); Nov 19–22 2014; New Delhi, India
Royse DJ, Baars J, Tan Q (2017) Current overview of mushroom production in the world. In: Zied DC, Pardo-Gime'nez A (eds) Edible and medicinal mushrooms: Technology and applications, 1st edn. Wiley

Royse DJ, Chalupa W (2009) Effects of spawn, supplement, and phase II compost additions and time of recasing second break compost on mushroom (*Agaricus bisporus*) yield and biological efficiency. Bioresour Technol 100:5277–5282
Royse DJ, Sanchez JE (2008a) Supplementation of first break mushroom compost with hydrolyzed protein, commercial supplements and crystalline amino acids. World J Microbiol Biotechnol 24:1333–1339
Royse DJ, Sanchez JE (2008b) Supplementation of 2nd break mushroom compost with isoleucine, leucine, valine, pheylalanine, Fermenten and Soy Plus. World J Microbiol Biotechnol 24:2011–2017
Royse DJ, Sanchez JE, Beelman RB, Davidson J (2008) Re-supplementing and recasing mushroom (*Agaricus bisporus*) compost for a second crop. World J Microbiol Biotechnol 24:319–325
San Antonio JP (1966) Effects of injection of nutrient solutions into compost on the yield of mushrooms (*Agaricus bisporus*). Proc Am Hortic Soc 89:415–422
Sánchez C (2010) Cultivation of *Pleurotus ostreatus* and other edible mushrooms. Appl Microbiol Biotechnol 85:1321–1337
Sarikurkcu C, Tepe B, Yamac M (2008) Evaluation of the antioxidant activity of four edible mushrooms from the Central Anatolia, Eskisehir—Turkey: *Lactarius deterrimus, Suillus collitinus, Boletus edulis, Xerocomus chrysenteron*. Bioresour Technol 99:6651–6655
Shah ZA, Ashraf M, Ishtiaq CM (2004) Comparative study on cultivation and yield performance of oyster mushroom (*Pleurotus ostreatus*) on different substrates (wheat straw, leaves, saw dust). Pak J Nutr 3:158–160
Stamets P (2000) Growing gourmet and medicinal mushrooms. Ten Speed Press, Berkeley
Savoie JM, Salmones D, Mata G (2007) Hydrogen peroxide concentration measured in cultivation substrates during growth and fruiting of the mushrooms *Agaricus bisporus* and *Pleurotus* spp. J Sci Food Agric 87:1337–1344
Tan KK (1981) Cotton waste is a good substrate for cultivation of (*Pleurotus ostreatus*). The Oyster Mushroom. Mush Sci 11(1):705–710
United States Department of Agriculture (USDA) (2015) Mushrooms. National agricultural statistics service, agricultural statistics board. 17p. http://usda.mannlib.cornell.edu/usda/current/Mush/Mush 08-20-2014.pdf. Accessed 2 Nov 2015
Upadhyay RC, Singh M (2010) Production of edible mushrooms. In: Hofrichter M (ed) The Mycota: a comprehensive treatise on Fungi as experimental systems forbasic and applied research. Industrial Applications X, Springer ISBN 978-3-642-11457-1
Wakchaure GC (2011) Production and marketing of mushrooms: global and national scenario. In: Singh M, Vijay B, Kamal S, Wakchaure GC (eds) Mushrooms-cultivation, marketing and consumption. Directorate of Mushroom Research, Solan
Wasser SP (2010) Medicinal mushroom science: History, current status, future trends, and unsolved problems. Int J Med Mushrooms 12(1):1–16
Zied CD, Savoie JM, Pardo-Giménez A (2011) Soybean the main nitrogen source in cultivation substrates of edible and medicinal mushrooms. In: El-Shmy H (ed) Soybean and nutrition. In Tech Open Access Publisher, Rijeka

Chapter 14
Medicinal Mushrooms: Cultivation and Pharmaceutical Impact

Peter Chiew Hing Cheong, Chon Seng Tan, and Shin Yee Fung

14.1 Introduction

Approximately 430 million years ago, giant fungal spires Prototaxites Dawson measuring 8 m high and 1 m wide dotted the ancient landscape (Hueber 2001; Schultz 2013). Today, a different fungal coverage is taking over with its presence still humongous and massive, consisting of difference species produced and used worldwide as food, medicine, industrial bio-products and bioremediation tools (Chang 1980; Kozarski et al. 2015; Narsing Rao et al. 2017; Wasser 2014). China, Italy, the United States and the Netherlands are the top producers of mushrooms (Fig. 14.1) (Misachi 2017) with *Lentinula edodes* (Shiitake mushroom) leading the world production by 22% as of 2013, followed by *Pleurotus* (Oyster mushroom) (19%), *Auricularia* (Jelly ear mushroom) (18%), *Agaricus* (button mushroom) (15%), *Flammulina* (Enokitake mushroom) (11%), *Volvariella* (Straw mushroom) (5%) and other mushrooms (10%) (Royse et al. 2017). Only about 20 types of mushrooms are cultivated in mass (Reddy 2015). China dominated the production

P. C. H. Cheong
Medicinal Mushroom Research Group (MMRG), Department of Molecular Medicine, University of Malaya, Kuala Lumpur, Malaysia

C. S. Tan
Ligno Research Foundation, Balakong Jaya, Selangor, Malaysia

S. Y. Fung (✉)
Medicinal Mushroom Research Group (MMRG), Department of Molecular Medicine, University of Malaya, Kuala Lumpur, Malaysia

Centre for Natural Products Research and Drug Discovery (CENAR), University of Malaya, Kuala Lumpur, Malaysia

University of Malaya Centre for Proteomics Research (UMCPR), University of Malaya, Kuala Lumpur, Malaysia
e-mail: syfung@um.edu.my

B. P. Singh et al. (eds.), *Biology of Macrofungi*, Fungal Biology,
https://doi.org/10.1007/978-3-030-02622-6_14

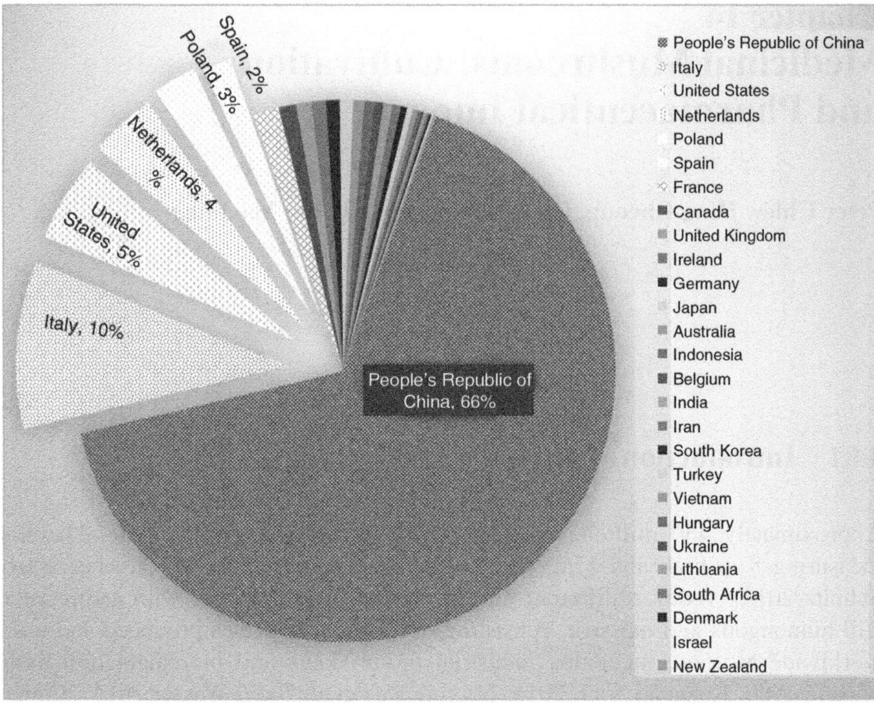

Fig. 14.1 The world's top producers of mushrooms (production in tons)(adapted from Misachi 2017)

of mushrooms with an annual production of about five million tons (Misachi 2017). China is also the major producer and exporter of non-edible mushroom *Ganoderma* (polypores) with capacity over 110,000 ton/year (Li et al. 2016). The worldwide market value of medicinal mushrooms was US $6.0 billion in 1999 and $18.0 billion in 2014 (Chang 2008; Wasser 2014). The global mushroom market is projected to rise to more than $50 billion by 2023, with the medicinal mushroom extract market growing at a compounded annual rate of 6.3 percent (McCall 2018). The increasing demand among health-conscious consumers is one of the main growth drivers of the mushroom market.

14.2 Man, Mushroom and Pharmaceuticals

Evidence of mushroom use as medicine went back as far as 5300 years ago when a mummified man found from Val Senales glacier in Italy was well-preserved with a *Piptoporus betulinus* mushroom in his medicine kit, as natural laxative and antibiotics (Capasso 1998). In Egypt and Aztecs, medicinal mushrooms were depicted as 'sons and flesh of Gods' and secured high respects among the priests who only served them to Pharaohs (Berlant 2005). Likewise, ancient Greece and Rome were fascinated by mushrooms and regarded them sacred (Ruck 2011). The mighty Vikings were said to consume mushrooms to enter a state of trance, fearless, berserk mode for battle (Fabing 1956). Yet some communities impoverished mushrooms and regarded them as ugly harbingers of death or illness, supposedly evil and often referred to as 'toadstools (Cheung 2008). Even in the 1600s, the English still remained distrustful of mushrooms and rarely used them neither for food nor medicine (Marley 2009). Elsewhere in France, Italy, Poland, Russia and eastern Europe regions, the acceptance of wild mushrooms as food gradually becomes a norm, with families picking mushrooms as pastimes and sources of income but not as medicine (Lemieszek and Rzeski 2012; Poucheret et al. 2006; Wasser and Weis 1999). Only countries in the Far East like China, Japan and Korea are mushrooms sought after and used both as food and medicine for centuries (Lemieszek and Rzeski 2012; Patel and Goyal 2012). Even in this modern era when Japan and China had incorporated medicinal properties from mushrooms to treat illness such as cancer; the western world is still rather sceptical on using mushrooms as medicine (Lemieszek and Rzeski 2012; Reis et al. 2017). The tide however is changing. Many European and western communities are changing their perception regarding medicinal mushrooms especially (McCall 2018) due to the influence of mass and social media compounded by growing evidence backed by clinical studies which demonstrated that both edible and non edible mushrooms contain outstanding curative properties that could ruffle several health and well-being aspects of modern lifestyles (Table 14.1).

Polysaccharides, beta glucan polymers, phenolic compounds, polyketides, triterpenoids, nucleotides, steroids and other secondary metabolites identified from medicinal mushrooms have many health promoting properties (Chen et al. 2015; Friedman 2016; Hetland et al. 2011; Ma et al. 2010; Tatjana and Marina 2013; Teplyakova and Kosogova 2016; Wasser 2002). The polysaccharides and beta glucan are reported to stimulate and boost the immune systems. Traditionally in Asian communities, dried nonedible medicinal mushrooms such as *Ganoderma lucidum, Trametes versicolor* and *Inonotus obliquus* are often steeped in boiling water and served as medicinal tonic for the sick and elderly due to their tough texture and bitterness (Wasser 2002). Edible mushrooms are usually added to the diet, cooked with other herbs and meat as soup or essence for flavour and perhaps the knowledge that mushrooms are good for health. Generally, mushrooms are regarded as safe and straight forward natural remedies of ailments (Lull et al. 2005). Several specific mushroom metabolites however, are gaining confidence and practical use following encouraging clinical trials conducted to treat a variety of diseases. In Japan and China, polysaccharide-K (PSK) and polysaccharide-peptide (PSP) from *T. versicolor* have been used clinically for almost 50 years as oral anticancer agents in surgical patients with breast cancers, gastric cancers and colorectal cancers and in patients with small cell lung cancers, with almost no severe adverse drug reactions (Katoh and Ooshiro 2007; Lemieszek and Rzeski 2012). Erinacines and hericenones identified from *Hericium erinaceus* demonstrate neuroprotective properties and improved mild cognitive impairment without toxicity in the experimental, animal or two clinical trials conducted (Ma et al. 2010; Mori et al. 2009; Spelman et al. 2017). Clinical trials of Ganopoly derived from *Ganoderma lucidum* for 12 weeks

Table 14.1 Medicinal mushrooms for general health and wellbeing

Lifestyle diseases, illness and well being		Mushroom	Bioactives	References
Health	Obesity	*Pleurotus ostreatus*	Polysaccharides, Lovastatin	(Friedman 2016; Geng et al. 2017; Gunde-Cimerman and Cimerman 1995; Lee et al. 2014a; Ng and Wang 2005; Schwartz and Hadar 2014; Tatjana and Marina 2013; Vitak et al. 2017; Yang et al. 2014; Zhou et al. 2005)
Metabolic	Hyperglycemia	*Grifola frondosa* *Schizophyllum commune*		
syndrome	Diabetes	*Ganoderma lucidum*	Eritadenin, Triterpenes, Diterpenoid derivative sterols	
	Hypercholesterolemia	*Lentinus edodes*		
Infectious diseases		*Sclerotium rolfsii*		
		Hericium erinaceus	Phenolic compounds	
	Hypertension	*Trametes versicolor*		
	Heart attack	*Agaricus brasiliensis*, *Ophiocordyceps militaris*,	Adenosine	
	Cardiovascular			
	Fatigue			
	Stroke			
	Inflammatory bowel disease			
Damage prevention (antioxidant)	Kidney	*H. erinaceus*, *O.sinensis*	Polysaccharide	(Ng and Wang 2005; Zhang et al. 2012)
	Liver			
	Cancer	*H. erinaceus*	Polysaccharides	(Chen et al. 2015; Daba and Ezeronye 2003; Kinoshita et al. 2010; Lee et al. 2012; Mizuno et al. 1999; Torisu et al. 1990; Torkelson et al. 2012; Wasser 2014; Yang et al. 2014)
		L. edodes	Low molecular weight secondary metabolites	
		G. lucidum		
		O. militaris,	Agaritine	
		Trametes versicolor	Lentinan	
		Agaricus blazei	Cordycepin	
		Lignosus rhinocerus		

Antiviral	Human immunodeficiency, herpes, West Nile, Influenza, hepatitis viruses, orthopoxviruses, variola virus	*Inonotus oblique, Lentinus edodes, Grifola frondosa, Ganoderma lucidum, Trametes versicolor, Agaricus brasiliensis*	Polysaccharides	(Adotey et al. 2011; Shibnev et al. 2015; Teplyakova and Kosogova 2016; Wang and Ng 2001; Zhou et al. 2005)
			Proteins, Glycoproteins, Terpenoids, Melanins, Nucleosides, Lectins	
Immunomodulatory	Nitric oxide production, phagocytosis, secretion of interleukins	*Ophiocordyceps militaris*	Polysaccharides	(Ng and Wang 2005; Zhu et al. 2014a)
		Ganoderma lucidum	β-D-glucan, Triterpenoids	
		Lentinus edodes, Trametes versicolor		
Wellbeing	Rheumatism	*Phellinus linteus*	Proteoglycan	(Kim et al. 2003; Lull et al. 2005)
	Body pain			
	MRSA	*Hericium erinaceus*	Polysaccharides Secondary	(Barneche et al. 2017; Jose Alves et al. 2013; Zhu et al. 2014b)
	Antibacterial	*Gymnopilus junonius*	metabolites	
		Oudemansiella canarii	Grifolin	
		Agaricus bisporus		
	Antifungal	*Albatrellus dispansus*		
	Anti-Parasitic *Helicobacter pylori*			
	Gastric & ulcer			
Respiratory	Allergy	*Lentinus edodes*	Lentinan	(Hwang et al. 2015; Jones 1998; Markova et al. 2003; Tuli et al. 2014)
	Influenza	*Ganoderma lucidum, Ophiocordyceps* sp. *Phellinus baumii*	Cordycepin polyphenols	
	Asthma			
	Tuberculosis			
Fertility	Ovulation	*Grifola frondosa*	Polysaccharides cordycepin	(Chen et al. 2010; Huang et al. 1987; Jiraungkoorskul and Jiraungkoorskul 2016; Lin et al. 2007)
		Ophiocordyceps sinensis		
	Sexual dysfunction	*Ophiocordyceps militaris*		

(continued)

Table 14.1 (continued)

Lifestyle diseases, illness and well being		Mushroom	Bioactives	References
Neurite outgrowth, nerve, addiction	Alzheimer	*Hericium erinaceus*	Erinacineshericenones secondary metabolites	(Carhart-Harris et al. 2017; Ma et al. 2010; Michael et al. 2015; Mori et al. 2009; Phan et al. 2015; Pitman et al. 2006; Sabaratnam et al. 2013; Spelman et al. 2017)
	Dementia	*Ganoderma lucidum, Grifola frondosa Sarcodon scabrosus, Lignosus rhinocerus Pleurotus giganteus*	FIP	
	Anxiety	Psychedelic mushrooms	Psilocybin	
	Depression			
	Insomnia (sleep promoting)			
	Alcoholism, tobacco addiction			
Beauty	Anti-aging	*Ganoderma lucidum, Agaricus bisporus, Lentinula edodes*	Ergosterol derivatives,	(Lam et al. 2012; Sun et al. 2009; Weng et al. 2010)
	Skin repair		Active hexose correlated compound, Polysaccharides	
	Hair loss due to chemotherapy			

showed **hypoglycemic** activity in Type II diabetes, improved signs of patients with coronary heart disease and produced some antiviral and liver protective effects in patients with chronic hepatitis B infection (Zhou et al. 2005). Wasser (2014) listed a nine steps pathway towards the production of new pharmaceuticals through medicinal mushrooms:

1. Mushroom cultivation and biomass production
2. **Biomass** extraction
3. Screening of mushroom extracts
4. Effect of selected extracts on a target of interest
5. Chemical fractionation of selected extracts
6. Elucidation of active fractions (compounds), mechanism of action, and potency
7. Effect on animal models
8. Preclinical drug development
9. Clinical drug development

Hence, large scale cultivation of mushrooms and new pharmaceuticals have become inextricably intertwined.

14.3 Mushroom Cultivation for Food then Medicine

Prior 1970, *Agaricus bisporus* the button mushroom, was the only mushroom species cultivated in the West, in open fields of France, then moved underground, into caves, catacombs, tunnels and quarries (Chang and Hayes 2013). Other edible mushrooms were collected from the wild. The cultivation of *Agaricus* was actually stumbled upon unintentionally. In the East however, *Lentinula edodes* (Shiitake), *Auricularia polytricha* (Wood Ear) and *Flammulina velutipes* (Enokitake) were cultivated on logs and woody substrate as early as 300–200 BC. In those days, the fruiting bodies of mushrooms were collected from the wild and used to cultivate fresh fruiting bodies on substrate. With time, the cultivation methods were refined, giving rise to a whole new wealth of knowledge; suitable substrates to grow certain mushroom species, optimal temperature, initial pH requirement, white light, influence of heavy metal, aseptic techniques and ways to curb diseases and pests to yield good harvest (Chen et al. 2011). Today, extensive literature reviews, patents and propriety methods related to mushroom cultivation have been established. The improved cultivation methods include inoculation using pure mushroom spawn of certain age. Optimal sources of nitrogen and mineral in growth substrate such as sterile compost waste material of various agro origins, sawdust in polypropylene bags and wood logs are used. For mycorrhizal species, mushroom mycelium is inoculated to the roots of living trees. Shake flasks and bioreactors are used to cultivate mushroom pellets in liquid state with temperature, pH, aeration and light monitored (Petre and Teodorescu 2011; Wu et al. 2013) (Table 14.2). The cultivation of mushrooms has shifted from backyard cottage farming to industrial based state-of-the-art methodologies, utilising knowledge and skills gained on cultivating

Table 14.2 Mushroom cultivation techniques, background, viability and bioactives

Mushroom	Cultivation	Year	Technique	Viability	Bioactives	Reference
Tuber melanosporum (Truffles)	US	1972	SSF artificially inoculated seedlings of oaks, hazelnuts and beeches	10 years	Polysaccharide	(Lefevre and Hall 2001; Liu et al. 2009)
	France	1987				
	Italy			5 years		
	NZ	1997		5 years		
	Australia	2009	Fed-batch	16–30 days		
Lentinula edodes (Shiitake)	China	1368–1644	Hardwood logsSawdust, millet, wheat bran	Viable	Lentinan	(Shiitake 1995)
Ophiocordyceps sinensis	China	2010	SSF presence of white light, rich based substrate, 20–25 °C, 70–80% humidity, addition of distilled water, initial pH inoculum 6 LSF flask initial pH 5–6, 5–6 days, 25 °C, 150 rpm	3–7 months to grow fruiting bodies via SSF 5–6 days	Cordycepin	(Kim et al. 2010; Yue et al. 2013)
	Korea				Adenosine	
Ophiocordyceps cardinalis	Malaysia				Polysaccharides	
Ophiocordyceps militaris						
Ganoderma lucidum	China	1960	SSF cut log sawdust pellets	6 months	Polysaccharides	(Fang and Zhong 2002)
	Japan	1970			Beta-glucan	
					Triterpenoids	
			LSF bioreactors	72–288 h	Ganoderic acids A and B	
			Shake flasks 23–25 °C, pH 5.5–6.5, rotary shaker 90–120 rev min^{-1}			
Lignosus rhinocerus	Malaysia	2009	SSF Sclerotia	6 months	Polysaccharides	(Abdullah et al. 2013; Kong et al. 2016; Lau and Abdullah 2016; Lau et al. 2015; Lau et al. 2014; Lee et al. 2014b; Tan et al. 2016; Wei et al. 2014; Yap et al. 2015; Yap et al. 2013)
Lignosus tigris			Fruiting body			
			LSF			
			Shake flasks	1–3 years	Exopolysaccharides	
				11–15 days	Low molecular weight molecules	
					Antioxidant	

mushrooms within the past 100 years. Only with sufficient supply can bioactivity validation work be carried out. In this era, cultivation of medicinal mushrooms utilizes advanced fermentation technology and aseptic techniques in enclosed sterile work stations equipped with high efficiency particulate air HEPA filters. High throughput analytical chemistry analysis and bioassays are applied to assess the quality and amount of nutrient and bioactive composition of the cultivated material. Mass production of medicinal mushrooms differs slightly from common cultivation of mushrooms depending on the stage of the growth that would produce the optimum amount of bioactive desired. Solid state fermentation (SSF) technique targets the fruiting bodies of mushrooms where stress condition is applied to the growing cultivar on substrate to produce fruiting bodies. Liquid state fermentation (LSF) is used to stimulate the mycelia to release the bioactive directly into the broth (culture filtrate) and shorten the fermentation time. Fermentation techniques can be selective as well, targeting specific stage of growth or compounds desired. To cultivate sclerotium-forming mushrooms, poor nutrient and specific growth conditions could trick the mycelia to keep retrieving nutrient from the surroundings and store in their sclerotia, a dormant stage which forbid them to germinate into fruiting bodies (Lau et al. 2015). Sclerotia are structures formed through dense aggregation of mycelia during adverse conditions; desiccation and depletion of nutrients in combination with unfavourable conditions for mycelial growth. Unlike mushroom cultivation for food, these medicinal mushrooms are cultivated to target optimum nutrient and biomass composition, monitoring the amount of target bioactives that could be produced. These bioactives would be the priority for quality control. Depending on the type of target bioactives, the cultivated material will be extracted to yield optimum amount of bioactives using a variety of solvents. These extracts and purified compounds are then subjected to different specific bioassays, pharmaceutical assessments and made into commercial products.

14.4 Brave New World, Solid State Fermentation SSF-*Ophiocordyceps sinensis*

Morchella esculenta, morel mushrooms can be cultivated but the cultivation method failed to sustain commercial distribution of the mushrooms. Truffle farmers in New Zealand and Australia are making profit US$300–1450 per kg 5 years earlier from the introduced species compared to farmers and founders in northern hemisphere who started the artificially inoculation of truffles in seedlings of oaks, hazelnuts and beeches. Immature cultivation techniques plagued cultivation of medicinal mushrooms at commercial level as well. In the early 1980s, many institutes in China started to cultivate medicinal mushroom *Ophiocordyceps sinensis* for fruiting bodies through artificial substrates. In the wild, *Ophiocordyceps* fungi parasitise larva of the ghost moth *Hepialus armoticamus* in the highland of Himalaya during winter and after 6 months during summer, stalk-like fruiting bodies would extend above ground from the heads of the insects to disperse infectious fungal spores.

Fig. 14.2 Fruiting bodies of *Ophiocordyceps sinensis* cultivated on rice based culture medium (Photo from Chon Seng Tan)

The fruiting bodies of *Ophiocordyceps* are highly sought after for their medicinal properties (Baral et al. 2015; Ng and Wang 2005) with market values hitting USD 20,000–40,000 kg^{-1} (Lo et al. 2013) due to their scarcity. On account of the immaturity of commercial artificial cultivation, *O. millitaris* was identified among the substitutes of *O. sinensis.* Uncertainty overshadowed previous work if it would be possible to grow *O. sinensis* fruiting bodies using artificial cultivation without any insect hosts. The next intermittent question; would the desired medicinal properties be produced? In 2010 however, progress on the cultivation of *Ophiocordyceps* fruiting bodies was reported via infected larvae reared in moist pits with plants (Yue et al. 2013). Fruiting bodies were also produced after tweaking the following parameters: (1) strains (2) culture temperature (3) light exposure (4) rich culture medium. With the right strain, understanding of the fungal biology, physiology, nutrient requirement, growth conditions; temperature, moisture content and stress triggers to develop fruiting bodies, several successes in cultivating fruiting bodies of *O. sinensis* are reported to date (Fig. 14.2) using rice based substrate. Chemical analysis, nutrient studies and assay conducted of the cultivated *O. sinensis* revealed minor disparaging differences when compared to wild types. On the contrary, the amount of bioactives and nutrient in fruiting bodies is higher in cultivated fruiting bodies of *O. sinensis* (Fung et al. 2018). A new mutagenesis technique called 'ion beam' could also be incorporated to mass produce cordycepins using mutated *Ophiocordyceps* species (Das et al. 2008).

14.5 Some Fairytale Bliss-Liquid State Fermentation LSF-*Ganoderma lucidum*

Traditionally *Ganoderma lucidum* is cultivated in the field. As the bioactivity was found higher in the mycelia than in the fruiting bodies and spores, LSF provided the way to accelerate mycelial growth for biomass, polysaccharides and triterpenoids with shortened fermentation duration and reduced contamination (Cui et al. 2015; Tang and Zhong 2002). Yeast extract, peptone and glucose combination at ratio 1:1:4 in LSF greatly affected the cell growth and product biosynthesis (Fang and Zhong 2002). Spores of *G. lucidum* consists several bioactive substances which are much abundant than in the fruiting body (Chaiyasut et al. 2010). Yet retrieving the bioactive using conventional approach and enzymes are expensive. *G. lucidum* is therefore cultivated for the spores and the spores fermented with *Lactobacillus plantarum* to break down the spores by the fifth day of fermentation.

14.6 Rescued from the Brink of Extinction –Sclerotial Technology- *Lignosus rhinocerus*

In the diary of John Evelyn (1664), a special mention was made of Tiger Milk Mushroom, referred to as *Lac Tygridis,* for looking like a mushroom but weighed like a metal and came from a hard coagulation of matter (Evelyn and De la Bédoyère 2004). Used by natives to treat diseases (Lau et al. 2015), Sir Henry Nicholas Ridley, the father of Malaya's rubber industry attempted to cultivate the wild solitary mushroom as early as 1890 but did not succeed. With forest clearing and over time, this mushroom is almost forgotten due to its rarity in the wild. Classified in the family of Polyporaceae, the mushroom has centrally stipitate pilei arising from sclerotia buried in the ground (Lau and Abdullah 2016). The sclerotium is the part with medicinal properties (Lau et al. 2015; Lee et al. 2009). The mushroom extracts of *Lignosus rhinocerus* inhibit breast, lung and human cervical cancer cell lines (Lee et al. 2012; Pushparajah et al. 2016). The scientific findings have so far verified some of its traditional applications (Fung and Tan 2017). The different degrees of potency and selectivity of the mushroom extract on cancer and their corresponding noncancerous cell lines could be attributed to the mushroom strain/cultivar/genetics, cultivation protocols and conditions, post-harvest processing, and extract preparation (Lau et al. 2015). To date there are three species of Tiger Milk Mushroom found in Malaysia; *L. rhinocerus*, *L. tigris* and *L. cameronensis*. High, intermediate and low molecular weight fractions from potent cold water extract from *L. rhinocerus* have been assessed for bioactivities such as anti-proliferation of cancer cells, immunomodulatory, antioxidant, anti-inflammatory, antimicrobial and neuritogenic activities (Eik et al. 2012; Lee et al. 2014b; Mohanarji et al. 2012; Wong et al. 2009; Yap et al. 2015; Yap et al. 2013).

The first report on artificial cultivation of *L. rhinocerus* through SSF was conducted on sawdust as substrate and sclerotia were produced (Huang 1999). In 2009, using propriety methodology, sclerotia of *L. rhinocerus* was successfully cultivated in 6 months to form sclerotia and 2–3 years to develop into fruiting bodies (Tan et al. 2012). Pilot cultivation of *L. rhinocerus* was carried out using optimised formulation consisting of sawdust, paddy straw and spent yeast at ratio 7.9:1:1 in bags (Abdullah et al. 2013). Sclerotia weighing between 80 and 120 g on fresh weight basis formed 3–4 weeks after burial, followed by fruiting bodies 8–12 months later.

The use of LSF; shake flasks and bioreactors to cultivate *L. rhinocerus* for exopolysaccharides and mycelial biomass was also reported (Rahman et al. 2012; Wei et al. 2014). Among the carbon sources tested, glucose yielded the highest mycelia growth and maximal exopolysaccharides production at 6.84 and 2.67 g/L, respectively after 11 days via bioreactor (Wei et al. 2014). Calcium nitrate was the most suitable for the growth of mycelial biomass (2.72 g/L) while potassium nitrate was the most favourable nitrogen source for the maximal production of exopolysaccharides (1.17 g/L) (Wei et al. 2014). Bioactives obtained through LSF shake flasks of *L. rhinocerus* was also found high in antioxidant content (Lau et al. 2014).

14.7 Quality Control and Commercial Claims

Bioactives from mushrooms could not be artificially synthesized due to their rather complex nature and structures. This is one of the reasons why commercial cultivation is crucial to prevent bottleneck in research and further pharmaceutical development. Still, without the bioactive ingredient to be singularly identified and gone through all stages of clinical trials, many could not be registered as pharmaceuticals. The supplementary market is therefore saturated with mushroom products claiming for a variety of health benefits with creative yet not proven statements all too confusing targeting patients and consumers in general. Wild, natural and precious medicinal mushrooms such as *Ophiocordyceps sinensis* often succumb to unscrupulous act of adulterations and heavy metal contamination due to pollution.

Cultivated edible mushrooms arena exception, if heavy pesticides were used during cultivation or if proper propagation techniques and sites were not monitored stringently and regulated. This will lead to questionable health value of the cultivated medicinal mushrooms is. Quality control is mandatory and the use of safe food grade material for medicinal mushroom cultivation has to be clear. HACCP monitoring during mushroom cultivation for medicinal purposes should be implemented for safety reasons. Cultivation using **lignocellulosic** residues and agro-industrial by-products is cost effective but if the substrate were previously treated with agrochemicals, the residues could pose issues as mushrooms in general absorb and retain chemicals hence their usage in bioremediation. Cultivated medicinal mushrooms on wood material should also be checked for the presence of harmful carcinogenic, mutagenic, and nephrotoxic toxins such as aristolochic acid.

Consumers will want to know the source of the raw materials used for cultivation, be it edible or non-edible, to produce the medicinal product. Amount of active ingredients, percentage of polysaccharides such as beta-glucans and/or starch, bioavailability, potency of extract, proof of authenticity (certificate of analysis) are also important information for consumers. Source of origin, the processes involved in producing the final product form, dosage to use for therapeutic effect are prime information for consumers with serious health issues.

The consumer protection association and regulatory bodies need to evaluate cultivation methods, together with high quality extraction protocols of mushrooms used for medicinal products. Only mushroom extract products with therapeutic effect demonstrated and passed the safety assessment in clinical studies should be allowed for sale.

14.8 Conclusion

Domesticating and cultivating medicinal mushrooms from the wild in controlled environment be it through SSF or LSF with HACCP implemented for pharmaceuticals is feasible. The health we could gain from medicinal mushrooms is dependent on how reliable and trustworthy the cultivators and manufacturers are to supply therapeutic mushroom extracts. The demands for safe and reliable products without adulterations are the ever-increasing attention of consumers these days. In the next 10–20 years, mass mushroom cultivation with close monitoring by relevant health authorities will increase production of mushroom bioactives and possibly make them more affordable, safe and bio-effective as mushroom pharmaceuticals.

References

Abdullah N, Haimi MZD, Lau BF, Annuar MSM (2013) Domestication of a wild medicinal sclerotial mushroom, *Lignosus rhinocerotis* (Cooke) Ryvarden. Ind Crop Prod 47:256–261

Adotey G, Quarcoo A, Holliday JC, Fofie S, Saaka B (2011) Effect of immunomodulating and antiviral agent of medicinal mushrooms (immune assist 24/7 TM) on CD4+ T-lymphocyte counts of HIV-infected patients. Int J Med Mushrooms 13(2):109–113

Baral B, Shrestha B, da Silva JAT (2015) A review of Chinese Cordyceps with special reference to Nepal, focusing on conservation

Barneche S, Alborés S, Borthagaray G, Cerdeiras MP, Vázquez A (2017) Anti-MRSA activity of fruiting body extracts of spectacular Rustgill mushroom, *Gymnopilus junonius* (Agaricomycetes). Int J Med Mushrooms 19(3):243–248

Berlant SR (2005) The entheomycological origin of Egyptian crowns and the esoteric underpinnings of Egyptian religion. J Ethnopharmacol 102(2):275–288

Capasso L (1998) 5300 years ago, the ice man used natural laxatives and antibiotics. Lancet 352(9143):1864

Carhart-Harris RL, Roseman L, Bolstridge M, Demetriou L, Pannekoek JN, Wall MB (2017) Psilocybin for treatment-resistant depression: fMRI-measured brain mechanisms. Sci Rep 7(1):13187

Chaiyasut C, Kruatama C, Sirilun S (2010) Breaking the spores of *Ganoderma lucidum* by fermentation with *Lactobacillus plantarum*. Afric J Biotechnol 9(43):7379–7382

Chang ST (1980) Mushrooms as human food. BioSci 30(6):399–401

Chang ST (2008) Overview of mushroom cultivation and utilization as functional foods. In: Cheung PCK (ed) Mushrooms as functional foods. Wiley, Hoboken, pp 1–33

Chang ST, Hayes WA (2013) The biology and cultivation of edible mushrooms. Elsevier Science, Saint Louis, p 2013

Chen J-T, Tominaga K, Sato Y, Anzai H, Matsuoka R (2010) Maitake mushroom (*Grifola frondosa*) extract induces ovulation in patients with polycystic ovary syndrome: a possible monotherapy and a combination therapy after failure with first-line clomiphene citrate. J Altern Complement Med 16(12):1295–1299

Chen P, Yong Y, Gu Y, Wang Z, Zhang S, Lu L (2015) Comparison of antioxidant and antiproliferation activities of polysaccharides from eight species of medicinal mushrooms. Int J Med Mushrooms 17(3):287–295
Chen Y-S, Liu B-L, Chang YN (2011) Effects of light and heavy metals on *Cordyceps militaris* fruit body growth in rice grain-based cultivation. Korean J Chem Eng 28(3):875–879
Cheung PCK (2008) Mushrooms as functional foods: Wiley, ISBN 978-0-470-05406-2
Cui M-L, Yang HY, He GQ (2015) Submerged fermentation production and characterization of intracellular triterpenoids from *Ganoderma lucidum* using HPLC-ESI-MS. J Zhejiang Univ Sci B 16(12):998–1010
Daba A, Ezeronye O (2003) Anti-cancer effect of polysaccharides isolated from higher basidiomycetes mushrooms. Afr J Biotechnol 2(12):672–678
Das S, Masuda M, Hatashita M, Sakurai A, Sakakibara M (2008) A new approach for improving cordycepin productivity in surface liquid culture of *Cordyceps militaris* using high-energy ion beam irradiation. Lett Appl Microbiol 47(6):534–538
Eik L-F, Naidu M, David P, Wong KH, Tan YS, Sabaratnam V (2012) *Lignosus rhinocerus* (Cooke) Ryvarden: a medicinal mushroom that stimulates neurite outgrowth in PC-12 cells. Evid-Based Complementary Altern Med 2012(1):7
Evelyn J, De la Bédoyère G (2004) The diary of John Evelyn. Boydell Press
Fabing HD (1956) On going berserk: a neurochemical inquiry. Am J of Psychiatry 113(5):409–415
Fang QH, Zhong JJ (2002) Submerged fermentation of higher fungus *Ganoderma lucidum* for production of valuable bioactive metabolites—ganoderic acid and polysaccharide. Biochem Eng J 10(1):61–65
Friedman M (2016) Mushroom polysaccharides: chemistry and antiobesity, antidiabetes, anticancer, and antibiotic properties in cells, rodents, and humans. Foods 5(4):80
Fung SY, Cheong PCH, Tan NH, Ng ST, Tan CS (2018) Nutrient and chemical analysis of the fruiting body of a cultivar of the Chinese caterpillar mushroom, *Ophiocordyceps sinensis* (Ascomycetes). Int J Med Mush 20(5):459–469
Fung SY, Tan CS (2017) The bioactivity of Tiger Milk mushroom: Malaysia's prized medicinal mushroom. In: Medicinal plants and fungi: Recent advances in research and development. Springer, Singapore, pp 111–133
Geng P, Siu KC, Wang Z, Wu JY (2017) Antifatigue functions and mechanisms of edible and medicinal mushrooms. Biomed Res Int 2017:1–16. https://doi.org/10.1155/2017/9648496
Gunde-Cimerman N, Cimerman A (1995) *Pleurotus* fruiting bodies contain the inhibitor of 3-Hydroxy-3-Methylglutaryl-coenzyme a reductase—lovastatin. Exp Mycol 19(1):1–6
Hetland G, Johnson E, Lyberg T, Kvalheim G (2011) The mushroom *Agaricus blazei* Murill elicits medicinal effects on tumor, infection, allergy and inflammation through its modulation of innate immunity and amelioration of Th1/Th2 imbalance and inflammation. Adv Pharmacol Sci 2011:157015
Huang N (1999) A new species of medicinal mushroom-cultivation of Hurulingzhi. Edible fungi China 18:8–9
Huang Y, Lu J, Zhu B, Wen Q, Jia F, Zeng S (1987) Prevention and improvement of adrenal glands and thymus hormones, and infertile sperm count improve by 300% after *Cordyceps* supplement. Zhong Cheng Yao Yan Jiu 10:24–25
Hueber FM (2001) Rotted wood–alga–fungus: the history and life of Prototaxites Dawson 1859. Rev Palaeobot Palynol 116(1–2):123–158
Hwang BS, Lee IK, Choi HJ, Yun BS (2015) Anti-influenza activities of polyphenols from the medicinal mushroom *Phellinus baumii*. Bioorganic Med Chem Lett 25(16):3256–3260
Jiraungkoorskul K, Jiraungkoorskul W (2016) Review of naturopathy of medicinal mushroom, *Ophiocordyceps sinensis,* in sexual dysfunction. Pharmacogn Rev 10(19):1–5
Jones K (1998) Reishi mushroom. Alt Complement Ther 4(4):256–266
Jose Alves M, CFR Ferreira I, Dias J, Teixeira V, Martins A, Pintado M (2013) A review on antifungal activity of mushroom (basidiomycetes) extracts and isolated compounds. Curr Top Med Chem 13(21):2648–2659

Katoh R, Ooshiro M (2007) Enhancement of antitumor effect of tegafur/uracil (UFT) plus leucovorin by combined treatment with protein-bound polysaccharide, PSK, in mouse models. Cell Mol Immunol 4:295-299

Kim GY, Kim SH, Hwang SY, Kim HY, Park YM, Park SK, Lee MK, Lee SH, Lee TH, Lee JD (2003) Oral administration of proteoglycan isolated from *Phellinus linteus* in the prevention and treatment of collagen-induced arthritis in mice. Biol Pharm Bull 26(6):823–831

Kim S-Y, Shrestha B, Sung GH, Han SK, Sung JM (2010) Optimum conditions for artificial fruiting body formation of *Cordyceps cardinalis*. Mycobiol 38(2):133–136

Kinoshita J, Fushida S, Harada S, Makino I, Nakamura K, Oyama K, Fujita H, Ninomiya I, Fujimura T, Kayahara M, Ohta T (2010) PSK enhances the efficacy of docetaxel in human gastric cancer cells through inhibition of nuclear factor-κB activation and survivin expression. Int J Oncol 36(3):593–600

Kong B, Tan N, Fung S, Pailoor J, Tan C, Ng S (2016) Nutritional composition, antioxidant properties, and toxicology evaluation of the sclerotium of Tiger Milk mushroom *Lignosus tigris* cultivar E. Nut Res 36(2):174–183

Kozarski M, Klaus A, Jakovljevic D, Todorovic N, Vunduk J, Petrović P, Niksic M, Vrvic MM, Van Griensven L (2015) Antioxidants of edible mushrooms. Molecules 20(10):19489–19525

Lam W, Wang C, Tsui T, Wai M, Tang H, Wong Y (2012) Extract of white button mushroom affects skin healing and angiogenesis. Microsc Res Tech 75(10):1334–1340

Lau BF, Abdullah N, Aminudin N, Lee HB, Tan PJ (2015) Ethnomedicinal uses, pharmacological activities, and cultivation of *Lignosus* spp. (Tiger Milk Mushrooms) in Malaysia–a review. J Ethnopharmacol 169:441–458

Lau BF, Abdullah N, Aminudin N, Lee HB, Yap KC, Sabaratnam V (2014) The potential of mycelium and culture broth of *Lignosus rhinocerotis* as substitutes for the naturally occurring sclerotium with regard to antioxidant capacity, cytotoxic effect, and low-molecular-weight chemical constituents. PLoS One 9(7):e102509

Lee ML, Tan NH, Fung SY, Tan CS, Ng ST (2012) The antiproliferative activity of sclerotia of *Lignosus rhinocerus* (Tiger Milk Mushroom). Evid Based Complement Alternat Med 2012: Article ID697603

Lee SS, Chang YS, Noraswati MNR (2009) Utilization of macrofungi by some indigenous communities for food and medicine in peninsular Malaysia. Forest Ecol Manag 257(10):2062–2065

Lee KF, Chen JH, Teng CC, Shen CH, Hsieh MC, Lu CC, Lee KC, Lee LY, Chen WP, Chen CC, Huang WS, Kuo HC (2014a) Protective effects of *Hericium erinaceus* mycelium and its isolated Erinacine A against Ischemia-injury-induced neuronal cell death via the inhibition of iNOS/p38 MAPK and nitrotyrosine. Int J Mol Sci 15(9):15073–15089

Lee SS, Tan NH, Fung SY, Sim SM, Tan CS, Ng ST (2014b) Anti-inflammatory effect of the sclerotium of *Lignosus rhinocerotis* (Cooke) Ryvarden, the Tiger Milk mushroom. BMC Complement Altern Med 14(1):359

Lefevre CK, Hall IR (2001) The status of truffle cultivation: a global perspective. International Society for Horticultural Science (ISHS), Leuven

Lemieszek M, Rzeski W (2012) Anticancer properties of polysaccharides isolated from fungi of the Basidiomycetes class. Contemp Oncol 16(4):285–289

Li S, Dong C, Ha W, Liu X (2016) Development of Ling-zhi industry in China – emanated from the artificial cultivation in the Institute of Microbiology, Chinese Academy of Sciences (IMCAS). Mycology 7(2):74–80

Lin WH, Tsai MT, Chen YS, Hou RCW, Hung HF, Li CH, Wang HK, Lai MN, Jeng KCG (2007) Improvement of sperm production in subfertile boars by *Cordyceps militaris* supplement. Am J Chi Med 35(04):631–641

Liu YW, Gao JL, Guan J, Qian ZM, Feng K, Li SP (2009) Evaluation of antiproliferative activities and action mechanisms of extracts from two species of Ganoderma on tumor cell lines. J Agric Food Chem 57:3087–3093

Lo TCT, Chang CA, Chiu KH, Tsay PK, Jen JF (2013) Correlation evaluation of antioxidant properties on the monosaccharide components and glycosyl linkages of polysaccharide with different measuring methods. Carbohydrate Poly 86:320–327

Lull C, Wichers HJ, Savelkoul HFJ (2005) Antiinflammatory and immunomodulating properties of fungal metabolites. Mediat Inflamm 2005(2):63–80
Ma B-J, Shen J-W, Yu H-Y, Ruan Y, Wu T-T, Zhao X (2010) Hericenones and erinacines: stimulators of nerve growth factor (NGF) biosynthesis in *Hericium erinaceus*. Mycol 1(2):92–98
Markova N, Kussovski V, Drandarska I, Nikolaeva S, Georgieva N, Radoucheva T (2003) Protective activity of Lentinan in experimental tuberculosis. Int Immunopharmacol 3(10–11):1557–1562
Marley G (2009) Mushrooms for health: Medicinal secrets of Northeastern fungi: Down East, ISBN-13: 9780892728084
McCall O (2018) Medicinal mushroom: A top food trend in 2018. Retrieved from https://born2invest.com/articles/medicinal-mushroom-top-food-trend-2018/
Michael PB, Alyssa AF, Jessica AP, Claire EW, Barbosa PCR, Rick JS (2015) Psilocybin-assisted treatment for alcohol dependence: a proof-of-concept study. J Psychopharmacol 29(3):289–299
Misachi J. The world's top producers of mushroom and truffle 2017. Available from: https://www.worldatlas.com/articles/the-world-s-top-producers-of-mushroom-and-truffle.html
Mizuno M, Ki M, Ito H, Kawade M, Terai H, Tsuchida H (1999) Anti-tumor polysaccharide from the mycelium of liquid-cultured *Agaricus blazei* mill. IUBMB Life 47(4):707–714
Mohanarji S, Dharmalingam S, Kalusalingam A (2012) Screening of *Lignosus rhinocerus* extracts as antimicrobial agents against selected human pathogens. J Pharm Biomed Sci 18(11):1–4
Mori K, Inatomi S, Ouchi K, Azumi Y, Tuchida T (2009) Improving effects of the mushroom Yamabushitake (*Hericium erinaceus*) on mild cognitive impairment: a double-blind placebo-controlled clinical trial. Phytother Res 23(3):367–372
Narsing Rao MP, Xiao M, Li W-J (2017) Fungal and bacterial pigments: secondary metabolites with wide applications. Front Microbiol 8:1113
Ng TB, Wang HX (2005) Pharmacological actions of *Cordyceps*, a prized folk medicine. J Pharm Pharmacol 57(12):1509–1519
Patel S, Goyal A (2012) Recent developments in mushrooms as anti-cancer therapeutics: a review. 3 Biotech 2(1):1–15
Petre M, Teodorescu A (eds) (2011) Medicinal mushrooms cultivation through the solid-state and submerged fermentations of agricultural wastes
Phan C-W, David P, Naidu M, Wong K-H, Sabaratnam V (2015) Therapeutic potential of culinary-medicinal mushrooms for the management of neurodegenerative diseases: diversity, metabolite, and mechanism. Crit Rev Biotechnol 35(3):355–368
Pitman JL, McGill JJ, Keegan KP, Allada R (2006) A dynamic role for the mushroom bodies in promoting sleep in *Drosophila*. Nature 441(7094):753–756
Poucheret P, Fons F, Rapior S (2006) Biological and pharmacological activity of higher fungi: 20-year retrospective analysis. Cryptogam Mycol 27:311–333
Pushparajah V, Fatima A, Chong CH, Gambule TZ, Chan CJ, Ng ST (2016) Characterisation of a new fungal immunomodulatory protein from Tiger Milk mushroom, *Lignosus rhinocerotis*. Sci Rep 6:30010
Rahman NA, Daud F, Kalil MS, Ahmad S (2012) Tiger milk mushroom cultivation by using submerged culture technique. WSEAS Trans Biol Biomed 3(9):83–92
Reddy S (2015) Diversity and applications of mushrooms. In: Plant biol Biotechnol. Springer, New Delhi, pp 231–261
Reis FS, Martins A, Vasconcelos MH, Morales P, Ferreira ICFR (2017) Functional foods based on extracts or compounds derived from mushrooms. Trends Food Sci Technol 66:48–62
Royse D, Baars JJP, Tan Q (2017) In: Zied DC, Pardo-Gim ÃA (eds) Current overview of mushroom production in the world: technology and applications. Wiley, pp 5–13
Ruck C (2011) Mushrooms, myth and Mithras: the drug cult that civilized Europe. City Lights Publishers, San Francisco
Sabaratnam V, Kah-Hui W, Naidu M, David PR (2013) Neuronal health–can culinary and medicinal mushrooms help? J Tradit Complement Med 3(1):62–68
Schultz C (2013) Long before trees overtook the land, earth was covered by giant mushrooms. https://www.smithsonianmag.com/smart-news/long-before-trees-overtook-the-land-earth-was-covered-by-giant-mushrooms-13709647/

Schwartz B, Hadar Y (2014) Possible mechanisms of action of mushroom-derived glucans on inflammatory bowel disease and associated cancer. Ann Transl Med 2(2):19

Shibnev V, Garaev T, Finogenova M, Kalnina L, Nosik D (2015) Antiviral activity of aqueous extracts of the birch fungus *Inonotus obliquus* on the human immunodeficiency virus. Vopr Virusol 60(2):35–38

Shiitake JK (1995) The healing mushroom US: inner traditions/bear. ISBN 9780892814992

Spelman K, Sutherland E, Bagade A (2017) Neurological activity of Lion's Mane (*Hericium erinaceus*). J Restorative Med 6(1):19–26

Sun B, Wakame K, Sato E, Nishioka H, Aruoma OI, Fujii H (2009) The effect of active hexose correlated compound in modulating cytosine arabinoside-induced hair loss, and 6-mercaptopurine- and methotrexate-induced liver injury in rodents. Cancer Epidemiol 33(3):293–299

Tan CS, Ng ST, Yap YHY, Lee SS, Lee ML, Fung SY et al (2012) Breathing new life to a Malaysia lost national treasure the tiger-milk mushroom (*Lignosus rhinocerotis*). In: Proceedings of the 18th congress of the International Society for Mushroom Science, pp 66–71

Tan NH, Fung SY, Pailoor J, Tan CS, Ng ST (2016) Nutritional composition, antioxidant properties, and toxicology evaluation of the sclerotium of Tiger Milk mushroom *Lignosus tigris* cultivar E. Nut Res 36(2):174–183

Tang YJ, Zhong JJ (2002) Fed-batch fermentation of *Ganoderma lucidum* for hyperproduction of polysaccharide and ganoderic acid. Enzym Microb Technol 31(1–2):20–28

Tatjana K, Marina K (2013) Therapeutic properties of mushrooms in managing adverse effects in the metabolic syndrome. Curr Top Med Chem 13(21):2734–2744

Teplyakova TV, Kosogova TA (2016) Antiviral effect of Agaricomycetes mushrooms (review). Int J Med Mushrooms 18(5):375–386

Torisu M, Hayashi Y, Ishimitsu T, Fujimura T, Iwasaki K, Katano M, Yamamoto H, Kimura Y, Takesue M, Kondo M et al (1990) Significant prolongation of disease-free period gained by oral polysaccharide K (PSK) administration after curative surgical operation of colorectal cancer. Cancer Immunol Immunother 31(5):261–268

Torkelson CJ, Sweet E, Martzen MR, Sasagawa M, Wenner CA, Gay J, Putiri A, Standish LJ (2012) Phase 1 clinical trial of *Trametes versicolor* in women with breast cancer. ISRN Oncology 2012:1–7. https://doi.org/10.5402/2012/251632

Tuli HS, Sandhu SS, Sharma A (2014) Pharmacological and therapeutic potential of *Cordyceps* with special reference to Cordycepin. 3 Biotech 4(1):1–12

Vitak T, Yurkiv B, Wasser S, Nevo E, Sybirna N (2017) Effect of medicinal mushrooms on blood cells under conditions of diabetes mellitus. World J of Diabetes 8(5):187–201

Wang HX, Ng TB (2001) Examination of lectins, polysaccharopeptide, polysaccharide, alkaloid, coumarin and trypsin inhibitors for inhibitory activity against human immunodeficiency virus reverse transcriptase and glycohydrolases. Planta Med 67(07):669–672

Wasser SP (2002) Medicinal mushrooms as a source of antitumor and immunomodulating polysaccharides. Appl Microbiol Biotechnol 60(3):258–274

Wasser SP (2014) Medicinal mushroom science: Current perspectives, advances, evidences, and challenges. Biomed J 37(6):345–356

Wasser SP, Weis AL (1999) Medicinal properties of substances occurring in higher basidiomycetes mushrooms: current perspectives. Int J Med Mushrooms 1(1):31–62

Wei H, Mohd Salleh S, Daud F, Zainal Z, Mazni Othman A, Mohd Saleh N (2014) Optimization of submerged culture conditions for the production of mycelial biomass and exopolysaccharides from Lignosus rhinocerus pp 73–80

Weng Y, Xiang L, Matsuura A, Zhang Y, Huang Q, Qi J (2010) Ganodermasides A and B, two novel anti-aging ergosterols from spores of a medicinal mushroom *Ganoderma lucidum* on yeast via UTH1 gene. Bioorganic Med Chem 18(3):999–1002

Wong K-H, Lai CKM, Cheung PCK (2009) Stimulation of human innate immune cells by medicinal mushroom sclerotial polysaccharides. Evid-based Complement Altern Med 11(3):215–223

Wu FC, Chen YL, Chang SM, Shih L (2013) Cultivation of medicinal caterpillar fungus, *Cordyceps militaris* (ascomycetes), and production of cordycepin using the spent medium from Levan fermentation. Int J Med Mush 15(4):393–405

Yang S, Jin L, Ren X, Lu J, Meng Q (2014) Optimization of fermentation process of *Cordyceps militaris* and antitumor activities of polysaccharides in vitro. J Food Drug Anal 22(4):468–476
Yap HYY, Tan NH, Fung SY, Aziz AA, Tan CS, Ng ST (2013) Nutrient composition, antioxidant properties and antiproliferative activity of *Lignosus rhinocerus* Cooke sclerotium. J Sci Food Agric 93(12):2945–2952
Yap HYY, Fung SY, Ng ST, Tan CS, Tan NH (2015) Shotgun proteomic analysis of tiger milk mushroom (*Lignosus rhinocerotis*) and the isolation of a cytotoxic fungal serine protease from its sclerotium. J Ethnopharmacol 174:437–451
Yue K, Ye M, Lin X, Zhou Z (2013) The artificial cultivation of medicinal caterpillar fungus, *Ophiocordyceps sinensis* (Ascomycetes). Int J Med Mushrooms A review:425–434
Zhang Z, Lv G, Pan H, Pandey A, He W, Fan L (2012) Antioxidant and hepatoprotective potential of endo-polysaccharides from *Hericium erinaceus* grown on tofu whey. Int J Biol Macromol 51(5):1140–1146
Zhou S, Gao Y, Chan E (2005) Clinical trials for medicinal mushrooms: experience with *Ganoderma lucidum* (W.Curt.:Fr.) Lloyd (Lingzhi mushroom). Int J Med Mushrooms 7(1&2):111–118
Zhu LN, Tang Q, Zhou S, Liu Y, Zhang Z, Gao X, Wang S, Wang Z (2014a) Isolation and purification of a polysaccharide from the caterpillar medicinal mushroom *Cordyceps militaris* (ascomycetes) fruit bodies and its immunomodulation of RAW 264.7 macrophages. Int J Med Mushrooms 16(3):247–257
Zhu Y, Chen Y, Li Q, Zhao T, Zhang M, Feng W, Takase M, Wu X, Zhou Z, Yang L, Wu X (2014b) Preparation, characterization, and anti-*Helicobacter pylori* activity of Bi3+-*Hericium erinaceus* polysaccharide complex. Carbohydr Polym 110:231–237

Chapter 15
Biological Control of Microbial Pathogens in Edible Mushrooms

Gail M. Preston, Jaime Carrasco, Francisco J. Gea, and María J. Navarro

15.1 Introduction

Mushrooms are commercially cultivated through an intensive agricultural practice performed under controlled environmental conditions using selective substrates specifically designed for the growth of the target host. Production is susceptible to losses due to the effect of harmful abiotic and biotic disorders. The **biotic disorders** are caused by microbial pathogens and pests, mainly bacterial or fungal parasites and flies, which provoke mushroom diseases and ultimately yield losses. To fight parasites and pests the mushroom industry implements strict hygiene measurements in addition to the use of preventive **phytosanitary** products, especially selective **fungicides** and **insecticides**. Environmental issues, the occurrence of resistant strains and consumer interest in the perceived health benefits of organic food are driving industry to seek out alternative approaches for pest and disease control. This chapter is focused on the biological control implemented to fight the harmful parasites and pests in mushroom crops. Biocontrol in the form of selected bacterial strains and **entomopathogenic** nematodes, or **bioactive compounds** showing antagonism against parasites and pests, have the potential to address these issues. In addition, the implementation of integrated management programs based on thorough cleaning between crops, use of effective air filters to prevent disease and pest dispersion, early detection of disease symptoms, and atmospheric control through ventilation provide the foundation for reducing chemical use in **mushroom cultivation**. Pest and diseases have been reported to affect different varieties of cultivated mushrooms. Since the biotic disorders affecting the white button mushrooms (*Agaricus bisporus*) are better described, and some of them affect other cultivated

G. M. Preston · J. Carrasco (✉)
Department of Plant Sciences, University of Oxford, Oxford, UK
e-mail: jaime.carrasco@plants.ox.ac.uk

F. J. Gea · M. J. Navarro
Centro de Investigación Experimentación y Servicios del Champiñón (CIES), Quintanar del Rey, Cuenca, Spain

B. P. Singh et al. (eds.), *Biology of Macrofungi*, Fungal Biology,
https://doi.org/10.1007/978-3-030-02622-6_15

species, such as *Auricularia mesenterica*, *Flammulina velutipes*, *Ganoderma lucidum* or *Pleurotus ostreatus* (Carrasco et al. 2017), we focus our discussion primarily on *A. bisporus* as a model organism.

15.2 Mushroom Pests and Diseases

A number of biotic disorders are responsible for yield losses in mushroom crops, among them bacterial and fungal diseases are the most renowned causative agents of mushroom diseases. The fungal disturbances can be subdivided into **competitors** and **parasites** (Geels et al. 1988). The first group is constituted by those fungi that fight with the cultivated mushroom for resources and space, being able to interfere with the growth of the mushroom mycelium during the colonization of the compost or the casing. Examples of competing fungi include plaster moulds such as *Scopulariopsis fimicola* or *Papulaspora byssina* (Fletcher and Gaze 2008). On the other hand, **parasitic fungi** infect the mycelium and/or the **fruiting bodies** causing damage, and in some cases becoming lethal. Among the most harmful fungal diseases are dry and wet bubble (*Lecanicillium fungicola* and *Mycogone perniciosa*), green mould (*Trichoderma aggressivum*) and cobweb (*Cladobotryum mycophilum*) (Fig. 15.1) (Fletcher and Gaze 2008; Largeteau and Savoie 2010).

Fig. 15.1 Examples of fungal diseases affecting *A. bisporus*: (**a**) Dry bubble (*Lecanicillium fungicola*); (**b**) Wet bubble (*Mycogone perniciosa*); (**c**) Green mould (*Trichoderma aggressivum*); (**d**) Cobweb disease (*Cladobotryum mycophilum*)

Symptoms of disease include the occurrence of undifferentiated masses of fungal tissue called **bubbles**, the reduction of the available crop surface due to colonisation of the casing by the parasite, conidial masses engulfing diseased mushrooms and causing rot, and the common symptom of spotting on caps due to the **germination** of spores that land on the cap surface and make the mushrooms unsalable (Fig. 15.1 and 15.2) (Gea and Navarro 2017).

A range of diseases caused by bacterial parasites affect mushroom crops (Fletcher and Gaze 2008). Of these, brown blotch disease, caused by *Pseudomonas tolaasii* and other *Pseudomonas* sp. and characterized by the occurrence of light brown, irregular and often sunken spots, on the surface of **mushroom caps** (Navarro et al. 2018), and the internal stipe necrosis caused by the parasite *Ewingella americana,* which provokes dark brown discolouration of the central tissue of the stalk, are the most renowned bacterial diseases (Fig. 15.2 and 15.3) (Fletcher and Gaze 2008).

Fig. 15.2 Symptoms of cap spotting on *A. bisporus* due to microbial pathogens: (**a**) *Lecanicillium fungicola*; (**b**) *Trichoderma aggressivum*; (**c**) *Cladobotryum mycophilum*; (**d**) *Pseudomonas tolaasii*

Fig. 15.3 Symptoms of bacterial diseases: (**a**) Brown blotch caused by *Pseudomonas* sp. in button mushrooms; (**b**) Infection by *Pseudomonas* sp. in oyster mushrooms, *Pleurotus ostreatus,* causing yellowing and stunting; (**c**) Internal stipe necrosis of button mushrroms caused by *Ewingella americana*

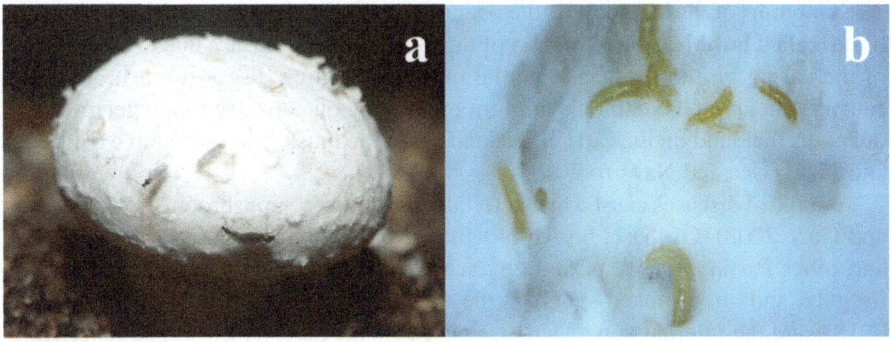

Fig. 15.4 Mushroom flies: (**a**) Adults of sciarid flies on mushroom cap; (**b**) Cecid larvae feeding on mushroom tissue

In addition, certain pests including flies, mites and nematodes condition mushroom yield (Fletcher and Gaze 2008). Among them, insects belonging to the order Diptera, including forids, sciarids and cecids, constitute the most serious pests. Fly larvae feed on the **vegetative mycelium** and the basidiomes while adults are attracted by the organic **volatile compounds** secreted by the mushroom and represent an important dispersal vector for fungal diseases such as *Lecanicillium fungicola* (Fig. 15.4) (Berendsen et al. 2010).

15.3 Biocontrol of Fungal Diseases

Mycoparasites are the biotic agents responsible for the largest crop losses in mushroom crops, and their control is currently based on hygiene and routine application of fungicides (Fletcher and Gaze 2008; Berendsen et al. 2010; Gea et al. 2010; Carrasco et al. 2016). Contamination of the casing is considered the primary source of infection (Carrasco et al. 2017). However, the germination of *L. fungicola* has been reported to be inhibited by the **microbiota** inhabiting the casing layer (Berendsen et al. 2010), and the presence of *A. bisporus* is required to release these mycoparasites from fungistasis. This fact has led researchers look for alternative biocontrol agents within the casing layer that inhibit germination or directly inhibit **mycoparasite** growth.

Currently, *Bacillus velezensis* QST 713 (Serenade®) is the only biocontrol agent registered for mushroom use (Pandin et al. 2018). The commercial biocontrol agent Mycostop® (*Streptomyces griseoviridis*) has been also shown to be as effective as chemical treatment against *L. fungicola* in cropping experiments (Beyer et al. 2016).

A number of bacterial strains from the genera *Bacillus* and *Pseudomonas*, that naturally inhabit the compost or the casing, including species such as *B. pumilus, B. licheniformis, B. amyloliquefaciens* or *B. subtilis*, have been found to show a selective suppressive effect over mycoparasites *in vitro* (Berendsen et al. 2012; Liu et al. 2015; Stanojević et al. 2016; Milijašević-Marčić et al. 2017). *Bacillus* sp. has

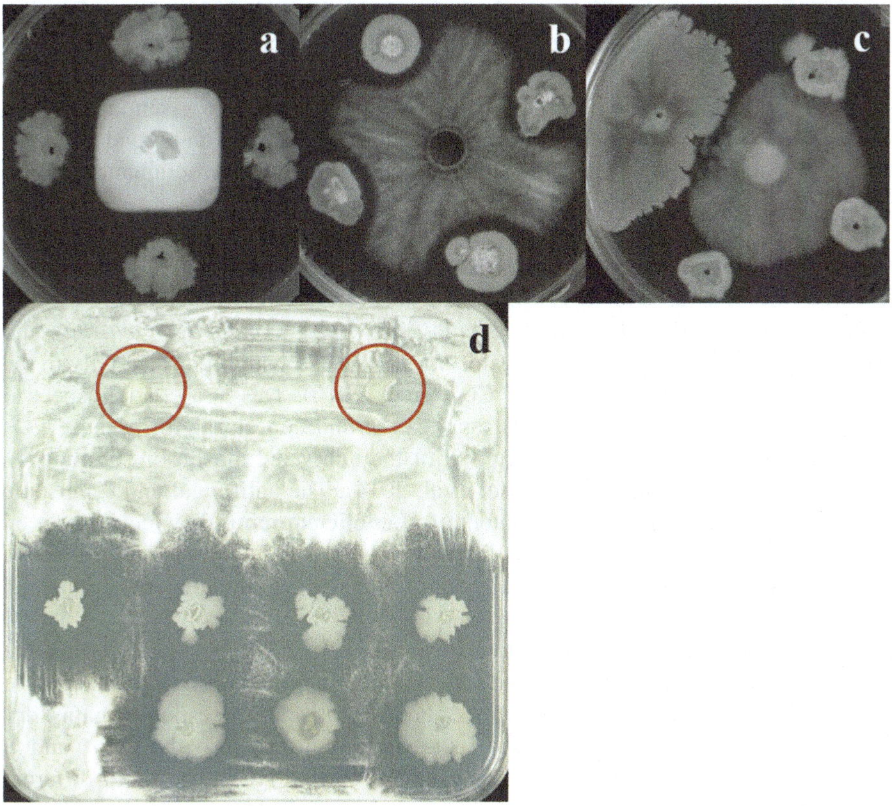

Fig. 15.5 Effect of biocontrol agents (*Bacillus* sp.) co-cultured with mycoparasites in LBA: (**a**) Inhibition of *L. fungicola*; (**b**) Inhibition of *T. aggressivum*; (**c**) Inhibition of *C. mycophilum*; (**d**) Inhibition of conidial germination (*L. fungicola*) compared to non-active strains (red circles). In a-c the mycoparasite is inoculated in the centre of the plate, and candidate biocontrol agents (CBAs) are inoculated adjacent to the growing mycelium. In d the conidial suspension is spread on the agar and the CBAs inoculated afterwards

been observed to inhibit mycelium growth and spore germination of mycoparasites *in vitro* (Fig. 15.5).

However, effective *in vitro* results on tested strains do not often translate efficiently to the commercial crop cycle. Therefore, both the preliminary selection of the most effective candidate and timely application appear to be essential conditions to success on crop trials (Milijašević-Marčić et al. 2017; Aslani et al. 2018). For instance, the introduction of biocontrol agents, such as *Bacillus velezensis* QST 713 and *Bacillus subtillis* B-38, on the casing during cropping show to prevent losses derived from green mould (Pandin et al. 2018; Milijašević-Marčić et al. 2017).

In order to reduce fungicide dependence, environmentally friendly biomolecules have also been tested, showing variable efficacy to inhibit **fungal diseases** in crop (Gea et al. 2014; Mehrparvar et al. 2016; Dos Santos et al. 2017). Essential oils (EOs) from aromatic plants have been employed since the middle ages for their antiseptic, bactericidal or fungicidal effects (Bakkali et al. 2008). Some EOs act synergistically as fungicides and as germination inhibitors against mycoparasites (Soković et al. 2006; Glamočlija et al. 2007, 2009). Among EOs from different aromatic plants and their active compounds, carvacrol and the essential oil from *Origanum vulgare* (wild marjoram) have shown higher activity and broad spectrum efficacy against mycoparasites (Soković et al. 2006). EOs from cinnamon, clove, thyme, "tea tree" (*Melaleuca alternifolia*), *Satureja hortensis* or *Zataria multiflora* and the volatile fraction of oregano (carvacrol and thymol) and geranium (citronellol and genariol) showed antifungal activity *in vitro* against bubble diseases and cobweb (Mehrparvar et al. 2016; Dos Santos et al. 2017; Tanović et al. 2006, 2009; Potočnik et al. 2010). EOs from *Lippia citriodora*, *Cymbopogon citratus* and *Thymus vulgaris* showed toxicity against *M. perniciosa* with high selectivity to the parasite in crop trials (Regnier and Combrinck 2010). This is a crucial aspect to attend in order to determine correct application and concentration, because EOs have been reported to inhibit the growth of the cultivated mushroom **hyphae**, as other chemical fungicides do (Geösel et al. 2014). The antifungal and fungistatic activity of EOs are based on the content of **phenolic compounds** which impact on the structure and function of the cellular membrane. In addition, these bioactive compounds are mostly volatiles, which could facilitate the formulation of commercial products while minimizing chemical residues (Soković et al. 2006).

Compost teas (CT), aqueous extracts from fermented composted materials that are used to control plant diseases and for crop fertilisation, have been evaluated for their use in the cultivation of *A. bisporus*, showing satisfactory response against dry bubble diseases and no inhibitory effects on button mushroom both *in vitro* and *in vivo* (Gea et al. 2012, 2014). Non-aerated compost teas (NCT) from four different composts obtained from agricultural wastes and aerated compost teas (ACT) from grape marc compost showed good results to control *L. fungicola in vitro* (Diánez et al. 2006; Gea et al. 2009). The analysis of the cultivable microbiome suggests that the inhibitory effect of compost teas is due to the activity of the native microbiota (Marín et al. 2013). In addition, the treatment with compost tea obtained from spent mushroom compost (waste material from mushroom cultivation) was an effective treatment both *in vitro* and *in vivo* to control *L. fungicola* and bubble disease while showing a reduce fungitoxic effect on the host (Gea et al. 2012; Dos Santos et al. 2017), which suggests an alternative to recycle this abundant by-product. However, since spent mushroom compost is susceptible of being a reservoir of mushroom parasites, the waste material must be previously re-composted to eliminate any source of disease inoculum.

15.4 Biocontrol of Bacterial Diseases

Bacterial diseases are important biotic disorders that provoke significant yield losses. The most important bacterial disease in mushroom crops is brown blotch caused by *Pseudomonas tolaasii* and *Pseudomonas* sp., and the majority of reported efforts to exercise biological control have been designed to minimize blotch symptoms.

Different *Agaricus* strains differ on sensitivity to bacterial diseases such as **bacterial blotch** (Soler-Rivas et al. 1999), so breeding for resistance is one of the strategies available to control bacterial diseases (Savoie et al. 2013). Comparative mapping of quantitative trait loci involved in resistance to the bacterial blotch, *Pseudomonas tolaasii*, affecting *Agaricus bisporus*, partially characterized loci involved is resistance to brown blotch (Moquet et al. 1999). However, as far as the authors are aware, currently there are no commercial varieties of white button mushroom resistant to brown blotch broadly cultivated, and therefore biocontrol agents have been explored as an alternative form of disease control.

Antagonistic bacteria have been reported as biological control agents of bacterial blotch by certain authors (Largeteau and Savoie 2010; Kim et al. 2011; Tajalipour et al. 2014; Aslani et al. 2018). For example, *P. reactans* produces the **lipopeptide** WLIP (the white line inducing principle), which interacts with and limits the toxicity of the lipopeptide toxin tolaasin that is thought to be largely responsible for the symptoms caused by *P. tolaasii* (Largeteau and Savoie 2010). The interaction between these two biomolecules can be observed as a white precipitate in agar (Fig. 15.6). Although the use of *P. reactans* or purified WLIP as a control agent for bacterial blotch disease has been evaluated (Largeteau and Savoie 2010), to date none commercial product is available.

The microbiota inhabiting fruiting bodies has been described as a potential source for biocontrol agents, since cultivated mushrooms produce volatiles with

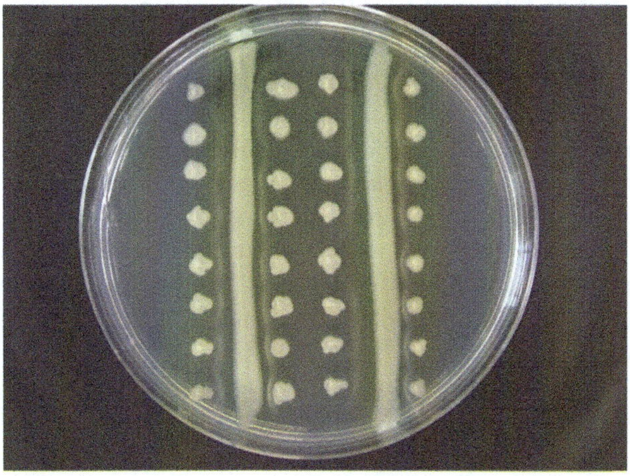

Fig. 15.6 "White line test" for tolaasin and WLIP production. Bacteria are inoculated at both sides of the line of the test bacterium (either *Pseudomonas reactans* or *Pseudomonas tolaasii*). Colonies that result in a white precipitate in between the two bacteria, after 48 h incubation at 28 °C on King's B medium are identified as potential producers of tolaasin and related molecules if *P. reactans* is the test bacterium, and as potential producers of WLIP and related molecules if *P. tolaasii* is the test bacterium. In the image shown *P. reactans* NCPPB387 is the test bacterium

antimicrobial effects associated to endomicrobes (Kanchiswamy et al. 2015; Kües et al. 2018). For instance, *Pseudomonas taiwanensis* Bi1 and *Bacillus thuringiensis* De4 isolated from wild mushrooms have been reported to show *in vitro* antagonistic properties against the bacterial parasites *Pseudomonas tolaasii* and *Ewingella americana*, and the ability to reduce brown blotch and internal stipe necrosis symptoms in crop (Aslani et al. 2018). The bacterial strains *Pseudomonas putida* A1, *Pseudomonas reactants* A2 and A6, *Pseudomonas fluorescens* A3 and A4, and *Bacillus subtilis* A5, isolated from compost, casing and fruit bodies, were described to inhibit symptoms of brown blotch both in mushroom tissue and during cultivation (Tajalipour et al. 2014).

Experiments conducted to screen the phytochemical control of bacterial diseases, using plant extracts, have reported variable effects (Soković et al. 2006; Malpani et al. 2012). Nowadays, the adequate management of environmental condition together with the regular drench application of chlorinated water to the casing layer are common methods employed to prevent the impact of bacterial diseases (Fletcher and Gaze 2008; Navarro et al. 2018; Diamantopoulou et al. 2012; Lomax et al. 2007). The development of the disease has been described to be associated to high relative air humidity. Moisture condenses on caps if there is no adequate ventilation. Environmental condition must be adequately managed to prevent condensation on mushroom caps and facilitate the water drying over their surface prevent in an effort to prevent occurrence and dispersion of bacterial blotch disease (Navarro et al. 2018).

15.5 Biological Control of Mushroom Pests

Historically mushroom flies have been treated though preventive chemicals but evidence of resistance has been reported (Bartlett et al. 1997; Smith 2002). Furthermore, the use of these phytosanitary products provokes adverse effects on mushroom mycelium that derives into yield losses or quality shortcuts and the presence of residues has been detected in the harvested mushrooms (Shamshad 2010). Alternative methods for the control of mushroom flies include physical barriers such as nets to prevent the entrance of adults in the farm (Coles 2002); plant extracts and biocontrol organisms, including mites, bacteria or (Jess and Bingham 2004; Shamshad et al. 2008; Erler et al. 2009a, b).

As regards the biocontrol organisms, nematodes are often applied to control mushroom flies. *Steinernema* spp. nematodes track and infest the fly larvae, causing the death of infested individuals by releasing bacteria (*Xenorhabdus* sp. or *Photorhabdus* sp.) (Kirk and Keil 2001). The nematodes employed require compatibility, in case that insecticide treatment has been previously applied (Koppenhöfer and Grewal 2005), and their effectiveness relies on the environmental factors applied during the crop cycle (Kirk and Keil 2001). The application of *Steinernema feltiae* against phorid flies has shown varied results (Erler et al. 2009b; Navarro and Gea 2014), whereas *Steinernema carpocapsae* treatments have shown favourable results to control phorids (Jess and Bingham 2004). *S. feltiae* has been postulated as a good biological agent for the control of sciarids (Shamshad et al. 2008), since *S. carpocapsae* showed lower infectivity than *S. feltiae* (Kim et al. 2004). Both *S. carpocapsae* and *S. feltiae* treatments reduced the population of the mushroom fly *Lycoriella auripila* in crops, with *S. feltiae* reported to be most effective treatment. However, no decrease in the population of the mushroom phorid fly *Megaselia halterata* was detected with either nematode treatment (Navarro et al. 2014).

Important aspects concerning dosage rate and time of application must be monitored to maximise the efficiency of the treatment and to reduce their impact on crop, since a detrimental effect on mushroom mycelia has been reported depending on the value of nematode applied (Grewal et al. 1992). Certain pathogenic bacteria have been also reported as potential biological control for mushroom flies. The bacterial larvicide *Bacillus thuringiensis* var. *israelensis* Berliner (Bti) was shown to reduce the incidence of fruit damage by *M. halterata* larvae and resulted in significantly lower losses (Koppenhöfer and Grewal 2005). However, *Bacillus thuringiensis* ssp. *israelensis* was ineffective against *Lycoriella ingenua* (Shamshad et al. 2008).

Among a series of different plant extracts evaluated against *M. halterata*, neem products and hot-water extracts of *Origanum onites* (pot marjoram) and *Pimpinella anisum* (aniseed) were identified as potential alternatives to conventional pesticides for the control of mushroom phorid fly since their application reduced the number of emerging adults and carpophore damages (Erler et al. 2009a).

15.6 Concluding Remarks

The control of microbial diseases currently relies on the application of integrated management programs in addition to the use of pesticides, mostly fungicides (Gea and Navarro 2017). The management of the environmental conditions in the facilities together with the use of air circulation systems, overpressure in the growing chambers and the possibility of applying "cooking-out" at the end of the cycle, provide an adequate environment to develop a program for integrated pest and disease control. However, an increasing demand for organic products and the reported evidence of resistant strains to fungicides or insecticides advocate for the development of novel sustainable strategies for mushroom crop protection.

Microbial biocontrol agents have significant potential for biocontrol, if formulated and applied appropriately. However, the mechanism of parasite inhibition in mushroom crops is still unknown. Mechanisms of parasite inhibition might be related to the production of germination inhibitors, various hydrolytic enzymes and secondary metabolites, or the formation of **biofilms** around the host hyphae. They may be based on antagonistic relationships involving for instance **mycophagy** of parasites or even elicitation of the immune response of the cultivated species (Leveau and Preston 2008; Payapanon et al. 2011; Pandin et al. 2018). The understanding of these mechanisms will facilitate the development of new and effective biological approaches to control diseases.

Bioactive compounds from aromatic plants show promising antifungal, bactericidal and insecticidal activity to be incorporated as alternative products although more work is required to understand their action and optimise their application.

Ultimately, mushroom science is a young discipline that could certainly benefit from the development of a range of approaches for the biological control of mushroom disease while developing resistant varieties to target parasites. The development of resistant cultivars appears as one of the most effective, economical, and environmentally friendly approach to the management of disease control (Savoie et al. 2013). Independent **quantitative trait loci** (QTL) mapping studies investigating the genetics of the resistance to dry bubble or green mould disease have been performed evaluating the QTL involved in the expression of disease symptoms (Foulongne-Oriol et al. 2011, 2012). In addition, resistance to fungal parasites such as *M. perniciosa* have been reported among wild and commercial strains of *A. bisporus* (Fu et al. 2016) suggesting a real possibility for exploring breeding as a potential alternative to design resistant strains to mycoparasites (Sonnenberg et al. 2017). However, the genetic resources available in the mushroom field are still scarce and some cultivated varieties such as the white button mushroom present technical difficulties due to the nature of their life cycle which hinders breeding (Sonnenberg et al. 2017). Therefore, further research into alternative approaches for control of mushroom diseases, including biocontrol, should continue to be a priority to ensure sustainable mushroom production in the future.

Acknowledgement This article has been funded by the European Union's Horizon 2020 research and innovation programme under the Marie Sklodowska-Curie grant agreement No 742966. Some of the pictures attached in the text have been kindly loaned by CIES (Antonio Martínez Carrasco).

References

Aslani MA, Harighi B, Abdollahzadeh J (2018) Screening of endofungal bacteria isolated from wild growing mushrooms as potential biological control agents against brown blotch and internal stipe necrosis diseases of *Agaricus bisporus*. Biol Control 119:20–26

Bakkali F, Averbeck S, Averbeck D, Idaomar M (2008) Biological effects of essential oils-a review. Food Chem Toxicol 46(2):446–475

Bartlett GR, Keil CBO (1997) Identification and characterization of a permethrin resistance mechanism in populations of the fungus gnat *Lycoriella mali* (Fitch) (Diptera: Sciaridae). Pest Biochem Physiol 58:173–181

Berendsen RL, Baars JJ, Kalkhove SI, Lugones LG, Wösten HA, Bakker PA (2010) *Lecanicillium fungicola*: causal agent of dry bubble disease in white-button mushroom. Mol Plant Pathol 11(5):585–595

Berendsen RL, Kalkhove SI, Lugones LG, Baars JJ, Wösten HA, Bakker PA (2012) Effects of fluorescent *Pseudomonas* spp. isolated from mushroom cultures on *Lecanicillium fungicola*. Biol Control 63:210–221

Beyer DM, Pecchia JA, Paley K (2016) Evaluation of bio-fungicides for the control of fungal diseases on *Agaricus bisporus*. In: Baars JJP, Sonnenberg AMS (eds) Proceedings of the 19th international society for mushroom science (ISMS) conference, pp 86–90

Carrasco J, Navarro MJ, Santos M, Diánez F, Gea FJ (2016) Incidence, identification and pathogenicity of *Cladobotryum mycophilum*, causal agent of cobweb disease on *Agaricus bisporus* mushroom crops in Spain. Ann Appl Biol 168(2):214–224

Carrasco J, Navarro MJ, Gea FJ (2017) Cobweb, a serious pathology in mushroom crops: a review. Span J Agric Res 15(2):19

Coles P (2002) Integrated pest management in mushroom production: specific control techniques: exclusion. In: MUSHROOM integrated pest management. The Pennsylvania State University, State College, pp 21–26

Diamantopoulou P, Philippoussis A, Lahouvaris L, Parissopoulos G (2012) The effect of calcium chloride irrigation on yield and quality of *Agaricus bisporus*. In: Proceeding of the 15th congress of ISMS. Rotterdam, Netherlands. pp. 390–393

Diánez F, Santos M, Boix A, De Cara M, Trillas I, Avilés M, Tello JC (2006) Grape marc compost tea suppressiveness to plant pathogenic fungi: role of siderophores. Compost Sci Util 14(1):48–53

Dos Santos TL, Belan LL, Zied DC, Dias ES, Alves E (2017) Essential oils in the control of dry bubble disease in white button mushroom. Cienc Rural 47, 05:e20160780

Erler F, Polat E, Demir H, Cetin H, Erdemir T (2009a) Control of the mushroom phorid fly, *Megaselia halterata* (wood), with plant extracts. Pest Manag Sci 65:144–149

Erler F, Polat E, Demir H, Cetin H, Erdemir T (2009b) Evaluation of microbial products for the control of the mushroom phorid fly, *Megaselia halterata* (wood). J Entomol Sci 44:89–97

Fletcher JT, Gaze RH (2008) Mushroom pest and disease control:A color handbook. Manson Publishing Ltd. Academic Press, San Diego, p 192

Foulongne-Oriol M, Minvielle N, Savoie JM (2011) QTL for resistance to Trichoderma lytic enzymes and metabolites in *Agaricus bisporus*. In; Proceedings of the 7th international conference on mushroom biology and mushroom products (ICMBMP7), Arcachon, France. vol 2. pp 17–25

Foulongne-Oriol M, Rodier A, Savoie JM (2012) Relationship between yield components and partial resistance to *Lecanicillium fungicola* in *Agaricus bisporus* assessed by QTL mapping. Appl Environ Microbiol 78:2435–2442

Fu Y, Wang X, Li D, Liu Y, Song B, Zhang C et al (2016) Identification of resistance to wet bubble disease and genetic diversity in wild and cultivated strains of *Agaricus bisporus*. Int J Mol Sci 17(10):1568

Gea FJ, Navarro MJ (2017) Mushroom diseases and control. In: Cunha D, Pardo-Gimenez A (eds) Edible and medicinal mushrooms: Technology and applications. Wiley Blackwell Ltd, Chichester, pp 239–259

Gea FJ, Navarro MJ, Tello JC (2009) Potential application of compost teas of agricultural wastes in the control of the mushroom pathogen *Verticillium fungicola*. J Plant Dis Protect 116(6):271–273

Gea FJ, Tello JC, Navarro MJ (2010) Efficacy and effects on yield of different fungicides for control of wet bubble disease of mushroom caused by the mycoparasite *Mycogone perniciosa*. Crop Prot 29(9):1021–1025

Gea FJ, Santos M, Diánez F, Tello JC, Navarro MJ (2012) Effect of spent mushroom compost tea on mycelial growth and yield of button mushroom (*Agaricus bisporus*). World J Microbiol Biotechnol 28(8):2765–2769

Gea FJ, Carrasco J, Diánez F, Santos M, Navarro MJ (2014) Control of dry bubble disease (*Lecanicillium fungicola*) in button mushroom (*Agaricus bisporus*) by spent mushroom substrate tea. Eur J Plant Pathol 138:711–720

Geels FP, Van de Geijn J, Rutjens AJ (1988) Pests and diseases. In: Griensven V, Interlingua LJLD (eds) The cultivation of mushrooms, Sussex, pp 361–422

Geösel A, Szabó A, Akan O, Szarvas J (2014) Effect of essential oils on mycopathogens of *Agaricus bisporus*. In: Proceedings of 8th international conference on mushroom biology and mushroom products (ICMBMP8), New Delhi, pp 530–535

Glamočlija J, Sokovic M, Vukojevic J, Milenkovic I, Tesevic V, Van Griensven LJLD (2007) Effect of oregano essential oil on infection of *Agaricus bisporus* by *Mycogone perniciosa in vitro* and in the mushroom growing unit. Int J Med Mushrooms 9(3/4):300

Glamočlija J, Sokovic M, Grubisic D, Vukojevic J, Milinekovic I, Ristic M (2009) Antifungal activity of *Critmum maritimum* essential oil and its components against mushroom pathogen *Mycogone perniciosa*. Chem Nat Compd 45(1):96–97

Grewal PS, Richardson PN, Collins G, Edmondson RN (1992) Comparative effects of *Steinernema feltiae* (Nematoda: Steinernematidae) and insecticides on yield and cropping of the mushroom *Agaricus bisporus*. Ann Appl Biol 121:511–520

Jess S, Bingham JFW (2004) Biological control of sciarid and phorid pests of mushroom with predatory mites from the genus Hypoaspis (Acari: Hypoaspidae) and the entomopathogenic nematode *Steinernema feltiae*. Bull Entomol Res 94:159–167

Kanchiswamy CN, Malnoy M, Maffei ME (2015) Bioprospecting bacterial and fungal volatiles for sustainable agriculture. Trends Plant Sci 20:206–211

Kim HH, Choo HY, Kaya HK, Lee DW, Lee SM, Jeon HY (2004) *Steinernema carpocapsae* (Rhabditida: Steinernematidae) as a biological control agent against the fungus gnat *Bradysia agrestis* (Diptera: Sciaridae) in propagation houses. Biocon Sci Technol 14:171–183

Kim MH, Park SW, Kim YK (2011) Bacteriophages of *Pseudomonas tolaasii* for the biological control of brown blotch disease. J Korean Soc Appl Biol Chem 54(1):99–104

Kirk DJ, Keil CB (2001) Factors influencing efficacy of two- entomopathogenic nematodes used for fly control in commercial mushroom crops. Mush News 49:4–17

Koppenhöfer AM, Grewal PS (2005) Compatibility and interactions with agrochemicals and other biocontrol agents. In: Nematodes as biocontrol agents, CABI Eds, pp 363–381

Kües U, Khonsuntia W, Subba S, Dörnte B (2018) Volatiles in communication of Agaricomycetes. In: Anke T, Schüfflerpp A (eds) Physiology and genetics. Springer, Cham, pp 149–212

Largeteau ML, Savoie JM (2010) Microbially induced diseases of *Agaricus bisporus*: biochemical mechanisms and impact on commercial mushroom production. Appl Microbiol Biotechnol 86(1):63–73

Leveau JH, Preston GM (2008) Bacterial mycophagy: definition and diagnosis of a unique bacterial–fungal interaction. New Phytol 177(4):859–876

Liu C, Sheng J, Chen L, Zheng Y, Lee DYW, Yang Y et al (2015) Biocontrol activity of *Bacillus subtilis* isolated from *Agaricus bisporus* mushroom compost against pathogenic fungi. J Agric Food Chem 63:6009–6018

Lomax KM (2007) Dew point temperature related to wet mushroom caps. Mushroom News 55(1):4–10

Malpani MO, Rajput PR, Ghodile NG (2012) Effect of medicinally potential plant extracts and isolated ingredients on pathogen damaging mushroom cultivation in Vidarbha region of India. In: Proceeding of the 18th congress of ISMS. Beijing, China, pp 390–393

Marín F, Santos M, Diánez F, Carretero F, Gea FJ, Yau JA, Navarro MJ (2013) Characters of compost teas from different sources and their suppressive effect on fungal phytopathogens. World J Microbiol Biotechnol 29(8):1371–1382

Mehrparvar M, Goltapeh EM, Safaie N, Ashkani S, Hedesh RM (2016) Antifungal activity of essential oils against mycelial growth of *Lecanicillium fungicola* var. *fungicola* and *Agaricus bisporus*. Ind Crop Prod 84:391–398

Milijašević-Marčić S, Stepanović M, Todorović B, Duduk B, Stepanović J, Rekanović E, Potočnik I (2017) Biological control of green mould on *Agaricus bisporus* by a native *Bacillus subtilis* strain from mushroom compost. Eur J Plant Pathol 148(3):509–519

Moquet F, Desmerger C, Mamoun M, Ramos-Guedes-Lafargue M, Olivier JM (1999) A quantitative trait locus of *Agaricus bisporus* resistance to *Pseudomonas tolaasii* is closely linked to natural cap color. Fungal Genet Biol 28:34–42

Navarro MJ, Gea FJ (2014) Entomopathogenic nematodes for the control of phorid and sciarid flies in mushroom crops. Pesq Agropec Bras 49(1):11–17

Navarro MJ, Carrasco J, Gea FJ (2014) Chemical and biological control of diptera in Spanish mushroom crops. In: Proceedings of 8th international conference on mushroom biology and mushroom products (ICMBMP8), New Delhi, pp 549–5556

Navarro MJ, Gea FJ, Gonzalez AJ (2018) Identification, incidence and control of bacterial blotch disease in mushroom crops by management of environmental conditions. Sci Hort 2018 229:10–18

Pandin C, Le Coq D, Deschamps J, Védie R, Rousseau T, Aymerich S, Briandet R (2018) Complete genome sequence of *Bacillus velezensis* QST713: a biocontrol agent that protects *Agaricus bisporus* crops against the green mould disease. J Biotechnol 278:10–19

Payapanon A, Suthirawut S, Shompoosang S, Tsuchiya K, Furuya N, Roongrawee P et al (2011) Increase in yield of the straw mushroom (*Vovariella volvacea*) by supplement with *Paenibacillus* and *Bacillus* to the compost. J Fac Agric Kyushu U 56:249–254

Potočnik I, Vukojević J, Stajić M, Rekanović E, Stepanović M, Milijašević S, Todorović B (2010) Toxicity of biofungicide Timorex 66 EC to *Cladobotryum dendroides* and *Agaricus bisporus*. Crop Prot 29:290–294

Regnier T, Combrinck S (2010) *In vitro* and *in vivo* screening of essential oils for the control of wet bubble disease of *Agaricus bisporus*. S African J Bot 76:681–685

Savoie JM, Foulongne-Oriol M, Barroso G, Callac P (2013) Genetics and genomics of cultivated mushrooms, application to breeding of agarics. In: Kempken F (ed) The mycota, agricultural applications, vol 11. Springer, Berlin, pp 3–33

Shamshad A (2010) The development of integrated pest management for the control of mushroom sciarid flies, *Lycoriella ingenua* (Dufour) and *Bradysia ocellaris* (Comstock), in cultivated mushrooms. Pest Manag Sci 66:1063–1074

Shamshad A, Clift AD, Mansfield S (2008) Toxicity of six commercially formulated insecticides and biopesticides to third instar larvae of mushroom sciarid, *Lycoriella ingenua* Dufour (Diptera: Sciaridae), in New South Wales, Australia. Aus J Entomol 47:256–260

Smith JE (2002) Dimilin resistance in mushroom sciarids. Mushroom J 626:15

Soković M, Van Griensven LJ (2006) Antimicrobial activity of essential oils and their components against the three major pathogens of the cultivated button mushroom, Agaricus bisporus. Eur J Plant Pathol 116(3):211–224

Soler-Rivas C, Jolivet S, Arpin N, Olivier JM, Wichers HJ (1999) Biochemical and physiological aspects of brown blotch disease of *Agaricus bisporus*. FEMS Microbiol Rev 23(5):591–614

Sonnenberg AS, Baars JJ, Gao W, Visser RG (2017) Developments in breeding of *Agaricus bisporus* var. *bisporus*: progress made and technical and legal hurdles to take. Appl Microbiol Biot 101(5):1819–1829

Stanojević O, Milijašević-Marčić S, Potočnik I, Stepanović M, Dimkić I, Stanković S, Berić T (2016) Isolation and identification of *Bacillus* spp. from compost material, compost and mushroom casing soil active against *Trichoderma* spp. Arch Biol Sci 68:845–852

Tajalipour S, Hassanzadeh N, Khabbaz Jolfaee H, Heydari A, Ghasemi A (2014) Biological control of mushroom brown blotch disease using antagonistic bacteria. Biocontrol Sci Tech 24:473–484

Tanović B, Potočnik I, Stanisavljević B, Đorđević M, Rekanović E (2006) Response of *Verticillium fungicola* var. *fungicola*, *Mycogone perniciosa* and *Cladobotryum* sp. mushroom pathogens to some essential oils. Pestic Phytomed 21(3):231–237

Tanović B, Potočnik I, Delibašić G, Ristić M, Kostić M, Marković M (2009) *In vitro* effect of essential oils from aromatic and medicinal plants on mushroom pathogens: *Verticillium fungicola* var. *fungicola*, *Mycogone perniciosa*, and *Cladobotryum* sp. Arch Biol Sci 61(2):231–237

Chapter 16
Cordycepin: A Biotherapeutic Molecule from Medicinal Mushroom

Mohammad Soltani, Roslinda Abd Malek, Nagib A. Elmarzugi, Mohamad Fawzi Mahomoodally, Davin Uy, Ong Mei Leng, and Hesham A. El-Enshasy

16.1 Introduction

For many years, mushrooms have been considered as important food based on its rich nutritional value. In addition, different mushrooms showed high potential application in medical applications. (El Enshasy et al. 2013) and pharmacological (Wasser 2002; Elenshasy 2010; El Enshasy et al. 2013). The "Chinese caterpillar

M. Soltani · R. A. Malek
Institute of Bioproduct Development (IBD), Universiti Teknologi Malaysia (UTM), Skudai, Malaysia

N. A. Elmarzugi
Department of Industrial Pharmacy, Faculty of Pharmacy & Biotechnology Research Center, Tripoli, Libya

M. F. Mahomoodally
Department of Health Sciences, Faculty of Science, University of Mauritius, Réduit, Mauritius

D. Uy
Research and Innovation Center, Tumnuptek, Chamkarmorn, Phnom Penh, Cambodia

O. M. Leng
Harita Go Green Sdn Bhd, Johor Bahru, Johor, Malaysia

H. A. El-Enshasy (✉)
Institute of Bioproduct Development, Universiti Teknologi Malaysia, Johor Bahru, Johor, Malaysia

School of Chemical Engineering, Faculty of Engineering, Universiti Teknologi Malaysia, Johor Bahru, Johor, Malaysia

City of Scientific Research and Technology Application, New Burg Al Arab, Alexandria, Egypt
e-mail: henshasy@ibd.utm.my

B. P. Singh et al. (eds.), *Biology of Macrofungi*, Fungal Biology,
https://doi.org/10.1007/978-3-030-02622-6_16

fungus", is one of the conventional medicinal mushrooms. This fungus attacks and expands inside a particular insect and kills the host in the later stage. During the time of winter passing inside the insect, the fruituning body grows on the surface of the corpse. However, there are numerous known types of Chinese caterpillar growths, which fit in with numerous distinctive such as *Cordyceps*, *Torrubiella* and *Paecilomyces*. One of the well-knownCordyceps species which is the, *Cordyceps sinensis*, become a parasite on the larvae of *Hepialus armoricanus*, been utilized to initiate the longevity, soothe fatigue and kill various ailments as Chinese traditional medicine (Pegler et al. 1994; Zhu et al. 1998; Buenz et al. 2005). Later contemplates showed that different species showed numerous pharmacological activities and showed potential application as **anticancer** (Leung et al. 2006; Wu et al. 2007; Lin and Chiang 2008; Lee et al. 2009; Lee et al. 2010; Choi et al. 2011; Jeong et al. 2011) **antioxidan**t (Wang et al. 2005; Ramesh et al. 2012; Ren et al. 2012;), **anti-diabetic** (hypoglycemic) (Yun et al. 2003; Ma et al. 2015), **anti-inflammatory** (Mizuno 1999; Won and Park 2005; Jeong et al. 2010; Choi et al. 2014), and **immunomodulator** (Ng and Wang 2005; El Enshasy & Hatti-Kaul, 2013; Jeong et al. 2013; Yao et al. 2014; Yang et al. 2015; Lee et al. 2015; Peng et al. 2015; Zhang et al. 2015;). Further investigations for isolation of bioactive molecules from this type of fungus showed that that low molecular weight compound, cordycepin is one of the potent active molecules with many **biotherapeutic** activities (Cunningham 1951; Suhadolnik and Cory 1964; Fujita et al. 1994; Kiho and Ukai 1995; Bok et al. 1999; Li et al. 2004; Yu et al. 2004a; Yu et al. 2004b; Yalin et al. 2006).

Cordycepin (3′-deoxyadenosine) well known as nucleoside analogue that have different type of bioactivities. The cordycepin will be converted into 5′-mono, di and triphosphates and thus hinder the movement of ribose-phosphate pyrophospho kinase and 5-phosphoribosyl-1-pyrophosphate amidotransferase in the de novo purines biosynthesis and/or the nucleic acids synthesis causing the antimetastatic, antitumor and antimicrobial results (Overgaard-Hansen 1964a; Rottman and Guarino 1964; Cory et al. 1965; Rich et al. 1965; Ahn et al. 2006; Yoshikawa et al. 2004; Nakamura et al. 2005). In addition, cordycepin with its antileukemic ability normally join with adenosine deaminase inhibitor and this will cause the inhibitoryn effect to take place which help to analogues of 2′, 5′- oligoadenylate towards the human immunodeficiency virus infection (Muller et al. 1991; Kodama et al. 2000;). Large scale culturing of mycelial through synthetic can be used a s a new source of cordycepin due to its limited amount in natural source. Tow stage control of dissolved oxygen or addition of NH4+ to the **submerged medium** can help to improve the production of cordycepin (Mao and Zhong 2004, 2006). In addition to production using fermentation technology, cordycepin could be also produced chemically. However, chemical synthesis has some limitation such as complication of the process and the utilization of large volume of organic solvents which decrease the attractiveness of this process (Hansske and Robins 1985; Aman et al. 2000).

The purpose of chapter is to give focus on the recent research for the production and therapeutic applications of cordycepin. However, weak and strong points of previous studies can be recognized through this review. Thus, it will be useful for further assessments to improve the production process and future medical applications of cordycepin in the treatment of different diseases.

16.2 Cordycepin Specifications

16.2.1 Chemical Structure, Molecular and Physical Properties

Cordycepin, or 3′-deoxyadenosine, is a subsidiary of the nucleoside adenosine, contrasting from the recent by the nonattendance of oxygen in the 3′ position of its ribose part. It was at first concentrated from fungi, family Cordyceps, yet is presently processed chemically. In light of the fact, as cordycepin is like adenosine, a few steps cannot separate between the two. Along these lines, it can take part in certain biochemical reactions (for instance, be joined into a RNA molecule, therefore initiating the untimely end of its amalgamation). Despite of other unknown bioactive ingredients (e.g. polysaccharides) extracted from various species of cordyceps with different molecular weights and biotherapeutic activities (Soltani et al. 2013). The molecular structure of adenosine, its analogues deoxyadenosine and cordycepin are presented in Fig. 16.1. Moreover, scientific data of cordycepin is indicated in Table 16.1.

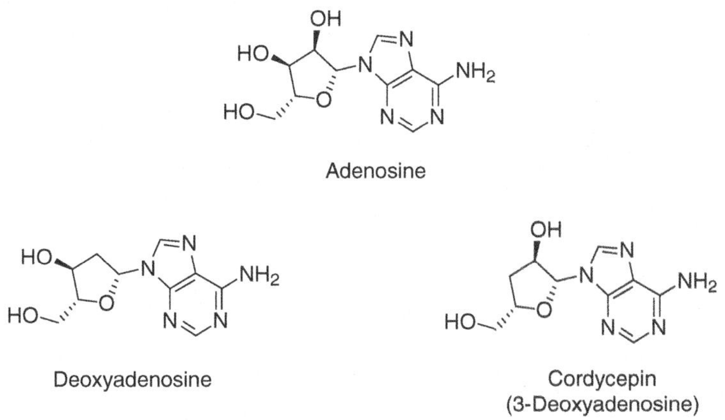

Fig. 16.1 Molecular structure of adenosine, deoxyadenosine and cordycepin (3′-deoxyadenosine)

Table 16.1 Cordycepin molecular information

Name	IUPAC name	9-(3-Deoxy-β-D-ribofuranosyl)adenine
	Other names	Cordycepine, 3′-Deoxyadenosine
Identifiers	CAS number	73–03-3
	PubChem	6303
	ChemSpider	6064
	ChEMBL	CHEMBL305686
	InChIKey	OFEZSBMBBKLLBJ-
Properties	Molecular formula	$C_{10}H_{13}N_5O_3$
	Molar mass	251.24 g mol^{-1}
	Melting point	225.5 °C, 499 K, 438 °F

Data are given for materials in their standard state (at 25 °C, 100 kPa)

16.2.2 Codycepin Biosynthesis

A **biosynthesis pathway** is frequently starts with a promptly accessible precursor particle that is like the product. The cell then joins together this antecedent with other small molecules, artificially adjusting the product along the way. At every step, the substrate will dynamically look like the final product. A multi-step biosynthesis pathway can have many steps along the way, experiencing consistent change by enzymes until the last compound is shaped. Concentrating on biosynthesis can yield numerous viable experiences into cures for human ailments. Comprehension the science of the human body plainly helps when a sickness results because of failing biosynthesis. Sometimes, on the other hand, concentrating on the biosynthetic pathways of different living beings can also turn up important hints for improving new drugs.

By picking up a comprehension of the biosynthesis pathway of a cordycepin, we can find that how it is synthesized and possibly mimic the synthesis in the laboratory. At last, it will be worthwhile to clone these genes to generate transgenic living organism, which might be designed to prepare natural product at more terrific concentration and purity at a division of the cost. Thus, in order to understand the synthesis of cordycepin deeply, the biosynthesis pathway of cordycepin from mushroom *Cordyceps militaris* is presented in detail on Fig. 16.2 (Zheng et al. 2012).

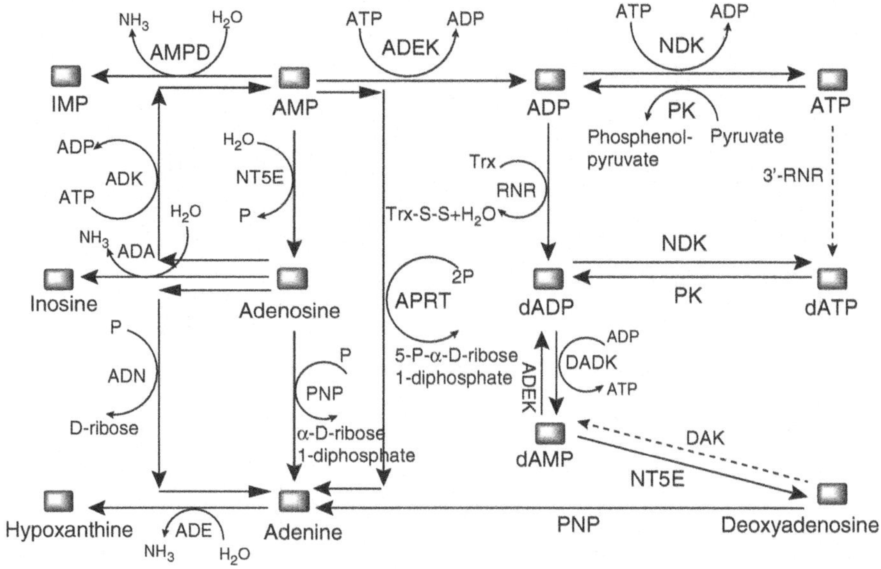

Fig. 16.2 *Cordyceps militaris* adenine metabolic pathway. Abbreviations for different enzymes: *ADA* adenosine deaminase, *ADE* adenine deaminase, *ADEK* adenylate kinase, *ADK* adenosine kinase, *ADN* adenosine nucleosidase, *AMPD*, *AMP* deaminase, *APRT* adenine phosphoribosytransferase, *DADK* deoxyadenylate kinase, *DAK* deoxyadenosine kinase, *NDK* nucleoside-diphosphate kinase, *NT5E*, 5′-nucleotidase, *PK* pyruvate kinase, *PNP* purine nucleoside phosphorylase, 3'-RNR, ribonucleotide triphosphate reductase. The dashed lines exhibit metabolic pathways present in other organisms but absent in *Cordyceps militaris* (Zheng et al. 2012)

It can be clearly seen that several enzymes as significant agents are present during the metabolic pathway of cordycepin incorporating process until the final product is achieved.

16.3 Coducepin: Mechanism of Action

The structure of cordycepin is truly comparable with cellular nucleoside, adenosine (Fig. 16.1) and demonstrates like a nucleoside analogue.

16.3.1 Hinderance of Purine Biosynthesis Pathway

Inside the cells, Cordycepin get changed over into 5' mono-, di- and tri-phosphate that can decrease the catalyst action like ribose-phosphate pyrophosphokinase and 5-phosphoribosyl-1-pyrophosphate amidotransferase that being utilized within **de novo biosynthesis of purines** (Fig. 16.3) (Klenow 1963; Overgaard-Hansen 1964b; Rottman and Guarino 1964).

16.3.2 Cordycepin Incites RNA Chain End

Cordycepin fails to offer 3′-hydroxyl group in its molecular form (Fig. 16.1), which is the main distinction from adenosine. Adenosine is a nitrogenous base and function as cell nucleoside, which is needed for the different molecular procedures in cells such as synthesis of RNA or DNA. Throughout the procedure of transcription (RNA combination), a few enzymes are not being recognize the adenosine and cordycepin, that prompts joining of 3′-deoxyadenosine, or cordycepin, in place of typical nucleoside avoiding further fuse of nitrogenous bases (A, U, G, and C), prompting untimely end of **transcription** (Chen et al. 2008; Holbein et al. 2009).

16.3.3 Cordycepin Meddles in mTOR Signal Transduction

Cordycepin has been used for its abbreviation of the **poly A tail** of m-RNA, which influences the strength within the cytoplasm. It was watched that restraint of polyadenylation by cordycepin of some m-RNAs made them touchier than alternate **mRNAs**. At higher dosages, Cordycepin represses cell connection and lessens focal attachment. Further rise in the usage of cordycepin may terminate **mTOR** (mammalian focus of rapamycin) signaling pathway (Fig. 16.4) (Wong et al. 2009). The name mTOR has been determined from the medication rapamycin, on the grounds that this medicine represses

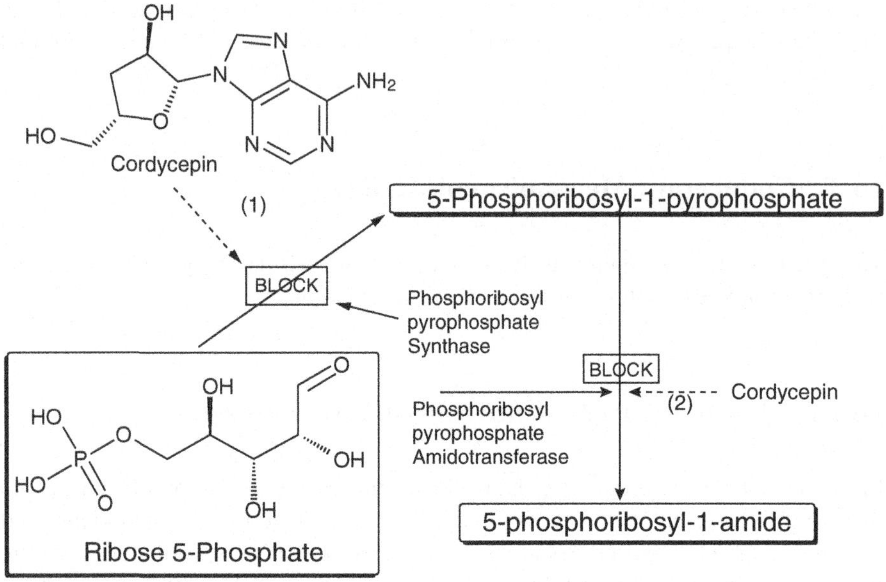

Fig. 16.3 The hindrance effect of cordycepin in mono- and tri- phosphate states on the catalyst enzymes, phosphoribosyl pyrophosphate synthase and phosphoribosyl pyrophosphate amidotransferase contain in biosynthesis pathway of purine (Tuli et al. 2014b)

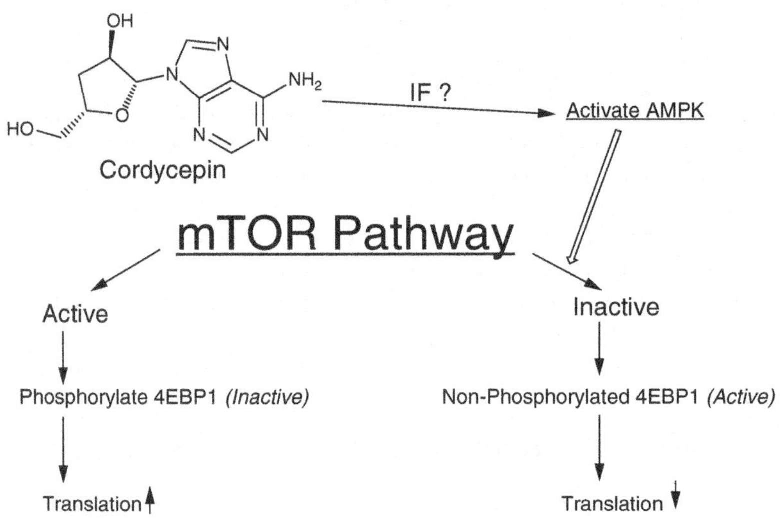

Fig. 16.4 Cordycepin presumably activates the AMPK by some unknown mechanism, which further negatively regulates the translation of mTOR signaling transduction pathway by the formation of a translational repressor, 4-E-binding protein-1 (4EBP1) (Tuli et al. 2014b)

mTOR action. The mTOR inhibitors, for example, rapamycin and CCI-779 have been tried as anticancer medicines, on the grounds that they repress or block mTOR pathway. mTORcan be defined as 298 kDa serine/threonine **protein kinase** from the family **PIKK** (Phosphatidylinositol 3-kinase-related kinase). The mTOR assumes an extremely imperative part to direct proteins production. Be that as it may, mTOR itself is controlled by different sorts of cell indicators such as factors of growth, nutritional environment, hormones, and energy level for cellular. As growth factor tie with cell receptor, Phosphatidyl inositol 3 kinase (PI3K) gets initiated, changes over phosphatidyl inositol bisphosphate (PIP2) to phosphatidyl inositol trisphosphate (PIP3). PIP3 further initiates PDK1 (phosphoinositide subordinate protein kinase 1). The actuated PDK1 then phosphorylates AKT 1 kinase and makes it somewhat initiated which is further made completely enacted by mTORC2 complex. The activated AKT 1 kinase now actuates mTORC1 complex that prompts the phosphorylation of 4EBP1 (translational repressor) and makes it inactive, exchanging on the protein production (Wong et al. 2009). The study confirmed that during the low level of nutrutional stress affirmed, Cordycepin actuates AMPK, which hinders the action of mTORC1 and mTORC2 by some obscure component. The inactivated mTORC2 complex cannot enact AKT 1 kinase completely, which inhibits mTOR signal transduction hindering translation and more cell expansion and development (Fig. 16.4) (Tuli et al. 2014b).

16.4 Pharmacokinetic of Cordycepin

The cordycepin effects at molecular level represent comprehensive analysis on the cordycepin kinetics as well as ADME (Absorption, Distribution, Metabolism and Elimination at site of action of the biological target) studies. Therefore, promising investigations has been carried out in order to enhance the kinetic and the cordycepin quality of action at different active sites such as plasma.

In terms of cordycepin effects on the RNA synthesis, the termination of RNA transcription was recognized as a result of the absence of 3′ hydroxyl moiety (Holbein et al. 2009). Therefore, the impact of cordycepin investigated on the RNA metabolism of the yeast. The results presented that Cordycepin-triphosphate (CoTP) acted as a toxic and limiting factor on the growth of cells via RNA synthesis inhibition. However, the modulation of 3′ end heterogeneity of ASC1 and ACT1 mRNAs as well as rapid extended NEL025c loci and CYH2 transcript were another obtained findings (Holbein et al. 2009). In addition, the amelioration of poly (A) polymerase mutants growth defects together with pap1–1 mutation was reported to neutralize the effects of gene expression of cordycepin. It exhibited the epistatic relation of cordycepin function and poly (A) polymerase activity along with its potency to drop the efficiency of 3′ formations independently (Holbein et al. 2009). Advantageous impacts of cordycepin on the metabolic profiles of plasma and liver specifically in hyperlipidemic hamsters were studied thoroughly. Thus, ^{1}H NMR spectroscopy was applied on the intact liver tissues and plasma resulted high lipid level in the hyperlipidemic hamsters. At the end, the lipid-regulating activities and also the protective effects of cordycepin on the plasma and liver especially in the fatty liver condition were recorded respectively (Sun et al. 2011).

Effects from cordycepin on the activation of Wnt/ß-catenin survival pathway to develop leukemia stem cells (LSCs), whether cordycepin regulates expression of ß-catenin in leukemia cells or not, was investigated (Ko et al. 2013). Cordycepin exhibited positive effects on the **malignant cancer** cells such as HepG2, A549, MCF7 and SK-Hep1 in a dose-dependent manner. Unexpectedly, reverse action with reduction on the levels of ß-catenin in the specific cancer cell line including THP1, K562 and U937 was resulted. GSK-3ß as pharmacological inhibitor was used in order to restore the negative effects of cordycepin on the above-mentioned group. The findings represented that cordycepin selectively reduced stability of **ß-catenin** in leukemia cells but not in other solid tumor cells (Ko et al. 2013). Another study by Wu et al. (2014), showed interaction of cordycepin with the γ1 subunit was found as activator for the AMP-activated protein kinase (AMPK) protein. The results clarified the mechanism and the role of cordycepin on the activation of AMPK in HepG2 cells. Therefore, circular and fluorescent dichroism as well as molecular docking measurements performed. The results showed the direct interaction of cordycepin with AMPK γ1 subunit. However, knock down of AMPKγ1 by siRNA and inhibition of AMPKγ1 expression has reported as significant factor leading abolishment of cordycepin actions in the lipid regulation especially in hyperlipidemia. Finally, cordycepin was introduced as promising agent to inhibit intracellular lipid accumulation by activating AMPK through the interaction with γ1 subunit (Wu et al. 2014).

16.5 Biological Activities of Cordycepin

16.5.1 Antioxidant Activity

In 2005, the **Supercritical carbon dioxide** (SC-CO2) was utilized as the elution dissolvable for fractioning ethanolic concentrate (E) of *Cordyceps sinensis* (CS), a customary Chinese natural cure, into R, F1, F2, and F3 parts. This extractive fractionation strategy was as an agreeable with extensive scale and as a nontoxic procedure. These four fractions were described regarding sum **polysaccharides** and cordycepin fixations; scavenge the free radicals and **antitumor** functions. Exploratory effects showed that fractionation changed the conveyances of aggregate polysaccharides and cordycepin in parts. Fraction R was the most dynamic portion to rummage free radicals and repress the expansion of carcinoma cells, emulated by the fraction F1 and the extract E. The impact of scavenging on 1, 1-diphenyl-2-picryl hydrazyl (**DPPH**) of *Cordyceps sinensis* concentrate and fractions at 2 mg/ml was R (93%), E (66%), F1 (75%), F2 (47%), and F3 (27%). The IC_{50} (half cell growth inhibitory concentration) of tumor cell expansion and colony structuring on human colorectal (HT-29 and HCT 116) and hepatocellular (**Hep 3b** and Hep G_2) carcinoma cells by fraction R were around 2 μg/ml. Then again, R did not influence the development of ordinary isolating human peripheral blood mononuclear cells (**PBMC**) by showing an extensive esteem of IC_{50} over 200 μg/ml. Tumor cells accumulation at sub-G_1 stage and the fracture of DNA, regular characteristics of customized cell expiration, were watched in a time and manner of dose dependent. Totally, the discoveries recommended that fraction R containing cordycepin and polysaccharide, acquired by SC-CO_2 liquid extractive fractionation, indicated solid scavenging capability and specifically restrained the development of colorectal and hepatocellular growth cells by the procedure of apoptosis (Wang et al. 2005).

The most significant consequence led to aging process was introduced as **oxidative damage** induced by free radicals (Ramesh et al. 2012). The capacities and activities of antioxidant actions of cells decrease with increase in age. In addition, main reason for aging is steady loss of antioxidant/pro-oxidant equilibrium resulting considerable increase in oxidative stress. Current study was concentrated on cordycepin impacts on antioxidation and lipid peroxidation in elder rats. By comparison between old and young rats, decline in activities of catalase (**CAT**), superoxide dismutase (**SOD**), glutathione peroxidase (**GPx**), glutathione-S-transferase (**GST**), glutathione reductase (**GR**), vitamin C and vitamin E, reduced glutathione (**GSH**), and elevated levels of malondialdehyde (MDA) were entirely detected in the aged rats' lungs, heart, kidneys and liver. Moreover, the elevation of serum aspartate aminotransferase (**AST**), urea, alanine aminotransferase (**ALT**), and creatinine were significantly observed in old rats compared to younger ones. Cordycepin treatment on aged rats by increasing the activity of CAT, SOD, GST, GPx and GR and elevation of vitamins E, C and GSH level like the values of these parameters found in younger rats, demonstrated promising effects of cordycepin in antioxidant activities. In addition, decrease in the level of creatinine, ALT, AST and MDA was reported due to using cordycepin on the aged rats. Authors suggested cordycepin as effective agent for decreasing **lipid peroxidation** and restoring antioxidant status in the aged rats (Ramesh et al. 2012).

Furthermore, NF-κB signaling pathway suppression by cordycepin was investigated (Ren et al. 2012). Dose-dependent decrease in activity of NF-κB induced by TNF-α in HEK-293 T cells by cordycepin was detected. Although cordycepin in its high concentration decreased the NF-κB transcriptional activities and DNA-binding, cordycepin did not limit nuclear translocation of p65 induced by TNF-α. Moreover, in order to suppress IκBα degradation, cordycepin inhibited **phosphorylation** of IκBα. Additional examination showed that activation of NF-κB mediated by IKKs suppressed by cordycepin. Also, inhibition of IKKγ ubiquitination was reported as the effect of cordycepin. Researchers proposed cordycepin as an agent, which strongly inhibits NF-κB signaling through IKK, IκB and NF-κB activity suppression. In addition, They suggested potential medical applications of cordycepin in treating disorders associated with inflammation and cancer therapy (Ren et al. 2012).

16.5.2 Anticancer Activity

A research study with the aim at examination and exhibition of high chemical constituents and antitumor effects of Cs-HK1 mycelium were conducted. When compared with natural *Cordyceps sinensis*, the mycelium of fungal had much higher contents of cordycepin (65/7 μg vs 20/8 μg) (per gram dry weight). Low cytotoxic effect was shown by the fungal extract from the hot water extraction towards the B16 melanoma cells in culture (about 25% inhibition) but significant antitumor effect in animal tests, that cause 50% inhibition on the B16 cell-induced tumor growth in mice. Major bioactive compounds like cordycepin and polysaccharides of *Cordyceps sinensis*, can be found in mycelium biomass and the mycelium extract had significant antitumor activity. The Cs-HK1 fungus was introduced as a new and promising medicinal fungus and as an effective and good substitute of economical of the wild *Cordyceps sinensis* for health care (Leung et al. 2006). One year later, the cultured mycelium of a *Cordyceps sinensis* (CS) organism was successively removed by petroleum ether (PE), ethyl acetic acid derivation (EtOAc), ethanol (EtOH) and heated water. All dissolvable concentrates with the exception of high temperature water extract demonstrated a huge and dose-dependent inhibitory impact on the cell growth of four cancer cell lines, MCF-7 breast growth, Hl-60 human premyelocytic leukemia, B16 rodent melanoma and Hepg2 human hepatocellular carcinoma, with IC_{50} values underneath 132 mg/ml. The EtOAc extract, specifically, had the most intense impact against every one of the four cancer cell lines, with IC_{50} between 12 mg/ml (on B16) and 45 mg/ml (on MCF-7). Conversely, it had lowest cytotoxicity on the ordinary rodent bone marrow cells. The EtOAc extract included significant amount of cordycepin cooperating the *in vitro* cytotoxicity. In an animal experiment, the EtOAc extract demonstrated noteworthy hindering impact on B16-induced melanoma in C57BL/6 mice, bringing on around the range of 60% abatement of tumor size over 27 days. The outcomes proposed that the EtOAc concentrate of *Cordyceps sinensis* mycelium has solid antitumor activity and can be positively applied as a potential anticancer and/or antitumor product (Wu et al. 2007).

In 2008, the study utilized Radix Astragali (RA) as the media for cultivation of *Cordyceps militaris* to explore the antitumor function of the fermentation stock. The cordycepin production through the culturing of *Cordyceps militaris* in Radix Astragali medium was found higher in antitumor activity than that cultured in artificial or synthesized medium. The broth used for fermentation purpose hindered the development of four different tumor cells incorporating human breast cancer MCF-7 cells, human gastric cancer AGS cells, murine colorectal adenocarcinoma CT26 cells and human hepatocellular carcinoma Hep G2 cells with IC50 37 mg/ml, 465 mg/ml, 20 mg/ ml, and 25 mg/ml, respectively. Despite the fact that cordycepin as the main bioactive segment with the strong antitumor functions in the fermentation medium of *Cordyceps militaris* in RA culture, there were different constituents, which improved the antitumor function of fermentation synergistically. To assess and accept the antitumor function, the BALC/c mice were embedded with CT26 cells and after that encouraged with different measurements of the fermentation broth. Clearly it was reported that 20-mg/kg-body weight (Bw)/day group had no noteworthy antitumor movement as contrasted with the control group. The measurements of 100 mg/kg Bw/day and 200 mg/kg Bw/day hindered the tumor size by 43.81% and 48.89%. Tumor weight was additionally decreased by 31.21% and 39.48% contrasted with the control bunch. Moreover, the fermentation broth demonstrated low cytotoxicity against essential rodent hepatocytes, and did not cause any genuine symptom or any side effects on the key organs of the mice as contrasted with the chemotherapeutic medicine 5-FU (Lin and Chiang 2008).

Other research on antitumor activity of cordycepin mechanism have been studied in two various bladder cancer cell lines, T-24 and 5637 cells (Lee et al. 2009). Important and dose-dependent tumor growth inhibition by cordycepin action at a dose of 200 μM (IC50) through cell-cycle progression, was reported as a main result of G2/M-phase arrest and led to up-regulation of p21WAF1 expression, free of the p53 pathway as well as cordycepin-induced phosphorylation treatment of JNK (c-Jun N- terminal kinases). **Small interfering RNA** (si-JNK1) and SP6001259 (JNK-specific inhibitor) blocked the JNK function and saved cell growth inhibition, cordycepin-dependent p21WAF1 expression and decreased cell cycle proteins. It demonstrated cordycepin as a positive and effective healing agent for the treatment of bladder cancer (Lee et al. 2009).

Regarding the molecular mechanism and molecular targets of cordycepin, a research study was explored promising molecular systems for the antitumor impacts of cordycepin on the human colon malignancy HCT116 cells. After using cordycepin to treat the cells, the dose-dependent cell was watched at an IC50 esteem of 200 μm. Cordycepin medicine demonstrated G2/m-stage cell cycle arrest, which was connected with expanded p21WAF1 levels and decreased measures of cyclin Cdc2, B1, and Cdc25c in a p53-free pathway. Additionally, treatment by cordycepin instigated enactment of JNK (c-Jun N-terminal kinases). SP600125 pretreatment as a JNK-particular inhibitor, annulled cordycepin-intervened p21WAF1 expression, cell development restraint, and diminished proteinscell cycle. Besides, JNK1 restraint by small meddling RNA (siRNA) handled comparative results: concealment of cordycepin-actuated p21WAF1 expression, diminished cell growth, and decreased cell cycle proteins. Together, these effects prescribed a basic role for G2/m-stage arrest in human colon growth cells and JNK1 activation in cordycepin-instigated restraint of cell development (Lee et al. 2010).

Other investigation was also carried out to determine the mechanism of cell death induced by cordycepin in the breast cancer cells. Reduction in cell viability and dose-responsive cell growth inhibition were resulted due to contact of both MCF-7 and MDA-MB-231 breast cancer cells with cordycepin. Moreover, Some particular properties of mitochondria-mediated apoptotic pathway confirmed by biochemical assays, TUNEL and **DNA fragmentation** was associated with cell death induced by cordycepin in MDA-MB-231 cells (Choi et al. 2011).

In addition, **Caspases-3** and **Caspases -9** activation and cytosolic release of cytochrome c was triggered by cordycepin-associated dose-dependent growth of mitochondrial translocation of Bax. Interestingly, in cytoplasm, large membranous vacuole ultrastructure morphology and autophagosome-specific protein were detected which confirmed the autophagy-associated cell death of MCF-7 cells. Regardless of ER response, MCF-7 cells with apoptotic defects can be treated by employing cordycepin-induced autophagic cell death. While survival roles of autophagy in tumorigenesis of other cancer cells have been confirmed, the significant function of autophagy on the cordycepin-induced MCF-7 cell death is also observed. Finally, results showed the strong killing of MCF-7 and MDA-MB-231 human breast cancer cells by the cordycepin. Further study was suggested to evaluate more about clinical usage, ER dependency and therapeutic actions of cordycepin on human breast cancer (Choi et al. 2011).

However, the recent research of (Pan et al. 2015) reported on the strong antitumor activity of cordycepin when treating MA-10 cells in different combined immuno-deciency (SCID) mice *in vivo*. The outcome of studyshows that cordycein is significantly selective treatment highly effective in inductionof MA-10 cells apoptosis through p38 MAPKs signaling.

16.5.3 Anti-Inflammatory Activity

In 2005, a study was carried out to illustrate **anti-inflammatory** functions of *Cordyceps militaris*. The 70% ethanolic extract concentrates of fruiting bodies (FBE) and cultivated mycelia (CME) of *Cordyceps militaris* were ready. Both FBE and CME demonstrated topical anti-inflammatory in the croton oil-prompted ear edema in mice. CME was reported to hold intense anti-inflammatory movement, which was assessed by the carrageenan-instigated edema, and additionally solid antinociceptive action in writhing experiment. FBE and CME hold powerful inhibitory action on the chick embryo chorioallantoic membrane (CAM) angiogenesis in a dose-dependent manner. Besides, cordycepin as famous product from *Cordyceps militaris* was reported to be responsible for only part of anti-angiogenic and anti-inflammatory activities (Won and Park 2005).

Kim et al. (2006) investigated the anti-inflammatory effect of cordycepin on lipopolysaccharide (LPS)-stimulated RAW 264.7 macrophage cells. The key component of endotoxin, Lipopolysaccharide (LPS), which is formed by lipid A (phosphoglycolipid) is connected to hydrophilic heteropolysaccharide covalently. Cytokines production such as TNF-α, IL-1γ, GM-CSF, and nitric oxide was increased by considerable LPS-induced macrophage activation and modulated by up-regulation of inducible NOS2 (**nitric oxide synthase**). In addition, butanol fraction of *Cordyceps militaris* was inhibited by the production of nitric oxide (NO) in LPS-activated macrophage. Moreover, Cordycepin was reported as main component of *Cordyceps militaris* identified by high performance liquid chromatography. The activation of MAP and Akt kinases in LPS-activated macrophage was evaluated to elucidate the mechanism of inducible nitric oxide synthase (iNOS) expression and NO production inhibition by cordycepin. Furthermore, the phosphorylation of p38 and Akt in dose-dependent manners LPS-activated macrophage strong inhibition by cordycepin was reported obviously. Tumor necrosis factor (TNF-α) expression, translocation of nuclear factor- B (NF-κB) and IκB alpha phosphorylation were reported by considerable suppressing property of cordycepin. In contrast, Cordycepin demonstrated decreasing effect on the expressions of inducible nitric oxide synthase (iNOS) and cycloxygenase-2 (COX-2) in RAW 264.7 cell significantly.

Results presented that inhibition of NO production was occurred by cordycepin by down-regulation of COX-2 and iNOS gene expression via p38 and Akt phosphorylation and suppression of NF-κB activation as antioxidant inhibitor. Hence, the conclusion drawn by authors is that the cordycepin may afford a potential therapeutic action for inflammation-associated illnesses (Kim et al. 2006).

Recently, cordycepin anti-inflammatory action on the inflammatory mediator's production in murine BV2 microglia stimulated by **lipopolysaccharide** (LPS) was assessed by Jeong et al. (2010). In addition, the cordycepin secretion results on phosphorylation of **mitogen-activated protein kinases** (MAPKs) and nuclear factor-kappaB (NF-κB) activation induced by LPS was studied. After LPS stimulation, Pro-inflammatory cytokine production, nitric oxide (NO) and prostaglandin E2 (PGE2) was sensed in BV2 microglia. However, significant inhibitory effect of cordycepin on the extreme production of PGE2, pro-inflammatory cytokines and NO in a concentration-dependent manner without producing cytotoxicity was reported. In addition, cordycepin illustrated the suppressing effect on NF-κB translocation by blocking the degradation of IkappaB-α (IκB-α) and inhibition of Akt, ERK-1/2, JNK, and p38 kinase phosphorylation. Results presented that cordycepin inhibition on LPS-stimulated inflammatory mediator production in BV2 microglia is related to Akt, NF-κB, and MAPK signaling pathway suppression. Therefore, authors proposed cordycepin as useful agent for the treatment of neurodegenerative diseases by inflammatory mediator production inhibition in activated microglia (Jeong et al. 2010).

16.5.4 Hypoglycemic (Anti-Diabetic) Activity

Anti-diabetic impact of different fractions of *Cordyceps militaris* was compared in pure compounds and crude extracts. In addition, the effect of the various fractions, CMESS (ethanol soluble supernatant), CCCA (crude cordycepin containing adenosine) and cordycepin were studied in diabetic mice (Yun et al. 2003). Regarding to starch-loaded mice, Potential inhibitory activity of 34.7% and decreased blood glucose level by 35.5% was reported for the CMESS. Nevertheless, cordycepin and CCCA presented no difference. It is reported that administration of cordycepin and CMESS for a period of 7 days, considerably decreased the level of blood glucose. But, the reduction of blood glucose did not occurred by utilizing of CCCA with high concentration of cordycepin. T-lymphocyte proliferation was expressively reduced; while NO production was augmented above than two-fold in the cordycepin-administered group. Additionally, NO production and macrophage proliferation were meaningfully reduced in the group administrated by CMESS. Authors suggested that cordycepin and CMESS might be useful kits use to control the blood glucose level in diabetes and encouraging new medicine as an anti-hyperglycemic instrument without the deficiencies of decreased immune responses and other side effects (Yun et al. 2003).

16.5.5 Immunomodulatory and Protective Effects of Codycepin

A research conducted on medicine by Zhou et al. (2002), which resulted in the discriminating up-regulation of interleukin-10. Authors focused on the immunoregulatory effects of cordycepin derived from Cordyceps spp. measuring the Interleukin-2 and interleukin-10 secretion of human peripheral blood mononuclear cells, which were incubated through cordycepin was performed. Furthermore, the expression of surface markers on T lymphocytes, the proliferative response and the effect of cordycepin on the expression of interleukin-10 mRNA were evaluated completely. In addition, cytotoxicity of cordycepin, interleukin-10-secreting cells and the impact of anti-interleukin-10 neutralizing antibody were assessed. Results showed that cordycepin has an expressively up-regulative impact on the interleukin-10 production and interleukin-10 **mRNA** expression. Moreover, Cells of CD8+, CD4+, CD56+, CD14+ and CD19+ were confirmed as Interleukin-10-producing cells. In parallel, Proliferation of peripheral blood mononuclear cells cordycepin and **phytohaemagglutinin**-induced interleukin-2 production was inhibited considerably. It is reported that decrease in expression of the surface markers CD45PRO, CD25, CD71, CD54 and HLA DR also reflected restricted T lymphocyte activation. Furthermore, Anti-interleukin-10 neutralizing antibody was reported, as component not completely hindered the cordycepin suppressive effect on production of interleukin-2. Authors stated an effective concentration (above 24 μg/ml) of cordycepin, which presented minor cytotoxicity but did not raise apoptosis. In contrast, high cordycepin concentration was stated as it may have exerted widely inhibitory effect on DNA synthesis or caused in cytotoxicity remarkably. However, in range of cordycepin concentration between 0 and 24 μg/ml, the strong up-regulation of interleukin-10 expression and immediate down-regulation of interleukin-2 expressions signified that an inhibitory result on **DNA synthesis** or a **cytotoxicity** mechanism might not be essential. Results indicated that cordycepin uses immunoregulativeoutcomes. Additional study on the method for the improvement of novel immunomodulatory medicines, which precisely adjust the excretion of cytokines, was suggested (Zhou et al. 2002).

Cho et al. (2007) carried out a research on the new effect of collagen stimulated human platelet accumulation resulted by cordycepin (3′-deoxyadenosine). Inhibition of measure-dependently collagen-induced platelet aggregation by the cordycepin in existanceof different wide range of concentrations of CaCl2 was reported thoroughly. Of thromboxane A2 (TXA2) and cytosolic freeCa^{2+} ($[Ca^{2+}]i$), as two aggregation-inducing molecules, up to 74% of up-regulation of $[Ca^{2+}]i$ was blocked with cordycepin (500 μM) while for TXA_2 production, it was suppressed by 46%. Afterward, phosphorylation of Ca^{2+}- for 20-kDa and 47-kD aproteins in collagen treated platelets was effectively reduced by by cordycepin. But, upstream routes such as formation of inositol1,4,5-triphosphate (IP3) and the activation of pphospholipase C- γ2 (PLC-γ2) (assessed by phosphotyrosine level) that normally used for producing these two inducers were not changed by cordycepin. Furthermore, the level of second messengers guanosine 3′, 5′-cyclic monophosphate (cGMP) and adenosine 3′,5′-cyclic monophosphate (cAMP) in collagen-stimulated platelets was increased in presence of cordycepin. However, cordycepin-induced up-regulation of cGMP did not changed by the NO-sensitive guanylyl cyclase inhibitor ODQ and the cAMP enhancement mediated by cordycepin was totally blocked by adenylyl cyclase inhibitor SQ22536, demonstrating the various cordycepin modes of action. Hence, Authors suggested that their results show the inhibitory effect of cordycepin on platelet aggregation could be related to the boost of cAMP/cGMP production and the down-regulation of $[Ca^{2+}]_i$ (Cho et al. 2007).

One of the common techniques for curing the **atherosclerosis** is Percutaneous Transluminal Coronary Angioplasty (PTCA). However, it has limited efficacy due to manifestation of restonesis during 3–6 months after angioplasty(Chang et al. 2008). Restenosis is stimulated in cells andextracellular matrix (ECM) inside the intimal layer and/or alteration of the vessel wall. Thus, one of the potential therapeutic agents for the atherosclerosis or restenosis treatment can be **matrix metalloproteinase** (MMP) system. In this study, the effect of cordycepin on the MMP system within vascularmuscle cells was investigated particularly. Cordycepin (20 μM/day, i.p) led to inhibition of rat aortic smooth muscle cells (RaoSMCs) proliferation. Thus, neointimal development in the carotid artery of a balloon injured-Spraque–Dawley (SD) rat was decreased considerably. In addition, the triggering of MMPapproachincollagentypeI activated RaoSMCs was evaluated in order to conduct on cordycepin inhibition mechanism on the buildupof both, cells and ECM and/or remodeling of thou thee vessel wall. The positive effect of cordycepin was reported as a strong agent to inhibit the activation of MMP-2 and extracellular matrix **metalloproteinaseinducer** (EMMPRIN) expression in a dose-dependent within RAoSMCs activated by collagen type I. Furthermore, the expression of cycloxygenase-2 (COX-2) correlated tohyperplasia of RAoSMCs was strongly suppressed by cordycepin. Taken together, anti-proliferation property of cordycepin in RAoSMCs through the varietypf vessel wall remodeling concluded by the authors. They also stated that cordycepin can be a probable use medicinal for restenosis treatment (Chang et al. 2008).

Previous investigation was also obtained by (Yao et al. 2011) to investigate the effects of cordycepin on CA1 pyramidal neurons exist in the rat hippocampal brain slices by whole-cell patch clamp method. The result exhibited the drop in the both evoked and spontaneous action potential (AP) firing frequencies. The hyperpolarization of neuronal membrane potential by cordycepin and the stability of either membrane input resistance or the amplitude of fast after hyperpolarization (fAHP) in presence of cordycepin was reported as opposed with AP, which spiked width. Collectively, Membrane hyperpolarization induced by cordycepin led to diminish neuronal activity presenting cordycepin as potential therapeutic approach to treat excitotoxic disorders such as ischemic.

Oxygen–glucose deprivation (OGD) injury in mice brain slices was investigated in the presence of cordycepin. Study outcomes illustrated that brain slice injury and especially postischemic neuronal degeneration can be prevented by cordycepin.

Also, two types of excitatory amino acids in brain homogenized supernatant, aspartate and glutamate, were enhancedin reperfusion/ischemiaset, were sensed by **HPLC**. The results also showed that the significant reduction in the extracellular level of aspartate and glutamate was the positive effect of cordycepin. Furthermore, Ameliorating the extent of oxidation, dropping malondialdehyde (MDA) level and increasing the superoxide dismutase (SOD) activity were reported as cordycepin activity on the system. The remarkable increase in the expression of matrix metalloproteinase-3 (MMP-3 as main enzyme involved in inflammation) after ischemia reperfusion was also inhibited by the protective activity of cordycepin. It was concluded that the potential neuroprotective action of cordycepin, in both *in vivo* and *in vitro* studies, was exerted subsequently after cerebral reperfusion/ ischemia (Cheng et al. 2011).

Drugs, mostly viable throughout the initial phase of trypanosomosis, however crudely enter the blood brain restraint, are ineffectual when parasites spread to the brain and reason cencephalitis. Therefore, a study with intended to assess the weakness of species*T. evansi* to cordycepin invitro and in rats tentatively affected was carried out. In vitro, a critical reduction ($P < 0.01$) in living trypanosomes in the doping between 5.0 and 10 mg/ ml was recorded one hour afterward the start of the investigation, and also at three, six, nine and twelve hours in all concentrations contrasted with control. Despite the fact that no remedial impacts were found in the in vivo assessments in the most of groups, the treatment was fit to keep the parasitemia at low levels, along these lines expanding the life span of rats when contrasted with positive control. Rats that gained cordycepin only or in combination by adenosimedeaminase inhibitor (ADA: EHNAhydrochloride), did not indicate trypomastigote manifestations of the parasite in the bloodstream 24 h after the consumption. Averages of 8 days were assigned to remain animals negative in blood smears, yet from that point had a repeat of parasitemia. Around all the tainted creatures, just 3 rats cured with the mixture of EHNA hydrochloride (2 mg/kg) and cordycepin (2 mg/kg) persisted negative throughout the trial period. The adequacy of 42.5% was affirmed by PCR utilizing *Trichoderma evansi* particular primers. Subsequently, authors concluded that cordycepin requires natural impact to against *T. evansi*. The cordycepin treatment, when secured by an inhibitor of ADA, be able to delay the *T. evansi*-contaminated rats survival then furnish therapeutic viability (Da Silva et al. 2011).

Jeong et al. (2011) carried out a research on the specific polyadenylation inhibitory effect of cordycepin as an operative compound in *Cordyceps militaris*. Authors believed that cordycepin has possessed many immunological activities, which is efficient for **cancer therapy**. However, anticancer mechanism still needs more investigation. In this study, cordycepin apoptotic effects were conducted on human leukemia cells. Apoptosis induction (not necrosis) by cordycepin treatment led to celldevelopment inhibition in a suitable concentration manner. Poly (ADP-ribose) polymerase protein cleavage, caspases activation, dysfunction of mitochondrial and initiation of **reactive oxygen species** (ROS) were reported as activities related to induction process. However, significant inhibitory effect of caspases, in the responses of cordycepin, attenuated cordycepin-induced apoptosis. N-acetyl-L-cysteine, a ROS scavenger, also inhibited caspases activation and apoptosis induced by cordycepin. Results supported a mechanism whereby human leukemia cells apoptosis was induced by cordycepin via a signaling cascade containing a caspase pathway mediated by ROS (Jeong et al. 2011).

As knowledge goes far, *Cordyceps sinensis* is a parasite utilized as conventional Chinese drug as a tonic to alleviate the lung for the medication of exhaustion and respiratory ailments. Idiopathic pulmonary fibrosis is a persistent, irreversible and incapacitating lung infection demonstrating fibroblast/myofibroblast development and extreme deposition of extracellular framework in the interstitium causing breathing issues. Recently, some results uncovered an incomplete help of lung fibrosis in patients experiencing intense respiratory syndrome (SARS). Subsequently, the theory that cordyceps has some benefits for lung fibrosis was expressed and the idea of more current study was pointed at investigating the target(s) of cordyceps in the alleviation of lung fibrosis in rats and cell unit models and understanding of its mechanisms. A rodent model of bleomycin (BLM)-incited lung fibrosis and a fibrotic cell pattern with converting growth factor beta −1 actuation were used in the studies. Lessening of invasion of inflammatory cells, accumulation of fibroblastic loci and collagen, creation of reactive oxygen species (ROS), and cytokines production, together with recuperation from unbalance of MMP-9/TIMP-1, were seen in fibrotic rats after medication with cordyceps in prophylactic (from the day of BLM consumption) and remedial (from 14 days after BLM) diets. In a fibrotic cell pattern with changing growth factor beta-1 infusion, the human lung epithelial A549 gained a mesenchymal phenotype with an expansion of vimentin manifestation with an associative lessening of E-cadherin. This epithelial–mesenchymal move could be returned in part by cordycepin as a significant component of Cordyceps. The results give an understanding of the prophylactic and remedial possibilities of Cordyceps for the cure of lung fibrosis (Chen et al. 2012).

Cordycepin exhibited an anti-atherogenic impact on exploratory animals. Notwithstanding, the impacts of cordycepin on the signaling pathway and cell-cycle regulation in **vascular smooth muscle units** (VSMC) maintain largely obscure; hence, in an alternate study, surprising function of cordycepin-actuated hindrance in VSMC development were examined. Cordycepin treated VSMC mechanisms were explored through a MTT test, an uptake trial of thymidine, FACS investigation, immunoblot examination, kinase assay, immunoprecipitation test, and transient transfection evaluation. Cordycepin repressed cell growth, actuated G1-phase cell-cycle arrest, down the cyclins and **cyclin-subordinate kinase** (CDK) secretion, and up-regulated p27KIP1 expression in VSMC. Cordycepin actuated JNK activity, ERK1/2 and p38MAPK. Obstructing of the ERK activity by either ERK1/2-particular inhibitor U0126 or a little meddling RNA (si-ERK1) restraint of cell development, switched p27KIP1 expression, and diminished cell-cycle proteins in VSMC treated by cordycepin. Ras initiation was expanded by cordycepin. Cells transfection with overwhelming negative Ras (RasN17) mutant genes expanded p27kip1 declaration, saved cordycepin-instigated ERK1/2 movement, decreased cell cycle proteins and restrained cell multiplication. In conclusion, Authors stated that findings of research demonstrated that Ras/ERK1 pathways take part in p27KIP1-intervened G1-phase cell cycle arrest incited by cordycep in through cyclin/CDK complexes reduction in VSMC (Jung et al. 2012).

Recently, novel research was accepted to examine the cordycepin impact on the normal sleep in rats, and intervened by **adenosine receptors** (ARs). Thus, **electroencephalogram** (EEG) was applied in order to record the sleep about 4 h later than cordycepin oral administration by rats. Analyses on the sleep structure and EEG effect spectra were performed. As a result, Cordycepin diminished wake cycles of sleep and expanded non rapideye movement (NREM) rest. Fascinatingly, cordycepin expanded (theta) waves force density throughout NREM rest. Furthermore, AR subtypes (A2A, A1 and A2B) protein levels were risen later the cordycepin consumption, particularly in the rodent hypothalamus which assumes a vital part in regulation of sleep. Subsequently, authors prescribed that cordycepin grows theta waves force density throughout NREM rest through nonspesifiv AR in rats. Authors declared that results were presented to give fundamental proof that cordycepin may be accommodating for sleep-bothered issues (Hu et al. 2013).

Lately, a research study with focuson the *in vitro* cordycepin effects on osteoclastogenesis as well asthe*in vivo* effects on ovariectomized (OVX) mice were conducted (Dou et al. 2016). An in vivo result showed the significant role of cordycepin on prevention of several diseases such as bone loss and bonemicroarchitecture as well as fixes bone mineral in OVX mice. The authors proposed the cordycepin asanosteoclastinhibitor and have big ability therapeutic effect especially effective inavoidingbone loss between postmenopausal osteoporosis victims.

16.6 Production of Cordycepin

16.6.1 Cultivation Conditions for Cordycepin Production

Optimal cultural environments for the creation of cordycepin through submerged cultures of *Cordyceps sinensis* and *Cordyceps militaris*was investigated by Kim and Yun (2005) in shake flask level cultivation and in 5 literstirred tankbioreactors. The mycelial biomass concentration of cordycepin attained in the submerged culture conditions of *Cordyceps militaris* was greater than *Cordyceps sinensis.*

In same year, production of cordycepin (3′-deoxyadenosine) in submerged cultivationof a *Cordyceps militariss* was performed and the effect of different types of carbon source and carbon: nitrogen (C:N ratio) on cordycepin was studied by Mao et al. (2005). Glucose (carbon source) was discovered to be the suitable for cordycepin production. The maximum cordycepin production of 245.7 was found in medium of 40 gL^{-1} glucose. In addition, response surface analysis as well as central composite design was used to study the impact of carbon: nitrogen ratio on cordycepin production. The highest cordycepin production of 345.4 mg/l was reached in culture media of 42.0 g/l glucose and 15.8 g/l peptone. (Mao et al. 2005).

One year later, *Cordyceps militaris* NBRC 9787 surface culture was used to study the production of cordycepin (Masuda et al. 2006). In this study, 98% of extracted cordycepin from *Cordyceps militaris* was excrete into medium (extracellular). Mixture of yeast extract and peptone was the best for the production of cordycepin. Glucose was utilized as the carbon source with the optimum carbon: nitrogen ratio of 2:1 (w/w). The maximum concentration and productivity of cordycepin in the medium under the optimal condition were reported as 640 mg/l and 32 mg/day, respectively (Masuda et al. 2006).

Mao et al. (2005) described that Peptone was the best nitrogen source involved in complex medium for the production of cordycepin. In addition, NH_4^+has significant role in cordycepin production. To improve production of cordycepin, in this, NH_4^+ was added during fed-batch culture along with complex medium in the presence of peptone. Furthermore, Maximum concentration of cordycepin (420.5 ± 15.1 mg/l) was obtained by optimizatation of feeding and ammonium feeding rate.

Different compounds such as glycine, l-glutamine, l-aspartic acid, adenosine and adenine were also supplemented to the base medium to enhance the cordycepin production, which illustrated positive effects on the production rate. The mixture of 16 g/l glycine and 1 g/l of adenine was the most suitable concentration and supported cordycepin production up to 2.5 g/l, which was4.1 times significant than the one in the basal medium. The results showed that guanine and cordycepin production may be connected to each other and also about 97% of *Cordyceps militaris* synthesized cordycepin was excreted into the medium (Masuda et al. 2007).

Fan et al. (2012) investigated the cordycepin productionin submerged cultivation of *Cordyceps militaris*, in shake flasks level. Adding of ferrous sulfate in concentration of 1 g/l at zero time supports volumetric production of cordycepin up to596.59 ± 85.5 mg/L (70% greater than control without addition of ferrous sulfate). Meanwhile consumption of a potential cordycepin precursor, inosine 5′-monophosphate (IMP), was dropped significantly. In addition, study on transcription levels of key genes encrypting IMP cyclohydrolase (purH), IMPdehydrogenase (guaB) and adenylosuccinate synthetase (purA) in the purine nucleotide biosynthetic pathway were done. In comparison with controlled sample, the transcription level of guaB and purH were slowly down-regulated, while in ferrous sulfate supplemented cultures, purA was expressively up-regulated (Fan et al. 2012). Recent research showed that the optimum culture condition for production of cordycepin were pH at 5.5, temperature 25 °C, inoculum size 8% v/v, inoculum age 72 h, incubation time 24 d. The optimum culture medium in the same study was composed of: 1.5% dextrose, 0.8% yeast extract, potassium phosphate dibasic (K_2HPO_4) 0.3%, Potassium phosphate monobasic (KH_2PO_4) 0.1%, sodium chloride (NaCl) 0.05%, Magnesium Sulfate ($MgSO_4$) 0.05% and NaCl 0.05%. The maximal cordycepin production in this medium was 846 mg/L (Tuli et al. 2014c).

16.7 Synthesis and Delivery of Cordycepin

A novel study on layered double hydroxides (LDHs) as nanocarriers for delivering of cordycepin was investigated. Using **XRD, CZE, TEM, FT-IR** and electrophoretic mobility confirmed negatively charged biomolecule-cordycepin intercalated in the gallery spaces of charge-compensating species, [Mg–Al–NO3]. Decomposition of cordycepin by adenosine deaminase was prevented by new bio-LDH nanohybrid particles strongly. This new preparation suggested as a new form cordycepin arterial injection (Yang et al. 2006).

In 2008, a research study on the preparation of 4′-benzoyloxy precursor (4′-substituted cordycepin from adenosine) was carried out. In order to make the precursor, an electhrophilic supplementation (iodo-benzoyloxylation) to the 4′, 5′-unsaturated byproducts was performed. Furthermore, radical mediated removal of the 3′-iodine atom of the consequential adducts was done. The 4′-substituted cordycepin effectiveness was concisely proved through the synthesis of 4′-cyano and 4′-allyl cordycepin analogues (Kubota et al. 2008).

Moreover, for easing off the rapid metabolic velocity and also expanding the cordycepin bioavailability, four N-acyl-(octanoyl-, propionyl-, stearoyl- and lauroyl-) cordycepin formatives were produced artificially and their pharmacokinetic profiles were examined. The outcomes demonstrated that half-life time ($t_{1/2}$) and maximum concentration time (T_{max}) might be stretched with the alkyl chain length expansion, however concentration time curve (AUC) and maximum concentration (C_{max}) enhanced at first, then diminished when the number of alkyl carbon surpassed eight. The C_{max}, T_{max} and AUC of N-octanoyl-cordycepin were about 30, 4 and 68 times, respectively, greater than that of cordycepin. It showed that N-octanoyl modification could expand the cordycepin bioavailability and diminish the metabolic activity (Wei et al. 2009).

Recently, an investigation on region selective and extremely efficient cordycepin acyltion with an immobilized species of *Candida antartica* lipase B (Novozym 435)-catalyzedvinylacetateinthesolventconsumed,whichis2-methyltetrahydrofuran (MeTHF) was carried out. High operational stability and excellent region selectivity during the transformation displayed by Novozym 435 as biocatalyst, and 96.2% isolated yield of cordycepin acetate 25-g scale syntheses by biocatalyst introduced Novozym 435 as useful and effective agent on production of cordycepin. In addition, 5′-substituted cordycepin derivative was described as the mainly product from acylation effects. Finally, Recycling of Novozym 435, as biocatalyst, for the cordycepin derivative synthesis on a 25 g scale and maintaining of 63% of its unique activity after reusing for 7 batches production (Chen et al. 2013). Moreover, the identification of genes involved in formation of cordycepin within the fungus *Cordyceps militaris* was investigated through bioinformatic analysis method. The results showed important role of glucose methanol choline oxido reductase and telomerase reverse transcriptase on the development of cordycepin in fruting body of fungus *Cordyceps militaris* (Zheng et al. 2015).

In recent years, several methods have been applied in command to enhancethe rate of cordycepin creation within the cell as first level biofactory unit (Lia et al. 2015) or in total manufacturing rate including the coupling of *Cordyceps militaris* cells with various additives in medium culture (Das et al. 2009; Leung and Wu 2007), repeated batch culture method (Das et al. 2016), optimization of bioprocess parameters in order to achieve higher rate of production (Tang et al. 2007; Jiapeng et al. 2014), using mixed static-liquid culture technique (Kang et al. 2014), and improved feeding strategy (Velut et al. 2007) as well as cordycepin extraction process enhancement (Wang et al. 2014; Park 2015).

Another research was conducted on the optimization of solvent-solvent extraction method in order to efficiently extract the cordycepin from fermented broth. The hexane, chloroform and n-butanol were used in different parameters such as solvent-solvent ratio, extraction time and extraction temperature. Results showed the maximum yield of extraction (95% cordycepin) was establish at solvent-solvent ratio (1:2 v/v), 90 min of extraction at 40 °C (Tuli et al. 2014a).

16.8 Extraction, Separation, Purification and Determination of Cordycepin

Hsu et al. (2002) investigated on characterization of the bioactive components containing cordycepin and adenosine from *Cordyceps sinensis*. The mycelium of *Cordyceps sinensis* was cultured in potato dextrose broth (PDB) for 5 day incubation period at 22 °C. The fermentation process was performed in a 5-L fermenter with a working volume of 3-L and 10 percent of inoculum at 22 °C for 5 days. The overall levels of amino acids were found considerably diverse, varying from 4–17% obtained by an automatic amino acid analyzer device. Aspartic and glutamic acid were found as two major amino acids significantly in all samples verified. The amount of overall amino acids in the mycelium was determine only half that found in the natural type. HPLC analysis proved that no bioactive components adenosine and cordycepin found in the mimic and counterfeit types. Regarding to Adenosine, it was exist rich in the fruiting body, more than in the mass of the ordinary type and also in the *Cordyceps sinensis* mycelium. In terms of cordycepin, it was found in both fermented and natural *Cordyceps* extracts. Results suggested that cordycepin and adenosine can be applied as indexing elements for distinguishing cordyceps from the mimic and counterfeit (Hsu et al. 2002).

In the same year, **Capillary zone electrophoresis** (CZE) method was applied to separate adenosine and cordycepin by Ling et al. (2002). In addition, determination of concentrations of both adenosine and cordycepin in the stroma of Cordyceps spp. was made by CZE. Outcomes presented that the CZE technique is a rapid, simple and sensitive system to quantify the cordycepin and adenosine concentration with upright repeatability (Ling et al. 2002).

Regarding to cordycepin determination, an innovative capillary electrophoresis (CE) technique with UV detection at 254 nm was advanced and enhanced (Rao et al. 2006). Ideal settings initiate were 20 mM sodium borate buffer plus 28.6% methanol, separation voltage20 kV, pH 9.5, temperature 25 °C and hydrodynamic injection time 10s. Above the 20–100 μg/mLvarieties of cordycepin concentration, linearity was found clearly. It should be noted that enhanced technique has already been practicalfor cordycepin determination in different pharmaceutical products. By comparing the results obtained by either CE or high performance liquid chromatography (HPLC) methods, it can be found that both techniques were offered comparable outcomes regarding cordycepin concentration in either natural types purchased from China and Taiwan or cultured fruiting body within the solid-state fermentation. At the end, Authors claimed that low running cost and short analysis time are two main advantages of CE technique were proposed (Rao et al. 2006).

In addition, Quantitative and qualitative evaluation of bases, nucleosides and their analogues in cultured and natural *Cordyceps* was done utilizing high performance liquid chromatography electrospray ionization tandem mass spectrometry (HPLC-ESI-MS/MS) and rapid pressurized liquid extraction (PLE) (Fan et al. 2006). Extractions were done using PLE technique while ZORBAX Eclipse XDB-C18column with 5 mM aqueous ammonium acetate as mobile phase and gradient elution of methanol was used for separation purposes. Precursor ions, product ions, and retention times were characterized to identify target compounds. Compounds were analyzed quantitatively operating time programmed selective reaction monitoring (SRM) or selective ionmonitoring (SIM) with10segmentsinpositive (or negative foruridine) ion mode. The total numbers of 43 nucleosides, bases and their analogues were discovered in cordyceps, which among them 16compoundswere known separately. PLE and HPLC–ESI–MS/MS techniques explained above, were demonstrated good selectivity, linearity, recovery, precision, short analysis time plus LOQ and LOD in the ng/ml range (Fan et al. 2006).

In order to investigate pharmacological and physiological effects of bioactive ingredients from mushrooms, nucleoside determination and its metabolic compounds are important concepts should be considered initially. In other study, a fast-response **ultra-performance liquid chromatography** (UOLC) technique was created for the cordycepin determination. The detachment was done on Water AcquityUPLC framework with Acquity UPLC BEH-C18 column with 0.5 mm acetic acid gradient elution and acetonitrile (5 min). In addition, high value of analyte correlation coefficient was recorded at R2 > 0.9995. The LOQ and LOD were lower to 47.0 and 11.9 ng/ml along with 1 μl of injection volume, respectively (Yang et al. 2007).

For extracting the cordycepin from the solid state fermentation, a novel column chromatographic extraction (CCE) technique was developed and presented. Imbibition of dried wastes (4times its volume of water) for6 h, transferring to the columns and washing with water were the initial steps for preparations. Separation step was done using macro-porous resinDM130columnsfollowedby precipitation phenomenon, crystallization process, and polyamide column chromatography as purification steps. The results were 97% extraction rates of cordycepin gained by 4 volumes of water for washed materials distributed through 3 different columns and 12 volumes of water for a single column aimed to concentrate cordycepin (Ni et al. 2009).

In other study, cordycepin and adenosine were extracted from *Cordyceps kyushuensis* using **supercritical fluid extraction** (SFE) technique. In order to get the optimal extraction conditions anorthogonal array strategy (OAD) test, L9(3), including different parameter synch as temperature, pressure and flow rate of CO_2as critical factors (Ling et al. 2009). In addition, scaling up the procedure up to 30 times using a SFE system was successfully achieved under the following conditions

40 °C, pressure of 40 MPa and a flow rate of CO2 ($2.0lmin^{-1}$). The same study showed also that **High-speed counter-current chromatography** (HSCCC) with a two-phase solvent system constituted of collected fractions analyzed by HPLC and ethyl acetate–n-butyl alcohol–water at ratio of 1:4:5 (v/v/v) was used successfully for cordycepin separation (Ling et al. 2009).

Moreover, **capillary electrophoresis mass spectrometry (CE–MS)** technique was improved in order to concurrent resolution of cordycepin in wild and artificially cultured *Cordyceps* by 5 chlorocytosine arabinoside as standard. The CE division environments and MS related parameters were enhanced methodically for obtaining exceptional CE and MS results of the explored compound. The 100 mM formic acid including 10% (v/v) methanol was reported as the optimal CE electrolyte. In addition, the optimal MS parameters were explained as 75%(v/v)methanol including 0.3% formic acid with a rate of 3 μl/min was chosen (Yang et al. 2009).

Xie et al. (2010) established a consistent, selective and sensitive **liquid chromatography mass spectrometry** (LCMS) method using electrospray ionization interface technique to separate and determine bioactive compounds adenine, thymine, cordycepin and adenosine in mushroom *Cordyceps sinensis.* In this method, gradient elution technique using a 2.0X150mm VP-OD Scolumn was used for separation (Xie et al. 2010).

Recently, One-step separation and purification of cordycepin from *Cordyceps militaris*(L.) Link, in a preparative measure, has been applied using **high speed counter current chromatography** (HSCCC) method (Xie Huichun et al. 2011). A two-phase solvent method of combinations of n-hexane-n-butanol-methanol-water (23,80,30:155, v/v/v/v) was used to perform HSCCC separation with high efficiency. The identification of product was done using IR, UV, ^{1}H NMR, ^{13}C NMR and MS (Xie Huichun et al. 2011). Recent study showed simple optimized solvent system composed of: water (23%), ethanol (30%), methanoll (25%), ethylacetate (22%) was successfully used for cordycepin extraction (Soltani et al. 2017).

In order to get better overall look on the recent investigations on the cordycepin determination techniques as provided above, data were summarized and presented in Table 16.2 as follows.

16.9 Conclusion

As we have recently faced by various hard-treating diseases such as cancer among the people in different societies, the researchers were encouraged to find the alternate treatment with potential therapeutic activity. Chinese medicinal mushrooms especially *Cordyceps* spp. are one of those with long history background about 2000 years which are used to treat many type of these diseases. Cordycepin is low molecular weight bioactive compound isolated from the mushroom *Cordyceps militaris* and/or *Cordyceps sinensis* and possess different biotherapeutic activities. Therefore, the method of production and its therapeutic actions on the different types of cancer cell lines were investigated extensively. However, more studies are

Table 16.2 Cordycepin determination using different methods

Cordyceps species	Objectives	Cultivation mode	Methods	Results	References
Cordyceps sinensis	Characterization	Submerged (cultured in Potato Dextrose Broth, PDB)	Automatic amino acid analyzer, HPLC	Cordycepin found in both cultured and natural *Cordyceps sinensis*, Cordycepin was introduced as an indicator for *Cordyceps* from its mimic and counterfeit	Hsu et al. (2002)
Cordyceps sinensis	Separation and concentration determination	Submerged (shake flask)	Capillary Zone Electrophoresis (CZE)	Rapid, Simple and Sensitive system to quantify the cordycepin	Ling et al. (2002)
Cordyceps militaris	Cordycepin determination	Solid-state fermentation	Capillary Electrophoresis (CE) + UV detection, HPLC	Low running cost and short analysis time are obtained as advantages of this method	Rao et al. (2006)
Cordyceps sinensis	Quantitative and qualitative evaluation of cordycepin	Submerged	HPLC-ESI-MS/MS, PLE, SRM and SIM	Good selectivity, linearity, recovery, precision, short analysis time + LOQ and LOD in the ng/ml range	Fan et al. (2006)
Cordyceps sinensis	Determination of cordycepin	Submerged	Ultra-performance liquid chromatography (UPLC)	Fast determination of analytes in pharmaceutical products and biological fluids	Yang et al. (2007)
Cordyceps militaris	Cordycepin extraction from wastes, propose technique for purification and separation	Solid fermentation	Column chromatographic extraction method (CCE)	High effective extraction method, minimal volume of solvent consumption, energy-saving, simple, environmentally friendly and low operating cost	Ni et al. (2009)
Cordyceps kyushuensis	Optimal extraction of cordycepin	Submerged	Surface fluid extraction (SFE), HSCCC + HPLC	Considerable amount of cordycepin with purity of 98.5% was obtained	Ling et al. (2009)
Cordyceps militaris and *Cordyceps sinensis*	Cordycepin analysis in natural and cultured Cordyceps	Submerged	Capillary electrophoresis-Mass spectroscopy (CE-MS)	Cordycepin was found in both natural and cultured *Cordyceps Militaris* but solely detected in natural *Cordyceps sinensis* with very low content	Yang et al. (2009)

Cordyceps sinensis	Determination of cordycepin and its transformation product	Submerged	IP-RP-LC-MS	No transformation was found in both natural and commercial cultured *Cordyceps sinensis*	Yang et al. (2010)
Cordyceps sinensis	Determination of cordycepin	Submerged	Liquid chromatography-Mass spectroscopy (LCMS)	Cordycepin was recovered from 98.47–99.32%	Xie et al. (2010)
Cordyceps militaris	Separation and purification of cordycepin	Submerged	High-speed counter-current chromatography (HSCCC)	Cordycepin with purity of 98.6% and 91.7% was produced	Zhang et al. (2016)
Cordyceps militaris Mutant G81–3	Correlation determination between adenosine and cordycepin	Surface-liquid culture of Mutant	High-energy proton beam irradiation	Significant overproduction of cordycepin at 0.48 g/l/day	Masuda et al. (2011)
Cordyceps militaris Mutant G81–3	Cordycepin production using crystallization method	Surface-liquid culture of Mutant	High-energy proton beam irradiation	High level production of cordycepin at 14.3 g/l	Masuda et al. (2014)
Cordyceps militaris 3936	Optimization of cordycepin extraction	Submerged cultivation broth	Solvent-Solvent extraction	High yield extraction (95%)	Tuli et al. (2014a)

still required to improve the production process in both upstream and downstream. In addition, more investigations are still needed for scaling up and industrialization of this process. Finally, finding the correlation between cordycepin and other bioactive ingredients in the extract need further investigations to improve the production process.

Acknowledgement The financial support from the Institute of Bioproduct Development (IBD), is gratefully acknowledged. We would like also to acknowledge the support of MOHE and UTM-RMC through HICOE grant no.R.J130000.7846.4 J262.

References

Ahn YJ, Park SJ, Lee SG, Shin SC, Choi DH (2006) Cordycepin: selective growth inhibitor derived from liquid culture of *cordyceps militaris* against *clostridium* spp. J Agric Food Chem 48(7):2744–2748

Aman S, Anderson DJ, Connolly TJ, Crittall AJ, Ji G (2000) From adenosine to 3 '-deoxy-adenosine: development and scale up. Org Process Res Dev 4(6):601–605

Bok JW, Lermer L, Chilton J, Klingeman HG, Towers GN (1999) Antitumor sterols from the mycelia of *Cordyceps sinensis*. Phytochemistry 51(7):891–898

Buenz EJ, Bauer BA, Osmundson TW, Motley TJ (2005) The traditional Chinese medicine *Cordyceps sinensis* and its effects on apoptotic homeostasis. J Ethnopharmacol 96(1–2):19–29

Chang W, Lim S, Song H, Song BW, Kim HJ, Cha MJ, Sung JM, Kim TW, Hwang KC (2008) Cordycepin inhibits vascular smooth muscle cell proliferation. Eur J Pharmacol 597(1–3):64–69

Chen LS, Stellrecht CM, Gandhi V (2008) RNA-directed agent, cordycepin, induces cell death in multiple myeloma cells. Br J Haematol 140(6):682–391

Chen M, Cheung FW, Chan MH, Hui PK, Ip SP, Ling YH, Che CT, Liu WK (2012) Protective roles of *Cordyceps* on lung fibrosis in cellular and rat models. J Ethnopharmacol 143(2):448–454

Chen ZG, Zhang DN, Cao L, Han YB (2013) Highly efficient and region selective acylation of pharmacologically interesting cordycepin catalyzed by lipase in the eco-friendly solvent 2-methyltetrahydrofuran. Bioresour Technol 133:82–86

Cheng Z, He W, Zhou X, Lv Q, Xu X, Yang S, Zhao C, Guo L (2011) Cordycepin protects against cerebral ischemia/reperfusion injury in vivo and in vitro. Eur J Pharmacol 664(1–3):20–28

Cho HJ, Cho JY, Rhee MH, Park HJ (2007) Cordycepin (3′-deoxyadenosine) inhibits human platelet aggregation in a cyclic AMP-and cyclic GMP-dependent manner. Eur J Pharmacol 558(1–3):43–51

Choi S, Lim MH, Kim KM, Jeon BH, Song WO, Kim TW (2011) Cordycepin-induced apoptosis and autophagy in breast cancer cells are independent of the estrogen receptor. Toxicol Appl Pharmacol 257(2):165–173

Choi YH, Kim GY, Lee HH (2014) Anti-inflammatory effects of cordycepin in lipopolysaccharide-stimulated RAW 264.7 macrophages through Toll-like receptor 4-mediated suppression of mitogen-activated protein kinases and NF-κB signaling pathways. Drug Des Devel Ther 8:1941

Cory JG, Suhadolnik RJ, Resnick B, Rich MA (1965) Incorporation of cordycepin (3′-deoxy-adenosine) into ribonucleic acid and deoxyribonucleic acid of human tumor cells. Biochim Biophys Acta 103(4):646–653

Cunningham KG (1951) Cordycepin, a metabolic product from cultures of Cordyceps militaris (Linn) Link. Part I. Isolation and characterization. J Chem Soc 2:2299–2302

Da Silva AS, Wolkmer P, Nunes JT, Duck MR, Oliveira CB, Gressler LT, Costa MM, Zanette RA, Mazzanti CM, Lopes ST, Monteiro SG (2011) Susceptibility of Trypanosoma evansi to cordycepin. Biomed Pharmacother 65(3):220–223

Das S, Masuda M, Sakurai A, Sakakibara M (2009) Effects of additives on cordycepin production using a *Cordyceps militaris* mutant induced by ion beam irradiation. Afr J Biotechnol 8(13):3041–3047

Das SK, Fujihara S, Masuda M, Sakurai A (2016) Efficient production of anticancer agent cordycepin by repeated batch culture of *Cordyceps militaris* mutant. WCECS 2010, San Francisco, USA. World Congress on Engineering and Computer Science 2010

Dou C, Cao Z, Ding N, Hou T, Luo F, Kang F, Yang X, Jiang H, Xie Z, Hu M, Xu J (2016) Cordycepin prevents bone loss through inhibiting osteoclastogenesis by scavenging ROS generation. Nutrients 8(4):231

El Enshasy H, Elsayed EA, Aziz R, Wadaan MA (2013) Mushrooms and truffles: historical biofactories for complementary medicine in Africa and in the middle East. Evid Based Complement Alternat Med 2013:1–10

El Enshasy HE (2010) Immunomodulators. In: Hofrichter M (ed) Mycota, 2nd edn. Springer Verlag, Heidelberg, pp 165–194

El Enshasy HA, Hatti-Kaul R (2013) Mushroom immunomodulators: unique molecules with unlimited applications. Trends Biotechnol 31(12):668–677

Fan DD, Wang W, Zhong JJ (2012) Enhancement of cordycepin production in submerged cultures of *Cordyceps militaris* by addition of ferrous sulfate. Biochem Eng J 60:30–35

Fan H, Li SP, Xiang JJ, Lai CM, Yang FQ, Gao JL, Wang YT (2006) Qualitative and quantitative determination of nucleosides, bases and their analogues in natural and cultured *Cordyceps* by pressurized liquid extraction and high performance liquid chromatography–electrospray ionization tandem mass spectrometry (HPLC–ESI–MS/MS). Anal Chim Acta 567(2):218–228

Fujita T, Inoue K, Yamamot OS, Ikumoto T, Sasaki S, Toyama R, Chiba K, Hoshino Y, Okumoto T (1994) Fungal metabolites. Part 11. A potent immunosuppressive activity found in *Isaria sinclairii* metabolite. J Antibiot (Tokyo) 47(2):208–215

Hansske F, Robins MJ (1985) Regiospecific and stereoselective conversion of ribonucleosides to 3-deoxynucleosides. A high yield three-stage synthesis of cordycepin from adenosine. Tetrahedron Lett 26(36):4295–4298

Holbein S, Wengi A, Decourty L, Freimoser FM, Jacquier A, Dichtl B (2009) Cordycepin interferes with 3′ end formation in yeast independently of its potential to terminate RNA chain elongation. RNA 15(5):837–849

Hsu TH, Shiao LH, Hsieh C, Chang DM (2002) A comparison of the chemical composition and bioactive ingredients of the Chinese medicinal mushroom Dong Chong XiaCao, its counterfeit and mimic, and fermented mycelium of *Cordyceps sinensis*. Food Chem 78(4):463–469

Hu Z, Lee CI, Shah VK, Oh EH, Han JY, Bae JR, Lee K, Chong MS, Hong JT, Oh KW (2013) Cordycepin increases nonrapid eye movement sleep via adenosine receptors in rats. Evid Based Complement Alternat Med:2013

Jeong JW, Jin CY, Kim GY, Lee JD, Park C, Kim GD, Kim WJ, Jung WK, Seo SK, Choi IW, Choi YH (2010) Anti-inflammatory effects of cordycepin via suppression of inflammatory mediators in BV2 microglial cells. Int Immunopharmacol 10(12):1580–1586

Jeong JW, Jin CY, Park C, Hong SH, Kim GY, Jeong YK, Lee JD, Yoo YH, Choi YH (2011) Induction of apoptosis by cordycepin via reactive oxygen species generation in human leukemia cells. Toxicol In Vitro 25(4):817–824

Jeong MH, Lee CM, Lee SW, Seo SY, Seo MJ, Kang BW, Jeong YK, Choi YJ, Yang KM, Jo WS (2013) Cordycepin-enriched *Cordyceps militaris* induces immunomodulation and tumor growth delay in mouse-derived breast cancer. Oncol Rep 30(4):1996–2002

Jiapeng T, Yiting L, Li Z (2014) Optimization of fermentation conditions and purification of cordycepin from *Cordyceps militaris*. Prep Biochem Biotechnol 44(1):90–106

Jung SM, Park SS, Kim WJ, Moon SK (2012) Ras/ERK1 pathway regulation of p27KIP1-mediated G1-phase cell-cycle arrest in cordycepin-induced inhibition of the proliferation of vascular smooth muscle cells. Eur J Pharmacol 681(1–3):15–22

Kang C, Wen TC, Kang JC, Meng ZB, Li GR, Hyde KD (2014) Optimization of large-scale culture conditions for the production of cordycepin with *Cordyceps militaris* by liquid static culture. Sci World J 2014:2014

Kiho T, Ukai ST (1995) (Semitake and others), Cordyceps species. Food Rev Int 11(1):231–234

Kim HG, Shrestha B, Lim SY, Yoon DH, Chang WC, Shin DJ, Han SK, Park SM, Park JH, Park HI, Sung JM (2006) Cordycepin inhibits lipopolysaccharide-induced inflammation by the suppression of NF-κB through Akt and p38 inhibition in RAW 264.7 macrophage cells. Eur J Pharmacol 545(2–3):192–199

Kim HO, Yun JW (2005) A comparative study on the production of exopolysaccharides between two entomopathogenic fungi *Cordyceps militaris* and *Cordyceps sinensis* in submerged mycelial cultures. J Appl Microbiol 99(4):728–738

Klenow H (1963) Formation of the mono-, di-and triphosphate of cordycepin in Ehrlich ascites-tumor cells in vitro. Biochim Biophys Acta (BBA) 76:347–353

Ko BS, Lu YJ, Yao WL, Liu TA, Tzean SS, Shen TL, Liou JY (2013) Cordycepin regulates GSK--3β/β-catenin signaling in human leukemia cells. PLoS One 8(9):e76320

Kodama EN, McCaffrey RP, Yusa K, Mitsuya H (2000) Antileukemic activity and mechanism of action of cordycepin against terminal deoxynucleotidyl transferase-positive (TdT+) leukemic cells. Biochem Pharmacol 59(3):273–281

Kubota Y, Kunikata M, Ishizaki N, Haraguchi K, Odanaka Y, Tanaka H (2008) Synthesis of 4-benzoyloxycordycepin from adenosine. Tetrahedron 64(10):2391–2396

Lee DH, Kim HH, Lim DH, Kim JL, Park HJ (2015) Effect of cordycepin-enriched WIB801C from *Cordyceps militaris* suppressing fibrinogen binding to glycoprotein IIb/IIIa. Biomol Ther (Seoul) 23(1):60–70

Lee SJ, Kim SK, Choi WS, Kim WJ, Moon SK (2009) Cordycepin causes p21WAF1-mediated G2/M cell-cycle arrest by regulating c-Jun N-terminal kinase activation in human bladder cancer cells. Arch Biochem Biophys 490(2):103–109

Lee SJ, Moon GS, Jung KH, Kim WJ, Moon SK (2010) c-Jun N-terminal kinase 1 is required for cordycepin-mediated induction of G2/M cell-cycle arrest via p21WAF1 expression in human colon cancer cells. Food Chem Toxicol 48(1):277–283

Leung PH, Wu JY (2007) Effects of ammonium feeding on the production of bioactive metabolites (cordycepin and exopolysaccharides) in mycelial culture of a *Cordyceps sinensis* fungus. J Appl Microbiol 103(5):1942–1949

Leung PH, Zhang QX, Wu JY (2006) Mycelium cultivation, chemical composition and antitumour activity of a *Tolypocladium* sp. fungus isolated from wild *Cordyceps sinensis*. J Appl Microbiol 101(2):275–283

Li SP, Li P, Lai CM, Gong YX, Kan KK, Dong TT, Tsim KW, Wang YT (2004) Simultaneous determination of ergosterol, nucleosides and their bases from natural and cultured *Cordyceps* by pressurised liquid extraction and high-performance liquid chromatography. J Chromatogr A 1036(2):239–243

Lia J, Guanb M, Lic Y (2015) Effects of cooking on the contents of adenosine and cordycepin in *Cordyceps militaris*. In: The 7th World Congress on Particle Technology (WCPT7). Elsevier Ltd, China, pp 485–491

Lin YW, Chiang BH (2008) Anti-tumor activity of the fermentation broth of *Cordyceps militaris* cultured in the medium of Radix astragali. Process Biochem 43(3):244–250

Ling JY, Sun YJ, Zhang H, Lv P, Zhang CK (2002) Measurement of cordycepin and adenosine in stroma of *Cordyceps* sp. by capillary zone electrophoresis (CZE). J Biosci Bioeng 94(4):371–374

Ling JY, Zhang GY, Lin JQ, Cui ZJ, Zhang CK (2009) Supercritical fluid extraction of cordycepin and adenosine from *Cordyceps kyushuensis* and purification by high-speed counter-current chromatography. Sep Purif Technol 66(3):625–629

Ma L, Zhang S, Du M (2015) Cordycepin from *Cordyceps militaris* prevents hyperglycemia in alloxan-induced diabetic mice. Nutr Res 35(5):431–439

Mao XB, Eksriwong T, Chauvatcharin S, Zhong JJ (2005) Optimization of carbon source and carbon/nitrogen ratio for cordycepin production by submerged cultivation of medicinal mushroom *Cordyceps militaris* (2005). Process Biochem 40(5):1667–1672

Mao XB, Zhong JJ (2004) Hyperproduction of cordycepin by two-stage dissolved oxygen control in submerged cultivation of medicinal mushroom *Cordyceps militaris* in bioreactors. Biotechnol Prog 20(5):1408–1413

Mao XB, Zhong JJ (2006) Significant effect of NH4+ on cordycepin production by submerged cultivation of medicinal mushroom *Cordyceps militaris*. Enzym Microb Technol 38(3–4):343–350

Masuda M, Das SK, Fujihara S, Hatashita M, Sakurai A (2011) Production of cordycepin by a repeated batch culture of a *Cordyceps militaris* mutant obtained by proton beam irradiation. J Biosci Bioeng 111(1):55–60

Masuda M, Das SK, Hatashita M, Fujihara S, Sakurai A (2014) Efficient production of cordycepin by the *Cordyceps militaris* mutant G81-3 for practical use. Process Biochem 49(2):181–187

Masuda M, Urabe E, Honda H, Sakurai A, Sakakibara M (2007) Enhanced production of cordycepin by surface culture using the medicinal mushroom *Cordyceps militaris*. Enzym Microb Technol 40(5):1199–1205

Masuda M, Urabe E, Sakurai A, Sakakibara M (2006) Production of cordycepin by surface culture using the medicinal mushroom *Cordyceps militaris*. Enzym Microb Technol 39(4):641–646

Mizuno T (1999) Medicinal effects and utilization of Cordyceps (Fr.) Link (Ascomycetes) and Isaria Fr. (Mitosporic Fungi) Chinese Caterpillar Fungi, "Tochukaso". Int J Med Mushrooms 1(3):251–261

Muller WE, Weiler BE, Charubala R, Pfleiderer W, Leserman L, Sobol RW (1991) Cordycepin analogues of 2′, 5′ -oligoadenylate inhibit human immunodeficiency virus infection via inhibition of reverse transcriptase. Biochemistry 30:2027–2033

Nakamura K, Konoha K, Yoshikawa N, Yamaguchi Y, Kagota S, Shinozuka K, Kunitomo MN (2005) Effect of cordycepin (3′-deoxyadenosine) on hematogenic lung metastatic model mice. In Vivo 19(1):137–141

Ng TB, Wang HX (2005) Pharmacological actions of *Cordyceps*, a prized folk medicine. J Pharm Pharmacol 57(12):1509–1519

Ni H, Zhou XH, Li HH, Huang WF (2009) Column chromatographic extraction and preparation of cordycepin from Cordyceps militaris waster medium. J Chromatogr B 877(22):2135–2141

Overgaard-Hansen K (1964a) The inhibition of 5-phosphoribosyl-1-pyrophosphate formation by cordycepin triphosphate in extracts of Ehrlich ascites tumor cells. Biochim Biophys Acta 80(3):504–507

Overgaard-Hansen K (1964b) The inhibition of 5-phosphoribosyl-1-pyrophosphate formation by cordycepin triphosphate in extracts of Ehrlich ascites tumor cells. Biochim Biophys Acta 80:504–507

Pan BS, Wang YK, Lai MS, Mu YF, Huang BM (2015) Cordycepin induced MA-10 mouse Leydig tumor cell apoptosis by regulating p38 MAPKs and PI3K/AKT signaling pathways. Sci Rep 25(5):13372

Park HJ (2015) Ethanol extract of *Cordyceps militaris* grown on germinated soybeans inhibits 2, 4-dinitrophenolfluorobenzene-induced allergic contact dermatitis. J Funct Foods 1(17):938–947

Pegler DN, Yao YJ, Li Y (1994) The Chinese 'caterpillar fungus'. Mycologist 8(1):3–5

Peng J, Wang P, Ge H, Qu X, Jin X (2015) Effects of cordycepin on the microglia-overactivation-induced impairments of growth and development of hippocampal cultured neurons. PLoS One 10(5):e0125902

Ramesh T, Yoo SK, Kim SW, Hwang SY, Sohn SH, Kim IW, Kim SK (2012) Cordycepin (3′-deoxyadenosine) attenuates age-related oxidative stress and ameliorates antioxidant capacity in rats. Exp Gerontol 47(12):979–987

Rao YK, Chou CH, Tzeng YM (2006) A simple and rapid method for identification and determination of cordycepin in *Cordyceps militaris* by capillary electrophoresis. Anal Chim Acta 566(2):253–258

Ren Z, Cui J, Huo Z, Xue J, Cui H, Luo B, Jiang L, Yang R (2012) Cordycepin suppresses TNF-α-induced NF-κB activation by reducing p65 transcriptional activity, inhibiting IκBα phosphorylation, and blocking IKKγ ubiquitination. Int Immunopharmacol 14(4):698–703

Rich MA, Meyers P, Weinbaum G, Cory JG, Suhadolnik RJ (1965) Inhibition of human tumor cells by cordycepin. Biochim Biophys Acta 95(2):194–204

Rottman F, Guarino AJ (1964) The inhibition of phosphoribosyl-pyrophosphate amidotransferase activity by cordycepin monophosphate. Biochem Biophys Acta 89:465–472

Soltani M, Kamyab H, El-Enshasy HA (2013) Molecular weight (Mw) and Monosaccharide composition (MC): Two major factors affecting the therapeutic action of polysaccharides extracted from *Cordyceps sinensis*. J Pure Appl Microbio 7(3):160111613
Soltani M, Abd Malek R, Ware I, Ramli S, Elsayed EA, Aziz R (2017) Optimization of cordycepin extraction from *Cordyceps militaris* fermentation broth. J Sci Ind Res 76:355–361
Suhadolnik RJ, Cory JG (1964) Further evidence for the biosynthesis of cordycepin and proof of the structure of 3-deoxyribose. Biochim Biophys Acta 91:661–662
Sun Y, Wang YH, Qu K, Zhu HB (2011) Beneficial effects of cordycepin on metabolic profiles of liver and plasma from hyperlipidemic hamsters. J Asian Nat Prod Res 13(06):534–546
Tang YJ, Zhu LW, Li HM, Li DS (2007) Submerged culture of mushrooms in bioreactors-challenges, current state-of-the-art, and future prospects. Food Technol Biotechnol 45(3): 221–229
Tuli HS, Sandhu SS, Kashyap D, Sharma AK (2014a) Optimization of extraction conditions and antimicrobial potential of a bioactive metabolite, cordycepin from *Cordyceps militaris* 3936. WJPPS 3:1525–1535
Tuli HS, Sandhu SS, Sharma AK (2014b) Pharmacological and therapeutic potential of *Cordyceps* with special reference to Cordycepin. 3 Biotech 4(1):1–2
Tuli HS, Sharma AK, Sandhu SS (2014c) Optimization of fermentation conditions for cordycepin production using *Cordyceps militaris* 3936. J Biol Chem Sci 1:35–47
Velut S, de Maré L, Hagander P (2007) Bioreactor control using a probing feeding strategy and mid-ranging control. Control Eng Pract 15(2):135–147
Wang B-J, Won S-J, Yu Z-R, Su C-L (2005) Free radical scavenging and apoptotic effects of *Cordyceps sinensis* fractionated by supercritical carbon dioxide. Food Chem Toxicol 43:543–552
Wang H-J, Pan M-C, Chang C-K, Chang S-W, Hsieh C-W (2014) Optimization of ultrasonic-assisted extraction of cordycepin from *Cordyceps militaris* using orthogonal experimental design. Molecules 19:20808–20820
Wasser SP (2002) Medicinal mushrooms as a source of antitumor and immunomodulating polysaccharides. Appl Microbiol Biotechnol 60:258–274
Wei HP, Ye XL, Chen Z, Zhong YJ, Li PM, Pu SC, Li XG (2009) Synthesis and pharmacokinetic evaluation of novel N-acyl-cordycepin derivatives with a normal alkyl chain. Eur J Med Chem 44:665–669
Won S-Y, Park E-H (2005) Anti-inflammatory and related pharmacological activities of cultured mycelia and fruiting bodies of *Cordyceps militaris*. J Ethnopharmacol 96:555–561
Wong YY, Moon A, Duffin R, Barthet-Barateig A, Meijer HA, Clemens MJ, de Moor CH (2009) Cordycepin inhibits protein synthesis and cell adhesion through effects on signal transduction. J Biol Chem 23:jbc-M109
Wu C, Guo Y, Su Y, Zhang X, Luan H, Zhang X, Zhu H, He H, Wang X, Sun G, Sun X, Guo P, Zhu P (2014) Cordycepin activates AMP-activated protein kinase (AMPK) via interaction with the γ1 subunit. J Cell Mol Med 18:293–304
Wu JY, Zhang QX, Leung PH (2007) Inhibitory effects of ethyl acetate extract of *Cordyceps sinensis* mycelium on various cancer cells in culture and B16 melanoma in C57BL/6 mice. Phytomedicine 14:43–49
Xie Huichun YL, Lao D, Zhang T, Ito Y (2011) Preparative separation of high-purity cordycepin from *Cordyceps militaris*(L.) link by high-speed countercurrent chromatography. J Liq Chromatogr Relat Technol 34:491–499
Xie J-W, Huang L-F, Hu W, He Y-B, Wong KP (2010) Analysis of the Main Nucleosides in *Cordyceps Sinensis* by LC/ESI-MS. Molecules 15:305–314
Yalin W, Cuirong S, Yuanjiang P (2006) Studies on isolation and structural features of a polysaccharide from the mycelium of a Chinese edible fungus (*Cordyceps sinensis*). Carbohydr Polym 63:251–256
Yang F-Q, Ge L, Yong JWH, Tan SN, Li S-P (2009) Determination of nucleosides and nucleobases in different species of *Cordyceps* by capillary electrophoresis–mass spectrometry. J Pharm Biomed Anal 50:307–314

Yang FQ, Guan J, Li SP (2007) Fast simultaneous determination of 14 nucleosides and nucleobases in cultured *Cordyceps* using ultra-performance liquid chromatography. Talanta 73:269–273
Yang QZ, Yang J, Zhang CK (2006) Synthesis and properties of cordycepin intercalates of Mg–Al–nitrate layered double hydroxides. Int J Pharm 326:148–152
Yang FQ, Li DQ, Feng K, Hu DJ, Li SP (2010) Determination of nucleotides, nucleosides and their transformationproducts in Cordyceps by ion-pairing reversed-phase liquid chromatography-mass spectrometry. J Chromatogr A.34:5501–5510. https://doi.org/10.1016/j.chroma.2010.06.062.
Yang X, Li Y, He Y, Li T, Wang W, Zhang J, Wei J, Deng Y, Lin R (2015) Cordycepin alleviates airway hyperreactivity in a murine model of asthma by attenuating the inflammatory process. Int Immunopharmacol 26:401–408
Yao L-H, Li C-H, Yan W-W, Huang J-N, Liu W-X, Xiao P (2011) Cordycepin decreases activity of hippocampal CA1 pyramidal neuron through membrane hyperpolarization. Neurosci Lett 503:256–260
Yao L-H, Meng W, Song R-F, Xiong Q-P, Sun W, Luo Z-Q, Yan W-W, Li Y-P, Liu X-P, Li H-H, Xiao P (2014) Modulation effects of cordycepin on the skeletal muscle contraction of toad gastrocnemius muscle. Eur J Pharmacol 726:9–15
Yoshikawa N, Nakamura K, Yamaguchi Y, Kagota S, Shinozuka K, Kunitomo M (2004) Antitumor activity of cordycepin in mice. Clin Exp Pharmacol Physiol 31:S51–S53
Yu R, Song L, Zhao Y, Bin W, Wang L, Zhang H (2004a) Isolation and biological properties of polysaccharide CPS-1 from cultured *Cordyceps militaris*. Fitoterapia 75:465–472
Yu R, Wang L, Zhang H, Zhou C, Zhao Y (2004b) Isolation, purification and identi- fication of polysaccharides from cultured *Cordyceps militaris*. Fitoterapia 75:662–666
Yun YH, Han SH, Lee SJ, Ko SK, Lee CK, Ha NJ, Kim KJ (2003) Anti-diabetic effects of CCCA, CMESS, and cordycepin from *Cordyceps militaris* and the immune responses in Streptozotocin-induced diabetic mice. Nat Prod Res 9:291–298
Zhang DW, Deng H, Qi W, Zhao GY, Cao R (2015) Osteoprotective effect of cordycepin on estrogen deficiency-induced osteoporosis in vitro and in vivo. Biomed Res Int 2015:423869
Zhang Z, Tudi T, Liu Y, Zhou S, Feng N, Yang Y, Tang C, Tang Q, Zhang J (2016) Preparative isolation of cordycepin, N 6 -(2-hydroxyethyl)-adenosine and adenosine from Cordyceps militaris by macroporous resin and purification by recycling high-speed counter-current chromatography. Journal of Chromatography B 1033-1034:218–225
Zheng P, Xia Y, Xiao G, Xiong C, Hu X, Zhang S, Zheng H, Huang Y, Zhou Y, Wang S, Zhao GP (2012) Genome sequence of the insect pathogenic fungus *Cordyceps militaris*, a valued traditional Chinese medicine. Genome Biol 12(11):R116
Zheng ZL, Qiu XH, Han RC (2015) Identification of the genes involved in the fruiting body production and cordycepin formation of *Cordyceps militaris* fungus. Mycobiology 43:37–42
Zhou X, Meyer CU, Schmidtke P, Zepp F (2002) Effect of cordycepin on interleukin-10 production of human peripheral blood mononuclear cells. Eur J Pharmacol 453:309–317
Zhu JS, Halpern GM, Jones K (1998) The scientific rediscovery of an ancient Chinese herbal medicine: *Cordyceps sinensis*: part I. J Altern Complement Med 4:289–303

Chapter 17
Biosynthesis of Nanoparticles Using Mushrooms

Anu Kalia and Gagandeep Kaur

17.1 Introduction

Nanoparticles (NPs) are the zero dimensional nanomaterials which exhibit diverse morphologies, dimensions and chemistry. This novel form of matter possesses unusual physicochemical and optoelectronic properties owing to phenomena of 'Quantum confinement' i.e. confinement of electrons within particles that are smaller than the bulk electron delocalization length (Foldbjerg et al. 2015). Therefore, nanoparticles have found wide applications in diagnostics, therapeutics, catalysis, electronics and many other fields and numerous commercial products contain engineered nanomaterials (Zhang et al. 2016; Matsoukas et al. 2015).

Nanoparticles can be generated by several physical, wet chemistry and other techniques. However, in the quest to design environmentally benign procedures for synthesis of NPs, the focus of researchers has been greatly shifted from chemical methods towards biological systems (Khan et al. 2018). Over the evolutionary period, living organisms have acquired specialized traits that mechanistically involve a range of escaping or counter-acting processes such as bio-accumulation/bio-mineralization and precipitation to evade the cellular toxicity due to exposure of **xenobiotics** and heavy metal compounds (Banerjee and Rai 2017). Thus, such organisms possess resistance machinery against the respective metal/semiconductor ions/compound (s), provided that their concentration does not exceed the threshold limit. The components of the resistance machinery include different primary and secondary metabolites as well as the cellular and extracellular macromolecules which work in a consorted manner within or on surface of the living cell to accomplish the **detoxification** process. This leads to formation of NPs inside, on the cell

A. Kalia · G. Kaur (✉)
Punjab Agricultural University, Ludhiana, Punjab, India
e-mail: gagandeep.21511@lpu.co.in

B. P. Singh et al. (eds.), *Biology of Macrofungi*, Fungal Biology,
https://doi.org/10.1007/978-3-030-02622-6_17

surface or outside the cell through process called 'nanobiosynthesis' (Siddiqi and Husen 2016).

The diversity of these macromolecules of a living cell can be harnessed to generate different morphologies, size dimensions and consequently varied functionalities of NPs. Biogenesis of NPs comprehensively relies on reduction of metal/semi conductor ions by various biomolecules such as proteins particularly the biocatalytic proteins (enzymes), and other proteins, peptides and even amino acids; cellular sugars (polysaccharides), and hetero−/homocyclic aromatic compounds such as vitamins of cellular origin. Therefore, the reduction potential and capacity of the system determines the effectiveness of the biogenerators (Siddiqi and Husen 2016).

Another specialized aspect of this clean, environmentally benign **nanobiosynthesis** process is the utilization of the diversity of the cell types, prokaryotic to eukaryote besides the complexity of tissue and organ systems which further intensifies the multiplicity for obtaining different types of NPs. Prokaryotes such as bacteria and actinomycetes have been elaborately evaluated for the synthesis of various NPs. Likewise, unicellular eukaryotes such as single cell algae and multicellular eukaryotes, fungi, devoid of organ-system hierarchy, have also shown a great promise for the synthesis of a variety of metal/metal oxide, non-metal oxide and semiconductor NPs. The fungal cell derived NP synthesis, mycosynthesis, has caught the attention of researchers worldwide due to few peculiar characteristics of fungi such as secretion of copious amount of diverse **extracellular enzymes** and proteins which act as the reducing as well as capping agents thereby mediating the extracellular NP synthesis rendering ease in the downstream isolation and purification processes (Banerjee and Rai 2017). Moreover, the several fold quicker growth compared to other microbes leads to generation of huge fungal biomass. This further enhances the NP yield per unit biomass as the cylindrical fungal mycelia offer a large surface area for interaction with the dissolved ionic forms of the metal/semiconductor elements leading to greater extent of availability of the interface for the reduction of ionic forms and generation of NPs (Mukherjee et al. 2001).

Among fungi, basidiomycetous macrofungi or mushrooms exhibit enormous variability regarding their habitat and morphology. These fungi exist in the wild and have been cultivated on farms (Atkins 1972) for their nutritional benefits due to presence of a range of bioactive compounds spanning over saccharides, proteins to complex terpenes, flavonoids, vitamins and minerals (Cheung 2010). The immunomodulatory and immune-stimulating properties of the medicinal mushrooms further magnify the nutraceutical benefits of intake of mushroom extracts, reductions, and concoctions. In recent few years, focus of nanoparticle research has intensified on exploring the possibilities of NP synthesis by utilizing various genera of edible and medicinal mushrooms. Presence of innumerable bioactive compounds in these mushrooms offers a huge potential for NP synthesis. Numerous proteins and polysaccharides found in mushrooms have been employed for the synthesis of both intracellular and extracellular organic (carbon NPs) and inorganic NPs (metal and non-metal NPs). The mushroom derived NPs exhibit unique features such as high stability, extended shelf-life, water solubility and good dispersion properties.

Therefore, nanobiosynthesis using mushroom hyphae and fruiting bodies seem to be very promising technology for the green synthesis of probably non-toxic, eco-friendly and stable nanomaterials (Banerjee and Rai 2017).

17.2 How Mushroom Derived Inorganic Nanoparticles can be Synthesized?

Both fruiting body and the mycelium of the mushrooms have been used for NP synthesis. Edible and medicinal mushrooms have been utilized for the synthesis of a variety of inorganic NPs by the use of metallic compounds such as mineral salts *viz.*, silver nitrate ($AgNO_3$), chloroauric acid ($HAuCl_4$), sodium selenite (Na_2SeO_3), zinc sulfide (ZnS), cadmium sulfide (CdS), ferrous sulfate ($FeSO_4$), palladium chloride and ($PdCl_2$), alumina (Al_2O_3) (Vaseghi et al. 2018). The generalized protocol starts with thorough washing of the basidiocarps with distilled water to remove any dirt, dust and other impurities. This is followed by blending or crushing of the cleaned fruit bodies and then suspending the resulting paste in known volume of distilled water. The cell extract can be obtained easily by either filtering or centrifugation of the suspension. The resultant filtrate/supernatant can be utilized for generation of NPs by supplementation of metal salt solution followed by incubation with or without electromagnetic radiation exposure. Certain noble metal NP synthesis can be figured out as change of color in the reaction mixture while the metal oxide NPs such as ZnO or TiO_2 may exhibit shift from near transparent to white suspension on incubation (Fig. 17.1).

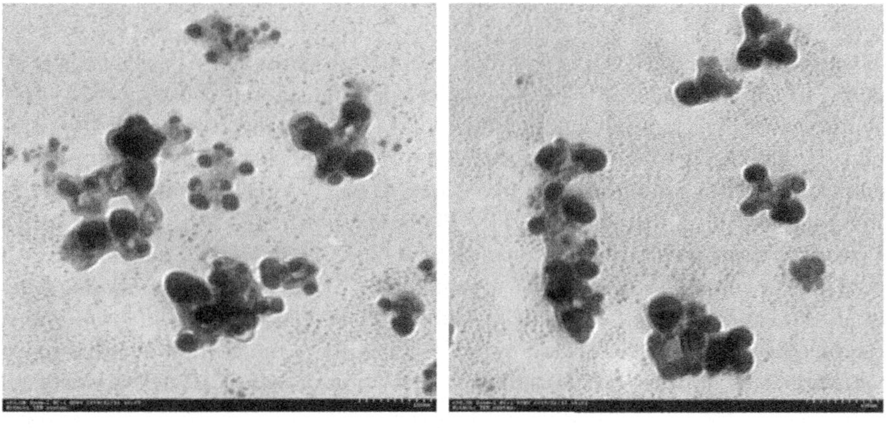

Fig. 17.1 Silver nanoparticles synthesized by photocatalysis using *Pleurotus florida* fruit body extract. The AgNPs can be seen embedded in a matrix which was found to be proteinaceous in nature. (Image: Electron Microscopy and Nanoscience Laboratory, Punjab Agricultural University, Ludhiana)

17.3 Mechanism for Synthesis of Nanoparticles by Mushrooms

Riboflavin is a water soluble compound that is sensitive to light and functions in the bound coenzyme forms **flavin mononucleotide** (FMN) and **flavin adenine dinucleotide** (FAD), which catalyze various cellular oxidation-reduction reactions. Mushrooms serve a s a rich source of riboflavin and its presence in mushroom extracts has been hypothesized to be responsible for the reduction of metal ions into NPs. When the reaction mixture is exposed to sunlight, the flavins present in the reaction mixture may get excited and act as electron donors or oxidizers. This gives a signiificant insight into the mechanism for the conversion of metal salts to nanometals (Bhat et al. 2013).

Fungal phenol oxidases (laccases, tyrosinases, and Mn-peroxidases) have also been hypothesized to play a role in recovery of gold, silver, selenium, and silicon nanoparticles using *Lentinus edodes*, *Pleurotus ostreatus*, *Ganoderma lucidum*, and *Grifola frondosa* (Vetchinkina et al. 2017). The study also introduced a hypothetical mechanism of gold (III) reduction from $HAuCl_4$ to gold (0) by phenol oxidases that resulted in formation gold nanoparticles of different shapes and sizes. Analysis of **FT-IR** spectra supported the involvement of proteins and heterocyclic compounds in the culture filtrate of *Schizophyllum commune* for the synthesis of silver nanoparticles.

17.4 Types of Mushroom Derived Engineered Nanoparticles (ENPs)

17.4.1 Silver Nanoparticles (AgNPs)

Mycosynthesis of AgNPs was first reported by Vigneshwaran (Vigneshwaran et al. 2007) that involved the role of proteins of spent mushroom substrate (SMS) for synthesis. Variable morphologies of the AgNPs (different shapes, sizes and topographies) have been reported from mycelial and fruit body extracts of various popular edible and medicinal mushrooms like *Agaricus bisporus* (Narasimha et al. 2011), *Pleurotus florida* (Bhat et al. 2011), *Pleurotus sajor-caju* (Musa et al. 2017), *Ganoderma neo-japonicum* (Gurunathan et al. 2013), *Ganoderma lucidum* (Karwa et al. 2011), *Inonotus obliquus* (Nagajyothi et al. 2014), *Tricholoma crassum* (Ray et al. 2011), *Tricholoma matsutake* (Anthony et al. 2014), and *Calocyble indica* (Gurunathan et al. 2015). Invariably, size variability is more often over the shape variabilities as most reports documented formation of spherical or near spherical morphologies though rare reports of triangular (Shankar et al. 2003) and hexagonal NP synthesis have also been in scientific literature. As both size and shape affect the activity potential of the synthesized **AgNPs**, maintenance of the monodisper nature of the NP suspension is desirable. The mushroom derived AgNP suspensions exhibit better stability which can be unanimously attributed to the capping action of the mushroom macromolecules (Narayanan and Sakthivel 2010).

17.4.2 Applications of Mushroom Derived AgNPs

Since the dawn of Nanotechnology, AgNPs have been lauded for excellent **anti-microbial** activities. Therefore, majority reports advocated the evaluation of the synthesized AgNPs against a range of plant and human pathogenic bacterial and fungal cultures. Moreover, a synergistic enhancement in potential of **antibiotics** against **biofilm forming** bacteria has been reported by (Al-Hamadani and Kareem 2017) on application of antibiotics in combination with AgNPs synthesized using mushrooms. The anti-cellular effects are not restricted to microorganisms rather significant anti-tumorogenic properties of multi-shaped AgNPs derived from *Calocybe indica* have been observed against breast cancer cells (Gurunathan et al. 2015). The possible mechanism of action potential of AgNPs regarding the cytotoxicity and apoptosis may be attributed to oxidative stress induced by AgNPs (Gurunathan 2015). Sen et al. (2013) reported use of *Pleurotus florida* glucans for the synthesis of AgNP-glucan conjugates that exhibited antibacterial activity against multiple antibiotic resistant (MAR) bacterium *Klebsiella pneumoniae* YSI6A. The antibacterial activity was hypothesized to be due to generation of reactive oxygen species (ROS) that damaged the cellular macromolecules. The hypothesis was supported by observed degradation of bacterial DNA. Further evidence came from introduction of ROS scavenger histidine that resulted in dose dependent decrease of bactericidal activity of AgNPs-glucan.

17.4.3 Gold Nanoparticles (Au-NPs)

Use of **gold nanoparticles** (AuNPs) for a wide variety of biomedical applications, like bio-imaging, lateral flow assays, environmental detection and purification, data storage, drug delivery, biomarkers, catalysis, chemical sensors, and DNA detection makes them one of the most sought after nanomaterials. Mycosynthesis of AuNPs has been reported using a variety of genera such as *Agaricus bisporus, Pleurotus sapidus, Inonotus obliquus, Ganoderma spp.* and *Flammulina velutipes* (Khan et al. 2014, Lee at al. (2015), Sarkar et al. (2013). Extract of *Volvariella volvacea* has been used for extracellular synthesis of Au, Ag and Au-Ag NPs in water (Philip 2009).

The dielectric heating technique is yet another alternative technique for the synthesis of AuNPs that utilized edible mushroom *Agaricus bisporus* extract as both reducing and stabilizing agent (Eskandari-Nojehdehi et al. 2016). This method has been used for synthesis AuNPs of different sizes (20–150 nm) and shapes (triangular nanoprisms to nearly spherical and hexagonal). The AuNPs morphology particularly the size and shape largely depended on extract temperature. Gold nanoparticles were also synthesized using glucans isolated form an edible mushroom *Pleurotus florida*, that acted to reduce chloroauric acid to AuNPs (Sen et al. 2013). The fungal glucan acted as reducing as well as stabilizing agent for AuNPs. The resulting AuNPs-glucan bio-conjugates were able to efficiently catalyze the reduction of 4-nitrophenol (4-NP) to 4-aminophenol (4-AP), in the presence of sodium borohydride.

17.4.4 Applications of Mushroom Derived Gold NPs

Flammulina velutipes derived intracellular AuNPs have been utilized to catalytically reduce organic pollutants methylene blue and 4-nitrophenol (Narayanan et al. 2015). While biocompatibility analysis of AuNPs synthesized from *Ganoderma* sp. showed no **cytotoxicity** towards breast cancer cell lines (Gurunathan et al. 2014), biofunctionalized AuNPs synthesized from *Pleurotus florida* showed significant anti-cancer activity against four different cancer cell lines and no lethal effect was observed in Vero cell lines (Bhat et al. 2013).

17.4.5 Selenium Nanoparticles (Se-NPs)

Elemental Se (Se^0) is an insoluble metalloid which can be synthesized both chemically and biologically (Yang et al. 2008) and is proposed to be a good replacement for selenate (Se^{+2}) or selenite (Se^{+4}) ions in clinical practices due to its lower toxicity (Zhang et al. 2008, 2001). Vetchinkina et al. (2013) reported that the medicinal mushroom *Lentinula edodes* can be used to synthesize spherical nanoparticles of slemental Se by reducing selenium from inorganic sodium selenite (Se^{IV}) and from the organo-selenium compound 1,5-diphenyl-3-selenopentanedione-1,5 (DAPS-25). *Pleurotus ostreatus*, *Ganoderma lucidum*, and *Grifila frondosa* also have the capability of formation of NPs on cultivation on selenium and germanium-containing media (Na_2SeO_3 and GeO_2, respectively) (Vetchinkina et al. 2016). Mushroom polysaccharides and proteins have been used extensively for bio-functionalization of Se-NPs. Xiao et al. (2017) reported generation of *Cordyceps sinensis* exopolysaccharide (EPS)-conjugated SeNPs through reduction of SeO_3^{2-}. Formation of C–O···Se bonds between the -OH groups of EPS and SeNPs could have resulted in good dispersion of SeNPs in the EPS matrix. The EPS-SeNPs conjugates exhibited significant scavenging ability against superoxide anion radical (O_2^-) and **ABTS** radical cation ($ABTS^+$) when compared to pure EPS, indicating that the conjugated SeNPs reinforced antioxidant effect of EPS.

17.4.6 Applications of Mushroom Derived Selenium NPs

Pleurotus tuber-regium polysaccharide functionalized nano-selenium hydrosol has been reported to exhibit enhanced anti-tumor effects because of synergistic activity of mushroom polysaccharide and Se-NPs (Chen et al. 2015). The SeNPs decorated by the water-soluble derivative of *Ganoderma lucidum* polysaccharides showed immune-modulatory activity (Wang et al. 2014). Mushroom polysaccharides-protein complexes were used as capping agent for synthesis of size controllable and highly stable **selenium nanoparticles** (SeNPs) (Wu et al. 2012, 2013).

The polysachharide-protein complex capped the SeNPs through physical adsorption of hydroxyl groups of polysaccharides and imino groups of proteins on the SeNP surface. These bio-functionalized SeNPs were found to exhibit enhanced cellular uptake through endocytosis and were able to significantly inhibit human breast carcinoma cells.

17.4.7 CdS Nanoparticles (Quantum Dots)

The increasing application of semiconductor nanoparticles like **Cadmium Sulfide** in biosensors, photocatlysts, solar cells, diodes and quantum dots for targeted drug delivery and therapy has intensified the research on their biological synthesis. Macrofungus *Coriolus versicolor* has been used for extracellular synthesis of CdS nanoparticles under ambient conditions. The sulfure required for the transformation of toxic Cd to non-toxic CdS was found to be provided by thiol groups of fungal proteins. The resulting CdS NPs were found to be highly stable and auto-capped (Sanghi and Verma 2009). In another study, *Pleurotus ostreatus* was used for synthesis of CdS quantum dots. The obtained quantum dots were of a spherical shape with predominant size ranging from 4 to 5 nm (Borovaya et al. 2015). Cadmium Sulfide NPs were synthesized by using Reishi mushroom (*Ganoderma lucidum*) aqueous extract and were found to have antibacterial and antifungal properties (Raziya et al. 2006). Senapati and Sarkar (2014) reported synthesis of ZnS nanoparticles using *Pleurotus ostreatus*. The **XRD** analysis of the synthesized NPs confirmed that the ZnS nanoparticles had cubic structure. Further, the analysis also revealed a relationship between smaller particle size and increase of dislocation density and strain with increase of amount of mushroom extract used for the synthesis.

17.4.8 Iron Nanoparticles (Fe-NPs)

Synthesis of Fe-NPs using *Pleurotus* sp. has been reported by Mazumdar and Haloi (2017). UV-Vis spectra showed SPR peaks at 226 nm and 276 nm wavelength, confirming the formation of nanoparticles. The study hypothesized use of **siderophores** by mycelium to transport the iron inside.

17.5 Future Prospects

Use of mushrooms for synthesis of nanoparticles has been a relatively new venture in nanotechnology. However, the trend has been catching on due to many advantages like easy downstream processing, high protein content in mushrooms, cost effectiveness and easy scale up of processing. Extracellular and intracellular reductases have been found to play major role in nanoparticle synthesis. While nanoparticles synthesized through mushrooms are already finding applications in fields of therapeutics and biomedical engineering, it only appears like we have only scratched the surface yet. For future research, mushrooms can be genetically engineered to produce more of the enzymes involved in synthesis of nanoparticles. Elucidation of mechanisms involved in mycosynthesis of nanoparticles in of utmost importance at this stage as variation in mechanism in various mushroom species can help identify different strategies of synthesis and help us modulate the size, shape and yield of nanoparticles for their application in medicine, agriculture and technology.

References

Al-Hamadani AH, Kareem AA (2017) Combination effect of edible mushroom – sliver nanoparticles and antibiotics against selected multidrug biofilm pathogens. Iraq Med J 1(3):68–74

Anthony KJP, Murugan M, Jeyaraj M, Rathinam NK, Sangiliyandi G (2014) Synthesis of silver nanoparticles using pine mushroom extract: a potential antimicrobial agent against *E. coli* and *B. subtilis*. J Ind Eng Chem 20(4):2325–2331

Atkins FC (1972) Mushroom growing today. Faber and Faber, London

Banerjee K, Rai VR (2017) A review on mycosynthesis, mechanism, and characterization of silver and gold nanoparticles. Bio NanoSci 8(1):17–31. https://doi.org/10.1007/s12668-017-0437-8

Bhat R, Deshpande R, Ganachari SV, Huh DS, Venkataraman A (2011) Photo-irradiated biosynthesis of silver nanoparticles using edible mushroom *Pleurotus florida* and their antibacterial activity studies. Bioinorg Chem Appl 2011:1–7. https://doi.org/10.1155/2011/650979

Bhat R, Sharanabasava VG, Deshpande R, Shetti U, Sanjeev G, Venkataraman A (2013) Photo-bio-synthesis of irregular shaped functionalized gold nanoparticles using edible mushroom *Pleurotus florida* and its anticancer evaluation. J Photochem Photobiol B 125:63–69

Borovaya M, Pirko Y, Krupodorova T, Naumenko A, Blume Y, Yemets A (2015) Biosynthesis of cadmium sulphide quantum dots by using *Pleurotus ostreatus* (Jacq.) P. Kumm. Biotechnol Biotechnol Equip 29(6):1156–1163

Chen T, Wong KH, Wu HL, Zheng W, Wong MK, Man WYC (2015) U.S. patent no. 9,072,669. U.S. Patent and Trademark Office, Washington, DC

Cheung PCK (2010) The nutritional and health benefits of mushrooms. Nutr Bull 35:292–299

Eskandari-Nojehdehi M, Jafarizadeh-Malmiri H, Rahbar-Shahrouzi J (2016) Optimization of processing parameters in green synthesis of gold nanoparticles using microwave and edible mushroom (*Agaricus bisporus*) extract and evaluation of their antibacterial activity. Nanotechnol Rev 5(6):537–548

Foldbjerg R, Jiang X, Miclaus T, Chen C, Autrupa H, Beer C (2015) Silver nanoparticles-wolves in sheep's clothing? Toxicol Res 4(3):563–575

Gurunathan S, Raman J, Malek SNA, John PA, Vikineswary S (2013) Green synthesis of silver nanoparticles using *Ganoderma neo-japonicum* Imazeki: a potential cytotoxic agent against breast cancer cells. Int J Nanomedicine 8:4399

Gurunathan S, Han J, Park JH, Kim JH (2014) A green chemistry approach for synthesizing biocompatible gold nanoparticles. Nanoscale Res Lett 9(1):248

Gurunathan S, Park JH, Han JW, Kim JH (2015) Comparative assessment of the apoptotic potential of silver nanoparticles synthesized by *Bacillus tequilensis* and Calocybe indica in MDA-MB-231 human breast cancer cells: targeting p53 for anticancer therapy. Int J Nanomedicine 10:4203

Karwa AS, Gaikwad S, Rai MK (2011) Mycosynthesis of silver nanoparticles using Lingzhi or Reishi medicinal mushroom, *Ganoderma lucidum* (W. Curt.: Fr.) P. karst. And their role as antimicrobials and antibiotic activity enhancers. Int J Med Mushrooms 13(5):483–491

Khan AK, Rashid R, Murtaza G, Zahra A (2014) Gold nanoparticles: synthesis and applications in drug delivery. Trop J Pharm Res 13(7):1169–1177

Khan AU, Malik N, Khan M, Cho MH, Khan MM (2018) Fungi-assisted silver nanoparticle synthesis and their applications. Bioprocess Biosyst Eng 41(1):1–20

Lee KD, Nagajyothi PC, Sreekanth TVM, Park S (2015) Eco-friendly synthesis of gold nanoparticles (AuNPs) using *Inonotus obliquus* and their antibacterial, antioxidant and cytotoxic activities. J Ind Eng Chem 26:67–72

Matsoukas T, Desai T, Lee K (2015) Engineered nanoparticles and their applications. J Nanomater 2015:1–2. https://doi.org/10.1155/2015/651273

Mazumdar H, Haloi N (2017) A study on biosynthesis of iron nanoparticles by *Pleurotus sp.* J Microbiol Biotechnol Res 1(3):39–49

Mukherjee P, Ahmad A, Mandal D, Senapati S, Sainkar SR (2001) Bioreduction of $AuCl_4^-$ ions by the fungus, *Verticillium* sp. and surface trapping of the gold nanoparticles formed. Angew Chem Int Ed 40:3585–3588

Musa SF, Yeat TS, Kamal LZM, Tabana YM, Ahmed MA, El Ouweini A, Sandai D (2017) *Pleurotus sajor-caju* can be used to synthesize silver nanoparticles with anti-fungal activity against *Candida albicans*. J Sci Food Agri 98(3):1197–1207

Nagajyothi PC, Sreekanth TVM, Lee JI, Lee KD (2014) Mycosynthesis: antibacterial, antioxidant and antiproliferative activities of silver nanoparticles synthesized from *Inonotus obliquus* (Chaga mushroom) extract. J Photochem Photobiol B 130:299–304

Narasimha G, Praveen B, Mallikarjuna K, Deva Prasad Raju B (2011) Mushrooms (*Agaricus bisporus*) mediated biosynthesis of sliver nanoparticles, characterization and their antimicrobial activity. Int J Nano Dimens 2:29–36

Narayanan KB, Sakthivel N (2010) Biological synthesis of metal nanoparticles by microbes. Adv Colloid Interf Sci 156(1–2):1–13

Narayanan KB, Park HH, Han SS (2015) Synthesis and characterization of biomatrixed-gold nanoparticles by the mushroom *Flammulina velutipes* and its heterogeneous catalytic potential. Chemosphere 141:169–175

Philip D (2009) Biosynthesis of Au, Ag and A–Ag nanoparticles using edible mushroom extract. Spectrochim Acta A Mol Biomol Spectrosc 73(2):374–381

Ray S, Sarkar S, Kundu S (2011) Extracellular biosynthesis of silver nanoparticles using the mycorrhizal mushroom *Tricholoma crassum* (Berk.) Sacc: its antimicrobial activity against pathogenic bacteria and fungus, including multidrug resistant plant and human bacteria. Dig J Nanomater Biostruct 6:1289–1299

Raziya S, Durga B, Rajamahanthe SG, Govindh B, Annapurna N (2006) Synthesis and characterization of CdS nanoparticles using Reishi mushroom. Intl J Adv Technol in Eng Sci 4:220–227

Sanghi R, Verma P (2009) A facile green extracellular biosynthesis of CdS nanoparticles by immobilized fungus. Chem Eng J 155(3):886–891

Sarkar J, Roy SK, Laskar A, Chattopadhyay D, Acharya K (2013) Bioreduction of chloroaurate ions to gold nanoparticles by culture filtrate of *Pleurotus sapidus* Quel. Mater Lett 92:313–316

Sen I K, Maity K, and Islam SS (2013) Green synthesis of gold nanoparticles using a glucan of an edible mushroom and study of catalytic activity. Carbohydr polym 91(2): 518-528.

Senapati US, Sarkar D (2014) Characterization of biosynthesized zinc sulphide nanoparticles using edible mushroom *Pleurotuss ostreatus*. Indian J Phys 88(6):557–562

Shankar SS, Ahmad A, Pasricha R, Sastry M (2003) Bioreduction of chloroaurate ions by geranium leaves and its endophytic fungus yields gold nanoparticles of different shapes. J Mater Chem 13(7):1822–1826

Siddiqi KS, Husen A (2016) Fabrication of metal nanoparticles from fungi and metal salts: scope and application. Nanoscale Res Lett 11:98. https://doi.org/10.1186/s11671-016-1311-2

Vaseghi Z, Nematollahzadeh A, Tavakoli O (2018) Green methods for the synthesis of metal nanoparticles using biogenic reducing agents: a review. Rev Chem Engg 34(4):529–559

Vetchinkina E, Loshchinina E, Kursky V, Nikitina V (2013) Reduction of organic and inorganic selenium compounds by the edible medicinal basidiomycete *Lentinula edodes* and the accumulation of elemental selenium nanoparticles in its mycelium. J Microbiol 51(6):829–835

Vetchinkina EP, Loshchinina EA, Kurskyi VF, Nikitina VE (2016) Biological synthesis of selenium and germanium nanoparticles by xylotrophic basidiomycetes. Appl Biochem Microbiol 52(1):87–97

Vetchinkina EP, Loshchinina EA, Vodolazov IR, Kursky VF, Dykman LA, Nikitina VE (2017) Biosynthesis of nanoparticles of metals and metalloids by basidiomycetes. Preparation of gold nanoparticles by using purified fungal phenol oxidases. Appl Microbiol Biotechnol 101(3):1047–1062

Vigneshwaran N, Kathe AA, Varadarajan PV, Nachane RP, Balasubramanya RH (2007) Silver–protein (core– shell) nanoparticle production using spent mushroom substrate. Langmuir 23(13):7113–7117

Wang J, Zhang Y, Yuan Y, Yue T (2014) Immunomodulatory of selenium nano-particles decorated by sulfated *Ganoderma lucidum* polysaccharides. Food Chem Toxicol 68:183–189

Wu H, Li X, Liu W, Chen T, Li Y, Zheng W, Wong KH (2012) Surface decoration of selenium nanoparticles by mushroom polysaccharides–protein complexes to achieve enhanced cellular uptake and anti-proliferative activity. J Mater Chem 22(19):9602–9610

Wu H, Zhu H, Li X, Liu Z, Zheng W, Chen T, Yu B, Wong KH (2013) Induction of apoptosis and cell cycle arrest in A549 human lung adenocarcinoma cells by surface-capping selenium nanoparticles: an effect enhanced by polysaccharide–protein complexes from *Polyporus rhinocerus*. J Agric Food Chem 61(41):9859–9866

Xiao Y, Huang Q, Zheng Z, Guan H, Liu S (2017) Construction of a *Cordyceps sinensis* exopolysaccharide-conjugated selenium nanoparticles and enhancement of their antioxidant activities. Int J Biol Macromol 99:483–491

Yang LB, Shen YH, Xie AJ, Liang JJ, Zhang BC (2008) Synthesis of Se nanoparticles by using TSA ion and its photocatalytic application for decolorization of cango red under UV irradiation. Mater Res Bull 43(3):572–582

Zhang JS, Gao XY, Zhang LD and Bao YP (2001) Biological effects of a nano red elemental selenium. Biofactors 15(1): 27–38.

Zhang J, Peng D, Lu H and Liu Q (2008) Attenuating the toxicity of cisplatin by using selenosulfate with reduced risk of selenium toxicity as compared with selenite. Toxicol and Appl Pharmacol 226(3): 251–259.

Zhang Y, Leu YR, Aitken RJ, Riediker M (2016) Inventory of engineered nanoparticle-containing consumer products available in the Singapore retail market and likelihood of release into the aquatic environment. Int J Environ Res Public Health 12:8717–8743. https://doi.org/10.3390/ijerph120808717

Chapter 18
Bioconversion and Biotransformation Efficiencies of Wild Macrofungi

Aparecido Almeida Conceição, Joice Raisa Barbosa Cunha, Vandinelma Oliveira Vieira, Rubén Darío Romero Pelaéz, Simone Mendonça, João Ricardo Moreira Almeida, Eustáquio Souza Dias, Euziclei Gonzaga de Almeida, and Félix Gonçalves de Siqueira

18.1 Introduction

Macrofungi are vital for the maintenance of life on earth, meanly because of their ability to biodegrade organic matter, such as all the components of wood. When these microorganisms utilize organic matter as a substrate to produce a valuable by-product, the process is called bioconversion. Purified enzymes from a macrofungus can be used for the production of new, valuable by-products from a specific

A. A. Conceição
Graduate Program of Biosciences, Federal University of Bahia, Vitória da Conquista, Bahia, Brazil

Embrapa Agroenergy, Distrito Federal, Brasília, Brazil

J. R. B. Cunha
Graduate Program Microbial Agricultural, Federal University of Lavras, Lavras, Minas Gerais, Brazil

Embrapa Agroenergy, Distrito Federal, Brasília, Brazil

V. O. Vieira
Graduate Program of Biotechnology and Biodiversity, Federal University of Mato Grosso, Mato Grosso, Cuiabá, Brazil

Embrapa Agroenergy, Distrito Federal, Brasília, Brazil

R. D. R. Pelaéz · J. R. M. Almeida
Graduate Program Microbial Biology, University of Brasilia, Distrito Federal, Brasília, Brazil

Embrapa Agroenergy, Distrito Federal, Brasília, Brazil

S. Mendonça
Embrapa Agroenergy, Distrito Federal, Brasília, Brazil

E. S. Dias
Graduate Program Microbial Agricultural, Federal University of Lavras, Lavras, Minas Gerais, Brazil

B. P. Singh et al. (eds.), *Biology of Macrofungi*, Fungal Biology,
https://doi.org/10.1007/978-3-030-02622-6_18

substrate, which is a process called biotransformation (Collins and Kennedy 1999). The fungi need to obtain energy from a nutrient source to create a new product. Complex substrates generally need to be degraded to produce sugars; this processes is called biodegradation, which is applied to disintegration of any matter by a biological process. The biodegradation occurs through the action of specialized enzymes. Some of these enzymes, called promiscuous enzymes, are able to degrade several analogous substrates. Through the action of these enzymes, the fungi can degrade/remove some toxic/xenobiotic substances by a process called biodetoxification. When the elimination of the xenobiotic compounds occurs in contaminated media, including water, soil and subsurface material, the fungi/microorganisms achieve the bioremediation or myco-bioremediation. When the macrofungi degrade complex organic matter into mineral samples, this process is called mineralization (Coleman and Raiswell 2015).

Macrofungi are important agents that are responsible for mineralization, and they are specialized in the process of deconstructing complex structures such as lignin, for example (Boyle et al. 1992). With the capacity to produce different enzymes, macrofungi can efficiently degrade a wide range of substances so they are used as biodetoxification and bioremediation agents. Because of the ability of fungi to adapt to the most variable and extreme conditions, different biomasses can be used as substrates for the growth of macrofungi and the production of by-products with great interest for industry.

18.2 Vegetal Biomass Transformation to Value By-Products

The bioconversion efficiency of a macrofungus is closely linked to its ability to produce an enzymatic arsenal capable of efficiently degrading the most diverse types of substrates. Through the production of an enormous variety of enzymes, the so-called white-rot fungi (WRF) are naturally-occurring organisms that cause the decay of fallen trees by degrading the lignin present. Under laboratory conditions, these macrofungi produce enzymes and metabolites that are used in human activities, such as medicine, food, textile, paper, and feedstock (Chen and Chiu 2005).

E. G. de Almeida
Graduate Program of Biotechnology and Biodiversity, Federal University of Mato Grosso, Mato Grosso, Cuiabá, Brazil

F. G. de Siqueira (✉)
Graduate Program of Biosciences, Federal University of Bahia, Vitória da Conquista, Bahia, Brazil

Graduate Program of Biotechnology and Biodiversity, Federal University of Mato Grosso, Mato Grosso, Cuiabá, Brazil

Embrapa Agroenergy, Distrito Federal, Brasília, Brazil
e-mail: felix.siqueira@embrapa.br

Pre-treatment is essential for the reduction of recalcitrant biomass, and it functions by destroying the lignin structure and, thereby, increasing the accessibility of enzymes to cellulose. During its development in the plant material, the fungus recognizes its environment and can adapt to the present conditions by secreting exoenzymes or substances that help to modulate a propitious condition or to compete and survive in the presence of other organisms (Rossi et al. 2013). During this process, the enzymes catalyze the release of small portions of the complex macromolecules present in the plant cell wall. This process acts as a foundation for the biorefinery technology technology that is based on the sugar platform, and it is an effective way to utilize the raw materials from lignocelluloses (Su et al. 2018).

Filamentous fungi produce two principal groups of enzymes that are important for the degrading of plant cell walls: holocellulotic (cellulases and hemicellulases) and lignolytic (oxidative) enzymes. Oxidative enzymes include peroxidases (lignin and manganese peroxidase) and laccases. These enzymes are capable of oxidizing smaller molecules (eg, organic acids), and they form unstable intermediates by decomposing the lignin matrix when they are in contact with the phenolic and aromatic amine components of lignin (Tian et al. 2012). The deconstruction of lignin is one of the crucial processes for reducing the recalcitrance of vegetal cell walls.

Three classes of enzymes are required for the hydrolysis of cellulose: β-1,4-endoglycanases, exoglycanases and β-glycosidase. Endoglycanases catayze the hydrolysis of glycosidic bonds in the interior of the cellulose chain, mainly in the amorphous regions, to release units that are later degraded by the other two classes of enzyme. Exoglycanases act on the terminal bonds of cellulose by cleaving it into oligomers and cellobiose; the cellobiose is finally cleaved into glucose monomers by the action of β-glycosidase (Kantharaj et al. 2017).

The degradation of hemicellulose is a much more complex process than that of cellulose because of its greater structural complexity. Thus, different enzymatic classes are required for the degradation of the different types of hemicellulose: xylanases (β-1,4-endoxylanase, β-1,4-xylosidase) for the degradation of xylan. β-1,4-endoxylanase catalyses the hydrolysis of the β-1,4 bonds of the xylan chain to produce oligosaccharides, which are converted to xylose by β-1,4-xylosidase. Xyloglucanase and β-glucosidases are necessary for the decomposition of xyloglucans (de Souza 2013).

Wild or commercially cultivated macrofungi are able to bio-transform the vegetal biomass, rich in carbon and nitrogen, into valuable commercial substances, for example, enzymes (hydrolases and oxidative enzymes), carbohydrates (β-glucans), proteins (lectin) and secondary metabolites (lovastatin) (Kozarski et al. 2015). The interest in these potential products from macrofungi is due to the fact that these molecules are also useful for human activities. For example, it is necessary to make the sugar bound in the form of cellulose and hemicellulose from the plant cell wall available to obtain second generation ethanol. For this purpose, the recalcitrance of the vegetal biomass, conferred principally by the lignin, must be reduced. Macrofungi from groups of white-rot fungi are great producers of an arsenal of enzymes that can deconstruct the lignin complex and degrade the cellulose and hemicellulose chains into oligomers and monomers.

The importance of enzymes from macrofungi goes beyond the ability of degrade lignocellulosic material; fungal enzymes are required in many industrial areas (food, textile, paper, medicine). Wild macrofungi are sources of different enzyme classes such as laccase produced by *Lactarius* sp (Khaund and Joshi 2014a), *Lentinula boryana* and *Pycnoporus* sp. (Díaz-Godínez et al. 2016). Other wild macrofungi are able to produce enzymes that are used in medicine, such as metalo-endo-peptidases from *Tricholoma saponaceum*. These enzymes are able to hydrolyze fibrinogen and fibrin, for example (Kim and Kim 2001).

In general, domestic and wild macrofungi are true enzymes factories. These macrofungi have the impressive ability to use different carbon-nitrogen sources to develop so that several biomasses can be used to cultivate these fungi and then use the hydrolases and oxidases to convert the lignocellulosic material. Wild strains such as *Trametes versicolor* (CMU- TA01), *Irpex lacteus* (CMU-84/13) and *Phlebiopsis* sp. (CMU-47/13) have been reported to be potentially useful for the degradation of lignin because of their oxidative activities; thus, they can be useful for the paper industry (Damián-Robles et al. 2017).

The same rule that governs the production of enzymes by wild macrofungi using different carbon sources also works very well for other fungal substances, such as carbohydrates. The difference is that the carbohydrates are usually part of the fungus cell wall and membrane. These carbohydrates have different industrial interests, such as the use of β-glucans in medicine. Several wild mushrooms are sources of glucans. Some species are better producers than commercial species, for example, *Cortinarius violaceus* (L. ex Fr.) Gray, *Leucocybe connata* (Schumach.) Vizzini, *Laccaria amethystina* (Cooke), *Trametes versicolor*, *Piptoporus betulinus* and *Phlebia tremellosa* (Sari et al. 2017).

Other commercially interesting molecules include linoleic acid (*Boletus reticulatus*, *Flammulina velutipes* var. *velutipes*, *Lactarius salmonicolor*, *Pleurotus ostreatus*, *Polyporus squamosus*, and *Russula anthracina*) (Günç Ergönül et al. 2013), lectin (Singh et al. 2015) and bioactive secondary metabolites. For example *Hymenogaster aromaticus* has been reported to be a great producer of antioxidant agents. It contains a large quantity of phenolic components such as (+)-catechin and p-hydroxybenzoic acid (Zengin et al. 2017). Some studies that have obtained valuable by-products from wild macro fungus are summarized in Table 18.1.

The white-rot fungi are able to degrade lignin into water-soluble components and CO_2 by enzymatic action so that they have total access to cellulose and hemicellulose (Boyle et al. 1992). The bioconversion process process is basically based on the utilization of the energy from the sugar present in cellulose and hemicellulose to form the macrofungi's own metabolites that are necessary for growth and survival.

Wild mushrooms are expected to be a great source of a wide range of secondary metabolites and enzymes. Secondary metabolites are produced to help the macrofungi compete and better adapt to a rough environment; these metabolites are also used for human activities (antibiotics, antifungals, nematicides, vitamins, anti-inflammatory agents, polysaccharides). To grow, the macrofungi need to obtain their nutrients mainly from dead wood, which has a recalcitrant matrix protecting

Table 18.1 By-product from bioconversion of vegetal biomass by wild macrofungi.

Wild macrofungi	Substrate	By-product	References
Coprinus cinereus	Sisal waste + cow dung manure	Laccase, lignin peroxidase, carboxy-methylcellulase, xylanase	Raymond et al. (2015)
Coriolus versicolor f. antarcticus BAFC 266, *Pycnoporus sanguineus* BAFC 2126 and *Phlebiabre vispora* BAFC 633, *Ganoderma applanatum*	Malt extract agar	Phenoloxidase, laccase, peroxidase	Fonseca et al. (2015)
Trametes versicolor (CMU- TA01), *Irpex lacteus* (CMU-84/13), *Phlebiopsis* sp. (CMU-47/13)	Kraft pulp + Potato dextrose broth	Laccase, manganese peroxidase, lignin peroxidase	Damián-Robles et al. (2017)
Trametes sp, *T. cingulata, T. elegans, T. poca*	Synthetic medium	Lignin peroxidase, manganese peroxidase, laccase activities	Teere et al. (2001)
Russula sp., *Pycnoporus cinnabarinus*	Sabouraud dextrose agar	Antimicrobial metabolite	Alofe et al. (2005)
Cerrena unicolor	Synthetic medium	Antioxidant and antimicrobial bioactive molecules	(Jaszek et al. 2013)
Rigidoporus microporus (Sw)	–	Tannin, saponin, terpenoid, alkaloid, steroid, phlobatannin, anthraquinone and cardiac glycoside	Falade et al. (2017)
Lactarius sp.	–	laccase	Khaund and Joshi (2014b)
Phlebia sp.	Hardwood kraft pulp	ethanol	Kamei et al. (2012)

the cellulose and hemicellulose macromolecules. However, the macrofungi, white-rot fungi, have an enzymatic arsenal that is able to biodegrade and biotransform the organic components of the wood (lignin) so that they have access to the cellulose and hemicellulose chains. The enzymes produced that degrade lignocellulosic material into sugar monomers for the production of second generation ethanol by a fermentation process using yeast are of industrial interest (Fig. 18.1).

18.3 Macrofungi as Decomposers

The biodegradation of lignocellulolytic materials is important in the carbon- and nitrogen-cycling process because of the abundance of these materials in most terrestrial ecosystems. Lignocellulolytic materials are predominantly composed of approximately 50% cellulose, 25% hemicellulose and 25% lignin (Cosgrove and Jarvis 2012).

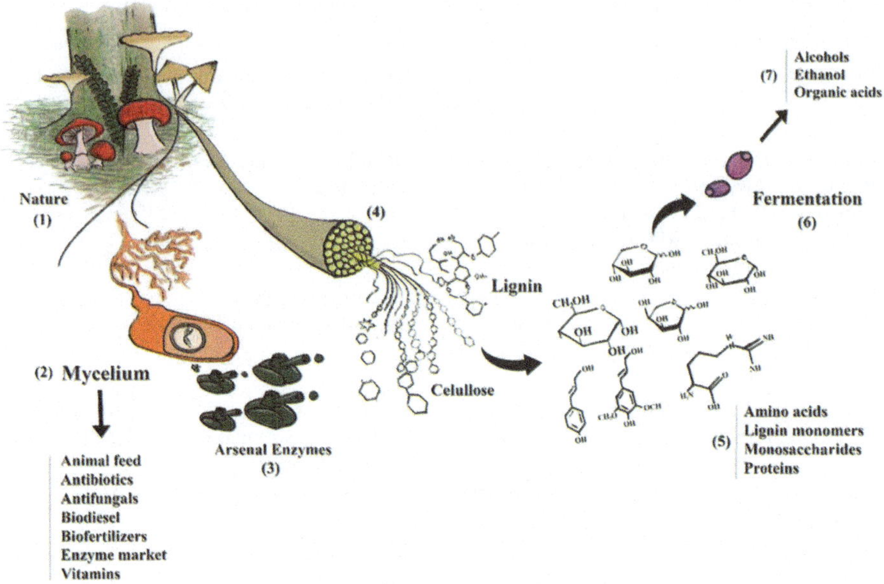

Fig. 18.1 Process of bioconversion to valuable by-products by wild mushrooms. Legend: Wild mushrooms are natural decomposers responsible for the decay of wood (1). Wild mushrooms are able of bioconvert the nutrients in the dead wood into important bioactive molecules for human activities (2). For that purpose, they produce an enzymatic arsenal (3) that is capable of biotransforming the complex plant cell wall components-lignin, cellulose, and hemicellulose (4) into oligomers and monomers (5). The sugar monomers can be further employed in the fermentation process (6) for the production of other valuable substances (ethanol, organic acids) (7)

Fungi participate in crucial processes of maintenance of tropical forests because they act as one of the main decomposers of organic matter in the soil (Shah et al. 2017). These organisms are considered to be natural recyclers because they use an enzymatic complex with a great capacity for degrading several substrates and transforming them into nutrients for their survival (Balaes et al. 2017).

Many of these decomposing fungi belong to the phylum Basidiomycota and participate directly in the degradation of major plant compounds such as cellulose and lignin (Baldrian and Valášková 2008). Even though there is some controversy with regard to the separation of macrofungi into two major groups on the basis of their ability to degrade different components of plant cell walls (Riley et al. 2014), it has still been didactically useful to divide them into white-rot and brown-rot macrofungi. White-rot fungi degrade cellulose, hemicellulose and lignin. This action is possible because of the presence of an extracellular enzyme system involving cellulases, ligninases and hemicellulases. They are of notable ecological importance because they play an important role in the decomposition of woody materials in forests. In addition, among the fungi, the white-rot group is the only one capable of totally degrading lignin. This degradation occurs as a result fo the presence of enzymes that produce hydrogen peroxide (Vasina et al. 2017). The fungi that cause brown-rot only degrade cellulose and hemicellulose; the lignin remains intact, and it is responsible for the brown color of the substrate (Monrroy et al. 2011). The ability to adapt to different carbon and nitrogen sources means that basidiomycetes can occupy diverse ecological niches, such as soil, wood and organic waste materials. Wild macro fungi have the potential of efficiently bioconverting these substrates into valuable products for the fungi and for human activities.

The ecological benefits of decomposing mycobiota go far beyond the decomposition of organic matter and recycling of nutrients for plants. These fungi can attract and interact with various groups of organisms, especially insects such as beetles and fly larvae. They can also facilitate the creation of a fertile environment for the emergence of algae, mosses and arachnids. They serve as food, environment and substrate for the reproduction of these organisms, feeding an entire chain within an ecosystem. This interaction results in a dynamic and vigorously growing environment (Mycena, 2017).

Fungi can degrade other types of organic molecules such as waxes, rubbers, phenol, benzene, toluene, xylene and xenobiotic substances, which are also present in forest ecosystems. They are the main agents responsible for the decomposition of those organic substances that impregnate the plant cells, fibers and vessels, and prevent the entry and displacement of external agents (Prenefetaboldú, 2002).

18.4 Soil Mycoremediation by Wild and Commercial Macrofungi

The use of fungi in soil remediation is also called Mycoremediation and has been reported to be a form of bioremediation. Many fungi have inhabited soil polluted by crude oil, and they have been reported to be capable of producing enzymes that help to predigest, degrade, utilize and mineralize various hydrocarbon pollutants.

White-rot fungi are considered to be efficient biodegraders of contaminants, including pesticides, as the result of an enzyme production system that is composed of lignin peroxidase, manganese peroxidase and laccase. White-rot fungi have been reported to have an important role in the biodegradation of various xenobiotic compounds, including chlorinated phenols, dioxins, polycyclic aromatic hydrocarbons, polychlorinated dibenzofuran, polychlorinated biphenyls, polychlorinated dibenzo-p-dioxin, lignin, explosives, pesticides and dyes. Different strains of white-rot fungi, such as *Phanerochaete chrysospoium*, *Trametes hirsutus*, *P. ostreatus*, *Phanerochaete sordida*, *Pleurotus* (*florida*, *sajorcaju*, *eryngii*, *ostreatus*), and *Cyathus bulleri* have been reported because of their ability to degrade lindane, diuron and other recalcitrant pesticides (Tortella et al. 2005).

Similarly, the fungal "waste" product from as edible mushroom farm (commercial) – Spent Mushroom Substrate (SMS) – has also been reported to be a useful tool for bioremediation. Several studies have shown that SMS can bind heavy metals and prevent them from migrating to water resources or being absorbed by plants from the soil. iI eliminates preservatives from wood and pesticides, chlorinated and non-chlorinated hydrocarbons from contaminated soils (Okerentugba et al. 2015). The compounds by SMS also play an important role in establishing and enabling plant growth on various contaminated or degraded soils, as well as improving biostimulation, nutrients, soil water content and retention (Asemoloye et al. 2017).

Many studies have been performed on microorganisms with the potential for decontamination processes. Chan-Cupul et al. (2016) analyzed the degradation of atrazine in a soil microcosm system by white-rot fungus (*Trametes maxima*) and its co-cultivation with *Paecilomyces carneus* and determined the absorption-desorption kinetics of atrazine in a clay soil. They concluded that both the fungal enzyme extract from *T. maxima* monoculture and that from co-cultivation with *P. carneus* were able to degrade atrazine in a short time (12 h) in a clay soil and that atrazine was highly adsorbed by clay soil. Degradation of atrazine by enzymatic extracts of fungi cultivated under different soil conditions revealed that both the monoculture of *T. maxima* and the co-cultivation with *P. carneus* degraded 100% of the atrazine in the soil, but the degradation time was shorter in the co-culture.

Huang et al. (2015) performed a market research and risk assessment for traceability in edible fungi and the role of the substrate in the accumulation of heavy metals, and they showed that there were significant positive correlations between the concentrations of metals (Cd, Pb and As) in mushrooms and their substrates. Martirani et al. (1996) studied the reduction of phenol content and toxicity in effluents from the olive oil industry by the ligninolytic fungus *Pleurotus ostreatus*, and the initial ability of this macrofungi to survive in the presence of wastewater from the olive oil industry (OMW) was observed in mycelia capable of growing in OMW up to a 20% dilution. This fungus was selected for use in transforming diluted crude OMW. *Pleurotus ostreatus* removed a detoxified OMW phenotype diluted to 10% in the absence of nutrients; dilute wastewater was also decontaminated, and this action occurred over a relatively short time (100 h). Bash et al (1999) investigated different factors associated with the use of white-rot fungi for bioremediation using SMS and substrate before fungal production. The contaminants were 16 polycyclic aromatic hydrocarbons (PAHs). Compounds with three rings were similarly degraded by the two sources; however, a significant difference was observed for compounds that were more difficult to degrade (4 and 5 rings). In addition, four- and five-ring compounds were degraded more rapidly when the soil and substrate were homogenized. There was an improvement in the degradation process when fish oil was added. The following removals occurred after seven weeks of incubation: 86% of the total of 16 PAHs, 89% of PAHs with three rings, 87% of PAHs with four rings and 48% of PAHs with five rings.

One of the advantages of this technology is that fertilizer components can be produced without leaking into groundwater (in contrast to inorganic salts). With this technology, micronutrients are bound to the surface of biological material that is completely biodegradable (Tuhy et al. 2015). SMS, like other agricultural residues, has been shown to contain several organic compounds and functional groups, which makes it a good biosorbent (Putra et al. 2014) with promising properties for use in agriculture (Tuhy et al. 2014).

Chiu et al. (1998) tested tested nine fungi (*Armillaria gallica, A. mellea, Ganoderma lucidum, Lentinula edodes, Phanerochaete chrysosporium, Pleurotus pulmonarius, Polyporus sp., Coprinus cinereus* and *Volvariella volvacea*) and the SMS of *Pleurotus pulmonarius* for the removal of the biocide pentachlorophenol; and, concluded that the use of SMS for the bioremediation of sites contaminated by biocides was promising because the performance of the substrate was better than that of many mushroom mycelia for the removal of the biocide. Jia et al. (2017) studied the bioremediation of soil co-contaminated with cadmium cadmium (Cd) and dichlorophene (DCP) by SMS of *Lentinus edodes* and the effects on the activity and biochemical properties of the soil. They concluded the study with promising values and an attractive option for reuse of SMS from bioremediation.

Studies over time have defined the mechanisms used by fungi for degradation and transformation of the contaminating compounds into compounds that can be assimilated by other organisms. The fungi secrete one or more enzymes that are known to play key roles in the processes of breaking down potentially polluting molecules. The three main enzymes are two peroxidases and laccase, whereas tolerance to pesticides has also been demonstrated for the the catalase group. More recently, cytochrome P450 has been reported to be a potential remedy. Some systems are complementary, as was observed from the activity of the laccase enzyme in the early stage of biodegradation, whereas the peroxidase participated more actively in the biodegradation in the later phase. Pizzul et al. (2009) used two mediators ($MnSO_4$ and Tween 80) to evaluate the capacity of lignin peroxidase (LiP) (LiP), manganese peroxidase (MnP), horseradish peroxidase (HRP) and laccase for the degradation of glyphosate and other pesticides. The results revealed that, in the presence of $MnSO_4$ and Tween 80, complete degradation of glyphosate by MnP occurred with and without H_2O_2.

18.4.1 Peroxidases

Among the peroxidases of fungal origin, lignin peroxidase (LiP) and manganese peroxidase (MnP) are heme peroxidases, which require the presence of hydrogen peroxide and manganese for activity and are mainly reported for the degradation of toxic compounds by white-rot fungi. MnP catalyzes an H_2O_2-dependent oxidation of Mn^{2+} to Mn^{3+}, and the Mn^{3+} ions are stabilized by chelation with organic acids. These chelated ions act as diffusible redox mediators facilitating the attack on several molecules and giving MnP a versatile oxidative capacity (Hofrichter 2002). On the other hand, LiP, in the presence of H_2O_2, catalyzes the oxidation of veratryl alcohol (a natural secondary metabolite in fungi that serves as a redox mediator), which then oxidizes non-phenolic aromatic residues or their partial degradation products to generate aryl radicals. These radicals, after various reactions, subsequently undergo complete degradation (Pointing 2001). Studies have demonstrated the production of LiP and MnP under conditions of nitrogen sufficiency. LiP and MnP productions are generally optimum at high oxygen stresses, but they are repressed by the agitation of fungi cultured in submerged liquid culture (conversely, Lac production is enhanced by agitation).

18.4.2 Laccase

Laccases are copper-containing extracellular enzymes from the group of blue oxidases, which use copper as a cofactor and molecular oxygen as the co-substrate. Laccases are able to oxidize most of the phenolic and non-phenolic compounds produced during the degradation of pesticides, and their activity has been observed to be 20 times greater in white-rot fungi, such as *Trametes versicolor*, than those of other organisms (Margot et al. 2013). Viswanath et al. (2014) reported detoxification mediated by marine fungal laccase. The non-specific nature of their activity on a variety of substrates makes them ideal catalysts for the metabolism of a variety of insecticides (Donoso et al. 2008). In the presence of molecules that act as mediators of electron transfer, laccases are able to oxidize many compounds (Atalla et al. 2013), and some of these laccase mediators are produced during the normal metabolic activity of white rot fungi (Asgher et al. 2008). Several recent studies have demonstrated the combined involvement of peroxidase and laccase enzymes in the biodegradation of different pesticides (Donoso et al. 2008; Kadimaliev et al. 2011). For example, Donoso et al. (2008) documented the involvement of peroxidase and laccase activities in the degradation of tribromophenol by *Trametes versicolor*. Kadimaliev et al. (2011) also measured the biodegradation of phenol by *Lentinus tigrinus* in liquid medium with the combined action of laccase and peroxidase.

18.4.3 Catalase

The production of reactive oxygen species (ROS) is a general toxic response in many biological systems. The accumulation of ROS results in damage to cell macromolecules, which is detrimental to cell integrity. Experiments with many other fungi have suggested that catalase activity could be used as a monitoring tool to quantify the efficiency of bioremediation (Lin et al. 2009).

18.4.4 Citocromo P450

The cytochrome P450 (CYP) enzyme system is a large family of widely distributed cysteinatoheme enzymes in nature that participate in the oxidative transformation of many endogenous and exogenous molecules through the insertion of an oxygen atom into a substrate and the consequent reduction of the oxygen atom to a water molecule (Ichinose 2013). Fungal CYP possesses hydrolytic and an enzymatic hydrolytic-oxidative complex for the detoxification of compounds in the environment. In addition to these systems, certain fungi have intracellular networks consisting of cytochrome P450 monooxygenases and glutathione transferases, which constitute its genome, for dealing with various types of pollutants. A cytochrome P450 (CYP) monooxygenase has been shown to be involved in the transformation of organochlorine compounds. In addition, CYP from white rot fungi was also involved in the initial oxidation in the degradation of pyrene, anthracene, fluorene and dibenzothiophene, through the production of epoxidation products (Bezalel et al. 1996). Analysis of the genome of *P. chrysosporium* showed that cytochrome P450 monooxigenases represent the largest and most important group of P450 genes in any fungal species and that the induction of these clusters is differentially expressed, depending on the type of xenobiotic and nutrition (Yadav et al. 2006; Yadav and Loper 2000). Many other fungal enzymes are also involved in combination with the degradation and detoxification of insecticides. While studying the degradation of simazine, trifluralin and dieldrin from *T. versicolor* and *P. chrysosporium,* Fragoeiro and Magan 2008, observed the production of extracellular enzymes with higher cellulase/dehydrogenase activity. Similarly, the degradation of organophosphate monocrotophos insecticides by three fungal strains – *Aspergillus flavus, Fusarium pallidoroseum* and *Macrophomina sp.* – was bound to the extracellular release of alkaline phosphatase, inorganic phosphates and ammonia (Jain et al. 2014).

18.5 Biodegradation in the Plant-Soil-Microbiota Relationship

The rhizosphere is densely populated with a variety of organisms. The interactions between members of the root and rhizosphere community are mainly achieved through chemical communication (Dam and Bouwmeester 2016).

By nature, soil communities are extremely diverse. As a consequence, the rhizosphere' is populated by numerous organisms, including nematodes, fungi, bacteria, arthropods and herbivores (Bonkowski et al. 2009). Each of these organisms, alone and in combination, can interact with the plant. Soil organisms depend mainly on chemical communication because other forms of communication are not viable underground. In fact, plants secrete a wide variety of primary and secondary plant metabolites into the rhizosphere to facilitate interactions with their biotic and abiotic environment. Even the root exudate of a small plant species, such as arabidopsis (*Arabidopsis thaliana*), can contain more than 100 different metabolites (Strehmel et al. 2014).

The main mechanism in bioremediation processes is the production of enzymes that have the ability to degrade xenobiotics. The peroxidase, laccase, catalase and cytochrome p450 enzymes are capable of degrading a diverse group of environmental pollutants, including dioxins, polychlorinated biphenyls (PCBs), petroleum hydrocarbons, HAP, trinitrotoluenes, industrial dye effluents, herbicides and pesticides (Magan et al. 2010).

The challenge over time is the production of sufficient enzymes for industrial use. The co-cultures have shown a great potential for inducing the production of these enzymes. Schneider et al. (2018) evaluated the effect of different sources and concentrations of carbon and nitrogen on the secretion of ligninolytic enzymes from a basidiomycete *M. palmivorus* VE111. They concluded that the concentrations of carbon and nitrogen interfere in the production of ligninolytic enzymes by this fungus, and they found that lower glucose concentrations in combination with intermediate concentrations of casein yielded the highest activity of laccase and peroxidases in a bioreactor. The activities of laccase and peroxidases achieved in this study were among the highest found in the literature.

Despite all the studies favoring the use of co-cultures in bioremediation processes, there are some factors that still need to be considered and studied. The bioremediation process in field studies is conducted under non-sterile conditions. White rot fungi's ability to colonize non-sterile soil differs from species to species. This ability also depends on the antagonistic activity in mixed cultures and consortia. An antagonist interaction is a more likely occurrence; as such, one should consider the negative effects that resident flora might have on the growth of newly introduced fungal species. Therefore, an investigation of the antagonistic effect of fungal co-cultures is of great importance (Albert et al. 2011).

18.6 Omics Approaches to Wild Macrofungi

Great attention has been directed to macrofungi because they have been shown to be an important source of molecules and food with great nutritional value. Because only a small percentage of macrofungus species are known, compared to their real extension in nature, wild macrofungi can be a great and unprecedented source of nutritional food and a source of better bioactive molecules or species that can be used to biodegrade/biotransform different types of substances. To better understand the environmental relationship and potential of these unknown species, techniques that are part of the "omics" concept are required.

Omics is revolutionizing the way biology is studied and how life is perceived. Omics approaches can provide a global view of an entire living system, which can englobe the processes that happen in a single cell or the interaction between cells or two or more different organisms (https://omics.org). The omics field has been driven largely by technological advances that have made cost-efficient analyses intrinsic to biologic systems possible, especially disease studies (Hasin et al. 2017). Fungal omics are now extensively applied to assist in the comprehension of both the fundamental fungal biology and related applications, including understanding and evaluating the processes that lead to fungal biotransformation and bioremediation efficiency. Genomics, transcriptomics, metabolomics, and proteomics can be used to understand the microbial network in an effective way. These "omics" approaches play an important role in understanding the molecular mechanism of degradation because of the ability to bring new perspectives, and they have been discussed at both the single species and at community levels (Aydin et al. 2017).

Silva (2016) states that fungal genomics deals with the discovery and characterization of all the sequences in the entire genome. However, the determination of the genomic sequence is only the beginning. Once it has been achieved, further work should be performed to determine the entire process, such as addressing a function of the numerous genes and compare with other organisms, study the expression profile, study the protein "expression profiling" (proteome) and study the low molecular weight compounds in the cell (metabolomics).

Even though there are visible advances in wild macrofungi by the omics approaches, there is a lack of information regarding fungal lifestyle and evolution, fungi-fungi interactions and fungi-environmental relationships. All the genomic material of the cells from hyphae can be extracted from an isolated specimen. Certification of the right species of the fungi and screening of specific genes for potential enzymes, proteins and secondary metabolites can be achieved with the entire genome. Knowledge of the potential for the enzyme production of the fungi is important mainly for understanding how the species can degrade hard wood or degrade other xenobiotic substances, for example (Li et al. 2018).

When is working with pure cultures, some information remains missing when all the fungal information is obtained with the omics technique, except that from the genomic method. This happens because the fungi are grown in a purified environment. Generally, the medium is made with sugar, water, and other nutrients readily available for the fungi, which does not happen in the nature. Therefore, the fungal potential is supressed under some laboratory conditions (Bertrand et al. 2013).

Methods such us cultivation in complex biomasses and co-culture with different microorganisms, including other fungal species, might help to activate the potential of these macrofungi (mushrooms) to produce different metabolites in quality and quantity. Criation of new toolbox approch macrofungi "omics" are required to understand how these changes occur by modifying just the environment. Some strategies that are able to visualize a biological system are necessary, and this visualization can be achieved by a multi-omic approach.

References

Albert S, Chauhan D, Pandya B, Padhiar A (2011) Screening of *Trichoderma* Spp. As potential fungal partner in co-culturing with white rot fungi for efficient bio pulping. Glob J Biotechnol Biochem 6(3):95–101

Alofe SOB, Onawunmi FV, Ogundaini AO, Tiwalade TA (2005) Mycelial growth and antibacterial metabolite production by wild mushrooms. group, vol 8, pp 157–162

Asemoloye MD, Ahmad R, Jonathan SG (2017) Synergistic action of rhizospheric fungi with *Megathyrsus maximus* root speeds up hydrocarbon degradation kinetics in oil polluted soil. Chemosphere 187:1–10. https://doi.org/10.1016/j.chemosphere.2017.07.158

Asgher M, Bhatti HN, Ashraf M, Legge RL (2008) Recent developments in biodegradation of industrial pollutants by white rot fungi and their enzyme system. Biodegradation 19(6):771–783. https://doi.org/10.1007/s10532-008-9185-3

Atalla MM, Zeinab HK, Eman RH, Amani AY, Abeer AAEA (2013) Characterization and kinetic properties of the purified *Trematos phaeria* mangrovei laccase enzyme. Saudi J Biol Sci 20(4):373–381. https://doi.org/10.1016/j.sjbs.2013.04.001

Aydin S, Karaçay HA, Shahi A, Gökçe S, Ince B, Ince O (2017) Aerobic and anaerobic fungal metabolism and Omics insights for increasing polycyclic aromatic hydrocarbons biodegradation. Fungal Biol Rev 31(2):61–72. https://doi.org/10.1016/j.fbr.2016.12.001

Balaes T, Petre CV, Ungureanu C, Mardari C, Tănase C (2017) Ligninolytic enzyme system in ecological adaptation of lignicolous macrofungi. Appl Ecol Environ Res 15(1):207–224. https://doi.org/10.15666/aeer/1501_207224

Baldrian P, Valášková V (2008) Degradation of cellulose by basidiomycetous fungi. FEMS Microbiol Rev 32(3):501–521. https://doi.org/10.1111/j.1574-6976.2008.00106.x

Bash E, Ceci A, Pierro L, Riccardi C, Pinzari F, Maggi O et al (1999) Biodegradation of Hexachlorocyclohexane (HCH) Isomers by White Rot Fungus, *Pleurotus florida*. J Bioremed Biodegr 32(4):117–126. https://doi.org/10.1007/978-3-540-88258-9

Bertrand S, Schumpp O, Bohni N, Bujard A, Azzollini A, Monod M, Wolfender JL (2013) Detection of metabolite induction in fungal co-cultures on solid media by high-throughput differential ultra-high pressure liquid chromatography-time-of-flight mass spectrometry fingerprinting. J Chromatogr A 1292:219–228. https://doi.org/10.1016/j.chroma.2013.01.098

Bezalel L, Hadar Y, Fu PP, Freeman JP, Cerniglia CE (1996) Initial oxidation products in the metabolism of pyrene, anthracene, fluorene, and dibenzothiophene by the white rot fungus *Pleurotus ostreatus*. Appl Environ Microbiol 62(7):2554–2559

Bonkowski M, Villenave C, Griffiths B (2009) Rhizosphere fauna: The functional and structural diversity of intimate interactions of soil fauna with plant roots. Plant Soil 321(1–2):213–233. https://doi.org/10.1007/s11104-009-0013-2

Boyle CD, Kropp BR, Reid ID (1992) Solubilization and mineralization of lignin by white rot fungi. Appl Environ Microbiol 58(10):3217–3224

Chan-Cupul W, Heredia-Abarca G, Rodríguez-Vázquez R (2016) Atrazine degradation by fungal co-culture enzyme extracts under different soil conditions. J Environ Sci Health, Part B Pesticides, Food Contaminants, and Agricultural Wastes 51(5):298–308. https://doi.org/10.1080/03601234.2015.1128742

Chen S, Chiu M (2005) Evaluation of white-rot fungi for treatment of organic wastes without enviromental impact. In: The 2005 world sustainable building conference, Tokyo, pp 27–29

Chiu SW, Ching ML, Fong KL, Moore D (1998) Spent oyster mushroom substrate performs better than many mushroom mycelia in removing the biocide pentachlorophenol. Mycol Res 102(12):1553–1562. https://doi.org/10.1017/S0953756298007588

Coleman ML, Raiswell R (2015) Microbial mineralization of organic matter: mechanisms of self-organization and inferred rates of precipitation of diagenetic minerals, vol 344. The Royal Society, pp 69–87

Collins AM, Kennedy MJ (1999) Biotransformations and bioconversions in New Zealand: Past endeavours and future potential. Australas Biotechnol 9(2):86–94

Cosgrove DJ, Jarvis MC (2012) Comparative structure and biomechanics of plant primary and secondary cell walls. Front Plant Sci 3:1–6. https://doi.org/10.3389/fpls.2012.00204

Dam NMV, Bouwmeester HJ (2016) Metabolomics in the rhizosphere: tapping into belowground chemical communication. Trends Plant Sci 21(3):256–265. https://doi.org/10.1016/j.tplants.2016.01.008

Damián-Robles RM, Castro-Montoya AJ, Saucedo-Luna J, Vázquez-Garcidueñas MS, Arredondo-Santoyo M, Vázquez-Marrufo G (2017) Characterization of ligninolytic enzyme production in white-rot wild fungal strains suitable for kraft pulp bleaching. 3 Biotech 7(5):319. https://doi.org/10.1007/s13205-017-0968-2

de Souza WR (2013) Microbial degradation of lignocellulosic biomass. In: Chandel DA (ed) Sustainable degradation of lignocellulosic biomass – techniques, applications and commercialization, pp 135–152. InTech. https://doi.org/10.5772/67458

Díaz-Godínez G, Cervantes-Muñoz P, Acosta-Urdapilleta M, Villegas E, Gupta V, Téllez-Téllez M (2016) Enzymatic activity of three wild mushrooms. Mycosphere 7:1568–1575. https://doi.org/10.5943/mycosphere/si/3b/8

Donoso C, Becerra J, Martínez M, Garrido N, Silva M (2008) Degradative ability of 2,4,6-tribromophenol by saprophytic fungi *Trametes versicolor* and *Agaricus augustus* isolated from chilean forestry. World J Microbiol Biotechnol 24(7):961–968. https://doi.org/10.1007/s11274-007-9559-4

Falade OE, Oyetayo VO, Awala SI (2017) Evaluation of the mycochemical composition and antimicrobial potency of wild macrofungus, *Rigidoporus microporus* (Sw). J Phytopharmacol 6(2):115–125

Fonseca MI, Zapata PD, Villalba LL, Fariña JI (2015) Caracterización del potencial enzimático oxidativo de cepas nativas de hongos de pudrición blanca de la selva subtropical de Misiones (*Argentina*). Acta Biológica Colombiana 20(1):47–56. https://doi.org/10.15446/abc.v20n1.38322

Fragoeiro S, Magan N (2008) Impact of *Trametes versicolor* and *Phanerochaete chrysosporium* on differential breakdown of pesticide mixtures in soil microcosms at two water potentials and associated respiration and enzyme activity. Int Biodeterior Biodegrad 62(4):376–383. https://doi.org/10.1016/j.ibiod.2008.03.003

Günç Ergönül P, Akata I, Kalyoncu F, Ergönül B (2013) Fatty acid compositions of six wild edible mushroom species. Sci World J 2013:163964–163964. https://doi.org/10.1155/2013/163964

Hasin Y, Seldin M, Lusis A (2017) Multi-omics approaches to disease. Genome Biol 18(1):1–15. https://doi.org/10.1186/s13059-017-1215-1

Hofrichter M (2002) Review: Lignin conversion by manganese peroxidase (MnP). Enzym Microb Technol 30(4):454–466. https://doi.org/10.1016/S0141-0229(01)00528-2

Huang Q, Jia Y, Wan Y, Li H, Jiang R (2015) Market survey and risk assessment for trace metals in edible fungi and the substrate role in accumulation of heavy metals. J Food Sci 80(7):H1612–H1618. https://doi.org/10.1111/1750-3841.12923

Ichinose H (2013) Cytochrome P450 of wood-rotting basidiomycetes and biotechnological applications. Biotechnol Appl Biochem 60(1):71–81. https://doi.org/10.1002/bab.1061

Jain R, Garg V, Yadav D (2014) In vitro comparative analysis of monocrotophos degrading potential of Aspergillus flavus, *Fusarium pallidoroseum* and *Macrophomina* sp. Biodegradation 25(3):437–446. https://doi.org/10.1007/s10532-013-9672-z

Jaszek M, Osińska-Jaroszuk M, Janusz G, Matuszewska A, Stefaniuk D, Sulej J, Jarosz-Wilkołazka A (2013) New bioactive fungal molecules with high antioxidant and antimicrobial capacity isolated from *Cerrena* unicolor idiophasic cultures. Biomed Res Int 2013:1–11. https://doi.org/10.1155/2013/497492

Jia Z, Deng J, Chen N, Shi W, Tang X, Xu H (2017) Bioremediation of cadmium-dichlorophen co-contaminated soil by spent *Lentinus edodes* substrate and its effects on microbial activity and biochemical properties of soil. J Soils Sediments 17(2):315–325. https://doi.org/10.1007/s11368-016-1562-7

Kadimaliev DA, Revin VV, Atykyan NA, Nadezhina OS, Parshin AA (2011) The role of laccase and peroxidase of *Lentinus* (Panus) *tigrinus* fungus in biodegradation of high phenol concentrations in liquid medium. Appl Biochem Microbiol 47(1):66–71. https://doi.org/10.1134/S0003683811010066

Kamei I, Hirota Y, Mori T, Hirai H, Meguro S, Kondo R (2012) Direct ethanol production from cellulosic materials by the hypersaline-tolerant white-rot fungus *Phlebia* sp. MG-60. Bioresour Technol 112:137–142. https://doi.org/10.1016/j.biortech.2012.02.109

Kantharaj P, Boobalan B, Sooriamuthu S, Mani R (2017) Lignocellulose degrading enzymes from fungi and their industrial applications. Int J Curr Res Rev 9(21):1–13. https://doi.org/10.7324/IJCRR.2017.9211

Khaund P, Joshi SR (2014a) Enzymatic profiling of wild edible mushrooms consumed by the ethnic tribes of India. J Korean Soc Appl Biol Chem 550(1):123–130. https://doi.org/10.1016/j.gene.2014.08.027

Khaund P, Joshi SR (2014b) Enzymatic profiling of wild edible mushrooms consumed by the ethnic tribes of India. J Korean Soc Appl Biol Chem 57(2):263–271. https://doi.org/10.1016/j.gene.2014.08.027

Kim J-H, Kim YS (2001) Characterization of a metalloenzyme from a wild mushroom, Tricholoma saponaceum. Biosci Biotechnol Biochem 65(2):356–362

Kozarski M, Klaus A, Jakovljevic D, Todorovic N, Vunduk J, Petrović P et al (2015) Antioxidants of edible mushrooms. Molecules 20(10):19489–19525. https://doi.org/10.3390/molecules201019489

Li H, Wu S, Ma X, Chen W, Zhang J, Duan S et al (2018) The genome sequences of 90 mushrooms. Sci Rep 8(1):9982. https://doi.org/10.1038/s41598-018-28303-2

Lin X, Li X, Sun T, Li P, Zhou Q, Sun L, Hu X (2009) Changes in microbial populations and enzyme activities during the bioremediation of oil-contaminated soil. Bull Environ Contam Toxicol 83(4):542–547. https://doi.org/10.1007/s00128-009-9838-x

Magan N, Fragoeiro S, Bastos C (2010) Environmental factors and bioremediation of xenobiotics using white rot fungi. Mycobiology 38(4):238–248. https://doi.org/10.4489/MYCO.2010.38.4.238

Margot J, Maillard J, Rossi L, Barry DA, Holliger C (2013) Influence of treatment conditions on the oxidation of micropollutants by Trametes versicolor laccase. New Biotechnol 30(6):803–813. https://doi.org/10.1016/j.nbt.2013.06.004

Martirani L, Giardina P, Marzullo L, Sannia G (1996) Reduction of phenol content and toxicity in olive oil mill waste waters with the ligninolytic fungus *Pleurotus ostreatus*. Water Res 30(8):1914–1918. https://doi.org/10.1016/0043-1354(95)00330-4

Monrroy M, Ortega I, Ramírez M, Baeza J, Freer J (2011) Structural change in wood by brown rot fungi and effect on enzymatic hydrolysis. Enzym Microb Technol 49(5):472–477. https://doi.org/10.1016/j.enzmictec.2011.08.004

Okerentugba PO, Orgi FA, Ibiene AA, Elemo GN (2015) Spent mushroom compost for bioremediation of petroleum hydrocarbon polluted soil : A review. Glob Adv Res J 4(1):1–7

Pizzul L, Castillo M d P, Stenström J (2009) Degradation of glyphosate and other pesticides by ligninolytic enzymes. Biodegradation 20(6):751–759. https://doi.org/10.1007/s10532-009-9263-1

Pointing SB (2001) Feasibility of bioremediation by white-rot fungi. Appl Microbiol Biotechnol 57:20–33. https://doi.org/10.1007/s002530100745

Putra WP, Kamari A, Najiah S, Yusoff M, Ishak CF, Mohamed A et al (2014) Biosorption of Cu(II), Pb(II) and Zn(II) ions from aqueous solutions using selected waste materials: adsorption and characterisation studies. J Encapsulation and Adsorpt Sci 4(4):25–35. https://doi.org/10.4236/jeas.2014.41004

Raymond P, Mshandete AM, Kajumulo Kivaisi A (2015) Production of oxidative and hydrolytic enzymes by *Coprinus cinereus* (Schaeff.) Gray from sisal wastes supplemented with cow dung manure. Biotechnol Res Int 2015:1–9. https://doi.org/10.1155/2015/650543

Riley R, Salamov AA, Brown DW, Nagy LG, Floudas D, Held BW, Grigoriev IV (2014) Extensive sampling of basidiomycete genomes demonstrates inadequacy of the white-rot/brown-rot paradigm for wood decay fungi. Proc Natl Acad Sci 111(27):9923–9928. https://doi.org/10.1073/pnas.1400592111

Rossi A, Cruz AHS, Santos RS, Silva PM, Silva EM, Mendes NS, Martinez-Rossi NM (2013) Ambient pH sensing in filamentous fungi: Pitfalls in elucidating regulatory hierarchical signaling networks. IUBMB Life 65(11):930–935. https://doi.org/10.1002/iub.1217

Sari M, Prange A, Lelley JI, Hambitzer R (2017) Screening of beta-glucan contents in commercially cultivated and wild growing mushrooms. Food Chem 216:45–51. https://doi.org/10.1016/j.foodchem.2016.08.010

Schneider WDH, Fontana RC, Mendonça S, de Siqueira FG, Dillon AJP, Camassola M (2018) High level production of laccases and peroxidases from the newly isolated white-rot basidiomycete *Marasmiellus palmivorus* VE111 in a stirred-tank bioreactor in response to different carbon and nitrogen sources. Process Biochem 69:1–11. https://doi.org/10.1016/j.procbio.2018.03.005

Shah F, Nicolas C, Bentzer J, Ellstrom M, Smits M, Rineau F, Tunlid A (2017) Ectomycorrhizal fungi decompose soil organic matter using oxidative mechanisms adapted from saprotrophic ancestors Ectomycorrhizal fungi decompose soil organic matter using oxidative mechanisms adapted from saprotrophic ancestors. New Phytol 209:1705–1719. https://doi.org/10.1111/nph.13722

Silva RN (2016) Perspectives in genomics the future of fungi in 'omics' era. Curr Genomics 17(2):82–84

Singh SS, Wang H, Chan YS, Pan W, Dan X, Yin CM et al (2015) Lectins from edible mushrooms. Molecules 20(1):446–469. https://doi.org/10.3390/molecules20010446

Strehmel N, Böttcher C, Schmidt S, Scheel D (2014) Profiling of secondary metabolites in root exudates of *Arabidopsis thaliana*. Phytochemistry 108:35–46. https://doi.org/10.1016/j.phytochem.2014.10.003

Su Y, Yu X, Sun Y, Wang G, Chen H, Chen G (2018) Evaluation of screened lignin-degrading fungi for the biological pretreatment of corn stover. Sci Rep 8(1):1–11. https://doi.org/10.1038/s41598-018-23626-6

Teere M, Zvauya R, Read JS (2001) Ligninolytic enzyme production in selected sub-tropical white rot fungi under different culture conditions. J Basic Microbiol 41(2):115–129. https://doi.org/10.1002/1521-4028(200105)41:2<115::AID-JOBM115>3.0.CO;2-S

Tian X, Fang Z, Guo F (2012) Impact and prospective of fungal pre-treatment of lignocellulosic biomass for enzymatic hydrolysis. Biofuels Bioprod Biorefin 6:335–350. https://doi.org/10.1002/bbb

Tortella GR, Diez MC, Durán N (2005) Fungal diversity and use in decomposition of environmental pollutants. Crit Rev Microbiol 31(4):197–212. https://doi.org/10.1080/10408410500304066

Tuhy Ł, Samoraj M, Michalak I, Chojnacka K (2014) The application of biosorption for production of micronutrient fertilizers based on waste biomass. Appl Biochem Biotechnol 174(4):1376–1392. https://doi.org/10.1007/s12010-014-1074-0

Tuhy Ł, Samoraj M, Witkowska Z, Rusek P, Chojnacka K (2015) Conversion of spent mushroom substrate into micronutrient fertilizer via biosorption in a pilot plant. Ecol Eng 84:370–374. https://doi.org/10.1016/j.ecoleng.2015.09.032

Vasina DV, Moiseenko KV, Fedorova TV, Tyazhelova TV (2017) Lignin-degrading peroxidases in white-rot fungus Trametes hirsuta 072. Absolute expression quantification of full multigene family. PLoS One 12(3):1–16. https://doi.org/10.1371/journal.pone.0173813

Viswanath B, Rajesh B, Janardhan A, Kumar AP, Narasimha G (2014) Fungal laccases and their applications in bioremediation. Enzyme Research 2014:1–21. https://doi.org/10.1155/2014/163242

Yadav JS, Loper JC (2000) Cytochrome P450 oxidoreductase gene and its differentially terminated cDNAs from the white rot fungus *Phanerochaete chrysosporium*. Curr Genet 37(1):65–73. https://doi.org/10.1007/s002940050010

Yadav JS, Doddapaneni H, Subramanian V (2006) P450ome of the white rot fungus *Phanerochaete chrysosporium*: structure, evolution and regulation of expression of genomic P450 clusters. Biochem Soc Trans 34:1165–1169. https://doi.org/10.1042/BST0341165

Zengin G, Uren MC, Kocak MS, Gungor H, Locatelli M, Aktumsek A, Sarikurkcu C (2017) Antioxidant and enzyme inhibitory activities of extracts from wild mushroom species from Turkey. Int J Med Mush 19(4):327–336. https://doi.org/10.1615/IntJMedMushrooms.v19.i4.30

Chapter 19
Wild Macro-Fungi from Northwest Himalayas: Future Prospects and Challenges

Monika Thakur

19.1 Introduction

With the increasing population and fast depletion of natural resources, it became necessary to explore the possibilities of using newer indigenous untapped plant resources. There are many natural resources, still lying unexplored and underexploited. Therefore, there has been focused attention by the researchers on exploiting underutilized natural resources for multifarious use. India possesses a rich treasure of wild edible mushroom. In the classification of forest products the mushrooms are covered under "Minor Forest Products", along with resin, honey, bamboos and medicinal plants etc., though the role they play in nature is a major one (Lakhanpal 2000; Thakur 2015; Thakur and Lakhanpal 2015 and Lakhanpal and Sai 2016)

Before the era of cultivation and domestication began, the mushrooms were collected and consumed from the wild only. Wild edible mushrooms, till date continues to be an integral component of the culture of many ethnic groups across the Indian local villages. Besides the gastronomic delight that they offer, these wild mushrooms have been exploited for their potential curative properties and are part of some medicinal practices. But, while collecting mushrooms from the wild natural sources, where the edible and the toxic ones may grow close by, there is no clear identification for the edibility of these mushrooms. Thus mushroom hunting, despite being an age old practice, involves a high risk of poisoning, even for an experienced forager.

Mushrooms have been consumed from wild and a speciality food form times immemorial. Wild edible mushrooms have been the food in wilderness; which now has come to occupy a very popular place in modernistic regimen, because of nutritive and **medicinal value** (Bano 1976). The Greeks regarded mushrooms as

M. Thakur (✉)
Amity Institute of Food Technology, Amity University Uttar Pradesh, Noida, India
e-mail: mthakur1@amity.edu

B. P. Singh et al. (eds.), *Biology of Macrofungi*, Fungal Biology,
https://doi.org/10.1007/978-3-030-02622-6_19

providing strength to soldiers in wars, Romans considered wild edible mushrooms as 'Foods of the Gods' while Chinese regarded them as 'Elixir of Life' (Chang and Miles 1982). Before the era of industrialization, the mushroom were collected and consumed from the wild. Those species which still defy cultivation or for which the cultivation technology is eluding are collected and consumed. Many of them are highly prized and priced as well. The wild edible mushrooms comprise a very vast variety of macro-fungi (Wasson 1980). They occupy diverse habitats and occur in climates ranging from tropical, subtropical to temperate occupying niches on ground, grass, litter, soil, wood all over the world and an integral component of the forest ecosystem.

There have been around 15,000 mushroom species globally and amongst them around 7000 species are edible; amongst these about 2000 species are considered nutraceutically important and about 200 are known to be mycorrhizal. Till date only 200 of them have been cultured in lab, 100 cultivated economically, approximately 60 grown on commercial scale and 10 have been grown in industrial level (Chang and Miles 2004). The present review provides an account of wild edible mushrooms specifically pertains to N.W. Himalayas. Wild edible mushrooms have been also underutilized and untapped resources, as wide variety of mushrooms are still under-exploited (Lakhanpal 1994; Thakur and Lakhanpal 2015; Thakur 2015; Lakhanpal and Sai 2016). In last decades, the focus has been shifted from other herbal natural products to wild edible mushrooms and their use by native people in food and medicine (FAO 2004).

The chapter, highlights the ethno-mycological aspects (dietary and **therapeutic** uses), mycophagy, nutritional and **Nutraceutical** potential, marketing, production and sustainable conservation of the wile edible mushrooms, so that their under-exploited potential shall be fully exploited.

19.2 Ethnomycological Studies

Ethnomycology is the branch of Ethnobotany and study the interaction of people and fungi. The ethnobotanical studies on fungi have been termed as 'Ethnomycology' as, "The study of the use of fungi by the native or tribal people" (Wasson 1980). Mushrooms played an important role for mankind as food, poison, medicine, in folklore, and religion (Militoris 2001). FAO (2004) listed ethnomycological importance of wild edible fungi as a source of nutritional components; source of income; source of delicacy and many other health benefits. Ethnomycology includes the cultural, ceremonial and medicinal uses of wild fungi by native people. Previously very little attention has been paid for the use of wild edible mushrooms by the local inhabitants. But as time passed, with the scarcity of food from natural resources and increasing interest of public in new and different food and their medicinal values has built up and expanded the market of wild edible mushrooms.

Ethnic knowledge in mycology is still prevalent in tribal and local inhabitants (Singh 1999). If such work is undertaken in the right perspective, the ethnomycological studies disclose the traditional wisdom and are under the danger of extinction and are vanishing rapidly due to assault of modern civilization and changes in sustenance economy. Several mycologists have reported ethnomycological usage of this natural resource wealth from some regions of India (Pandey and Singh 1978; Lakhanpal and Shad 1986a, b; Lakhanpal and Kaisth 1987; Harsh et al. 1993, 1996; Rai et al. 1993; Lakhanpal 1994; Lakhanpal 1997; Boruah et al. 1997; Kamat 1999; Lakhanpal and Shad 1999; Rana 2006, 2002; Deshmukh 2004; Sagar et al. 2005; Lakhanpal and Rana 2005; Lakhanpal et al. 2010; Singh and Aneja 1999; Kumar and Sharma 2011; Thakur 2015; Thakur and Lakhanpal 2015; Lakhanpal and Sai 2016 and Paul et al. 2018). This has been very important that the traditional ethnomycological data is well documented so that the same can be passes from generation to generations.

1. **Wild Edible Mushrooms: As Underutilized Treasures of Nature**

In India, the mushroom cultivation has shown the progress in the field of food, medicine and employment at the grass root level. Their cultivation had shown improvement in sustainable livelihood but the nutraceutical potential is yet to be fully explored. Now days, mushroom production technologies have grown enormously but still the **cultivation technologies** are not successful in many wild edible species.

(i) ***Cultivated Mushrooms***

From amongst the edible mushrooms, about 21 species have been successfully cultivated artificially. Some of the most important species which have been commercially exploited all over the world are: *Agaricus bisporus* (button mushroom), *Pleurotus sajor caju*, *P. ostreatus* (Dhingri, oyster mushroom), *Volvariella volvacea* (straw mushroom), *Lentinus edodes* (Shiitake, Japanese mushroom), *Agrocybe* sp. (Black Popular Mushroom), *Calocybe indica* (Milky mushrooms), *Aurocularia* sp. (wood Ear mushrooms), *Ganoderma lucidum* (Reishi mushrooms) (Manjit 2011; Thakur 2015). There are enough challenges ahead. Still there is vast number of the species collected and consumed from the wild and the numbers of wild edible mushrooms are not explored.

(ii) ***Wild Edible Mushrooms***

Wild mushrooms are also important Non-Timber Forest Product (NTFP's) from the forests and are being used as food since time immemorial (Devkota 2008). From India, many species of edible fungi has been reported to be traditionally and regularly consumed by the local inhabitants without causality and fatality. Table 19.1 and (Figs. 19.1, 19.2) lists the wild edible species which have been regularly consumed by people having both nutritional and medicinal potential.

2. **Mycophagy**

Mycophagy is the study of consuming fungi (mushrooms) collected available in wild forests. This practice has been very common by the tribal population and local inhabitants. Many people consume these wild mushroom fruiting bodies on the

Table 19.1 Major Wild edible mushrooms from North West Himalayas

Sr. No.	Wild edible mushrooms	Use (Edible / Medicinal)
1	*Lactarius deliciosus; L. Sanguifuulus*	Edible
2	*Russula brevipes*	Edible
3	*Boletus edulis; B. Hoarkii*	Edible
4	*Cantharallus cibarius*	Edible
5	*Sparasis crispa; S. Radicata*	Edible
6	*Hericium coralloides*	Edible
7	*Trappeinda himalayansis*	Edible
8	*Albatrellus cristatus*	Edible
9	*Helvella crispa; H. Elastic*	Edible
10	*Ramaria bitrytoides*	Edible
11	*Hydnum repandum*	Edible
12	*Volvariella bombycina*	Edible
13	*Morchella angusticeps; M. Deliciosa; M. Esculenta; M. Crassipes; M. conica*	Edible & Medicinal
14	*Cordyceps sinensis*	Edible & Medicinal
15	*Termitomyces* sp.	Edible
16	*Lycoperdon pusitum*	Medicinal
17	*Rhizopogon* spp.	Edible
18	*Ramaria* spp.	Edible

Fig. 19.1 Wild Edible Mushrooms, **(a)** *Morchella* sp.; **(b)** *Lactarius delicious*; **(c)** *Rusulla brevipes*; **(d)** *Boletus edulis*; **(e)** *Volvariella bombycina*; **(f)** *Hericium coralloies*

Fig. 19.2 Wild Edible Mushrooms, (**g**) *Helvella crispa*; (**h**) *Sparassis crispa*; (**i**) *Termitomyces* sp.; (**j**) *Cordyceps sinensis*; (**k**) *Lycoperdon* sp.; (**l**) *Hydnum repandum*

basis of their aroma, earthy flavour, texture and taste. They even don't have idea about the nutritional and nutraceutical importance of the same (Ene-Obong and Camovale 1992; Osemwegie et al. 2006). This is a very common practice. But before consumption, one should have a proper knowledge of edibility and a fool-proof method to identify poisonous mushrooms. Unfortunately no such method exists and those which have been used in the past have proved to be just figments of superstition. It was believed for instance that if a silver spoon / silver can / an onion boiled with fungi turned black, this was an infallible sign that the mushrooms was poisonous. One should neither be afraid nor careless with the consumption of wild mushrooms. Species that cannot satisfactorily identify should be used only after an expert has declared them edible (Thakur 2015).

19.3 Importance of Wild Edible Mushrooms

In the past decades, because of ethno-mycological data there has been visible upsurge in the collection, consumption and selling of wild edible mushrooms available in the forest areas. Historically, there has been no documentation of mushroom gathering and hunting by people, but the ancient wisdom of collection and consumption is reflected through the species collected and consumed. During the past few years, export market is also developing for other wild edible mushrooms. Aside their importance as a source of food and income, they are excellent sources of various **bioactive compounds** showing medicinal properties (Das 2010; Singh and Aneja 1999; Lakhanpal and Sai 2016). The collection of wild edible mushrooms by the local inhabitants can directly improve livelihoods through economic, nutritional, medicinal attributes and also ecologically beneficial as mentioned below:

(a) ***Nutritional Potential***

Mushrooms are the macro-fungi that have been used as traditional food since time immemorial. Nutritionally they are valuable health food, which is low in calories, rich in carbohydrates, proteins, essential amino acids, fibre, vitamins and minerals and many important bioactive components. The consumption of mushrooms can make a valuable addition to the often unbalanced diets of people in developing countries. The mushroom fruiting bodies have also been an excellent source of high quality proteinaceous food (Singh and Aneja 1999).

(b) ***Medicinal Potential***

Mushrooms **fruiting bodies** are also known as 'Health potentiators' and 'Elicitors of immune system'. In the last two decades there has been an upsurge on the use of wild edible mushroom fruiting bodies in the form of mycelium, powder or extract form in form of functional foods or nutraceuticals. As there has been much awareness now on the relationship between diet and disease, and this has evolved the concept of Myco-nutraceuticals. Myco-nutraceuticals are complete foods that are eaten not only to satisfy functional dietary needs but also elicit additional medicinal benefits. Mushrooms also have been variously used in traditional Indian medicine system. Many mushrooms like **puffballs** have been used on wounds to reduce inflammations and infection. Many species have also been traditionally used as tonics, aphrodisiacs and in several other preparations. The derivatives form the medicinally important wild edible mushrooms – the bioactive components have received diligence for improving biological function thus making people fitter, healthier and prosperous.

(c) ***Income Benefits***

Wild edible mushrooms having very good commercial value are species of Morchella, Helvella, Hericium, Sparassis, Hydnum, Trapeinda, Clavaria, Ramaria, Boletus, Albatrellus, Cordyceps, Lactarius and Rusulla (Lakhanpal 2000; Thakur 2015). All these mushrooms are consumed as well as sold fresh, collected and dried for sale. A few species of wild edible mushrooms dominate the world market with an estimated value of more than 2 billion dollars (Wang and Hall 2004).

In India the most valued and important mushroom species collected are *Morchella* spp. (Morels) and *Cordyceps sinensis*. Morels are presently the costliest and also most sought after edible mushrooms with great commercial potential. Till date, so many attempts have been made to domesticate morels, but all in vain and the process to domesticate morels is still incomplete (Singh and Aneja 1999; Thakur 2015). Himachal Pradesh, Jammu Kashmir and Uttrakhand are the main exporters of morels in India (Lakhanpal 1994). Every house sells around 10–15 kg/season and the annual return only from one village is Rs. 1–1.5 Lakhs/season. The main trading markets for the selling of morels are: Pathankot, Amritsar, Delhi and Bombay. From these centers the morels are exported to France, U.K., Switzerland, Dubai, Germany etc. and their export range from 150–170 million rupees (Iqbal 1993).

Some other species (other than morels) are also used in local trade and export. *Cordyceps sinensis* is one of the wild costliest mushrooms but mainly valued for medicinal values and nutraceutical potential. The cost of one kg of wild collected *Cordyceps sinensis* in the market varies from Rs. 60,000 to Rs. 1 lakh in India. Therefore, the collection of these wild edible mushrooms contributes to the income of the tribal population and local inhabitants.

19.4 Production and Trade of Wild Edible Mushrooms

(i) **Drying and storage**

Owing to a high moisture content of 80–85%, fresh mushrooms are perishable in nature and deteriorate very quickly. Freshly harvested morel fruiting bodies can only be stored at ambient temperatures for only 24 h, and after that the same are to be refrigerated, just to increase their shelf life for 2–3 days. Traditional drying has been the easiest, fastest and long lasting method of preservation. The collected mushrooms are cleaned with a damp cloth or a soft brush to remove mud, dirt, debris, bugs and insects, etc. These are then dried in the sun by spreading them on slate roofs or by making 'garlands'. The garlands may be dried indoors by hanging them above the fireplace where fuel wood is burnt for cooking. Drying on roof tops has been discontinued by many people because it causes dark spots to appear on the morels. Instead, they are first spread on gunny bags for 3–4 days to absorb surplus moisture and then are left to dry in the sun for 2–3 days.

The specimens to be marketed are preserved by wrapping in polythene's or paper bags and those kept for home consumption are hanged indoors. It is believed that drying enhances the flavor and aroma of these mushrooms. They are kept in closed chambers (with little ventilation) as in open they may absorb moisture and initiate the spoilage process. Almost identical data on drying and storage of mushrooms has been recorded by a number of researchers working in this field (Shad 1985; Shad 1989; Jandaik and Sharma 1995; Rathore 2001; Thakur 2015; Paul et al. 2018).

19.5 Production and Trade

So far the forests resources were considered only as a source of timber and timber products. Now it has been realized that forests are actually store houses of large number of other resources also. Mushroom though classified as MFP or NTFP in reality play a major role in ecological processes of the forest. The extraction for these MFP's and NATP's has been constructive while the extraction of timber can be destructive process. Marketing of wild edible mushrooms by the local inhabitants contributes to the food security, nutritional and medicinal benefits; generating additional employment and income (local, regional and national trade); and offering opportunities for various small village level entrepreneurs (Marshall and Nair 2009; Singh and Aneja 1999; Singh, 2011; Lakhanpal and Sai 2016). Picking and selling of mushrooms form wild supplements income of the people.

Production of mushrooms has been seasonal activity, but where the snow falls, the activity is both during the rainy season and spring season and in other areas it is only during rainy season. The ethno-mycological information for collection, consumption and trade of wild edible mushrooms is only speculative in India. Nevertheless, it is certain and observationally correct, that wild mushrooms are collected and consumed by local people, especially rural populations and almost all the tribal people for consumption and for trade as well. In North East India, in most of the places it is an organized activity and wild varieties of mushrooms are displayed in the markets and sold to the customers regularly (Thakur and Lakhanpal 2015). In India there has been large population which is **mycophilic** but the data is neither systematized nor categorized, which actually needs to be quantified and qualified.

19.6 Myco-Nutraceuticals Form Wild Edible Mushrooms

The medicinal mushrooms has emerged a new field of nutraceuticals in past two decades and it is interesting that most of the cultivated and wild edible mushrooms have been proven to be medicinally useful. Mushrooms though been used as food since time immemorial and their nutritional value is well recognized. In many Asian cultures, especially China, mycophagy is very common and is used to promote good health, vitality and to increase the adaptive capability of the individuals (Lombardi and Rountree 2002). The usage of mushrooms fruiting bodies for health benefits was most probably in the form of extracts, health tonics, beverages, tea's, soup concentrates, and arid healthful food dishes. The mushroom fruiting bodies are complete nutraceuticals (contain many bioactive components and then termed are now termed as 'Mushroom Nutraceuticals' (Smith et al. 2002). The term "**Mushroom Nutraceuticals**" has been coined by Chang and Buswell (1996) and is a refined / partially defined extractive from either mycelium or fruiting bodies, which is consumed in a capsule or tablet form as a dietary supplement (not as a regular food) with potential therapeutic applications (Lakhanpal and Sai 2016). Therefore, mushrooms are beneficial with dual role as, they have nutritional value plus accrue health benefits. The metabolites from the nutraceuticals may exhibit various properties such as anti-tumour, immune-modulating and hypo-chloestrolemic properties etc.

In the last decades, there has been an upsurge on the use of mushrooms as nutraceuticals. Many cultivated and wild edible mushroom species have been thoroughly investigated and authenticated for medicinal use. The species that have been properly analysed for medicinal value are: *Ganoderma lucidum* (Reishi), *Lentinus edodes* (Shiitake), *Grifola frondosa* (Maitake), *Agaricus blazei* (Himematsutake), *Cordyceps sinensis* (Caterpillar fungus), *Pleurotus ostreatus* (Oyster mushroom), *Morchella* species (morels) and *Hericium erinaceous* (Lions mane) etc. There are various bioactive components present in the fruiting bodies like polysaccharides, proteins, vitamins, and other phytochemicals. These have been good for the body as they provide some of the important health benefits like – immune enhancement, **anti-cancerous** effects, anti-HIV and anti-viral activity, and reduce the toxic effect of chemo- and radiotherapy (Wasser and Weis 1999; Wasser et al. 2000). Some of the fruiting bodies also show the aphrodisiac properties and many commercial application are also available in the markets (Lakhanpal and Rana 2005; Thakur 2012 and Thakur 2015).

(i) **Antioxidant Potential**

The **antioxidant** rich food reduces the plunge of various diseases (Liu et al. 2009; Huang et al. 2007). Mushrooms contain various biologically active components as phytochemicals and have antioxidant properties (Thakur and Sayeed 2014; Thakur and Lakhanpal 2015). The listed species which possess antioxidant properties are: *Agaricus bisporus, A. brasiliensis, A. silvaticus, Agrocybe cylindracea, Antrodia camphorate, Boletus edulis, Boletus badius, Geastrum saccatum, Grifola frondosa, Hypsizigus marmoreus, Inonotus obliquus, Lactarius sanguifluus, Laetiporus sulphureus, Lentinula edodes, Leucopaxillus giganteus, Lepista nuda, Morchella esculenta, Phellinus linteus, Polyporus squamosus, Russula delica, Suillus collitinus, Turbinaria conoids, Verpa conica* (Thakur and Sayeed 2014; FAO, 2004; Elmastas et al. 2007; Soares et al. 2009; Barros et al. 2008; Tsai et al. 2007; Shu and Lung 2008; Ribeiro et al. 2008; Sarikurkcu et al. 2008; Dore et al. 2007; Lee et al. 2007a, b, 2008; Turkoglu et al. 2007; Zheng et al. 2005; Song et al. 2003; Chattopadhyay et al. 2009; Thakur and Lakhanpal 2015).

(ii) **Anti-microbial Potential**

In addition to antioxidant activity, the mushrooms have been shown to display **antimicrobial** activity of a varying degree. The mushroom species showing antimicrobial activities are: *Cheimonophylum candissimum, Clitocybe cyathiformis; C. diatreta, Cordyceps militaris, Coprinus atrementarius, Crepidotus fulvotomentosus, Lentinus edodes, Lactarius sanguifluus Russula delica, Morchella esculenta, M. conica, Ganoderma lucidum; G. japonicum, Laetiporus sulphureus* (Thakur and Lakhanpal 2015; FAO 2004; Thakur and Sayeed 2014; Elmastas et al. 2007).

(iii) **Anti-tumoral Properties**

One of the most significant features of almost all the mushrooms analyzed is their **anti-tumor** activity. The species showing anti-tumoral activities are: *Agaricus brasiliensis, Agrocybe aegerita, Albatrellus confluens, Cordyceps sinensis, Coriolus versicolor, Fomes fomentarius, Ganoderma capense, Ganoderma lucidum, G. tsugae, Grifola frondosa, Inonotus obliquus, Poria cocos* (Peng et al. 2003; Ngai and Ng 2004; Kim et al. 2006; Cui et al. 2007; Hatvani 2001; Fan et al. 2007; Dyiabalanage et al. 2008; Harhaji et al. 2008; Chen et al. 2008; Rubel et al. 2008).

(iv) **Anti-inflammatory and Immuno-modulatory properties**

Mushrooms are also known to have a wide variety of bioactive components having a wide range of therapeutic effects and can act as **immuno-modulatory**, anticarcinogenic, antiviral, antioxidant, and anti-inflammatory agents (Lee et al. 2007a, b; Kim et al. 2006; Nishizawa et al. 2007). The concentration and efficacy of the bioactive compounds available will depends upon the type of species, cultivation trials, fruiting bodies, prevalent storage conditions, processing and cooking process. Many of the bioactive compounds found in mushrooms exhibit significant anti-inflammatory properties. The most of the well-known mushroom species and their anti-inflammatory and immunomodulatory properties, along with their bioactive compounds are: *Agrocybe cylindracea* (Agrocybin), *Amanita muscaria* Hot water, methanolic and ethanolic extracts, *Cordyceps militaris* (Aqueous and alkaline), *Fomitopsis pinicola (*Polysaccharides), *Ganoderma lucidum* (Triterpene), *Geastrum saccatum* (Glucans), *Inonotus obliquus* (Hot water), *Phellinus linteus* (Butanol fraction), *Poria cocos* (Polysaccharides), *Coriolus versicolo* (Polysaccharide), *Ganoderma lucidum* (Fractions, Polysaccharide), (Kim et al. 2004; Zheng et al. 2005; Lu et al. 2009, 2006; Ji et al. 2007; Gu and Belury 2005; Wu et al. 2006; Kim et al. 2006; Nishizawa et al. 2007; Yang et al. 2007; Dore et al. 2007; Smirdele et al. 2008; Cheng et al. 2008; Dudhgaonkar et al. 2009; Van et al. 2009).

(v) **Prebiotics**

Wild edible mushrooms are rich in dietary content and are the good source of carbohydrates and thus potential sources of the **prebiotics**. They are considered non-digestible food ingredients which stimulate the growth of beneficial bacteria (probiotics) in the gastrointestinal tract. They mainly consist of the dietary fibers and oligosaccharides (Thakur 2016). They are having the beneficial effects like gut health maintenance, cancer inhibition, immune-potentiation, cholesterol removal, prevention of obesity. Different mushroom species produces the different types of polysaccharides (soluble or insoluble) (Aida et al. 2009; Wani et al. 2010).

Various mushroom species which are widely used as sources of prebiotics viz. Agaricus bisporus, A. bitorquis, A. blazei, Auricularia auricular-judae, Boletus erythropus, Calocybe indica, Flammulina velutipes, Ganoderma lucidium, Geastrum accatum, Hericium erianaceus, Lentinus edodes, Phellinus linteus, Pleurotus eryngii, *P. florida*, P. ostreatus (Thakur and Lakhanpal 2015). The different types of polysaccharides and their benefits as prebiotic are: D-Glucan, β-1, 3 glucan; B-1, 6 glucan; Lentinan; Grifloan; Lentinan.

Hence, the polysaccharides extracted from mushroom species not only activates probiotic organisms but also sustains the gastrointestinal tract and shows various health benefits like tumor therapy, cardiovascular disease, anti-viral, antibacterial role. Hence, the mushroom species are considered potential source of prebiotics.

3. **Sustainable Conservation**

So far no measures have been undertaken for the conservation of mushrooms in wild. Almost all the wild edible mushroom has been cultivated, but still some are the uncultivated ones. Since, it is still a source from the wild, over exploitation can lead to the total depletion and then disappearance of this resource. Hence, proper conservation strategies are to be developed for the same. However, there is some traditional wisdom used for conservation. The traditional morel hunters pick morels very delicately by cutting the stipe with knife and leaving the base and underground mycelium which serves as inoculum for next season. There are other collectors who leave 1–2 fruiting bodies at some specific sites, believing that this will lead to increased production of *Guchhis* next year. Such traditional picking methods ensure propagation and can help in morels not becoming endangered. Some prominent researchers have also discussed similar methods for further propagation of fruiting bodies of morels (Lakhanpal and Sai 2016; Lakhanpal and Shad 1986a, b; Singh and Aneja 1999).

19.7 Future Prospects

Therefore, with the awareness in the use of mushroom as nutraceuticals have been up-surged and this has been named as a "**Non -green revolution**" (Thakur and Lakhanpal 2015). Wild edible mushrooms have bigger role in our lives as alleviating poverty, enhancing human health, and arresting environmental degradation etc. Mushrooms can serve as agents for promoting equitable economic growth in society. This chapter also aims to explore the possibilities of conservation and **sustainable development** of wild edible mushrooms and the indigenous knowledge associated with them as Wild edible fungi have great potential for generating a great socio-economic impact in human welfare. A system and policies for training local people must be developed to use the **ethnomycological** information for the wise exploitation of wild edible fungi (Christensen and Larsen 2005). The dissemination of knowledge about wild edible and poisonous mushrooms should also be there, so that no fear of poisoning will be there in the local inhabitants. The bioactive components of the medicinally important wild edible mushrooms have received great attention for being products of improving biological function thus making people fitter and healthier.

19.8 Conclusion

The wild edible mushroom sector is growing day by day in India is vast but also unexplored and unexploited one. The mushroom diversity supports and is an unexplorable part of the socio-economic culture of many traditional people in India. A lot many wild edible mushrooms have still been unexplored and their potential is still underutilized. The present chapter is aimed to summarize the importance of untapped natural resources – wild edible mushrooms and with the increasing; more ethnomycological studies are to be conducted so that they can be used as a potent source of nutraceuticals. Hence, by fully exploiting the unexplored potential, mushrooms can bring '**Non -green revolution**'.

References

Aida FMNA, Shuhaimi M, Yazid M, Maaruf AG (2009) Mushroom as a potential source of prebiotics: a review. Trends Food Sci Technol 20(11–12):567–575

Bano Z (1976) The nutritive value of mushrooms. In: Proceeding of first symposium on survey and cultivation of edible mushrooms of India. Regional Lab, Srinagar

Barros L, Falcão S, Baptista P, Freire C, Vilas-Boas M, Ferreira ICFR (2008) Antioxidant activity of *Agaricus* sp mushroom by chemical, biochemical and eletrochemical assays. Food Chem 111:61–66

Boruah P, Kailta P, Bordoloi D, Gogi P, Adhikary RK (1997) Some fleshy fungi of ethnobotanical use from north East India. Advan Fores Res India 16:165–171

Chang ST, Buswell JA (1996) Mushroom nutriceuticals. World J Microbiol Biotechnol 12:473–476

Chang ST, Miles PG (1982) Introduction to mushroom science. In: Chang ST, Quimio TH (eds) The tropical mushrooms: biological nature and cultivation method. The Chinese University Press, Hong Kong

Chang ST, Miles PG (2004) Mushrooms- cultivation, nutritional value, medicinal effect and environmental impact. CRC Press, Washington, DC

Chattopadhyay N, Ghosh T, Sinka S, Chattopadhyay K, Karmakar P, Ray B (2009) Polysaccharides from *Turbinaria conoides*: structural features and antioxidant capacity. Food Chem. https://doi.org/10.1016/jfoodchem.200905.069

Chen W, Zhao Z, Chen SF, Li YQ (2008) Optimization for the production of exopolysaccharide from *Fomes fomentarius* in submerged culture and its antitumor effect *in vitro*. Bioresour Technol 99:3187–3194

Cheng JJ, Lin CY, Lur HS, Chen HP, Lu MK (2008) Properties and biological functions of polysaccharides and ethanolic extracts isolated from medicinal fungus, *Fomitopsis pinicola*. Process Biochem 43:829–834

Christensen M, Larsen HO (2005) How can collection of wild edible fungi contribute to livelihood in rural areas of. Nepal J Forest & Live 44(2):50–55

Cui FJ, Tao WY, Xu ZH, Guo WJ, Xu HY, Ao ZH, Jin J, Wei YQ (2007) Structural analysis of antitumor heteropolysaccharide GFPS1b from the cultured mycelia of *Grifola frondosa* GF9801. Bioresour Technol 8(2):395–401

Das K (2010) Diversity and conservation of wild mushrooms in Sikkim with special reference to Barsey Rhododendron sanctuary. NeBIO 1(2):1–13

Deshmukh SK (2004) Biodiversity of tropical basidiomycetes as sources of novel secondary metabolites. In: Jain PC (ed) Microbiology and biotechnology for sustainable development. CBS Publishers and Distributors, New Delhi

Devkota S (2008) Distribution and status of highland mushrooms: a study from Dolpa. Nepal J Nat His Mush 23:51–59

Dore CMPG, Azevedo TCG, Souza MCR, Rego LA, Dantas JCM, Silva FRF, Rocha HAO, Baseia IG, Leite EL (2007) Anti-inflammatory, antioxidant and cytotoxic actions of ß-glucan-rich extract from *Geastrum saccatum* mushroom. Int J Immunopharmacol 7:1160–1169

Dudhgaonkar S, Thyagarajan A, Sliva D (2009) Suppression of the inflammatory response by triterpenes isolated from the mushroom *Ganoderma lucidum*. Int J Immunopharmacol 9:1272–1280. https://doi.org/10.1016/j.intimp.2009.07.011

Dyiabalanage T, Mulabagal V, Mills G, Dewitt DLM, Nair MG (2008) Health-beneficial qualities of the edible mushroom, *Agrocybe aegerita*. Food Chem 108:97–102

Elmastas M, Isildak O, Turkekul I, Temur N (2007) Determination of antioxidant activity and antioxidant compounds in wild edible mushrooms. J Food Compos Anal 20:337–345

Ene-Obong HN, Camovale E (1992) Nigerian soup condiments: traditional processing and potential as dietary fiber sources. Food Chem 43:29–34

Fan L, Soccol AT, Pandey A, Soccol CR (2007) Effect of nutritional and environmental conditions on the production of exo-polysaccharide of *Agaricus brasiliensis* by submerged fermentation and its antitumor activity. LWT Food Sci Technol 40:30–35

FAO (2004) In: Boa E (ed) Non wood forest products, wild edible fungi: a global overview of their use and importance. FAO Publication, Rome

Gu YH, Belury MA (2005) Selective induction of apoptosis in murine skin carcinoma cells (CH72) by an ethanol extract of *Lentinula edodes*. Cancer Lett 220:21–28

Harhaji L, Mijatovic S, Maksimovic-Ivanic D, Stojanovic I, Momcilovic M, Maksimovic V, Tufegdzie S, Marjanovic Z, Mostarica-Stojkovic M, Vucinic Z, Stosic-Grujicic S (2008) Antitumor effect of *Coriolus versicolor* methanol extract against mouse B16 melanoma cells: *in vitro* and *in vivo* study. Food Chem Toxicol 46:1825–1833

Harsh NSK, Rai BK, Ayachi SS (1993) Forest fungi and tribal economy- a case study in Baiga tribe of Madhya Pradesh. J Trop Forest 9:270–279

Harsh SK, Tiwari CK, Rai BK (1996) Forest fungi in the aid of tribal women of Madhya Pradesh. Sustain For 1:10–15

Hatvani N (2001) Antibacterial effect of the culture fluid of *Lentinus edodes* mycelium grown in submerged liquid culture. Intern J Antimicro Agents 17:71–74

Huang Q, Jin Y, Zhang L, Cheung PCK, Kennedy JF (2007) Structure, molecular size and antitumor activities of polysaccharides from *Poria cocos* mycelia produced in fermented. Carbohydr Polym 70:324–333

Iqbal M (1993) International trade in non-wood forest products. In: An Overview. FAO, Rome

Jandaik CL, Sharma SR (1995) Present status of *Morchella* in India. In: Chadda KL, Sharma SR (eds) Advances in horticulture. Malhotra Publishing House, New Delhi

Ji Z, Tang Q, Zhang J, Yang Y, Jia W, Pan Y (2007) Immunomodulation of RAW264.7 macrophages by GLIS, a proteopolysaccharide from *Ganoderma lucidum*. J Ethnopharmacol 112:445–450

Kamat N (1999) Ecotheological dimensions of termite hill. Govapuri Bull. Inst Menezes Braganza 1:3

Kim SH, Song YS, Kim SK, Kim BC, Lim CJ, Park EH (2004) Anti-inflammatory and related pharmacological activities of the n-BuOH subfraction of mushroom *Phellinus linteus*. J Ethnopharmacol 93:141–146

Kim YO, Park HW, Kim JH, Lee JY, Moon SH, Shin CS (2006) Anti-cancer effect and structural characterization of endo-polysaccharide from cultivated mycelia of *Inonotus obliquus*. Life Sci 79:72–80

Kumar S, Sharma YP (2011) Diversity of wild edible mushrooms from Jammu and Kashmir (India). In: Proceedings of 7th International Conference on Mushroom Biology and Mushroom Production, ICMBMP

Lakhanpal TN (1994) Prospects of mushrooms from the Wild. In: Nair MC, Gokulapalan G, Das L (eds) Advances in mushroom biotechnology. Scientific Publishers, Jodhpur

Lakhanpal TN (1997) Diversity of mushroom mycoflora in the North Western Himalayas. In: Sati SC, Saxena J, Dubey RC (eds) Recent research in ecology, environment and pollution, vol 10. Today and Tomorrow's Printers and Publishing House, New Delhi

Lakhanpal TN (2000) Wild and cultivated mushrooms. In: Verma LR (ed) Natural resources and development in Himalaya. Malhotra Publishing House, New Delhi

Lakhanpal TN, Kaisth KR (1987) Studies on wild edible mushrooms of Himachal Pradesh (N.W. Himalayas)-III. Ecology and Ethnomycology of two wild edible *Lactarii*. Indian J Mush 12–13:95–105

Lakhanpal TN, Rana M (2005) Medicinal and nutraceutical genetic resources of mushrooms. Plant Gen Res – Charac Utiliz 1:17

Lakhanpal TN, Shad OS (1986a) Studies on wild edible mushrooms of Himachal Pradesh (N.W . Himalayas)-I. Ethnomycology, production and trade of *Morchella* species. Indian J Mush 12–13:5–13

Lakhanpal TN, Shad OS (1986b) Studies on wild edible mushrooms of Himachal Pradesh (N.W. Himalayas)-II. Ecological relationships of *Morchella* species. Indian J Mush 12–13:15–20

Lakhanpal TN, Shad OS (1999) A Re-appraisal of the genus *Morchella* in India. In: Tewari JP, Lakhanpal TN, Singh J, Gupta R, Chamola BP (eds) Advances in microbial biotechnology. A.P.H. Publishing Corporation, New Delhi

Lakhanpal TN, Shad O, Rana M (2010) Biology of Indian morels. I K International Publishing House, New Delhi

Lakhanpal TN, Sai I, Thakur M (2016) Biodiversity of wild edible, mycorrhizal and nutraceutically important mushrooms. In: Sharma S, Singh M, Prasad, Rathore H (eds) Mushroom of India. Institute of Rural Development IIT Delhi, III Delhi 8–37

Lee CL, Yang X, Wan JMF (2006) The culture duration affects the immunomodulatory and anticancer effect of polysaccharopeptide derived from *Coriolus versicolor*. Enzyme Microb Technol 38:14–21

Lee IK, Kim YS, Jang YW, Jung JY, Yun B (2007a) New antioxidant polyphenols from medicinal mushroom *Inonotus obliquus*. Bioorg Med Chem Lett 17:6678–6681

Lee YL, Yen MT, Mau JL (2007b) Antioxidant properties of various extracts from *Hypsizigus marmoreus*. Food Chem 104:1–9

Lee JS, Park BC, Ko YJ, Choi MK, Choi HG, Yong CS, Lee JS, Kim JA (2008) *Grifola frondosa* (Maitake mushroom) water extract inhibits vascular endothelial growth factor-induced angiogenesis through inhibition of reactive oxygen species and extracellular signal-regulated kinase phosphorylation. J Med Food 11(4):643–651

Liu D, Zheng H, Liu Z, Yang B, Tu W, Li L (2009) Chemical composition and antimicrobial activity of essential oil isolated from the culture mycelia of *Ganoderma japonicum*. J Nanjing Med Univ 23(3):168–172

Lombardi R, Rountree R (2002). Mycological medicine . Functional Food and Nutraceuticals. https://doi.org/10.1080/mmy.40.5.507.517.

Lu MK, Cheng JJ, Lin CY, Chang CC (2009) Purification structural elucidation, and anti-inflammatory effect of a water-soluble 1, 6-branched 1, 3-α-d-galactan from cultured mycelia of *Poria cocos*. Food Chem. https://doi.org/10.1016/j.foodchem.2009.04.126

Marshall E, Nair NG (2009) Make money by growing mushrooms. In: FAO, pp 1–51

Militoris HP (2001) Mushrooms and man in medicine, myth and religion. Int J Med Mush 3:97–98

Manjit (2011). Mushroom production: an agribusiness activity. In: Singh M, Vijay B, Kamal S, Wakchaure G C (eds.) Mushroom: cultivation, marketing and consumption. Directorate of Mushroom Research, Solan, H.P.

Ngai PHK, Ng TB (2004) A ribonuclease with antimicrobial, antimitogenic and antiptoliferative activities from edible mushroom *Pleurotus sajor-caju*. Peptides 25:11–17

Nishizawa K, Torii K, Kawasaki A, Katade M, Ito M, Terashita K, Aiso S, Matsuoka M (2007) Antidepressant-like effect of *Cordyceps sinensis* in the mouse tail suspension test. Biol Pharm Bull 30:1758–1762

Osemwegie OO, Eriyaremu EG, Abdulmalik J (2006) A survey of macrofungi in Edo/Delta region of Nigeria, their morphology and uses. Global J Pure App Sci 2(2):149–157

Pandey G, Singh VK (1978) A note on the concept of mushrooms in ancient India. Indian Mush Sci I:383–388

Paul N, Slathia PS, Vaid A, Kumar R (2018) Traditional knowledge of Gucchi, *Morchella esculenta* (ascomycetes), in Doda District, Jammu and Kashmir, India. Intern J Med Mush 20(5):445–450

Peng Y, Zhang L, Zeng F, Xu Y (2003) Structure and antitumor activity of extracellular polysaccharides from mycelium. Carbohydr Polymer 54:297–303

Rai BK, Ayachi SS, Arvinder R (1993) A note on ethnomycomedicines from Central India. Mycologist 7:192–193

Rana M (2002) Ethno botanical Studies on the Morels of District Kinnaur (H.P.). M.Phil. Dissertation, Himachal Pradesh University, Shimla

Rana M (2006) Nutraceutical potential and genotypic Diversity of *Morchella* species from North West Himalayas. Ph.D. Thesis, Himachal Pradesh University, Shimla

Rathore M (2001) Studies on some wild edible mushrooms of Chamba (Himachal Pradesh). M. Phil. Dissertation, Himachal Pradesh University, Shimla

Ribeiro B, Lopes R, Andrade PB, Seabra RM, Gonçalves RF, Baptista P, Quelhas I, Valentão P (2008) Comparative study of phytochemicals and antioxidant potential of wild edible mushroom caps and stipes. Food Chem 110:47–56

Rubel R, Dalla Santa HS, Fernandes LC, Lima Filho JHC, Figueiredo BC, Di Bernardi R, Moreno AN, Leifa F, Soccol CR (2008) High immunomodulatory and preventive effects against sarcoma 180 in mice feed with Ling Zhi or Reishi mushroom *Ganoderma lucidum* (W. Curt.:Fr.) P. Karst. (Aphyllophoromycetideae) mycelium. Intern J Med Mush 10:37–48

Sagar A, Chauhan A, Sehgal AK (2005) Ethnomycological study of some wild edible mushrooms of tribal district of Kinnaur of Himachal Pradesh. Indian J Mush 23:1–8

Sarikurkcu C, Tepe B, Yamac M (2008) Evaluation of the antioxidant activity of four edible mushrooms from the Central Anatolia, Eikisehir – Turkey: *Lactarius deterrimus*, *Suillus collitinus*, *Boletus edulis*, *Xerocomus chrysenteron*. Bio/Technology 99:6651–6655

Shad OS (1985) Studies on some edible fungi of Himachal Pradesh. M.Phil. Dissertation, Himachal Pradesh University, Shimla

Shad OS (1989) Biological studies on *Morchella* species (morels) of Himachal Himalayas. Ph.D. Thesis, Himachal Pradesh University, Shimla

Shu C, Lung MY (2008) Effect of culture pH on the antioxidant properties of *Antrodia camphorata* in submerged culture. J Chin Inst Chem Eng 39:1–8

Singh J (1999) Ethnomycology and folk remedies: fact and fiction. In: Singh J, Aneja KR (eds) From ethnomycology to fungal biotechnology: exploiting fungi from natural resources for novel products. Kluwer Academic/Plenum Publishers, New York

Singh M (2011) Mushroom production: an agribusiness activity. In: Singh M, Vijay B, Kamal S, Wakchaure GC (eds) Mushroom: cultivation, marketing and consumption. Directorate of Mushroom Research, Solan

Singh J, Aneja KR (1999) From ethnomycology to fungal biotechnology: exploiting fungi from natural resources for novel products. In: Singh J, Aneja KR (eds) Proceedings of the international conference. Kluwer Academic/Plenum Publishers, New York

Smirdele FR, Olsen LM, Carbonero ER, Baggio CH, Freitas CS, Marcon R, Santos ARS, Gorin PAJ, Iacomini M (2008) Anti-inflammatory and analgesic properties in a rodent Modelo f a (1→3),(1→6)-liked ß-glucan isolated from *Pleurotus pulmonarius*. Eur J Pharmacol 597:86–91

Smith J, Rowan N, Sullivan R (2002) Medicinal mushrooms: their therapeutic properties and current medical usage with special emphasis on cancer treatments. University of Strathclyde, Glasgow

Soares AA, Souza CGM, Daniel FM, Ferrari GP, Costa SMG, Peralta RM (2009) Antioxidant activity and total phenolic content of *Agaricus brasiliensis* (*Agaricus blazei* Murril) in two stages of maturity. Food Chem 112:775–781

Song YS, Kim SH, Sa JH, Jin C, Lim CJ, Park EH (2003) Anti-angiogenic, antioxidant and xantine oxidase inhibition activities of the mushroom *Phellinus linteus*. J Ethnopharmacol 88:113–116

Thakur M (2012) Mushrooms: a prominent source of nutraceuticals. J Food Process Technol 3(10):177–178

Thakur M (2015) Wild mushrooms as natural untapped treasures. In: Chauhan AK, Pushpangadan P, George V (eds) Natural products: recent advances. Write & Print Publications, New Delhi

Thakur M (2016) Significance of prebiotics, probiotics and synbiotics as health potentiators. Food Sci Res J 7(2):327–334

Thakur M, Lakhanpal TN (2015) Mushroom cultivation and women entrepreneurship: prospects & challengers. In: Vasudevan P, Sharma S, Sharma VP, Verma M (eds) Women, technology and development. Narosa Publishers, New Delhi, pp 220–240

Thakur M, Sayeed R (2014) Qualitative phytochemical screening, Total phenolic content, *in – vitro* antioxidant activity and antimicrobial activities in methanolic extracts of *Lactarius sanguifluus* (Paulet) Fr. J Pure Appl Microbiol 8(6):4735–4741

Tsai SY, Tsai HL, Mau JL (2007) Antioxidant properties of *Agaricus blazei*, *Agrocybe cylindracea* and *Boletus edulis*. LWT Food Sci Technol 40:1392–1402

Turkoglu A, Duru ME, Mercan N, Kivrak I, Gezer K (2007) Antioxidant and antimicrobial of *Laetiporus sulphureus* (Bull.) Murril. Food Chem 101:267–273

Van Q, Nayak BN, Reimer M, Jones PJH, Fulche RG, Rempel CB (2009) Anti-inflammatory effect of *Inonotus obliquus*, *Polygala senega* L., and *Viburnum tribolum* in a cell screening assay. J Ethnopharmacol 125:487–493. https://doi.org/10.1016/j.jep.2009.06.026

Wang Y, Hall LR (2004) Edible ectomycorrhizal mushrooms: challenges and achievements. Canadian J Bot 82(2):1063–1073

Wani BA, Bodha RH, Wani AH (2010) Nutritional and medicinal importance of mushrooms. J Med Plants Res 4(24):2598–2604

Wasser SP, Weis AL (1999) Medicinal properties of substances occurring in higher basidiomycete mushrooms: current perspectives. Inter J Med Mush 1:31–62

Wasser PS, Nevo E, Sokolov D, Reshetnikov S, Timor-Tismenetsky M (2000) Dietary supplements from medicinal mushroom: diversity of types and variety of regulations. Intern J Med Mush 2(1):19

Wasson RG (1980) The wondrous mushrooms. Mc Graw Hill Book Co, New York

Wu MJ, Cheng TL, Cheng SY, Lian TW, Wang L, Chiou SY (2006) Immunomodulatory properties of *Grifola frondosa* in submerged culture. J Agric Food Chem 54:2906–2914

Yang BK, Gu YA, Jeong YT, Jeong H, Song CH (2007) Chemical characterics and immune-modulating activities of exo-biopolymers produced by *Grifola frondosa* during submerged fermentation process. Int J Biol Macromol 41:227–233

Zheng R, Jie S, Hanchuan D, Moucheng W (2005) Characterization and immunomodulating activities of polysaccharide from *Lentinus edodes*. Int J Immunopharmacol 5:811–820

Index

B. P. Singh et al. (eds.), *Biology of Macrofungi*, Fungal Biology,
https://doi.org/10.1007/978-3-030-02622-6

M

N

Q

R

The manufacturer's authorised representative in the EU is Springer Nature Customer Service Centre GmbH, Europaplatz 3, 69115 Heidelberg, Germany. If you have any concerns regarding our products, please contact ProductSafety@springernature.com

Printed and bound by CPI Group (UK) Ltd, Croydon, CR0 4YY
22/09/2026
02221048-0005